Zeolites: Science and Technology

NATO ASI Series

Advanced Science Institutes Series

A Series presenting the results of activities sponsored by the NATO Science Committee, which aims at the dissemination of advanced scientific and technological knowledge, with a view to strengthening links between scientific communities.

The Series is published by an international board of publishers in conjunction with the NATO Scientific Affairs Division

A	Life Sciences	Plenum Publishing Corporation
B	Physics	London and New York
C	Mathematical and Physical Sciences	D. Reidel Publishing Company Dordrecht and Boston
D	Behavioural and Social Sciences	Martinus Nijhoff Publishers The Hague/Boston/Lancaster
E	Applied Sciences	
F	Computer and Systems Sciences	Springer-Verlag Berlin/Heidelberg/New York
G	Ecological Sciences	

Series E: Applied Sciences – No. 80

Zeolites:
Science and Technology

edited by

F. Ramôa Ribeiro
Instituto Superior Técnico
Technical University of Lisbon, Portugal

Alírio E. Rodrigues
Department of Chemical Engineering
University of Porto, Portugal

L. Deane Rollmann
Mobil Research and Development Corporation
Central Research Division
Princeton, NJ, USA

Claude Naccache
Institut de Recherches sur la Catalyse
Villeurbanne, France

1984 **Martinus Nijhoff Publishers**
The Hague / Boston / Lancaster
Published in cooperation with NATO Scientific Affairs Division

Proceedings of the NATO Advanced Study Institute on Zeolites: Science and Technology, Alcabideche, Portugal, May 1-12, 1983

Library of Congress Cataloging in Publication Data

```
NATO Advanced Study Institute on Zeolites--Science
    and Technology (1983 : Alcabideche, Portugal)
    Zeolites--science and technology.

    (NATO advanced science institutes series. Series E,
Applied sciences ; no. 80)
    "Proceedings of the NATO Advances Study Institute on
Zeolites: Science and Technology, Alcabideche, Portugal,
May 1-12, 1983"--T.p. verso.
    "Published in cooperation with NATO Scientific Affairs
Division."
    1. Zeolites--Congresses.  I. Ribeiro, F. Ramoa.
II. Title.  III. Series.
TP245.S5N38  1983      660.2'8423      83-25486
```

ISBN 90-247-2935-1 (this volume)
ISBN 90-247-2689-1 (series)

Distributors for the United States and Canada: Kluwer Boston, Inc., 190 Old Derby Street, Hingham, MA 02043, USA

Distributors for all other countries: Kluwer Academic Publishers Group, Distribution Center, P.O. Box 322, 3300 AH Dordrecht, The Netherlands

Printed in The Netherlands

PREFACE

Zeolites have been the focus of intensive activity and growth in applications over the past 25 years in ion exchange, in adsorption and in catalytic process technology. Beginning with the synthetic zeolites A,X and Y, continuing into the emerging ZSM series, and including selected natural zeolites, applications span the range from large-scale purification and separation to such major petroleum and petrochemical processes as catalytic cracking and aromatics alkylation.

The future promises several new areas of signiciant use as our energy resource base is expanded. As a result, a NATO Advanced Study Institute on Zeolites was held in Alcabideche, Portugal, May 1-12, 1983. Its purpose was to summarize the state-of-the-art in zeolite science and technology, with particular emphasis on recent developments. This summary is intended to complement presentations of the latest research results at the 1983 International Zeolites Association meeting in Reno, Nevada - USA.

Both the fundamentals concepts and industrial applications are addressed in the lectures of the Institute. Individual chapters cover historical development, structure, crystallography and synthesis techniques. Basic principles of adsorption, diffusion, ion exchange and acidity are reviewed. A section on catalysis addresses shape selectivity, transition metals, bifunctional catalysis and "methanol-to-gasoline".

Included in the section on industrial applications are chapters on reactor and adsorber design, catalytic cracking, xylene and n--paraffins isomerization, as well as ion exchange and adsorption.

We would like to thank the members of the Advisory Board for the support given by them to the organization of this meeting; and all lecturers are to be thanked for their clear presentations and written contributions. We particularly wish to acknowledge the efforts of the local committee.

This NATO Advanced Study Institute was made possible through the NATO-ASI Programme. We are much indebted to this organization.

May 1983   F. Ramôa Ribeiro
Alcabideche   Alírio E. Rodrigues
                                  L. Deane Rollmann
                                    Claude Naccache

NATO ADVANCED STUDY INSTITUTE "ZEOLITES:SCIENCE AND TECHNOLOGY"

Director: Prof. F. Ramôa Ribeiro

Co-Directors: Dr. C. Naccache, Prof. A. Rodrigues, Dr. D. Rollmann

Advisory Board

Prof. L. Alves, Instituto Superior Técnico, Portugal

Prof. B. Delmon, Université Catholique de Louvain, Belgium

Dr. G. Martino, Institut Français du Pétrole, France

Prof. R. Maurel, Université de Poitiers, France

Prof. M. F. Portela, Instituto Superior Técnico, Portugal

Lectures

Prof. R.M. Barrer, Imperial College, London, UK

Dr. D. Barthomeuf, Exxon, Linden, USA

Prof. E. Derouane, Mobil, Princeton, USA

Dr. Z. Gabelica, Université de Namur, Belgium

Prof. M. Guisnet, Université de Poitiers, France

Prof. C. Kenney, University of Cambridge, UK

Dr. P. Jacob, Elf-Aquitaine, France

Dr. G. Kokotailo, University of Drexel, USA

Prof. H. de Lasa, University of Western, Ontario, Canada

Prof. H. Lechert, University of Hamburg, FRG

Dr. C. Naccache, Institut Recherches Catalyse, France

Dr. J. Rabo, Union Carbide, Tarrytown, USA

Prof. F. Ramôa Ribeiro, Instituto Superior Técnico, Portugal

Prof. A. Rodrigues, University of Porto, Portugal

Dr. D. Rollmann, Mobil, Princeton, USA

Dr. J. Sherman, Union Carbide, Tarrytown, USA

Local Committee

Eng. M. J. Pires, Instituto Superior Técnico, Portugal
Eng. J. M. Loureiro, University of Porto, Portugal
Eng. F. Freire, Instituto Superior Técnico, Portugal
Eng. F. Lemos, Instituto Superior Técnico, Portugal
Eng. M. L. Palha, Instituto Superior Técnico, Portugal

TABLE OF CONTENTS

X

PART I

HISTORY, STRUCTURE AND SYNTHESIS

MOLECULAR SIEVE ZEOLITE TECHNOLOGY: THE FIRST TWENTY-
FIVE YEARS*

Edith M. Flanigen

Union Carbide Corporation
Tarrytown Technical Center
Tarrytown, New York 10591, USA

ABSTRACT

    In twenty-five years molecular sieve zeolites have
substantially impacted adsorption and catalytic process
technology throughout the chemical process industries;
provided timely solutions to energy and environmental
problems; and grown to over a hundred million dollar
industry worldwide. The evolution in zeolite materials
with improved or novel properties has strongly influenced
the expansion of their applications, and provided new
flexibility in the design of products and processes.

INTRODUCTION

    The year 1979 marked the twenty-fifth anniversary
of the commercial birth of molecular sieve zeolites as
a new class of industrial materials. They were introduced
in late 1954 as adsorbents for industrial separations and
purifications. Since that time the fascination with and
the elegance of, this unique class of materials has
generated a masss of scientific literature describing
their synthesis, properties, structure and applications,
which probably now numbers well over 15,000 scientific
contributions and over 10,000 issued patents. The
molecular sieve industry has been projected to have grown
into an estimated quarter of a billion dollar market
{1} serving all of the major segments of the chemical
process industries including major applications in the
petroleum refining and petrochemical industries, and
has generated a myriad of other adsorption, catalytic,

* This article is used as reference material for J. A. Rabo's
  lecture on "Historical Aspects of Zeolites".

and most recently ion exchange applications.

Milton reviewed the beginnings and development of
molecular sieve zeolites in 1967 {2}. He traced the
early discoveries and synthesis of the new zeolites A,
X, and Y, which led to their commercial applications
as selective adsorbents and catalysts. It will be the
purpose of this paper to review the evolution of
molecular sieve materials, their synthesis, properties
and applications, over the span of 1954 to 1979, with
emphasis on the major milestones and trends in these
areas. There will be no attempt here to repeat the in
depth coverage of zeolite molecular sieves given by
Breck {3} or Barrer {4}, or recent up-to-date review
articles on the applications of molecular sieve zeolites
 as adsorbents {6}, catalysts {7,8}, and ion exchangeers
{9}, and on natural zeolites and their applications
{10,11,1,5}.

Success of molecular sieve zeolites has been due
primarily to the discovery of new materials whose properties have been engineered into improvements in known
processes and into the development of new ones. This
discussion will therefore emphasize the materials and
properties aspects of molecular sieve zeolites as they
evolved and developed into various application areas.

THE EVOLUTION IN MATERIALS

The theme in research on molecular sieve zeolite
materials over the twenty-five year period has been a
quest for new structures and compositions. Because zeolites are unique as crystalline porous materials, with
structure as well as composition controlling properties,
there arose the strong conviction that novel and useful
properties would result from the discovery of new compositions and structures. Let us now trace the web of
change in those discoveries over twenty-five years.

"Low-Silica" Zeolites or Aluminum-Rich Zeolites.
The discovery of zeolites A and X by Milton {12} at the
Union Carbide Coporation Laboratories represented a
fortunate optimum in composition, pore volume, and
channel structure, guaranteeing these two zeolites their
lasting commercial prominence out of more than 150
synthetic species known and discovered over the last
twenty-five years. Both zeolites are nearly "saturated"
in aluminum in the framework composition with a molar
ratio of Si/Al near one, the maximum aluminum content
possible in tetrahedral aluminosilicate frameworks if

one accepts Loewenstein's rule. As a consequence they
contain the maximum number of cation exchange sites
balancing the framework aluminum, and thus the highest
cation contents and exchange capacities. These compositional characteristics combined give them the most
highly heterogeneous surface known among porous materials,
due to exposed cationic charges nested in an aluminosilicate
framework which results in high field gradients. Their
surface is highly selective for water, polar and polarizable
molecules which serves as the basis for many of their
applications particularly in drying and purification.
Their pore volumes of nearly 0.5 cm^3/cm^3 are the highest
known for zeolites and give them a distinct economic
advantage in bulk separation and purifications where
high capacity is essential to and economic design. Their
3-dimensional channel structures allow the maximum in
diffusion characteristics. By a judicious selection of
cation composition achieved by facile ion exchange
reactions, nearly the entire spectrum of known pore
sizes in zeolites can be obtained. The pore sizes
achievable by cation exchange of types A and X span
the entire range from the smallest pore-sized zeolite
known, Cs-A at 0.2nm in size {13} through the 0.3nm
potassium A, the 0.4nm sodium A, the 0.5nm calcium A,
to the largest known which is about 0.8nm in sodium
X. This large pore size of zeolite X was a key to its
introduction as a catalytic cracking catalyst.

"Intermediate Silica" Zeolites. The next evolution
in zeolite materials was the impetus to synthesize more
siliceous zeolites, primarily to improve stability
characteristics, both thermal and acid. It was recognized
in the early 1950's by scientists at Union Carbide
Laboratories that the tetrahedral framework aluminum
provided a site of instability for attack by acid and
water vapor or steam. Also, the siliceous mineral zeolite mordenite was known with a Si/Al molar ratio of
5 and possessing superior stability characteristics.
Breck provided the first success in this quest with the
discovery of the third commercially important molecular
sieve zeolite type Y {14}, with an Si/Al ratio of from
1.5 to 3.0, and a framework topology like that of zeolite X and the rare zeolite mineral faujasite. Not only
was the desired improvement in stability over the more
aluminous X achieved, but also the differences in composition and structure had a striking, unpredicted
effect on properties that has led to the preeminence of
zeolite Y based catalysts in nearly all of the important
catalytic applications involving hydrocarbon conversion,

6

(i.e., cracking, hydrocracking and isomerization) since its initial commercial introduction in 1959.

The next commercially important synthetic zeolite introduced in the early 1960's was a large pore mordenite made by the method of Sand {15} and marketed as "Zeolon" by the Norton Co.{16}, which continued the progression toward higher Si/Al ratio, in this case a value near 5. Again, thermal, hydrothermal, and acid stability improvement was evident. This improved stability coupled with its specific structural and compositional characteristics found it a small but a significant commercial market as both an adsorbent and hydrocarbon conversion catalyst. Type L zeolite, discovered in the early 50's by Breck and Acara {17} with an Si/Al ratio of 3.0 and a unique framework topology, has only recently recieved attention as a commercial catalyst in selective hydrocarbon conversion reactions {18}.

Other zeolites with "intermediate" Si/Al compositions of from 2 to 5 and their own unique framework topologies which have achieved commercial status are the zeolite minerals mordenite, erionite, chabazite, and clinoptilolite, and the synthetic zeolite omega {19} with a typical Si/Al of 3 to 4. Their properties exhibit a common characteristic in terms of improved stability over the "low" silica zeolites. However, unique properties as adsorbents, catalysts and ion exchange materials are also exhibited which reflect their unique structural features. The surface of these intermediate silica zeolites is still heterogeneous and exhibits high selectivity for water and other polar molecules.

"High Silica" Zeolites. The most recent stages in the quest for more siliceous molecular sieve compositions was achieved in the late 1960's and the early 1970's with the syntehsis at the Mobil Research and Development Laboratories of the "high silica zeolites", compositions exemplified first by zeolite beta discovered by Wadlinger, Kerr and Rosinski {20}, and later ZSM-5 discovered by Argauer and Landolt {21}. Subsequently, ZSM-11 {22}, ZSM-21 {24}, and ZSM-34 {25} were described. These compositions are molecular sieve zeolites with Si/Al ratios from 10 to 100 or higher, and with unexpected, strikingly different surface characteristics. In contrast to the "low" and "intermediate" silica zeolites, representing heterogeneous hydrophilic surfaces within a porous crystal, the surface of the high silica zeolites approaches a more homogeneous characteristic with an organophilic-hydrophobic selectivity. They more strongly adsorb the less polar organic molecules and only weakly interact with water and other strongly polar molecules. In addition to this novel surface selectivity, the high silica zeolite compositions still contain a small concentration of aluminum in the framework and the accompanying stoichiometric cation exchange sites. Thus, their cation exchange properties allow the introduction of acidic OH groups via the well known zeolite ion exchange reactions, essential to the development of acid hydrocarbon catalysis properties.

Silica Molecular Sieves. The ultimate in siliceous molecular sieve compositions, and a much discussed aspiration of early workers in zeolite synthesis in the 1950's, was also achieved in the 1970's with the synthesis of the first pure silica molecular sieve, silicalite {26}, containing essentially no aluminum or cation sites. In the complete absence of strong field gradients due to framework aluminum and exchangeable metal cations which serve as hydrophilic sites, silicalite exhibits a high degree of organophilic-hydrophobic character, capable of separating organic molecules out of water-bearing streams {26}. Silicalite does however contain extraneous or defect hyroxyl groups which contribute a small concentration of hydrophilic sites capable of interacting with water and polar molecules. A related new composition, fluoride-silicalite {27}, completely free of hydroxyl groups, exhibits the ultimate in near perfect hydrophobicity, adsorbing less than 1 wt. % water at 20 torr and 25°C, and even exhibits bulk hydrophobicity: the crystals (d = 1.7 g/cm^3) actually float on water. Silicalite reportedly {26a} has the same framework topology as zeolite ZSM-5 {28}. Other silica molecular sieve compositions have been reported, including silicalite-2 {29}, and TEA-silicate {30}.

Chemically Modified Zeolites. An alternate method for producing highly siliceous zeolite compositions had its beginnings in the mid 1960's when thermochemical modification reactions that lead to framework dealumination were first reported {31,32}. These reactions include those described as "stabilization," or "ultrastabilization" involving high temperature steaming of ammonium exchanged or acid forms of the zeolite {31}, and framework aluminum extraction with mineral acids or chelates. Repetitive treatments in effect produce zeolites with framework Si/Al compositions and stability characteristics comparable to those observed in the synthesized high silica zeolite compositions. Such highly siliceous variants have been described by Scherzer for zeolite Y

{33}, by Chen {34} and others for mordenite, and by
Eberly et al. {35} and Patton et al. {36} for erionite.
The stabilized Y zeolite of McDaniel and Maher {31} and
the highly siliceous mordenite products of Chen {34},
were reported to be hydrophobic. Although the ultra-
stabilized   and other dealuminated forms of zeolites
emerged at the same time as the synthesized high silica
zeolite beta, focus in the former case in the late 60's
was on their improved stability characteristics and
catalytic applications, rather than on their surface
selectivity.

Other highly siliceous analogs or "pseudomorphs"
of clinoptilolite and mordenite prepared during the
same period by acid extraction {37} have Si/Al compositions like the high silica zeolites and silica molecular
sieves. However, their crystallinity, stability, and
hydrophobicity are substantially less than the thermo-
chemically derived ultrastabilized and dealuminated
compositions, presumably due to the presence of high
concentrations of hydroxyl defects {37b}, and the
absence of substantial silicon reinsertion into framework tetrahedral sites.

Natural Zeolites. In contrast to the development
of the synthetic zeolites which required their discovery
and successful synthesis in the laboratory, the evolution of the natural or mineral zeolites depended on
their availability in mineable deposits. The discovery
in 1957 of mineable deposits of relatively high purity
zeolite minerals in volcanic tuffs in the western
United States and in a number of other countries represents the beginning of the commercial natural zeolite era {10 }. Prior to that time there was no recognized
indication that zeolite minerals with properties useful
as molecular sieve materials occurred in large deposits.
Commercialization of the natural zeolites chabazite,
erionite, and mordenite as molecular sieve zeolites
commenced in 1962 with their introduction as new adsorbent materials with improved stability characteristics
in various acid natural gas drying applications. Their
improved stability over the then prevalent synthetic
zeolite adsorbents, types A and X, again reflects
their higher intermediate Si/Al ratio of 3-5. The
applications of clinoptiolite in radioactive waste recovery and in waste water treatment during the same
period of the 60's was based not only on superior
stability characteristics but also a high cation exchange selectivity for cesium and strontium, or for
the ammonium ion.

A summary of the evolution of molecular sieve materials as developed above is given in Table 1, with emphasis on the framework Si/Al variation.

TABLE 1

THE EVOLUTION OF MOLECULAR SIEVE MATERIALS

"Low" Si/Al Zeolites (1 to 1.5):
    A, X

"Intermediate" Si/Al Zeolites (~2 to 5):
    A) Natural Zeolites:
            erionite, chabazite, clinoptilolite,
            mordenite
    B) Synthetic Zeolites:
            Y, L, large pore mordenite, omega

"High" Si/Al Zeolites (~10 to 100):
    A) By thermochemical framework modification:
            highly siliceous variants of Y,
            mordenite, erionite
    B) By direct-synthesis:
            ZSM-5

Silica Molecular Sieves: Silicalite

During this period of twenty-five years of research and development over 150 species of synthetic zeolites have been synthesized and some 7 mineral zeolites have been found in substantial quantity and purity {38}. Yet commercially only twelve basic types are utilized. Table 2 lists these twelve types from Ref. 5. The major large volume commercial molecular sieve zeolites used in adsorption and catalysis remain after twenty-five years, the zeolites A, X and Y.

TABLE 2

ZEOLITE TYPES IN COMMERCIAL APPLICATIONS {5}

Zeolite Minerals	Synthetic Zeolites	
Mordenite	A	Na, K, Ca forms
Chabazite	X	Na, Ca, Ba forms
Erionite	Y	Na, Ca, NH_4, rare earth forms
Clinoptilolite	L	K, NH_4 forms
	Omega	Na, H forms
	"Zeolon", Mordenite	H, Na forms
	ZSM-5	Various forms
	F	K form
	W	K form

TRANSITION IN PROPERTIES

The transition in properties of molecular sieve materials is summarized in Table 3. The emphasis chosen in on the framework Si/Al increasing from aluminum saturated "Si/Al = 1" to infinity, as represented by the aluminum-free, pure silica molecular sieve, silicalite. The property transitions shown are somewhat generalized and should be considered only as trends.

TABLE 3

THE TRANSITION IN PROPERTIES

Transition in:

Si/Al, from 1 to ∞	"Acidity", increasing strength
Stability, from $\leq 700°C$ to ~1300°C	Cation concentration, decreasing
Surface selectivity, from hydrophilic to hydrophobic	Structure, from 4, 6, and 8-rings to 5-rings

The _stability_ characteristics vary substantially from a crystalline decomposition temperature near 700°C for the "low" silica zeolites, to above 1300°C for silicalite. The low silica zeolites are at best "fragile" in the presence of acid, whereas the high silica zeolites are completely stable even in boiling, concentrated mineral acids. In contrast, the high silica zeolites show decreased base stability.

The _surface selectivity_ changes from the highly polar or hydrophilic surface exhibited by the aluminum-rich zeolites to a more homogeneous or nonpolar organophilic or hydrophobic properties appears to occur at an Si/Al near 10.

Characterization of the surface selectivity of the high silica zeolites as compared to the low silica zeolites has been studied extensively in this laboratory {39}. In the adsorption of H_2O, O_2 and n-hexane, the hydrophilic NaX zeolite exhibits the near rectilinear type I isotherm shape typical of zeolite adsorption where the mechanism of volume filling of micropores controls isotherm shape. The hydrophobic silicalite similarly pore fills with oxygen and n-hexane at low relative pressure, but fills only about 25% of its pore volume with water near saturation {26a}. The hydrophilic NaX zeolite removes the water, and the hydrophobic silicalite removes the organic, from organic-water mixtures. As a consequence, separations and catalysis which usually require the absence of water with hydrophilic zeolites, can now be carried out in the presence of water with the hydrophobic zeolites.

The transition in hydrophobic properties within one zeolite structure type is illustrated in the case of the zeolite mordenite {"Zeolon"} by Chen {34} who shows a quantitative linear relationship of water capacity with increasing framework Si/Al ratio for a series of variously stabilized and dealuminated mordenite materials. There is only a small loss in crystal pore volume for cyclohexane over the Si/Al range of 5 to 50. The onset of change in surface selectivity from polar to nonpolar appears to occur here near an Si/Al ratio of 7.5.

The change in _cation concentration_ accompanying the change in Si/Al affects cation specific interactions in adsorption, catalysis and ion exchange, where the effect of crystal structure and resulting cation siting are also important. Cation exchange selectivity is perhaps most strongly affected. Its importance is observed in many adsorbent applications involving cation specific inter-

actions such as air separation involving specific interactions of the nitrogen quadrupole with cations {40}, in
ion exchange applications such as ammonium removal, and
perhaps equally strikingly in acid catalysis, where the
catalytic advantages of zeolite Y versus zeolite X are
well established.

The concept of increasing acidity as well as
stability in hydrogen and "decationized" forms of zeolites with increasing framework Si/Al has persisted
since the early work in catalysis by Rabo et al. {41}.
Recently several attempts to quantify and predict that
change have been published. Barthomeuf {42} showed a
linear decrease in infrared frequency of the acid
hydroxyl group with increasing framework Si/Al for a
large number of zeolite structural types, and related
this to increase in acid strength of the proton with
the change in framework charge density. Subsequently
{43} invoking the concept of acid activity coefficients
in zeolites, she noted that both acid strength and proton activity coefficients increased with decreasing
aluminum content in the zeolite framework, whereas
acid site concentration decreases. This suggested that
there should be a maximum in acid site activity and in
catalytic reaction rate at a specific aluminum content.
Vedrine et al. {44} report that the acid sites present
in H-ZSM-5 (Si/Al = 19.2) are similar to, but slightly
stronger than, those present in hydrogen mordenite
("Zeolon", Si/Al = 5). The concentration of acid sites
is substantially higher in the lower Si/Al mordenite.

In a somewhat different approach, Mortier {45}
applied the Sanderson electronegativity model to zeolites, and concluded that the acid strength of zeolitic
protons increases along with the calculated residual
hydrogen charge, with decreasing aluminum content.
Mortier, Jacobs and Uytterhoeven {46} confirmed that the
overall catalytic efficiency in the dehydration of
isopropanol increased linearly with the residual charge
on the proton for several zeolite structure types.

EVOLUTION IN SYNTHESIS

The early era of the discovery of zeolites in the
late 1940's and the early 1950's, which led to about 20
novel synthetic zeolites {47} in a short period of time,
flowed from the discovery of a new regime of chemistry
{2} involving highly reactive alkaline aluminosilicate
gels, metastable crystallization, and low temperature, low
pressure crystallization. It was later emphasized that

the cation played a dominant role in directing the formation of specific structures {48}. These early zeolites
resulted from nearly the same chemistry and the use of
only two alkali cations, sodium and potassium, or mixtures thereof. A second important variable in synthesis
as well as in properties was the Si/Al ratio. Increase
in the Si/Al in the reaction mixture resulted in synthesis of intermediate or transition Si/Al zeolites, such as
T {49} and L, still using the same two alkali cations.

The next major advance in synthesis of new zeolite
materials was the introduction of a new chemistry into
zeolite synthesis, that of the addition of alkylammonium
cations to synthesis gels. Barrer et al. {50} first
reported the synthesis of N-A, a siliceous analog of
zeolite A, by adding tetramethylammonium cations to
sodium aluminosilicate gels, and noted the effect of
the alkylammonium ion on increasing framework Si/Al composition. Nitrogenous analogs of zeolites B, X and Y
were also synthesized {50b}. Thus the first effect of
the addition of the alkylammonium cations was to generate
more siliceous framework compositions of previously known
structure types. Subsequently the addition of alkylammonium cations to sodium aluminosilicate gels led to
the crystallization of new zeolite structure types,
exemplified by zeolite ZK-5 {51}, omega {19,52}, and
zeolite N {53}. In the recent work by Mobil Research
and Development Corporation scientists {7} the addition
of some numbers of alkylammonium and other nitrogeneous
organic molecules {54}, such as TEA, TPA, TBA, and
pyrrolidine, to highly siliceous gels (Si/Al = 10 to
100) resutled in the high silica zeolite materials. These
compositions represent siliceous analogs of previously
known structure types, ZSM-21 {24}, a ferrierite-type,
and ZSM-34 {25}, an erionite-offretite type, as well as
new structure types in the case of ZSM-5 {21}, ZSM-11
{22}, and probably ZSM-12 {23} and zeolite beta {20}.
The addition of alkylammonium cations to pure silica
systems ultimately resulted in the silica molecular sieves:
silicalite {26} and fluoride-silicalite {27} with TPA;
silicalite-2, structurally related to the zeolite ZSM-11
{55}, with TBA {29}; and TEA-silicalite {30}, an apparent
structural analog of zeolite ZSM-12, with TEA.

Thus, variation intwo important parameters in synthesis, cation and Si/Al ratio, has resulted in the
spectrum of synthetic zeolites now known. Their synthesis still uses the basis reactive gel crystallization
method developed by Milton {2} in the late 40's.

The <u>synthesis</u> <u>mechanisms</u> of the low silica and high silica zeolites appear to differ. It is suggested here {56} that the nucleation mechanism in the low silica zeolites involves the formation of stabilized alkali metal cation aluminosilicate complexes and is primarily controlled by the aluminate and aluminosilicate solution chemistry. Four and six-membered rings and "cages" of aluminosilicate tetrahedra, stabilized by alkali metal cations, dominate the synthesis chemistry and appear in the structures. In the case of the highly siliceous molecular sieves, a true "templating" or clathration mechanism pervades wherein the alkylammonium cation forms complexes with silica via hydrogen bonding inter- actions {26 a}. These complexes template or cause re- lication of the structure via a stereo specific hydrogen- bonding interaction of the quaternary ammonium cation with the framework oxygens. Synthesis chemistry and structure is now dominated by silica, five-rings of tetrahedra, and the sterospecific interactions with alkylammonium ion. The concept of cation templating in zeolite synthesis has been discussed by Flanigen {48} and more recently developed and summarized by Rollman {57}.

It is interesting to note that recent theoretical work on molecular electrostatic potential by Mortier et al. {58} shows stabilization of a 6-ring containing aluminum due to the presence of metal cations near its center, in support of the mechanistic concept of a stabilized metal aluminosilicate species in alumina-rich synthesis. Also of note is the observation that the cross-over in zeolite surface selectivity from hydrophilic to hydrophobic at an Si/Al near 7.5 to 10, corresponds to the change in zeo- lite structural features from 4, 6, and 8-ring structures to those containing an increasing fraction of 5-rings (for example, $Y \rightarrow \Omega \rightarrow$ mordenite $\rightarrow$ ZSM-5). This suggests that as the fraction of Al decreases below one per 6-ring corresponding to an Si/Al of 5, 5-ring formation is favored.

Although the level of understanding of the synthesis chemistry and its relationship to the resulting structural features of zeolites has advanced substantially over the period of twenty-five years, the ability to execute chemical architecture in the laboratory has unfortunately eluded zeolite synthesis scientists for the same period. The synthesis of new zeolite has flowed from the innova- tive alteration of the chemistry of the synthesis system, rather than from the ability to design the chemistry to form a desired structure.

The manufacture of basic zeolite materials still remains similar to that developed in the initial work in the late 1940's by Milton {2}. The temperature range for crystallization has tended to increase with increasing Si/Al ratio in the zeolite, from 25 to 125°C for the aluminum-rich zeolites, to 100 to 150°C for the intermediate Si/Al zeolites, L, omega, and mordenite, and to near 125 to 200°C for the high silica zeolites as exemplified by ZSM-5. This is consistent with the suggested relationship of pore volume and synthesis temperature 2 , that the lower temperatures favor the highest pore volume materials (near 0.4 cm^3/g) such as A, X and Y, and the higher temperatures favor lower pore volumes (0.15-0.20 cm^3/g) for mordenite, L, omega and ZSM-5.

THE EVOLUTION IN APPLICATIONS

The molecular sieve behavior of crystalline zeolites and their large potential in performing molecular sieving separations was first demonstrated in the pioneering work of Barrer {4} and colleagues in England. With the commercialization of molecular sieve zeolites in late 1954 a new class of materials became available, capable of being tailor-made in terms of structure, composition and properties. Yet many of their early adsorbent and catalytic applications involved simple replacement of the then used adsorbent and catalyst materials in known adsorption and catalytic processes, based on the improved properties and performance of the molecular sieve zeolites. This is exemplified by the replacement of silica gel and activated alumina in drying and purification applications by zeolites A and X, due to the improved capacity and greater selectivity of the zeolites. Their introduction into catalytic cracking in 1962 was as a replacement for amorphous silica-alumina catalysts in existing moving bed and fluid bed catalytic cracking processing. Replacement followed the discovery of the higher activity of zeolite X in the cracking reaction and the higher selectivity to gasoline compared to silica-alumina.

Applications engineered specifically for zeolites were developed and continue to evolve over the twenty-five year span, especially in the adsorbent area, in such processes as isoparaffin/n-paraffin separation, xylene separation and olefin separation, and pressure swing adsorption air separation {6}. All of these adsorbent applications combine the unique adsorptive properties of a specific tailor-made molecular sieve adsorbent, and an adsorption process designed and engineered to optimize the product-process characteristics.

Interestingly, the original incentive of Milton to use molecular sieve zeolites was to separate oxygen from nitrogen as a new method for air separation. The commercialization of that application in oxygen and nitrogen production from air via the now used pressure swing adsorption processes {6,59}, occurred nearly twenty years later when a market in waste water purification and other applications requiring relatively low tonnage production of oxygen allowed it to finally compete successfully with the established cryogenic separation technology.

Adsorbent Applications. The use of molecular sieve zeolite absorbents to perform a host of separations and purifications has become firmly established in the chemical process industries. A summary list of major adsorbent applications adapted from Anderson's review at the last Molecular Sieve Zeolite Conference in Chicago {6} is given in Table 4.

TABLE 4

COMMERCIAL ADSORBENT APPLICATIONS OF MOLECULAR SIEVE ZEOLITES

A. Purification

I.   Drying
        natural gas (including
        LNG)
        cracked gas (ethylene
        plants)
        insulated windows
        refrigerant
II. CO_2 Removal
        natural gas
        cryogenic air separation
        plants
III. Sulfur Compound Removal
        sweetening of natural
        gas and liquified
        pretroleum gas
IV. Pollution Abatement
        removal of Hg, NO_x, SO_x

B. Bulk Separations

I.   Normal/iso-paraffin separation
II. Xylene separation
III. Olefin separation
IV. O_2 from air
V.   Sugar separation {60}

It can be seen that present day applications fall into
two categories, purification applications which in
general depend on surface selectivity for polar or polarizable molecules such as water, CO_2 or surface compounds;
and bulk separations many of which are based on molecular
sieving principles. Pressure swing adsorption in air
separation, originally envisaged by Milton as a molecular
sieving separation based on the slight difference in size
of the oxygen and nitrogen molecule, is rather based on
the strong specific interaction of the nitrogen molecular
quadrupole with the cation {40}. Many of the purification applications also involve molecular sieving in that
the selection of the zeolite adsorbent involves a pore
size designed to exclude potentially co-adsorbed molecules,
for example the use of type 3A molecular sieve in cracked
gas drying to prevent the co-adsorption of ethylene and
heavier unsaturated hydrocarbons. Refrigerant drying
and purification (of halogenated hydrocarbons), the
first broadly applicable commercial use of molecular
sieves {2}, still remains a major nonregenerative application.

Recently molecular sieve zeolite adsorbent separations have been extended to liquid phase aqueous systems
in the separation of fructose from fructose-dextrose-
polysaccharide mixtures {60}.

Catalyst Applications. Fundamental discoveries
in the use of zeolites in hydrocarbon catalysis were
made in the 50's primarily at the laboratories of Union
Carbide and Mobil Oil Corporations, and Esso Research
and Engineering Company {61}, with the recognition of the
acidic properties of hydrogen, multivalent metal cation,
and decationized forms of zeolites X and Y {41,62}, and
the novel shape selective properties of zeolite A {63}.
A commercial zeolite Y isomerization catalyst, introduced by Union Carbide in 1959 {2}, was the first·of a
series of molecular sieve based catalysts for the petroleum
industry.

The first major commercial catalytic application
resulted from the introduction of the use of zeolite X
in catalytic cracking of crude to produce liquid fuels
in 1962, based on the early work of Plank and Rosinski
{64}. The introduction of zeolite containing catalysts
caused a revolution in catalytic cracking {65}, because
of their increased catalytic activity and improved yields
to gasoline compared to amorphous silica-alumina catalysts.
Mechanistically this has been related by Weisz {66} and

others to the more efficient hydrogen redistribution between hydrocarbon molecules over zeolite catalysts, resulting in high rates of intermolecular hydrogen transfer, coupled with extremely high intrinsic cracking activity. Because of the very strong adsorption forces within zeolites they also concentrate hydrocarbon substrates to a much larger extent than other catalysts and favor bimolecular reactions {7} such as hydrogen transfer.

Developments since 1962 in zeolite catalytic cracking have occurred both in materials and process {67,65}. Zeolite X has been essentially replaced by the more stable and active zeolite Y. Metals resistant, and controlled combustion zeolite catalysts have been developed the former to allow handling of heavier feedstocks, and the latter for pollution control to convert CO emissions to CO_2. Zeolite content has been increased from 5-10% in 1964, to as high as 40% in 1979 {68}. Process innovations to utilize the unique properties of zeolites include concepts based on short contact riser cracking and have led to some number of proprietary zeolite engineered designs now in commercial use {67}. Recent developments have been strongly influenced by environmental requirements in pollution control and the need for higher octane unleaded gasoline (especially in the United States).

Other established industrial processes that utilize zeolite based catalysts in addtion to catalytic cracking, are hydrocracking, and paraffin isomerization {7}. All are based on the unique properties of zeolite catalysts which have in common, extremely high strength acid sites, and selectivities related to strong adsorptive forces within the zeolite. All use hydrothermally stable acid forms of large pore zeolites. Two examples of shape selective catalysis {63,69} utilizing, in addition to the above properties, a specific pore size and shape of zeolite, are in commercial use. The Selectoforming process of Mobil Oil Corp. selectively hydrocracks the normal paraffin components of catalytic reformate using an offretite-erionite type catalyst which excludes non-normal paraffin molecules from adsorption and reaction. A second commercial shape selective hydrocracking process is catalytic dewaxing which typically employs a large pore mordenite containing single channels approximately 0.7nm in diameter which contribute to selectivity {7}.

The catalytic properties of the high silica zeolite ZSM-5 have received much attention {70,7}. The initial commercial or near commercial applications reported {7} for ZSM-5 include: a) the isomerization of C_8 aromatics to produce isomerically pure xylenes, especially para-xylene for polyester manufacture; b) ethylbenzene synthesis for styrene production; and c) catalytic de-waxing {71}. The conversion of methanol to gasoline {72} as a new route from coal or synthetic or natural gas to motor fuel is under development {73}, with the first plant based on coal scheduled fro construction in New Zealand {74}. More recently Mobil workers have also reported {75} a similar conversion of oxygenated hydrocarbon com-pounds in biomass to gasoline.

These initial commercial applications for ZSM-5 appear to be elegant examples of shape selective catalysis {63,69} reflecting its unique crystal structure with 0.6nm pores outlined by 10-membered rings of oxygen. In addition they depend upon the other zeolite-specific properties of highly acidic sites as in the H-ZSM-5 catalyst, and the substrate or reactant concentration effect. The novel organophilic-hydrophobic selectivity also appears to contribute to the apparently unique selectivity of ZSM-5 for the conversion of oxygenated hydrocarbons to paraffins and aromatics.

The entrance into several of these commercial appli-cations by ZSM-5 reportedly {7} may involve retrofit of existing processes with a new catalyst. In the case of C_8 aromatic isomerization ZSM-5 replaces the platinum/silica-alumina catalysts developed for the Octafining process. In ethylbenzene synthesis ZSM-5 could replace the current technologies based on $AlCl_3$ and BF_3 supported on alumina. In the case of catalytic dewaxing ZSM-5 may replace the use of metal loaded tubular zeolite catalysts such as mordenite. Early reports of performance advan-tages of ZSM-5 in the process suggest that the H-ZSM-5 catalyst carries out the needed hydrogenation function without the addition of noble metal to the catalyst {71}.

TABLE 5

PRESENT AND PROJECTED APPLICATIONS OF ZEOLITES IN CATALYSIS {5}

Hydrocarbon conversion
    Alkylation
    Cracking
    Hydrocracking
    Isomerization

Hydrogenation and dehydro-
genation
Hydrodealkylation
Methanation

Shape-selective reforming
Dehydration
Methanol to gasoline
Organic catalysis
Inorganic reactions
    H_2S Oxidation
    NO Reduction of NH_3
    CO Oxidation
    $H_2O \rightarrow O_2 + H_2$

The first interesting application of a combined catalytic-adsorptive integrated process, named TIP (total isomerization process) for gasoline octane improvement weds the Union Carbide "IsoSiv" molecular sieve adsorption process for separating normal and isoparaffins employing 5A zeolite, with the "Hysomer" catalytic process of the Shell Oil Co. to isomerize normal paraffins to higher octane branched isomers using a highly acid, large pore zeolite catalyst based on a large pore mordenite {7}.

Thus, molecular sieve zeolites have found wide use in both catalytic conversion and adsorption separation processes over the last twenty-five years. In both adsorbent and catalytic applications the initial commercial introduction tended to involve replacement of existing non-zeolite adsorbents or catalysts because of improved zeolite performance. Zeolite engineered processes evolved subsequently and more facilely in adsorption applications where the capital costs are generally lower than those in major hydrocarbon conversion processes such as cracking and hydrocracking. In tha latter case the introduction of zeolites resulted in improved performance and production without expenditures of capital. After twenty-five years both kinds of processes are still in use. The retrofit processes have been extensively modified or redesigned to utilize unique properties of zeolites, and the sophistication or tailor-making of the zeolite designed or engineered processes have become more innovative and systematized to take maximum advantage of zeolite properties.

Ion Exchange Applications. The third unique
property of zeolite molecular sieves, that of selective
cation exchange, went through a quite different history
of development. Among the earliest application areas
explored in the Union Carbide laboratories by Thomas {76}
in the early 50's was the use of zeolites A, B and X in
cation exchange applications, including industrial and
domestic water softening. The synthetic polymeric ion
exchange resins, which themselves were replacements
for the amorphous metal aluminosilicate permutite-type
products, had only recently been introduced to the market-
place. At that time it appeared that the organic resins
had technical performance advantages in the regeneration
cycle over the synthetic zeolites A and X. Thus the
performance advantages of zeolites realized in adsorption
and catalysis did not apparently exist in ion exchange.

Cation exchange of zeolites is used routinely in
modifying the properties of zeolite products used in
adsorption and catalysis, and a large body of literature
on cation exchange selectivities, structural characteristics, and thermodynamics by Barrer, Sherry, and others
{77}, evolved very shortly after the commercial introduction of zeolites A and X in 1954. Yet, significant
use of the zeolites as ion exchangers has occurred only
recently. Their development as commercial ion exchangers
strongly followed the enactment of new environmental
pollution standards in the late 60's.

Sherman {9} recently reviewed the subject of ion
exchange separations with molecular sieve zeolites.
The summary of present and potential ion exchange applications reported by Breck {5} and adapted from Sherman {9}
is shown in Table 6.

TABLE 6

ION EXCHANGE APPLICATIONS {9,5}

Present Applications	Advantage
Removal of Cs^+ and Sr^{++} Radioisotopes - LINDE AW-500, mordenite, clinoptilolite	Stable to ionizing radiation Low solubility Dimensional stability High selectivity
Removal of NH_4^+ from waste water -LINDE F, LINDE W, clinoptilolite	NH_4^+ - selective over competing cations
Detergent builder Zeolite A, Zeolite X (ZB-100, ZB-300)	Remove Ca^{++} and Mg^{++} by selective exchange No environmental problem

Present Applications	Advantage
Radioactive waste storage	Same as Cs^+, Sr^{++} removal
Aquaculture AW-500, clinoptilolite	NH_4^+ - selective
Regeneration of artifical Kidney dialysate solution	NH_4^+ - selective
Feeding NPN to ruminant animals	Reduces NH_4^+ by selective exchange to nontoxic levels
Metals removal and recovery	High selectivities for various metals
Ion exchange fertilizers	Exchange with plant nutrients such as NH_4^+ and K^+ with slow release in soil.

The single potentially largest ion exchange application as builders in detergents {78} is ironically in
the water softening area, the original ion exchange'
application considered in the 50's. It became a commercial reality due to two changed factors. First, its
use to soften water as a builder in detergents in the
70's is non-regenerative and therefore the main earlier
disadvantage in regeneration as evaluated in the 50's
is absent. Second, there are currently a number of areas
in the world in which the use of phospate builders is
restricted for environmental reasons. After considerable
R and D effort, apparently no other suitable substitute
for phosphate has been found to date except the synthetic zeolites A and X.

Replacement of phosphates in detergents by zeolite ion exchangers is also based on performance and cost. The zeolites in powder form provide the same function as phosphates, the removal of hardness ions, Ca^{++} and Mg^{++}, as active ions in the wash water. The maximum ion exchange capacity of the aluminum saturated zeolite A, theoretically the highest possible in zeolites, coupled with a structurally controlled cation selectivity for Ca^{++}, give zeolite A a unique advantage in this application. Projections show a very large volume use and a bulk volume cost that is less than that of phosphates {78a}.

Natural zeolites have played an important role in the development of ion exchange applications. The use of zeolite minerals chabazite, mordenite and clinoptilolite for the removal of recovery of cesium and strontium radioisotopes in the nuclear industry was among the earliest applications of zeolites as ion exchangers. Their superior selectivity and stability characteristics spurred the development of other zeolite ion exchange applications. The high selectivity of clinoptilolite for ammonium ion in waste water treatment and other applications generated interest in developing synthetic zeolites such as Linde F and Linde W. Again these commercial applications were responsive to environmental problems.

The success of the zeolites both natural and synthetic in ammonium removal applications rests principally on their high cation selectivity. In this case their performance is far superior to the organic resin ion exchangers which show poor selectivity for ammonium ions, especially in competition with calcium and magnesium ion {9}.

Natural Zeolites. There have been a number of comprehensive and excellent reviews of the uses of natural zeolites in the last five years {1,10,11}. A summary adapted from these references is shown in Table 7. The major use of natural zeolites is in bulk mineral applications {1}: in Europe in the building and construction industry, where proximity to building location makes them cost effective; and in the Far East as filler in the paper industry, largely because of the unavailability of alternate mineral resources. As discussed previously, a modest market for zeolite minerals has developed as a molecular sieve adsorbent in acid gas drying in the natural gas industry, in NH_4 removal in water treatment

systems by ion exchange, and in the production of oxygen and nitrogen via adsorptive air separation, especia
lly in Japan {59}. In general, however, their penetration into molecular sieve applications has been quite
limited.

TABLE 7

SUMMARY OF USES OF NATURAL ZEOLITES {1,10,5}

Bulk Applications: Molecular Sieve Applications:

    Filler in Paper   Separation of Oxygen and Nitro-
    Pozzolanic Cements and   gen from Air
    Concrete   Acid-resistant Adsorbents in
    Dimension Stone   Drying and Purification
    Lightweight Aggregate   Ion Exchangers in Pollution
    Fertilizers and Soil   Abatement Processes
    Conditioners
    Dietary Supplement in
    Animal Nutrition

EVOLUTION IN COMMERCE

    The synthetic zeolite markets progressed from a re
latively small adsorbent market of the order of a
million dollars in the late 50's, through a rapid growth
which was influenced by the use of zeolites X and Y in
catalytic cracking to a published estimate {1} of $40MM
in 1970, and a projected $250MM in 1979. The projected
market in the large volume, bulk commodity detergent
area for zeolite builders in reported to be $25MM, or
approximmately 100MM lbs. of zeolite A in 1980 {79},
and an optimistic projection for growth to a 400MM lb.
market in 1982 {78a}.

    The introduction of zeolites into cracking catalysts
caused a commercial as well as a technical revolution
in catalytic cracking. It is reported that zeolite cra
cking catalysts have saved refiners over $250MM per
year, and increased gasoline capacity substantially
{80}. Today, type Y zeolite has 100% of the U.S. market,
and 75-80% of that outside of the United States {68}.
The total worldwide consumption of zeolites in catalytic
cracking in 1978 is estimated at between 70-90MM pounds
per year {68}, representing the single largest use of
synthetic molecular sieve zeolites to date. It is likely that bulk use of sinthetic zeolites in detergents
may surpass that volume if projections are realized.

In a recent estimate of the natural zeolite market, Leonard {1} suggests worldwide sales beginning in 1965 of 24MM lbs. at a value of $1MM, 160 MM lbs. at $8MM in 1970, and a 1979 projection of 560MM lbs. at a value of $35MM. Greater than 90% of the market is in bulk mineral applications, and only about 2% of that in North America. The major markets are in Europe, Russia, and the Far East, especially Japan

The synthetic molecular sieve zeolites as industrial materials are appropriately classified as specialty chemicals, or preferably as engineered products. The molecular sieve industry is technology and engineering intensive. The majority of molecular sieve zeolite applications are engineered in all respects, from the synthesis of the zeolite material, the modification of their properties, the selection of a tailormade zeolite product, to a process engineering design and execution that integrates and optimizes the material with the process. The successful growth of molecular sieve zeolites as an industrial chemical depended most strongly on their development as an engineered product.

As developed previously, the effect of outside forces in the changing world surrounding molecular sieves had a profound impact on their commercial growth and development. This is especially true in the case of the problems and crises in energy and environment which evolved in the 60's and erupted in the 70's, and which offered opportunities for their uniquely suited properties in separations and catalysis. This is seen in the natural gas purification area where large growth has recently occurred in the treatment of LNG in giant base load facilities in the Middle East {6}, in increased FCC production of gasoline and octane improvement for unleaded gasoline, in the fledging use of zeolite ion exchangers in waste water treatment, in insulated window adsorbents, in substitution of zeolites for phosphate in detergents, and possibly in the future, methanol or biomass to gasoline. The normal/iso-paraffin processes using molecular sieve zeolites were influenced by the need for biodegradable detergents. The xylene separation processes were influenced by the industrial need for raw material for polyester fiber production.

THE PAST AS INDICATOR OF THE FUTURE: THE NEXT TWENTY-FIVE YEARS

The inital discoveries of the synthetic molecular sieve zeolites A, X and Y in the late 40's and early 50's spawned an immense worldwide science and technology, utilizing the resources of many major industrial R and D organizations throughout the world.   Zeolites are "researched" and used commercially in every major country in the world.   (Recently China displayed molecular sieves as one of its industrial products at the Shanghai Industrial Exhibit {81}.)   Worldwide there are over a dozen manu- facturers of zeolites or zeolite-containing products. As an industrial material, their market has grown to hundreds of millions of dollars.   Since the mid-1960's, the later developing natural zeolites have reached the status of an important industrial mineral resource   82 with more than 300,000 tons mined worldwide, and used principally in bulk applications.

The Past.   What was necessary for the success of the molecular sieve zeolite industry?   The development of any new commercial product and process is usually the result of complex interactions among many contributing factors. I will attempt here to highlight some key factors that facilitated the zeolite "explosion".

The commercial use of zeolites depends on: 1) useful properties controlled by their structural chemistry; 2) availability; and 3) cost {5}.   The pioneering work of Barrer in the 40's in outlining the large number of molecular sieve separations possible gave the impetus to the industrial researchers, Milton and associates, who made initial key discoveries in novel synthetic zeolite compositions and a practical method for their manufacture. The scientific and engineering resources were then committed by a large, major chemical corporation, Union Carbide Corporation, with available technical and com- mercial resources.   As a result they became available to the industrial and scientific communities in late 1954.

The major early synthesis efforts in the late 40's and early 50's could not have been successful without the development of rapid, effective characterization techniques to evaluate the synthesized products {2}. Such techniques were developed to determine their struc- ture, chemical compostion, purity, and adsorption pro- perties.   As a result time efficient analysis of a very

large number of synthesis experiments was possible which
facilitated the discovery of twenty-some new zeolite
species and delineated their optimum synthesis system.

The general synthesis method developed by Milton
provided a simple, cost effective manufacturing process,
involving readily available, cheap raw materials such as
hydrated alumina, soluble alkali silicates, caustic, and
water, and process conditions of relatively low temperature
and pressure and short crystallization times. The unit
operations of batch crystallization, filtration and drying, were well established in the practices manufacturing
art. Thus, molecular sieve zeolites could be manufactured
to compete on a cost performance basis with other known
commercial adsorbent, catalyst, and ion exchange materials.

The development of formed or bonded zeolite products
necessary for commercial application in supported or moving bed systems, required extensive development of forming
technology. The as-synthesized 1-5μm zeolite crystals
are formed into beads, pellets, or mesh, typically by use
of a clay binder. Over the period of twenty-five years,
the development of an unsung and little discussed forming
technology has resulted in the ability to control particle
properties such as a strength and attrition resistance
and mass and heat transfer characteristics, and to
optimize the formed products' properties and performance.

As the new molecular sieve zeolites became known
and available and subsequently used, extensive research
effort in industrial and academic circles provided a
wealth of scientific information on the physical,
chemical, aned structural characteristics of this unique
class of materials. The resulting in-depth understanding
of properties allowed the selection of zeolite product
for a specific application, the identification of applications for the product, and the design and engineering
of the application process.

Their application as adsorbents required a major
development in adsorption process design and engineering
technology. The unit operation of adsorption has under-
gone a major development in the last twenty years mainly
as a result of the introduction of molecular sieves as
commercial adsorbents {6,2}. It is now a mature engineering practice that has brought adsorption to the forefront
as a major tool of the chemical process industry.

Indispensable to the commercial success of molecular sieve zeolites has been the dedication and contributions of the scientists and engineers who provided the key to their discovery and development, and subsequently unfolded their elegant structural and chemical architecture and novel properties. The zeal with which these molecular sieve "apostles" preached their gospel to the scientific world, to industrial technology management, and to the hard-to-sell chemical process industries, was essential to their success. Hundreds of these apostles and champions became committed, in addition to the original pioneers. The growth of molecular sieve science and technology has evolved a community of outstanding scientists and engineers, whose contributions include creative and practical scientific work, and the successful translation of R and D results to commercial manufacture, sale and utility. It is estimated at the onset of their second twenty-five years, that over five thousand scientists and engineers devote a substantial portion of their technical effort to molecular sieve zeolites.

The Future. The future trends in materials will no doubt see the development of new commercial zeolites selected from newly discovered compositions and structures, chemical modifications of present commercial products to generate new and useful properties, and a reevaluation of the host of known zeolites which never achieved commercial success. It seems likely that with the increasing number of laboratories devoting resources to the search for new structures and compositions, new classes of molecular sieve materials will be discovered. The modification chemistry of zeolites practiced to date, such as steaming and chemical extraction, leaves a vast area of chemical and structural modification of solids as yet unexplored with zeolites. Additional types of natural zeolites will probably not achieve commercial prominence since the large geological exploration efforts for zeolite deposits throughout the world during the last ten to fifteen years have probably identified all of the zeolite mineral species of commercial significance.

The commercialization of "stored" or "shelved" zeolites has largely been hampered by their lack of general availability and their apparent inability to compete performance-wise with the current commercial products. With the worldwide expansion of scientific zeolite centers with the capability of synthesizing non-commercial zeolites and determining their properties and potential applications, it is likely that several "old" zeolites will achieve commercial status.

It is likely that there will be more development and change in zeolite manufacturing processes during the next decade than during the last twenty years, due to the cost incentives of the bulk chemical and consumer markets, and the availability of the natural zeolites.

Breck has recently reviewed {5} a large number of new potential applications areas for zeolites. His compilation of proposed applications based on the reported literature is reproduced in Table 8.

TABLE 8

SOME PROPOSED APPLICATIONS OF ZEOLITES {5}

Adsorption

    New adsorbents for sieving
    Hydrophobic adsorbents
    Gas storage systems
    Carriers of chemicals

Nuclear Industry Applications

Environmental

    Weather modification
    Solar energy

Agricultural

    Fertilizers and soils
    Animal culture

Consumer Applications

    Beverage carbonation
    Laundry detergents
    Flame extinguishers
    Electrical conductors
    Ceramics
    New catalysts

The major trends in future commercial applications will probably comprise substantial growth in most of the presently existing separations and catalysis areas, and the development of new applications. The emergence of zeolite ion exchange applications could parallel that of zeolite adsorption technology over the last twenty-five years. The maturing of the developing ion exchange process design and engineering technology, with the capability of advanced systems design and engineering concepts, should stimulate the growth and acceptance of zeolite ion exchange separations in the chemical process industry alongside those based on adsorption and catalysis.

The growth of bulk chemical and consumer applications for synthetic as well as natural zeolites appears to be certain. In addition to those now prevalent for natural zeolites, Table 8 includes their use in agriculture, beverage carbonation, and raw materials for ceramics.

Natural zeolites should continue to grow as an important industrial mineral resource used principally in bulk application areas. The "engineering" of the mined zeolites, by beneficiation to upgrade purity and chemical modification to tailor properties, will no doubt emerge as the level of technically intensive effort on natural zeolites expands {1}. They should then enjoy a larger share of the "engineered" molecular sieve type applications. The extent of such growth will not likely be related to their lower cost but rather improved property and performance characteristics. The expansion will continue to be a relatively minor portion of the total molecular sieve applications market, largely because of the increasing capability in the manufacture and availability of a large number of synthetic zeolites, and the ability to control purity and properties during manufacture.

The trends in molecular sieve process design should see more compound, multistep process systems utilizing multiple or composite molecular sieve materials and combined unit operations such as integrated adsorption and catalytic systems. The energy savings possible in adsorption separations has received emphasis only recently {83}. It is likely that the energy efficiency in molecular sieve adsorption and catalytic processes will be more fully exploited in the process design.

Over the long range, molecular sieve zeolite technology should continue to be strongly influenced by the new emphasis on clear environment, and energy and renewable resource technology. Cost effective and novel separation and recovery processes will be required to meet pollution standards and material and energy resource limitations in thenext several decades. Development of adequate energy resources, especially the presently considered alternate synthetic fuel technologies based on synthesis gas, oil shale, coal and gasohol among others, all involve technically difficult and complex separations and catalytic problems.

Molecular seive zeolites are well positioned historically and offer a most appropriate technology
because of their unique properties which give them a
near-infinite flexibility to tailor product and process.
The recent extension of their shape and surface selectivity characteristics with the advent of high silica
zeolites and silica molecular sieves offers new opportunities in design parameters, as exemplified by the
methanol to gasoline process with ZSM-5. The availability of hydrophobic molecular sieve adsorbents opens
up a new application area in removing and recovering
organic molecules from aqueous systems. Combined hydrophobic and hydrophilic adsorbent systems would allow the
concentration or removal of an organic molecule from an
aqueous solution, and efficient drying of the recovered
organic. Similar separation schemes are now under investigation in the production of gasohol from grain {84}.

The zeolite future looks bright.

REFERENCES

1. D. W. Leonard, Preprint of Paper Presented at
   Soc. of Mining Engineers of A.I.M.E., Oct. 17-19,
   1979, Tucson, Ariz.
2. R. M. Milton, in "Molecular Sieves", Soc. Chem.
   Ind., London, (1968), p. 199.
3. D. W. Breck, "Zeolite Molecular Sieves", Wiley-
   Interscience, New York, (1974).
4. R. M. Barrer, "Zeolites and Clay Minerals as
   Sorbents and Molecular Sieves", Academic Press,
   London, (1978).
5. D. W. Breck, Proceedings of the Conference on The
   Properties and Applications of Zeolites, Soc.
   Chem Ind., London, April 18-20, 1979, to be
   published.
6. R. A. Anderson, ACS Symposium Series, 1977, 40,
   637.
7. J. A. Rabo, R. D. Bezman, M. L. Poutsma, Acta Phys.
   Chem., 1978, 24, 39.
8. J. A. Rabo, ed., "Zeolite Chemistry and Catalysis",
   Amer. Chem. Soc. Monograph 171, (1976).
9. J. D. Sherman, Adsorption and Ion Exchange
   Separations, AIChE Symposium Series, 1978, 74,
   No. 179 98.
10. F. A. Mumpton, in "Natural Zeolites, Occurrence,
    Properties, Use", L. B. Sand and F. A. Mumpton, eds.,
    Pergamon, London, (1978), p. 3.

32

11. K. Torii, in ref. 10, p. 441.
12. R. M. Milton, U.S. Patent 2,882,243 (1959); R. M.
     Milton, U.S. Patent 2,882,244 (1959).
13. D. Fraenkel and J. Shabtai, J. Am. Chem. Soc., 1977,
     99, 7074.
14. D. W. Breck, U.S. Patent 3,130,007 (1964).
15. L. B. Sand, U.S. Patent 3,436,174 (1969).
16. Anon: Chem. ⸗ Eng. News, 1962, Mar. 12, 62.
17. D. W. Breck and N. A. Acara, U.S. Patent 3.216,789
     (1965).
18. J.-R. Bernard and J. Nury, Belgian Patent 845,458
     (1976).
19. E. M. Flanigen and E. R. Kellberg, Dutch Patent
     6,710,729 (1967).
20. R. L. Wadlinger, G. T. Kerr and E. J. Rosinski,
     U.S. Patent 3,308,069 (1967).
21. R. J. Argauer and G. R. Landolt, U.S. Patenet
     3,702,886 (1972).
22. P. Chu, U.S. Patent 3,709,979 (1973).
23. E. J. Rosinski and M. K. Rubin, U.S. Patent
     3,832,449 (1974).
24. C. J. Plank, E. J. Rosinski and M. K. Rubin,
     U.S. Patent 4,046,859 (1977).
25. M. K. Rubin, E. J. Rosinski and C. J. Plank,
     U.S. Patent 4,086,186 (1978).
26. (a) E. M. Flanigen, J. M. Bennett, R. W. Grose,
     J. P. Cohen, R. L. Patton, R. M. Kirchner and
     J. V. Smith, Nature, 1978, 271, 512; (b) R. W. Grose
     and E. M. Flanigen, U.S. Patent 4,061,724 (1977).
27. E. M. Flanigen and R. L. Patton, U.S. Patent
     4,073,865 (1978).
28. G. T. Kokotailo, S. L. Lawton, D. H. Olson and
     W. M. Meier, Nature, 1978, 272, 437.
29. D. M. Bibby, N. B. Milestone and L. P. Aldridge,
     Nature, 1979, 280, 664.
30. R. W. Grose and E. M. Flanigen, U.S. Patent
     4,104,294 (1978).
31. C. V. McDaniel and P. K. Maher, in ref. 2, p. 186.
32. G. T. Kerr, J. Phys. Chem., 1967, 71, 4155.
33. J. Scherzer, J. Catalysis, 1978, 54, 285
34. N. Y. Chen, J. Phys. Chem., 1976, 80, 60.
35. P. E. Eberly, S. M. Laurent and H. E. Robson,
     U.S. Patent 3.506,400 (1970).
36. R. L. Patton, E. M. Flanigen, L. G. Dowell and
     D. E. Passoja, ACS Symposium Series, 1977, 40, 64.
37. (a) R. M. Barrer and M. B. Makki, Can. J. Chem., 1964,
     42, 1481; (b) R. M. Barrer and B. Coughlan, in ref.
     2, p. 141; (c) R. Beecher, A. Voorhies Jr. and
     P. Eberly Jr., Ind. Eng. Chem., Prod. Res. Develop.,
     1968, 7, 203; (d) M. M. Dubinin, G. M. Fedorova, D. M.
     Plavnik, L. I. Plguzova and E. N. Prokofeva, Iz. Akad.
     Nauk, SSR, Ser. Khim., 1968, 2429.

38. D. E. W. Vaughan, in ref. 10, p. 353.
39. R. W. Grose and E. M. Flanigen, Union Carbide laboratories, unpublished results.
40. D. W. McKee and R. P. Hamlen, Union Carbide laboratories, unpublished results; D. W. McKee, U.S. Patents 3,140,932 and 3,140,933 (1964); ref. 3, p. 694.
41. J. A. Rabo, P. E. Pickert, D. N. Stamires and J. E. Boyle, Actes Congr. Intern. Catalyse, 2e, Paris, 1961, 2, 2055.
42. D. Barthomeuf, J.C.S. Chem. Comm., 1977, 743.
43. (a) D. Barthomeuf, Acta Phys. Chem., 1978, 24, 71; (b) D. Barthomeuf, J. Phys. Chem. 1979, 83, 249.
44. J. C. Vedrine, A. Auroux, V. Bolis, P. Dejaifve, C. Naccache, P. Wierzchowski, E. G. Deroune, J. B. Nagy, J.-P. Gilson, J.H.C. van Hooff, J. P. van der Berg, and J. Wolthuizen, J. Catalysis, 1979, 59, 248-262.
45. W. J. Mortier, J. Catalysis, 1978, 55, 138.
46. P. A. Jacobs, W. F. Mortier and J. B. Uytterhoeven, J. Inorg. Nucl. Chem., 1978, 40, 1919.
47. D. W. Breck, W. G. Eversole, and R. M. Milton, J. Am. Chem. Soc., 1956, 78, 2338.
48. E. M. Flanigen, Adv. Chem. Ser., 1973, 121, 119.
49. D. W. Breck and N. A. Acara, U.S. Patent 2,950,952 (1960).
50. (a) R. M. Barrer and P. J. cenny, J. Chem. Soc., 1961, 971; (b) R. M. Barrer, P. J. Denny and E. M. Flanigen, U.S. Patent 3,306,922 (1962).
51. G. T. Kerr, Science, 1963, 140, 1412.
52. R. M. Barrer and H. Villiger, J. Chem. Soc., D, 1969, 659.
53. N. A. Acara, U.S. Patent 3,414,602 (1968).
54. Abbreviated organic base nomenclature described in ref. 48.
55. G. T. Kokotailo, P. Chu, S. L. Lawton, and W. M. Meier, Nature, 1978, 275, 119.
56. E. M. Flanigen and R. L. Patton, Union Carbide laboratories, unpublished results.
57. L. D. Rollman, Adv. Chem. Ser., 1979, 173, 387.
58. W. J. Mortier, P. Geerlings, C. Van Alsenoy and H. P. Figeys, J. Phys. Chem, 1979, 83, 855.
59. H. Minato and T. Tamura, in ref. 10, p. 509.
60. D. B. Broughton, Chem. Eng. Progr., 1977, 73, No. 10, 49; H. Odawara, Y. Noguchi and M. Ohno, U.S. Patent 4,014,711 (1977); R. W. Neuziland and J. W. Priegnitz, U.S. Patent 4,024,331 (1977); and H. Odawara, M Ohno, T. Yamazaki and M. Kanaoko, U.S. Patent 4,157,267 (1979).
61. C. N. Kimberlin, Jr. and E. M. Gladrow, U.S. Patent 2,971,903 (1961).

34

62. J. A. Rabo and J. E. Boyle, U.S. Patent 3,130,006 (1964); J. A. Rabo, P. E. Pickert and J. E. Boyle, U.S. Patent 3,236,762 (1966).
63. P. B. Weisz and V. J. Frilette, J. Phys. Chem., 1960, 64, 382.
64. C. J. Plank, E. J. Rosinski and W. P. Hawthorne, Ind. Eng. Chem., Prod. Res. Dev., 1964, 3, 165; C. J. Plank and E. J. Rosinski, Chem. Eng. Prog. Symp. Ser., 1967, 73 (63), 26.
65. P. B. Venuto and E. T. Habib, Jr., Catal. Rev.-Sci. Eng., 1978, 18, 1.
66. P. B. Weisz, Chemtech, 1973, 498.
67. J. S. Magee and J. J. Blazek, in ref. 8, p. 615.
68. D. P. Burke, Chem. Week, Mar. 28, 1979, 42.
69. S. M. Csicsery, in ref. 8, p. 680.
70. N. Y. Chen and W. E. Garwood, J. Catalysis, 1978, 52, 453.
71. N. Y. Chen, W. E. Garwood, W. O. Haag, and A. B. Schwartz, paper presented at the Symposium on Advances in Catalytic Chemistry, Oct. 3-5, 1979, Snowbird, Utah.
72. C. C. Chang and A. J. Silvestri, J. Catalysis, 1977, 47, 249.
73. S. L. Meisel, J. P. McCulloguh, C. H. Lechthaler and P. B. Weisz, Chemtech, 1976, 6, 86.
74. Anon: Oil and Gas Journal, Jan. 14, 1980, 95.
75. P. B. Weisz, W. O. Haag and P. G. Rodewald, Science, 1979, 206, 57.
76. T. L. Thomas, Jr., U.S. Patent 3,033,641 (1962).
77. Reviewed in ref. 9.
78. (a) Anon: Chem. and Eng. News, May 22, 1978, (b) P. Berth, G. Jakobi, E. Schmadel, M. J. Schwuger and C. H. Krauch, Angew. Chem. Internat. Edit., 1975, 14, 94; (c) H. G. Smolka and M. J. Schwuger, in ref. 10, p. 487.
79. Anon: Chem. Week, Jan. 2, 1980, 29.
80. J. S. Magee, ACS Symposium Series, 1977, 40, 650.
81. R. J. Seltzer, Chem. and Eng. News, Dec. 17, 1979, 21.
82. Cpt. on Zeolites, in "Industrial Minerals and Rocks", 4th ed., S. J. Lefond, ed., 1975, 1234-1274.
83. J.-P. Sicard and P. Richman, Oil & Gas Journal, Octo. 29, 1979, 145.
84. C. D. Chrisholm, Ames Laboratory, Iowa State University, unpublished results, 1979.

ZEOLITE STRUCTURES

R.M. Barrer

Chemistry Department
Imperial College of Science and Technology
London SW7 2AY
England

ABSTRACT

An account has been given of silicate anions from topological
and configurational aspects, with special reference to zeolites.
Different ways of constructing zeolite anions are described which
lead to known frameworks and to a large number of novel ones.
Brief consideration is also given to resultant channel systems
and to Al, Si order and disorder.

1. INTRODUCTION: SILICATE ANIONS

The structural side of zeolite chemistry is concerned with
their three-dimensional giant anions and the associated intrazeolite cations and water molecules. Much the most accurate
structural information relates to the anionic frameworks which
have geometrical elegance, provide numerous honeycomb structures
and determine important zeolite properties. However zeolite
anions are a part only of a remarkable array of silicate anions
in which many sub-units found in zeolites are also present. It
is therefore of considerable interest to give examples of these
anions in a build up to the main topic of zeolite anions. Zeolites
are usually synthesised from gels or from other silicates.
Because nucleation can be influenced by the starting materials
examples of the diverse silicate anions which could be chosen as
sources of silica acquire an additional significance.

Two aspects of the structure of silicate anions are their
topology and their configuration. By topology we will mean the

Table 1: Silicates grouped according to anion types

Type	Sub-types
Finite anions ("Island" anions)	Ortho- and pyro-silicates. Short unbranched chains. Single ring anions. Branched single rings. Double ring prisms. Structures with two or more island anions.
Infinite chain anions	Unbranched single chains. Open-branched single chains. Loop-branched single chains (linked rings). Multiple chains (two, three, four or five cross-linked chains). Hybrid chains. Tubular chains.
Infinite sheet anions	Unbranched single sheets. Branched single sheets. Double sheets. Hybrid triple sheets.
Infinite three-dimensional anions	Non-porous and porous tecto-silicates.

pattern of the Si–O–T bonds (T = Al or Si) and by configuration we will mean the spatial disposition of the bond pattern. For example, SiO_4 tetrahedra can join with other SiO_4 tetrahedra to give unbranched linear chains. These can, as we shall see, be variously puckered. These anions all have the same topology in terms of the bond pattern, but they have different configurations. Silicates can be grouped according to their topologies as in Table 1. In the following sections examples of the various categories of Table 1 will be given, leading up to and with emphasis on zeolites.

2. FINITE ANIONS

Some island anions are as follows (1):-

Single tetrahedron	SiO_4^{4-}	Orthosilicates
Two linked tetrahedra	$Si_2O_7^{6-}$	Disilicates
Chain of three tetrahedra	$Si_3O_{10}^{8-}$	Trisilicates

Ring of three tetrahedra (3-ring)	$Si_3O_9^{6-}$	Benitoite
4-ring	$Si_4O_{12}^{8-}$	Taramellite
6-ring	$Si_6O_{18}^{12-}$	Beryl
8-ring	$Si_8O_{24}^{16-}$	Muirite
9-ring	$Si_9O_{27}^{18-}$	Eudialite
12-ring	$Si_{12}O_{36}^{24-}$	Traskite
Double 3-ring (triangular prism)	$Si_6O_{15}^{6-}$	$(Ni(En)_3)_3[Si_6O_{15}]26H_2O$
Double 4-ring (cubic unit)	$Si_8O_{20}^{8-}$	Ekanite
Double 6-ring (hexagonal prism)	$Si_{12}O_{30}^{12-}$	Milarite
Branched chain of three tetrahedra	$Si_5O_{16}^{12-}$	Zunyite
Branched 4-ring	$Si_6O_{18}^{12-}$	Eakerite
Branched 6-ring	$Si_{18}O_{54}^{36-}$	Tienshanite

The line diagrams give the topologies of the last three anions. The dots represent the Si atoms and the lines show the linking. The oxygen atoms are not shown: they are between the pairs of Si atoms they link. The non-linking oxygens required to complete the tetrahedral groups are, for clarity, omitted. Additionally to the above examples a crystalline tetra-alkylammonium silicate has been reported which contains the anion $Si_{10}O_{25}^{10-}$, believed to be a pentagonal prism (2).

3. INFINITE CHAIN ANIONS

Unbranched single chains are found in a variety of meta-silicates. They can have the configurations illustrated in Fig. 1 (3), the configurations being characterised by the periodicities, i.e. the number of linked tetrahedra which must be counted along the chain before the arrangement repeats itself. The periodicities of Fig. 1 a to k and silicates to which they refer are as follows:

Letter in Fig. 1	Mineral	Periodicity
(a)	Pyroxenes	2
(b)	$Ba_2[Si_2O_6]$	2
(c)	Wollastonite	3
(d)	Krauskopfite	4

(e)	Haradite	4
(f)	Rhodonite	5
(g)	Stokesite	6
(h)	Pyroxyferroite	7
(i)	Ferrosilite III	9
(k)	Alamosite	12

Even-period chains tend to become less stretched with higher mean electronegativity and mean valence of the charge-balancing cations while for odd-period chains the extent of chain puckering is strongly correlated with mean electronegativity of the cations but less so with cation radius (4).

Instances of open- and loop-branched chains are shown in Fig. 2a to g (4). The examples and their periodicities are:

Letter in Fig. 2	Mineral	Periodicity
(a)	Aenigmatite	4
(b)	Astrophyllite	2
(c)	Deerite	4
(d)	Vlasovite	6
(e)	Lemoynite	6
(f)	Pellylite	8
(g)	Nordite	10

(a) and (b) are open-branched single chains; (c) to (g) are loop-branched single chains. The latter produce rings linked to other rings via Si-O-Si bonds.

Some double chains with different periodicities are shown in Fig. 3 a to i (5) for the minerals named below:

Letter in Fig. 3	Mineral	Periodicity
(a)	Sillimanite	1
(b)	Amphibole	2
(c)	Synthetic $Li_4\left[(SiGe_3)O_{10}\right]$	2
(d)	Xonotlite	3
(e)	Devitrite	3
(f)	Synthetic $Na_2Be_2H[Si_6O_{15}]OH$	3
(g)	Narsarsukite	4
(h)	Inesite	5
(i)	Tuhualite	6

Examples (a) (c) (f) and (i) are all ladder anions composed of 4-rings linked by shared edges, with the same topologies but different configurations. A synthetic aluminate, $Na_7[Al_3O_8]$ (6) has the same topology as (e) (devitrite) but with Al replacing Si.

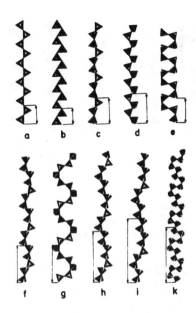

Fig. 1.   Configurations of some single chain unbranched anions
(3). For identification of species to which they refer
see text.

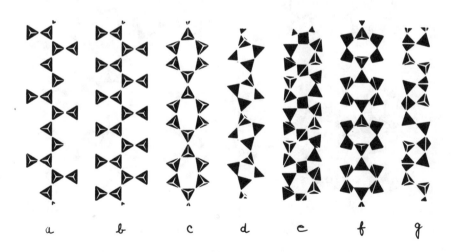

Fig. 2.   Some branched and loop-branched single chain anions (4).
Species to which they refer are identified in the text.

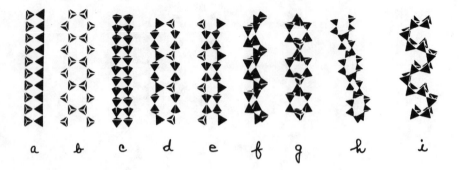

Fig. 3.   Some double chain anions with different periodicities
for species identified in the text (9).

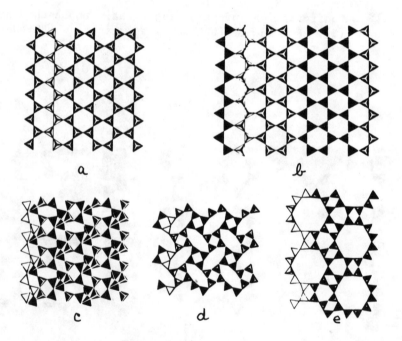

Fig. 4.   Several kinds of unbranched sheet anion for species
identified in the text (9).

As final examples of anions based on infinite chains one may include multiple, tube and hybrid chain anions (1,7):

Multiple chain anions	Example	Periodicity
Three cross-linked chains (8)	Synthetic $Ba_4[Si_6O_{16}]$	2
Four cross-linked chains, giving strip of hexagons, pattern as above, three hexagons wide (8)	Synthetic $Ba_5[Si_8O_{21}]$	2
Five cross-linked chains, giving strip of hexagons, pattern as above, four hexagons wide (8)	Synthetic $Ba_6[Si_{10}O_{26}]$	2
Tube anions (7)	Litidionite $NaKCu[Si_4O_{10}]$	3
	Miserite $KCa_5[Si_2O_7][Si_6O_{15}]OH,F$	3
Hybrid chain anion	Tinaksite $K_2NaCa_2Ti[Si_{17}O_{19}]OH$	3

Models of complex chain anions, such as tubular anions, when opened out and flattened yield open branched strip anions (7). The narsarsukite anion in this way opens to chain of branched 6-rings; the litidionite anion gives a strip hybrid anion of 4- and 8-rings; and the miserite anion gives a strip hybrid anion of 4-, 6- and 8-rings. Various hypothetical tubular anions and the strip anions they can yield when opened and flattened have also been considered (7). The strips can be extended to make infinite sheets composed of various combinations of rings. Some of these combinations are known to occur in sheet silicates, for example 4-, 6- and 12-rings in manganpyrosmalite, $(Mn, Fe)_8[Si_6O_{15}](OH, Cl)_{10}$. There seems no reason why complex chain anions of the kinds envisaged may not be found or synthesised in the rich field of silicate chemistry, and used as starting materials for zeolite synthesis.

4. INFINITE SHEET ANIONS

The next stage of polymerisation in silicate anions yields infinite sheets. Unbranched sheets of several kinds are shown in Fig. 4a to e (9). These are identified below:

(a) Hexagon based layers of tetrahedra found in mica, where all apices point in one direction.   Periodicity = 2

(b) Hexagon based layers in sepiolite; bands with apices pointing up alternate with bands having apices pointing down.   2

(c) The single layer in dalyite, $K_2Zr[Si_6O_{15}]$, (4-, 6- and 8-rings in layer).   3

(d) The single layer in apophyllite, $KCa_4[Si_8O_{20}]F.8H_2O$ (4- and 8- rings in layer).   4

(e) The manganpyrosmalite, $Mn_6[Si_{16}O_{30}]$ $(OH)_8Cl_2$, sheet, containing 4-, 6- and 12-rings.   6

Open branched single layers are also found, two such types being those identifiable in prehnite, $Ca_2(Al, Fe)[(Si_3, Al)O_{10}](OH)_2$ with periodicity 2, and in zeophyllite, $Ca_{13}[Si_5O_{14}]_2F_8(OH)_26H_2O$, with periodicity 4 and a sheet of 12-rings. A loop-branched single layer of periodicity 4 occurs in synthetic $NaPr[Si_6O_{14}]$.

Double layer sheet anions are illustrated in Fig. 5a, b and c, (10) in which

(a) is the unbranched double layer of hexacelsian, $Ba[Si_2Al_2O_8]$ of periodicity 2;

(b) is the loop-branched double layer of delhayelite, $Ca_4(Na_3, Ca)K_7[Si_{14}Al_2O_{38}]Cl_2F_4$ of periodicity 3; and

(c) is the loop-branched double layer of carletonite, $K_2Na_8Ca_8[Si_{16}O_{36}](CO_3)_8(OH,F)_22H_2O$, which has a periodicity of 6.

Two sub-layers of each kind are linked via the oxygen atoms marked with dots and lying on mirror or pseudo-mirror planes. The bottom parts of the diagram show the double layers edge on

5. SILICATES CONTAINING TWO ANIONS

In some silicates there may be more than one kind of anion. Examples of silicates of this type include the following (1):

Silicate	Anions Present	
Zoisite, epidote, ganomalite, vesuvian, serendibite, rustumite	$[SiO_4]^{4-}$	$[Si_2O_7]^{6-}$
Kilchoanite, ardennite	$[SiO_4]^{4-}$	$[Si_3O_{10}]^{8-}$
Meliphanite	$[SiO_4]^{4-}$	$[Si_7O_{20}]^{12-}$
Traskite	$[Si_2O_7]^{6-}$	$[Si_{12}O_{36}]^{24-}$
Miserite	$[Si_2O_7]^{6-}$	$[Si_{12}O_{30}]^{12-}$
Bavenite	$[Si_3O_{10}]^{8-}$	$[Si_6O_{16}]^{8-}$
Eudialite	$[Si_3O_9]^{6-}$	$[Si_9(O,OH)_{27}]^{\leqslant 18-}$
Chesterite	$[Si_4O_{11}]^{6-}$	$[Si_6O_{16}]^{8-}$
Reyerite	$[Si_8O_{20}]^{8-}$	$[Si_{14}Al_2O_{38}]^{14-}$
High pressure "garnet"	$[SiO_4]^{4-}$	$[SiO_6]^{8-}$
Synthetic $Si_5(PO_4)_6O$	$[SiO_4]^{4-}$	$[SiO_6]^{8-}$
High pressure $K_2Si_4O_9$	$[Si_3O_9]^{6-}$	$[SiO_6]^{8-}$

It is of interest that under high pressure conditions Si octahedrally co-ordinated with O appears. In stishovite, a high pressure form of crystalline silica, octahedra SiO_6 appear to be linked by edge sharing as much as by corner sharing. No example of face sharing of SiO_6 octahedra is known. Also, tetrahedra SiO_4 (or AlO_4) link only by corner sharing. The distances between central Si atoms are:

	corners	edges	faces
SiO_4 tetrahedra sharing: Si ... Si distance (Å)	3.24	1.87	1.08
SiO_6 octahedra sharing: Si ... Si distance (Å)	3.56	2.52	2.06

Edge and face sharing of SiO_4 tetrahedra and face sharing of octahedra draw Si^{4+} central ions too close for stability relative to corner sharing for SiO_4 pairs and edge and corner sharing for SiO_6 pairs.

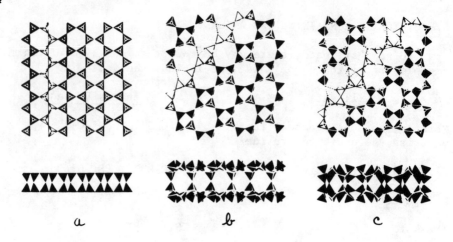

Fig. 5.   Examples of unbranched double layer anions identified
in the text (10).

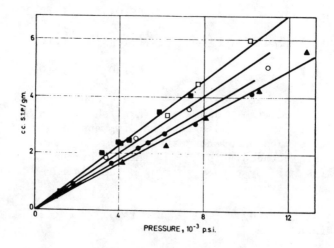

PRESSURE, 10⁻³ p.s.i.

Fig. 6.   Sorption of He in α- and in β_1-tridymite (12)

▪ 293 K		
☐ 375 K	α-tridymite	

	409.7 ○	
	433 ●	β_1-tridymite
	533 ▲	

Among the infinite single chain anions (Fig. 1) the periodicity appears to be determined or influenced by the associated
cations. Even-period chains are less stretched the higher the
mean electronegativity and mean valence of the cations. For odd-
period chains mean electronegativity more than mean radius of the
cations influences the chain puckering (11).

6. TECTOSILICATES

Three-dimensional frameworks (crystalline silicas, felspathoids, felspars and zeolites) have $O/(Al + Si) = 2$, indicating
the presence only of tetrahedra TO_4 ($T = Al$ or Si) where each
tetrahedron shares its four oxygens with four other tetrahedra.
In clay minerals on the other hand octahedral layers of AlO_6 or
MgO_6 are present attached either to one tetrahedral layer of
6-rings (kandites) or to two such layers one one each side of the
layer of octahedra (smectites, vermiculites and micas). The
resultant multiple sheets may be uncharged as with the kandites or
they may be anionic as with most of the three-layer sheet structures. In the latter case the charge-neutralising cations are
located between successive sheets. In tectosilicates there is
one equivalent of cations per g. atom of the Al replacing Si, and
the ions are distributed in three dimensions in the interstices of
the network.

The tectosilicates may be sub-divided into those which are not
and those which are porous (Table 1). The felspars, the denser
crystalline silicas such as quartz and various felspathoids
(egs. nepheline, kaliophilite, kalsilite and eucryptite) are non-
porous; in them the interstices are not large enough to contain
even the smallest guest molecules. Tridymite and cristobalite can
however take up considerable amounts of He and Ne, as shown for
tridymite in Fig. 6 (12) and so can just be considered as porous
tectosilicates. Other porous crystalline silicas are silicalites 1
and 2(4) dodecasil-3C (15) and melanophlogite (see §. 7). Some
felspathoids such as sodalite and cancrinite (Fig. 7) also have
frameworks with cavities or channels large enough to contain
various salts and/or zeolitic water.

The most important porous tectosilicates are found among the
large and growing number of zeolites, each type having its
individual configuration of windows, cavities and channels of
molecular dimensions. The pore volumes accessible to water, and
often to numerous other guest molecules, range from ~ 0.18 cm^3 per
cm^3 of crystal for the least porous (analcime) to ~ 0.50 cm^3 per
cm^3 of crystal for the most porous (such as faujasite, chabazite
and zeolites A, H-RHO, ZSM-2 and ZSM-3. The open structures allow ready migration of water molecules and of ions in
cation exchange. The exchange capacity of some zeolites is given
for the idealised compositions in Table 2. This capacity can of

Table 2: Exchange capacities in meq/100g of some zeolites

Zeolite	Idealised composition	Exchange capacity
Natrolite	$Na_2[Al\ Si_3O_{10}]2H_2O$	530
Analcime	$Na[AlSi_2O_6]H_2O$	450
Levynite	$Ca[Al_2Si_4O_{12}]6H_2O$	400
Chabazite	$((\frac{1}{2})Ca,Na)[AlSi_2O_6]3H_2O$	400
Gmelinite	$((\frac{1}{2})Ca,Na)[AlSi_2O_6]3H_2O$	400
Edingtonite	$Ba[Al_2Si_3O_{10}]4H_2O$	390
Faujasite	$(Ca,Na_2)[Al_2Si_5O_{14}]6.6H_2O$	390
Harmotome	$(K_2,Ba)[Al_2Si_5O_{14}]5H_2O$	390
Heulandite	$Ca[Al_2Si_6O_{16}]5H_2O$	330
Stilbite	$(Na,(\frac{1}{2})Ca)[AlSi_3O_8]3H_2O$	320
Mordenite	$((\frac{1}{2})Ca,Na)[AlSi_5O_{12}]3.3H_2O$	230

course vary with the Si/Al ratio which can usually be altered
according to synthesis conditions. Zeolites like ZSM-5 or -11
with Si/Al ratios varying between $\sim$25 and 1000 carry very much less
negative charge per 100g and are thus less polar than the most
aluminous zeolites of Table 2.

7. ZEOLITE GROUPS

Attempts have been made to group together zeolites which have
structural elements in common. For numerous synthetic zeolites
the structures are still unknown, but examples of grouping are
given in Table 3. This table contains instances of framework
anions which are variants of a single topology, i.e. isotypes
(egs. all members of the analcime group; heulandite and clinoptilolite in the heulandite group; stilbite, stellerite and barrerite, also in the heulandite group; phillipsite and harmotome in
the phillipsite group; gismondine and Na-P also in the phillipsite
group; and natrolite scolecite and mesolite in the natrolite group).
Variants of a given topology arise as a result of differing chemical
compositions. For examples analcime has a cubic unit cell of edge
13.72 Å and unit cell content $Na_{16}[Al_{16}Si_{32}O_{96}]16H_2O$. The Na^+ and

Table 3: Classification of some zeolites and porous tectosilicates

1. Analcine Group

   Analcime
   Wairakite
   Leucite (felspathoid)
   Rb-analcime (")
   Pollucite
   Viseite (aluminosilico-
   phosphate)
   Kehoeite (aluminophosphate)

2. Chabazite Group

   Chabazite
   Gmelinite
   Erionite
   Offretite
   Levynite
   Mazzite (zeolite Ω)
   Zeolite L
   Sodalite hydrate
   Cancrinite hydrate

3. Clathrate Group

   Melanophlogite
   Zeolite ZSM-39, Clathrasil-3C

4. Faujasite Group

   Faujasite (zeolites X and Y)
   Zeolite ZSM-2
   Zeolite ZSM-3
   Paulingite
   Zeolite A
   Zeolite RHO
   Zeolite ZK-5

5. Heulandite Group

   Heulandite
   Clinoptilolite
   Brewsterite

   Stilbite
   Stellerite
   Barrerite

6. Laumontite Group

   Laumontite
   Yugawaralite

7. Mordenite Group

   Mordenite
   Ferrierite
   Dachiardite
   Epistilbite
   Bikitaite

8. Natrolite Group

   Natrolite
   Scolecite
   Mesolite
   Thomsonite
   Gonnardite
   Edingtonite
   Metanatrolite

9. Pentasil Group

   Zeolite ZSM-5, Silicalite I
   Zeolite ZSM-11, Silicalite II

10. Phillipsite Group

    Phillipsite
    Harmotome
    Gismondine
    Zeolite Na-P
    Amicite
    Garronite
    Merlinoite
    Zeolite Li-ABW

48

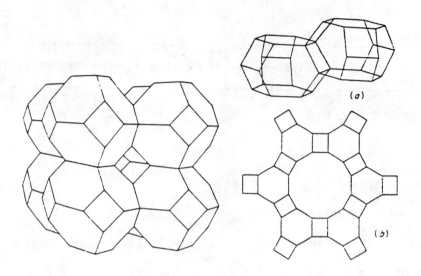

Fig. 7.   Frameworks of (i) sodalite (l.h.s.) and (ii) cancrinite
(r.h.s.). In (i) 14-hedral cages are identifiable.
(ii) shows (a) 11-hedra typical of cancrinite and (b)
a section normal to the c-direction, indicating a wide
channel circumscribed by 12-rings. Al or Si are
centred at each corner and O atoms are centred near
but not on the mid-point of each edge of Fig. 7 and
analogous subsequent framework representations. The
scales of the two drawings are not the same.

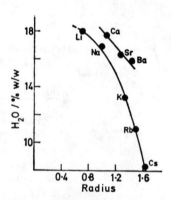

Fig. 8. Water contents of different cationic forms of phillipsite
plotted against cation radii (16).

the H_2O occupy different sub-lattices. Exchange of Na^+ by K^+ gives leucite with a tetragonal unit cell having a = 12.98 Å and c = 13.68 Å. The crystals are anhydrous and the K^+ ions are in the sub-lattice occupied by H_2O in analcime. The topology of the analcime anion is however unchanged. Another way in which variants of a given topology may arise is through changes in Si/Al ratios in the framework. This changes the cation density, which can also be altered by exchanges such as $2Na^+ \gtrless Ca^{2+}$. These two factors operate for the isotypes heulandite, $Ca_4[Al_8Si_{28}O_{72}]24H_2O$, (monoclinic, a = 17.72 Å, b = 17.90 Å, c = 7.43 Å, β = 116°25') and clinoptilolite, $Na_6[Al_6Si_{30}O_{72}]24H_2O$, (monoclinic, a = 17.64 Å, b = 17.90 Å, c = 7.40 Å, β = 116°22'). When one cation is replaced by another of different size or charge the distributions of the ions between sub-lattices may change as well as the water content under ambient conditions. Such changes can in turn cause minor or significant framework distortions without altering the topology. An example of the variation of water content with exchange cation is shown in Fig. 8 for phillipsite (16). For ions of the same charge water contents decrease as ion radius increases. A different curve is obtained for a series of divalent ions than for a series of univalent ones.

Guest species other than water can also result in variants of a given topology. Sodalite ($Na_6[Al_6Si_6O_{24}]2NaCl$; cubic with a = 8.87 Å) and nosean ($Na_6[Al_6Si_6O_{24}]Na_2SO_4$; cubic with a = 9.0 Å) both have the sodalite framework, slightly dilated in the case of nosean. In general, tectosilicate frameworks although strong and rigid can undergo small adjustments of crystallographic significance.

The majority of the frameworks in most zeolite groups are not variants of a given topology but possess different topologies, with structural similarities which justify their being grouped together. Such classifications are not without alternatives because on a basis of the possession of structural sub-units in common certain zeolites could equally well appear in more than one group. Thus in sodalite placed in the chabazite group 14-hedral cavities exist which are also found in some members of the faujasite group (faujasite and zeolite A). However sodalite and cancrinite are structurally related and cancrinite is well placed in the chabazite group and so is sodalite. The classification of zeolites into groups is thus somewhat arbitrary.

Zeolites can also be considered in terms of the densities of bonds Si-O-T (T = Al or Si) in the x, y and z directions. Certain network anions have bond densities comparable in all directions (e.g. groups 1, 2, 3, 4 and 9 of Table 3); others have bond densities greater in two directions than in the third (group 5); and others have these densities greater in one direction than in the remaining two ("fibrous" zeolites of the natrolite group). Bond density differences can influence thermal stability and rigidity

50

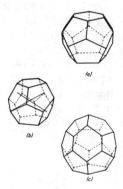

Fig. 9.   Dodecahedra (b), tetradecahedra (a) and hexadecahedra
          (c) found in melanophlogite ((a) + (b)) and in
          dodecasil-3C or zeolite ZSM-39 ((b) + (c)) (18).

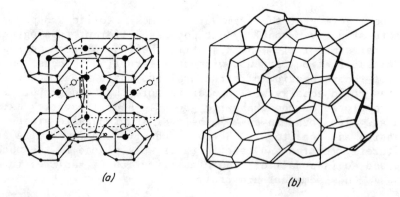

(a)   (b)

Fig. 10.   Stacking of pentagonal dodecahedra in (a) melanophlogite
           and (b) dodecasil-3C or zeolite ZSM-39 (18).

and the extent of deformation when undergoing modification by ion
exchange or in changes of the Si/Al ratios and in dehydration and
rehydration. In the fibrous zeolites the dense-bond chains are
cross-linked to like chains by single Si-O-T bonds and this appears
to make the frameworks readily deformable. On the other hand in
zeolite A 14-hedral sodalite cages are each linked through the six
4-ring faces to six other sodalite cages by four Si-O-T bonds
creating a cubic linkage unit. Cross-connecting sodalite cage sub-
units by multiple Si-O-T bonds produces a more rigid less easily
deformable framework than when, as in the natrolite group, cross-
linking involves single Si-O-T bonds. Similarly in faujasite each
sodalite cage is linked to four other such cages through four of
its eight 6-ring faces by six Si-O-T bonds, creating a hexagonal
prism linking unit. Rigidity and thermal stability is found for
such zeolites ·as A and faujasite (X and Y) despite the smaller
density, greater intracrystalline porosity, and smaller number of
(Al + Si) atoms (and hence of Si-O-T bonds) per unit volume. In
zeolite A, faujasite, natrolite and thomsonite these numbers are
as follows:

Zeolite	natrolite	thomsonite	zeolite A	faujasite
$(Al + Si)_o$ per 1000 $\mathring{A}^3$	14.5	14.4	12.9	12.7

Included in Table 3 are examples of porous crystalline silicas.
These are melanophlogite, rare in Nature and only recently synthesised (17), dodecasil-3C (15)and silicalites I (14) and II (15).
Melanophlogite has the structure of clathrate hydrate type I, in
which the pentagonal dodecahedra and tetradecahedra with twelve
5-ring and two 6-ring faces of Fig. 9 (18) b and a respectively (1)
are stacked to as to fill all space. There are two dodecahedra and
six tetradecahedra per cubic unit cell of 13.4 $\mathring{A}$ edge (Fig. 10a)
(18). In clathrasil-3C as in its isotype ZSM-39 (19) pentagonal
dodecahedra are stacked with hexadecahedra having twelve 5-ring and
four 6-ring faces to fill all space (Fig. 10b). There are 6 dodecahedra and 8 hexadecahedra per cubic unit cell of edge 19.4 $\mathring{A}$. If
there was one guest molecule, G, per void the limiting composition
of melanophlogite would be $23SiO_2.2G$ and that of clathrasil-3C would
be $13SiO_2.2G$. If only the larger voids contained a guest molecule
the limiting compositions would be $23SiO_2.3G$ and $17SiO_2.G$
respectively. Because of their analogy with clathrate hydrates
they have been termed the clathrate group in Table 3. Since a
range of clathrate hydrates is known synthesis may reveal more
porous silicas or zeolites in this group structurally like their
clathrate hydrate counterparts. All are rich in 5-rings and indeed
the clathrate, mordenite, heulandite and pentasil zeolite groups
are all related in this respect.

The pentasil zeolites ZSM-5 and ZSM-11 and their respective
end members silicalites I and II differ from melanophlogite and

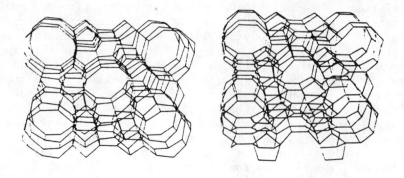

Fig. 11.   Frameworks of ZSM-5 (l.h.s.) and ZSM-11 (r.h.s.).

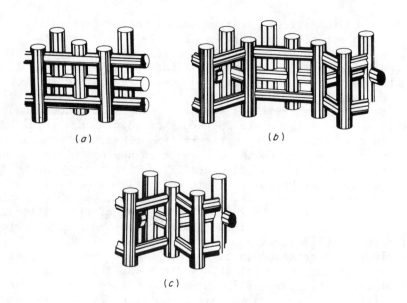

(a)   (b)

(c)

Fig. 12.   Formal representation of channel patterns in
            (a) ZSM-11, (c) ZSM-5 and (b) an intermediate
            structure (20).

clathrasil-3C in having relatively open continuous channels rather than semi-isolated voids permeating their structures. A view of their frameworks is shown in Fig. 11a and b, and a formal represent-ation of the three-dimensional channel patterns in Fig. 12c and a together with this pattern for an intermediate structure, of which many are possible (Fig. 12b) (20). These channels have minimum free dimensions from 5.1 to 5.6 Å, by contrast with the largest windows in melanophlogite or dodecasil-3C of ∿2.6 Å at most.

Table 3, with the exception of kehoeite, refers only to silica and aluminosilicates. AlPO$_4$ can crystallise in forms isostructural with quartz, tridymite and cristobalite and in addition, using organic bases as templates, a number of other AlPO$_4$ species have been made which include a structural analogue of sodalite and of erionite/offretite (21). Analogues of zeolites are known in which Ga replaces Al and Ge replaces Si (22). Table 3 is not intended to include such materials.

8. CONSTRUCTING 3-D FRAMEWORKS

In what follows several ways of constructing zeolite frame-works will be considered. These are most informative in that they show not only how known structures emerge but also many novel ones which point the way for further synthesis. This approach is preferred to describing individual crystallographic structures. Among the ways of visualising and constructing zeolite anions are the following:

1. Stacking of polyhedra
2. Use of the sigma transformation
3. Use of operators
4. Cross-linking chains of varied complexity
5. Cross-linking various kinds of layer

8.1. Frameworks as Assemblies of Polyhedral Voids

This method of building porous frameworks was illustrated in §. 7. for melanophlogite and dodecasil-3C. Polyhedral voids are very common in zeolite structures, as illustrated in Table 4 (23), and many frameworks can be built by stacking polyhedra of one or more different shapes and sizes so as to occupy all space available or to create open channels surrounded by linked polyhedra. The simplest example involves the sodalite cage (Fig. 7), i.e. the 14-hedron of type I. This is one of Federov's space-filling polyhedra which, stacked by face sharing will fill all space with its fellows, in the same orientation. The result is the sodalite structure. The forms of some of the polyhedra listed in Table 4 are shown in Fig. 13 (24).

TABLE 4

Some polyhedra in zeolites(23)

Polyhedron	Faces	Vertices[a]	Approximate free dimension (Å)	Examples
6-hedron (cube)	6 × 4-rings	8	–	zeolite *A*
8-hedron (hexagonal prism)	2 × 6-rings 6 × 4-rings	12	2·3 in plane of 6-rings	faujasite, zeolite *ZK-5*, chabazite, erionite, offretite, levynite
10-hedron (octagonal prism)	2 × 8-rings 8 × 4-rings	16	4·5 in plane of 8-rings	paulingite, zeolite *RHO*
11-hedron	5 × 6-rings 6 × 4-rings	18	4·7 along *c* axis 3·5 normal to *c*	cancrinite, zeolite *L*, erionite, offretite, zeolite losod
14-hedron Type I (truncated octahedron)	8 × 6-rings 6 × 4-rings	24	6·6 for inscribed sphere	sodalite, faujasite, zeolite *A*
14-hedron Type II	3 × 8-rings 2 × 6-rings 9 × 4-rings	24	6·0 along *c* 7·4 normal to *c*	gmelinite, offretite, mazzite (zeolite Ω)
17-hedron Type I	3 × 8-rings 5 × 6-rings 9 × 4-rings	30	9·0 along *c* 7 to 7·3 normal to *c*	levynite
17-hedron Type II	11 × 6-rings 6 × 4-rings	30	7·7 along *c* 6·4 normal to *c*	zeolite losod
18-hedron (oblate spheroidal form)	6 × 8-rings 12 × 4-rings	32	10·8 × 6·6 (6·6 is measured between centre planes of opposite 8-rings)	paulingite, zeolite *ZK-5*
20-hedron	6 × 8-rings 2 × 6-rings 12 × 4-rings	36	11 along *c* 6·5 normal to *c*	chabazite
23-hedron	6 × 8-rings 5 × 6-rings 12 × 4-rings	42	15 along *c* 6·3 normal to *c*	erionite
26-hedron Type I (truncated cubo-octahedron)	6 × 8-rings 8 × 6-rings 12 × 4-rings	48	11·4 for inscribed sphere	paulingite, zeolite *ZK-5*, zeolite *A*, zeolite *RHO*
26-hedron Type II	4 × 12-rings 4 × 6-rings 18 × 4-rings	48	11·8 for inscribed sphere	faujasite (zeolites X and Y)

[a] If *n* denotes the number of faces, $2n-4$ gives the number of vertices.

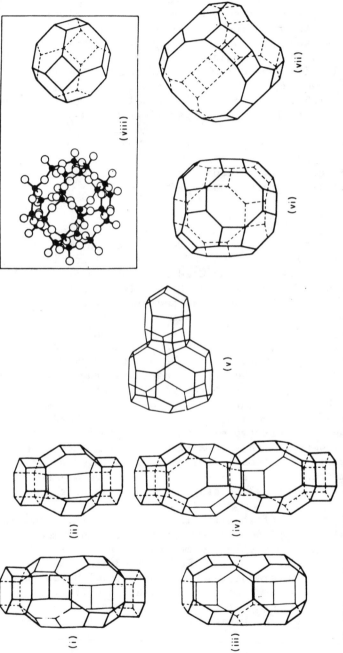

FIG. 13 Some of the polyhedral voids found in zeolite structures: (i) the chabazite 20-hedron, capped by hexagonal prisms; (ii) the gmelinite 14-hedron of Type II; (iii) the erionite 23-hedron; (iv) the levynite 17-hedron of Type I, with associated hexagonal prisms; (v) the losod 17-hedron of Type II, with associated 11-hedral canrinite cage; (vi) the zeolite *A* 26-hedron of Type I; (vii) the faujasite 26-hedron of Type II; (viii) the sodalite 14-hedron shown as in Fig. 3 and also as a "ball and stick" atomic model. Filled circles denote Al or Si and open circles denote oxygen. Part of the rear of the 14-hedron is, for clarity, omitted from the ball and stick model(2_h).

Next in simplicity to sodalite are frameworks made by stacking
two kinds of polyhedron only, as exemplified by melanophlogite and
dodecasil-3C in §. 7. Then one may consider structures composed of
assemblies of three kinds of polyhedron such as zeolite A (cubic
units, sodalite cages and 26-hedra of type I) and faujasite
(hexagonal prisms, sodalite cages and 26-hedra of type II). Next
come structures made by assembling polyhedra, often in columns,
which do not fill all space but which leave continuous channels in
the framework. Examples of these ways of building zeolites are
given in Table 5 (25). Wide parallel channels circumscribed by
12-rings arise in unfaulted, i.e. crystallographically ideal,
cancrinite hydrate, gmelinite, offretite and mazzite. Where such
channels do not occur windows of 8-rings (chabazite, erionite,
zeolite A, zeolite ZK-5), of octagonal prisms (zeolite RHO) or of
12-rings (faujasite) allow molecules to migrate, in these instances
in all three dimensions throughout the frameworks.

8.2. The Sigma Transformation

The sigma transformation is a purely conceptual device for
inter-relating and building known and hypothetical zeolite frameworks (26). A tetrahedrally connected structure is expanded by
imaginary fission of T atoms (T = Al or Si) lying on specified
planes running through the structure, and creating new oxygen
bridges connecting pairs resulting from the fission. As an
example Fig. 14, in the top half, shows how a single tetrahedron
becomes a pair, a 4-ring, a 6-ring and an 8-ring; and how these
three rings become respectively cube, hexagonal prism and octagonal
prism. The bottom half of the figure shows the stages of transformation of a sodalite cage into the 26-hedron of type I found
in zeolite A. The basic requirement is that every T atom lying
in the transformation plane must have two of its linkages lying
in the plane and the other two emerging from opposite sides of
the plane. An inverse of the sigma transformation may occur with
a plane containing no T atoms if every oxygen bridge cut by a
plane is a common edge of two 4-rings. This device may notionally
reduce double rings (prisms) to single rings or ladders to single
chains.

The transformation of sodalite into zeolite A (c.f. the bottom
half of Fig. 14) takes place via two hypothetical intermediate
zeolites. Starting with sodalite numerous other transformations
were effected, to zeolite RHO with two intermediate structures,
faujasite with three intermediates and chabazite with two intermediates. Transformations of cancrinite yielded offretite and
gmelinite; tridymite gave paracelsian and phillipsite; and cristobalite gave zeolite ABW and gismondine as well as two unknown
structures (26).

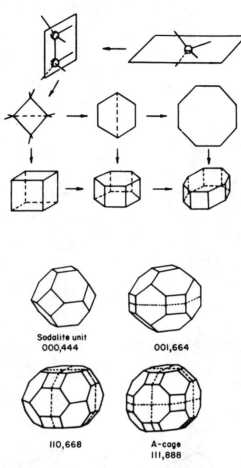

Sodalite unit
000,444

001,664

110,668

A-cage
111,888

FIG. 14 Examples of the "sigma transformation". The top half of the figure illustrates the transformation of a single tetrahedron to yield 4-, 6- and 8-rings and 4-4-, 6-6- and 8-8- prisms. The bottom half shows stages in transforming a sodalite cage (14-hedron of Type I) to the 26-hedral cage of zeolite $A(26)$.

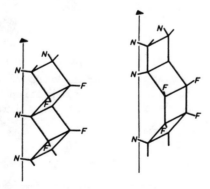

Fig. 15. Two configurations of ladders of 4-rings relative to a three-fold operator axis (27).

58

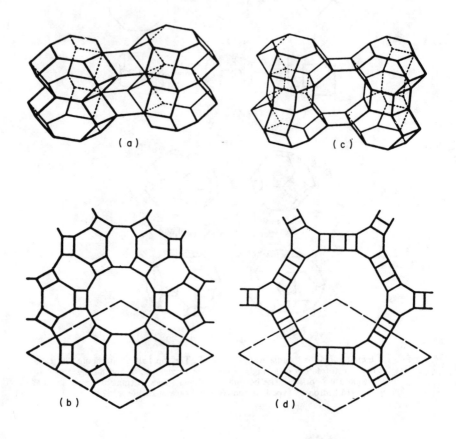

Fig. 16.   In (a) two connected NFNF ladders do not face each other
directly and generate the projection (b). In (c) these
connected ladders face each other directly and generate
the projection (d) (27).

...ed. They consisted of 4-rings each 4-ring linked to two others by sharing opposite edges. Such ladders variously buckled serve to illustrate the use of operators in producing zeolite frameworks (27). Fig. 15 shows two configurations of 4-ring ladders relative to a three-fold operator axis. The terminal unshared oxygen atoms in these ladders are designated N and F ("near" and "far" according to their distance from the axis. Thus on the left the chain is in an NFNF ... configuration and on the right an NNFF ... configuration.

The three-fold operator of Fig. 15 serves to link the ladders through the unshared terminal oxygens near the axis to produce columns of polyhedra characteristic of some zeolites of the chabazite group (see Table 5). There are then three ways of extending the column in the plane normal to the operator axis:

1. The chain considered is shared by two operators.
2. The chain is linked to a second chain belonging to another operator in such a way that
   (a) the two connected chains do not face each other directly (Fig. 16a), and
   (b) the two connected chains face each other (Fig. 16c)

The NFNF ... chains of Fig. 15 then in Fig. 16a and c lead respectively to the projections normal to the channel axis shown in Fig. 16b and d. These have channels circumscribed by puckered 12- and 24-rings. For various chains having periodicities n along their lengths (the c direction) the structures obtained for n < 8 are given in Table 6. All structures in column 4 have the projection normal to the c-direction shown in Fig. 16b, and all structures in column 5 have the projection seen in Fig. 16d. The large free diameters of zeolites of column 5 and the observation that they are not blocked by stacking faults would make such zeolites of special interest. The table includes 33 unknown zeolite frameworks.

Other members of the chabazite group such as sodalite, erionite, chabazite and levynite as well as additional unknown structures can be generated by applying 6_3 and $\bar{3}$ axes as operators (27).

8.4. Cross-linking of Chains

In §. 8.3 operators served to cross-link ladder-like chains and to form zeolites. Smith and Rinaldi (28) considered other aspects of the cross-linking of ladders based on 4-rings. Each vertex of a given ring may point up (U) or down (D), giving rise to the four types of ring shown in Fig. 17 (28). The last of these with another of like kind can form only the cubic unit shown in the

TABLE 5

Examples of combinations of polyhedra in zeolites(25)

Zeolite	Polyhedra and other features	Proportions	Mode of combination
Cancrinite hydrate	11-hedra (cancrinite cages). Wide channels 11^l to c	–	11-hedra in columns 11^l to c. Six linked columns surround and create each wide channel. Alternate columns are displaced by $c/2$.
Losod	11-hedra in columns A 17-hedra of Type II in columns B	1 1	A column of A is surrounded by six of B, 11^l to c. B-columns are alternately displaced by $c/2$.
Chabazite	Hexagonal prisms 20-hedra	1 1	Prisms and 20-hedra alternate in columns 11^l to c. A given column is surrounded by six like columns, three displaced by $c/3$ and three by $2c/3$.
Gmelinite	Hexagonal prisms 14-hedra of Type II Wide channels 11^l to c	1 1	Prisms and 14-hedra alternate in columns 11^l to c. Six linked columns surround and create wide channels. Alternate columns displaced by $c/2$.
Erionite	Hexagonal prisms 11-hedra 23-hedra	1 1 1	Prisms and 11-hedra alternate in columns A, 23-hedra form columns B, both A and B 11^l to c. Each column A is surrounded by six of B, where columns B are alternately displaced by $c/2$.
Offretite	Hexagonal prisms 11-hedra 14-hedra of Type II Wide channels 11^l to c	1 1 1	Prisms and 11-hedra alternate in columns A; 14-hedra form columns B; both A and B are 11^l to c. Three A and three B surround and create each wide channel.
Mazzite	14-hedra of Type II Wide channels 11^l to c. Narrow channels 11^l to c	–	Six linked columns of 14-hedra surround and create wide and narrow channels 11^l to c. Alternate columns displaced by $c/2$.
Unknown	14-hedra of Type II. Wide channels 11^l to c	–	Six linked columns of 14-hedra surround and create wide channels. All at same height.
Faujasite	Hexagonal prisms 14-hedra of Type I 26-hedra of Type II	2 1 1	A given 14-hedron is linked by four prisms arranged tetrahedrally on four of its eight hexagonal faces, to its four nearest 14-hedron neighbours. Arrangement of 14-hedra is as are the atoms in diamond. This creates 26-hedral voids, also arranged like atoms in diamond.

TABLE 5—*continued*

TABLE 5—*continued*

Zeolite	Polyhedra and other features	Proportions	Mode of combination
Zeolite *A*	Cubic units	3	A given 14-hedron is linked by cubic units through its six 4-ring faces to a 4-ring face of each of six other 14-hedra. This arrangement creates 26-hedra.
	14-hedra of Type I	1	
	26-hedra of Type I	1	
Zeolite *RHO*	Octagonal prisms	3	A given 26-hedron is linked by octagonal prisms through each of its six 8-ring faces to an 8-ring face of one of six other 26-hedra.
	26-hedra of Type I	1	
Zeolite *ZK-5*	Hexagonal prisms	4	A given 26-hedron is linked by hexagonal prisms through its eight 6-ring faces to a 6-ring face of each of eight other 26-hedra. This creates the 18-hedral voids.
	18-hedra	3	
	26-hedra of Type I	1	

TABLE 7

Schemes for interconnecting chains of Fig. 18(31).

Number	Direct	Unitary	Inverse	Rotational symmetry of scheme
1		$\parallel$ $=n=$ $\parallel$		4
2	$\parallel$ $\leftarrow n \rightarrow$ $\parallel$		$\parallel$ $\rightarrow n \leftarrow$ $\parallel$	2
3		$\downarrow$ $\leftarrow n \rightarrow$ $\uparrow$		2
4	$\parallel$ $=n\rightarrow$ $\parallel$		$\parallel$ $=n\leftarrow$ $\parallel$	1
5		$\downarrow$ $=n\rightarrow$ $\parallel$		1
6	$\uparrow$ $=n\leftarrow$ $\downarrow$		$\downarrow$ $=n\rightarrow$ $\uparrow$	1

Table 6: Actual and hypothetical structures based on ladders of 4-rings (27).

Periodicity n	Type of Chain	Three-fold operator		
		Chain shared	Chain not shared	
			not facing	facing
2	NF	cancrinite	unknown	unknown
3a	NNF ⎱	offretite	zeolite L	unknown
3b	FFN ⎰		unknown	unknown*
4	NNFF	gmelinite	unknown	unknown*
5a	NFFNF ⎱	unknown	unknown	unknown*
5b	FNNFN ⎰		unknown	unknown
6	NNFNFF	unknown	unknown	unknown*
7a	NFNFNFN ⎱	unknown	unknown	unknown
7b	FNFNFNF ⎰		unknown	unknown
7c	NNFFNFF ⎱	unknown	unknown	unknown*
7d	FFNNFNN ⎰		unknown	unknown*
8a	FNFFNFN	unknown	unknown	unknown*
8b	NFNFFNFF ⎱	unknown	unknown	unknown*
8c	FNFNNFNN ⎰		unknown	unknown
a(hex) in Å		12.8 to 13.7	$\sim$ 18.5	$\sim$ 22
c(hex) in Å		n x 2.5	n x 2.5	n x 2.5
Free diameter of Main channel in Å		$\sim$ 6.5	$\sim$ 7.5	$\sim$ 15
Ring circumscribing main channel		12-ring	12-ring	24-ring
Stacking faults		Block main Channel	Do not block	Do not block

*FF leads to heavily distorted 4-membered rings.

figure. From the other three kinds of 4-ring three types of chain can be made, as shown in Fig. 17. The UUDD chain resembles the NNFF chain of Table 6 and Fig. 15. From it 17 other frameworks were made by variously cross-linking the chains, in which repeat distances in the plane normal to the chain direction were less than 15 Å. These included felspar, paracelsian, phillipsite and a synthetic material termed phase A (29) ($Ba[AlSi_2O_6]Cl$, OH; group I4/mmm; a = 14.194 Å and c = 9.934 Å). Others were novel structures. Chains can be linked in two ways:

1. Flexible mode. Tetrahedra connected to one another along the chain are both connected to the same adjacent chain.

These include paracelsian, phillipsite, gismondine and phase A.

2. Inflexible mode. Tetrahedra connected to one another along the chain are linked to different adjacent chains. An example of this is felspar.

In the flexible mode 4-rings can rotate co-operatively but this is not possible in the inflexible mode.

The second kind of chain (UDUD) appears un-cross-linked as the anion in narsarsukite, $Na_4(TiO)_2[Si_8O_{20}]$, (see §. 3 and Fig. 3g). These chains could be cross-linked in four ways, one of which is found in the non-zeolite banalsite, $Ba,Na_2[Al_4Si_4O_{16}]$. The third kind of chain (UUUD) could also be cross-linked to yield frameworks of novel kinds (28).

The chains in Fig. 17 are examples only of those which may lead through cross-linking to tectosilicates. Other examples of single and double chain anions were referred to in §. 3. Additional chains from which zeolite anions may be constructed occur in the natrolite group and mordenite group. In the natrolite chain UDUD 4-rings are linked together by single tetrahedra as shown in Fig. 18 (30) for natrolite and thomsonite. These chains differ in Si/Al ratio, which is 3/2 for natrolite and 1 for thomsonite. The Al,Si distributions on tetrahedral sites are ordered and the periodicity in the c-direction is 6.6 Å. The heights of free vertices of tetrahedra are multiples of c/8. Albert and Gottardi (31) considered ways of cross-linking a chain to four other chains through its free vertices. Each chain can be coded by the height, n (nc/8) of the centre of its single tetrahedron. The linkage of a given chain to its neighbours is through the pair of U and the pair of D tetrahedra. For the latter the possibilities are:

1. The cross-links are both to D tetrahedra of two other chains (these chains being at height n).
2. One cross-link is to D and the other to U (the two adjacent chains being at heights n and (n-2)).
3. Both cross-links are to U (the adjacent chains being at heights (n-2)).

Three similar possibilities arise for the pair of U tetrahedra, namely both cross-links to U tetrahedra; one to U and one to D; and both to D. In U-D bonds the height of the adjacent chain is n + 2. Table 7 (31) gives the schemes for cross-linking. In it the symbols =, → and ← have the following meanings:

n = n for a cross-link with a chain at the same height
n → n + 2 for a cross-link with a chain at height n + 2
n ← n - 2 for a cross-link with a chain at height n - 2

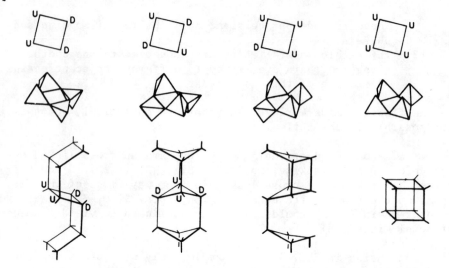

Fig. 17. Four types of 4-ring with apices of tetrahedra pointing
          up (U) or down (D). Below, three kinds of chain are
          shown based on UUDD, UDUD and UDUUD rings, together
          with the cubic unit made from two 4-rings of the fourth
          kind (28).

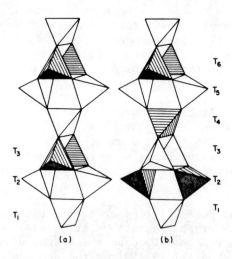

Fig. 18. The chains present in zeolites of the natrolite group
          (30). (a) refers to natrolite, (b) to thomsonite.
          Shaded tetrahedra denote AlO_4, unshaded denote SiO_4.
          There is ordering in the Al,Si distributions, and the
          periodicities in view of the ordering are 3 for (a) and
          6 for (b).

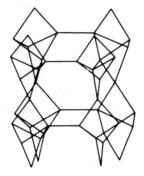

Fig. 19. The framework of edingtonite viewed along [110] (32).

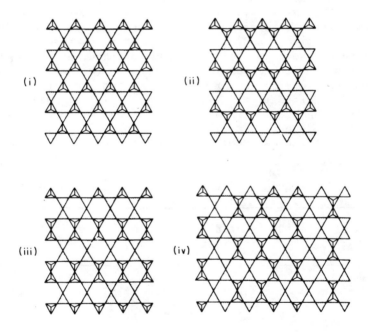

Fig. 20. Four kinds of hexagonal single sheet differentiated by patterns in which apices of the linked tetrahedra point up (▽) or down (▽) (35).

If corresponding inverse and direct schemes are considered as a single scheme the nine different possibilities in Table 7 are reduced to six. In fibrous zeolites two kinds of Si/Al ordering have been observed, as already noted for natrolite and thomsonite in Fig. 18. If possibilities of Si/Al ordering are combined with the six schemes then the 15 possibilities of Table 8 (31) are found. Ten of these refer to unknown structures. As an example, Fig. 19 (32) shows a portion of the edingtonite framework, viewed along [110].

The chain characteristic of some zeolites of the mordenite group is based on 5-rings (33). The chains run in the c direction with a periodicity 7.52 Å. Eight structures which can be made by cross-linking the chains are given in Table 9 (34). Only two of these refer to known minerals.

8.5. Frameworks made by cross-linking Sheets

Many kinds of sheet can be made from polygons sharing edges with like or unlike polygons, a number of which have been identified in silicates (§.4). These sheets were composed only of 6-rings (mica); of 4- and 8-rings (apophyllite); of 4-, 6- and 8-rings (dalylite); and of 4-, 6- and 12-rings (manganpyrosmalite). Each type of sheet can be further sub-divided, according to the patterns of un-linked vertices pointing up (U) and down (D). Four such patterns are shown for hexagonal sheets in Fig. 20 (35). There are accordingly many ways of cross-linking sheets, which may in addition be buckled in a variety of ways. Puckered sheets of type (i) are found, parallel to [001] in bikitaite, and such sheets also occur in the non-zeolites nepheline and carnegieite. In

Table 8: Space groups of the natrolite group of zeolites (31)

Structural Type	Disordered (Al,Si) distribution	Natrolite type order	Thomsonite type order
1	$P\bar{4}2_1m$	$P2_12_12$ Edingtonite	$P\bar{4}2_1c$
2	Pman Gonnardite(?)	P2/b	Pcnn Thomsonite
3	$I\bar{4}2d$ Tetragonal natrolite	Fdd2 Natrolite	Not possible
4	Pmma	P2/a	Pcca
5	Imma	B2/b	Not possible
6	Pmna	$P2_1/a$	Not possible

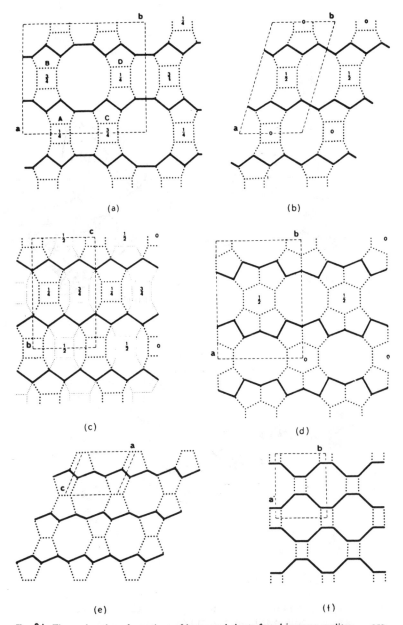

(a)

(b)

(c)

(d)

(e)

(f)

FIG. 21 The puckered conformations of hexagonal sheets found in some zeolites are shown edge on (full lines). The interconnections between sheets are shown as dashed lines. The zeolites concerned are: (a) mordenite; (b) dachiardite; (c) epistilbite; (d) ferrierite; (e) bikitaite; and (f) zeolite Li-ABW(34).

68

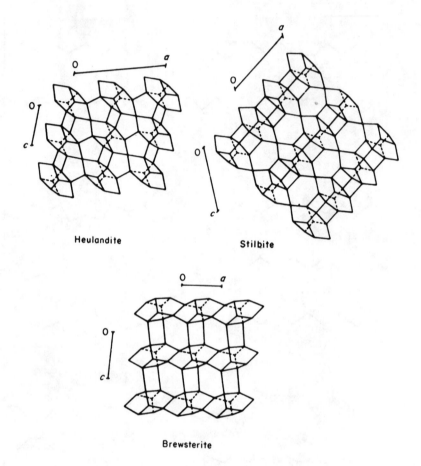

Heulandite

Stilbite

Brewsterite

Fig. 22. Layers with a common structural unit found in heulandite, stilbite and brewsterite (38).

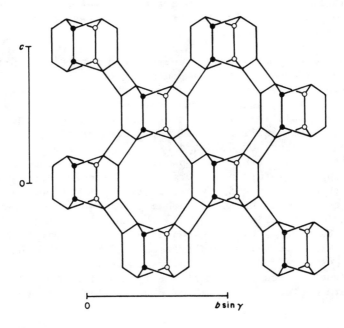

Fig. 23. The laumontite layer showing sub-units composed of four 6-rings and two 4-rings (39).

bikitaite the sheets are linked to one another through single
tetrahedron chains parallel to [010], whereas in nepheline and
carnegieite the sheets are linked directly to one another.

Sheets of type (ii) are directly connected to one another in
zeolite ABW. Those of type (iii) are found in dachiardite,
epistilbite and ferrierite. They are linked by single 4-rings for
the first two zeolites and by single 6-rings for ferrierite.
Type (iv) sheets connected through single 4-rings form the mordenite
structures. Fig. 21 (36) shows the puckered sheets edge on , as
the bold lines, while the connections between sheets are indicated
as the dashed lines, for the various zeolites referred to in this
and the preceding paragraph.

In the chabazite group sheets containing either 6-rings or
hexagonal prisms (6-6-rings) can be identified. These are linked
to similar sheets above and below in the various stacking sequences
of Table 10 (34), where layers containing single 6-rings are
denoted by small letters and those containing 6-6-rings by capital
letters. Novel zeolites involving sequences of A, B and C layers
analogous to those of a, b and c layers in losod, liottite,
afghanite and franzinite are possible, as well as sequences of
double and single 6-ring layers in addition to those giving
offretite, erionite and levynite. Two of these possibilities with
sequences ABCB and ABCACB have been examined (37). They have
hexagonal unit cells with a $\sim$ 13.7 Å and c $\sim$ 20 and 30 Å respectively. Each contains elongated 26-hedra of a third kind, with
nine 8-ring, two 6-ring and fifteen 4-ring faces. The free length
approaches 20 Å and the free diameter about 6.5 Å. In addition the
ABCB structure contains gmelinite type 14-hedra and the ABCACB
structure contains both gmelinite and 20-hedral chabazite type
cages.

Heulandite, stilbite and brewsterite can be represented in
terms of other layers each containing the same unit composed of
four 5-rings and two 4-rings, as shown in Fig. 22 (38). The bond
density in the layers is greater than that of the links between
layers, resulting in their platy character. Laumontite contains
sheets of the type shown in Fig. 23 (39) in which structural units
composed of four 6-rings and two 4-rings occur.

Zeolites can be constructed from layers of, or containing,
sodalite cages. In sodalite itself the layer consists of these
cages each linked to four other cages by sharing 4-ring faces, and
to a cage in an upper layer and one in a lower layer also by sharing
its remaining two 4-ring faces. In zeolite A a sodalite cage in
a layer is linked to four other cages by double 4-rings (cubic
units) and to a sodalite cage in the layer above and one in the
layer below, also by cubic units. From a topological viewpoint

71

Table 9: Some Structures Obtainable by Cross-linking Mordenite Chains (34).

Number	Space group	a(Å)	b(Å)	c(Å)	γ	Example
1	B2/m	18.73	10.30	7.52	107.9°	Dachiardite
2	B2/m	18.73	11.7	7.52	123°	Modified dachiardite
3	Pmnm	18.1	10.25	7.52		Unknown
4	Amam	18.1	20.5	7.52		Unknown
5	Cmcm	18.13	20.49	7.52		Mordenite
6	Immm	18.13	20.49	7.52		Modified mordenite
7	Bbcm	18.7	19.6	7.52		Unknown
8	Bmmm	18.7	19.6	7.52		Unknown

Table 10: Structures formed by linking layers containing single 6-rings (a,b,c) and layers containing 6-6-rings (A,B,C) (34).

No. of layers in repeat unit	Structural type	Space Group	a(Å)	c(Å)	Name
2	ab	P6₃	12.72	5.19	Canc
3	abc	P4̄3m	8.87	–	Sodalite
4	abac	P6₃/mmc	12.91	10.54	Losod
6	ababac	P6m2	12.85	16.10	Liottite
8	ababacac	P6₃mc	12.77	21.35	Afghanite
10	abcabcbacb	P3̄m1	12.88	26.76	Franzinite
2	Ab	P6̄m2	13.29	7.58	Offretite
4	AbAc	P6₃/mmc	13.26	15.12	Erionite
4	aBaC	P6₃/mmc	13.26	15.12	–
6	AbCaBc	R3̄m	13.34	23.01	Levynite
2	AB	P6₃/mmc	13.75	10.05	Gmelinite
3	ABC	R3̄m	13.78	15.06	Chabazite

sodalite cages may represent the positions occupied by spheres in cubic close-packing (40). An indefinite number of structures can be made from hexagonal layers of close-packed spheres in various sequences. The sequence ABCABC ... corresponds with cubic symmetry as in faujasite, in which the 14-hedra are joined by hexagonal prism units to four other 14-hedra, the ABC layer sequence being along [111]. The layer sequence ABAB ... corresponds with a hexagonal structure, the sodalite 14-hedra being linked, as in faujasite, by hexagonal prism units to four other 14-hedra. The resultant very open structure has cages and channels comparable with those in faujasite. This sequence is shown formally as Fig. 24a. Fig. 24b is that in faujasite and Fig. 24c is the sequence ABABCA of one of many possible novel structures. Zeolite ZSM-3 might, it was thought, be related to faujasite as one of these stacking arrangements of layers of sodalite cages (40). It had a hexagonal unit cell with a = 17.5 Å and a maximum possible value of c of 129 Å. Since the distance between adjacent sodalite cage layers is 14.3 Å, c = 129 Å represents a 9-layer sequence.

Moore and Smith (41) gave further consideration to structures based on sodalite cages and/or on 26-hedra of type I (as found in zeolite A). In a short-hand description of possibilities they used the symbols

H = hexagonal face (6-ring)
H' = hexagonal prism (6-6-ring)
S = square face (4-ring)
S' = cubic unit (4-4-ring)
O = octagonal face (8-ring)
O' = octagonal prism (8-8-ring)

The type of contact between polyhedra was indicated by the appropriate one of these letters and the letters following (in parentheses) denoted the types of polyonal faces opposing each other across the contact. The results are summarised in Table 11 in which frameworks of four so far unknown zeolites appear.

High resolution electron microscopy (HREM) has revealed a tendency to recurrent twinning in synthetic faujasite (zeolite Y) (42). This can generate a new structure within the zeolite with elliptical apertures along [110] and an elongated "hypercage" the length of which depends on the extent of twinning. If Δ denotes a building unit repeat along [111] ($\sim$ 14.3 Å repeat distance) and ∇ denotes a twin lamella, then ...$\Delta\Delta\Delta$... is the normal layer sequence and ... $\check{\nabla}/\Delta/\nabla/\Delta$... is a tunnel structure in which a twin lamella bounded by a pair of {111} twin planes, denoted by slashes, alternates every 14.2 Å with the regular faujasite lamellae. When there are n twin planes the tunnel in the structure is $\sim$14.3 (n+1) + 6.95 Å in length. The sequence ...$\Delta\Delta/\nabla/\Delta\Delta/\nabla$...

Table 11: Frameworks made by linking sodalite 14-hedra and/or 26-hedra, both of Type 1 (41).

(a) From 14-hedra

Arrangement	Space Group	a(Å)	c(Å)	TO$_4$ per unit cell	Example
S(H-H)	Pm3M	8.8	–	12	Sodalite
S'(H-H)	Pm3m	11.9	–	24	Zeolite A
H(H-S)	Fd3m	17.5	–	96	Unknown
H'(H-S)	Fd3m	24.7	–	192	Faujasite
H(H-S) and (H-H)	P6$_3$/mmc	12.4	20.5	64	Unknown
H'(H-S) and (H-H)	P6$_3$/mmc	17.5	28.5	128	Unknown

(b) From 26-hedra

Arrangement	Space Group	a(Å)	c(Å)	TO$_4$ per unit cell	Example
O(S-S) ≡ S'(H-H)	Pm3m	11.9	–	24	Zeolite A
O'(S-S) ≡ O'(S-S)	Im3m	15.1	–	48	Zeolite RHO
H(O-S) ≡ O'(S-S)	Im3m	18.7	–	96	Zeolite ZK-5
H'(O-S)					

(c) From 14-hedra and 26-hedra

Arrangement	Space Group	a(Å)	c(Å)	TO$_4$ per unit cell	Example
H(S-S) = S'(H-H)	Pm3m	11.9	–	24	Zeolite A
H'(S-S)	Fm3m	31.1	–	384	Unknown

indicates a twin lamella between flanking pairs of regular repeat
units and gives a cage sequence in which each cage is 49.6 Å long
and varies between 13 and 7.4 Å in free diameter. This particular
sequence has been observed (42). Recurrent twinning at the unit
cell level could also convert other known zeolite structures into
novel ones.

9.   INTRAZEOLITE CHANNELS

Each different zeolite topology has a different system of
channels and cavities. Nevertheless a broad classification is
possible, from the viewpoint of migration of guest molecules,
into three divisions:

1. Intracrystalline channels are parallel and are not interconnected (1-D diffusion).
2. Channels are inter-connected in two dimensions but not in
   the third (2-D diffusion).
3. Channels are inter-connected in three dimensions (3-D
   diffusion).

The geometry of channel and cavity systems is defined as
precisely as are the positions of framework atoms. In addition
to varied spatial arrangements of the open pathways, as in the
above ,three categories, in different topologies cross-sectional
areas and shapes are specific to each topology and configuration.
Cations may also be located in the diffusion pathways and if they
are present at strategic points, such as the windows or narrowest
points along a pathway they can act as partial or complete
barriers to the movement of guest molecules. The location and
numbers of the cations can be modified by exchanges such as
$2Na^+ \rightleftarrows Ca^{2+}$ and the blocking effects of cations can be dramatically
changed by such means.

Some examples of zeolites with 1-D, 2-D and 3-D channel
systems are:

1-D. Cancrinite hydrate (12, 6.2); laumontite (10, 4.0x5.6);
      mazzite, zeolite Ω (12, 7.4); mordenite (12, 6.7x7.0);
      zeolite L (12, 7.1).
2-D. Dachiardite (10, 3.7x6.7 and 9, 3.6x4.8); ferrierite
      (10, 4.3x5.5 and 8, 3.4x4.8); levynite (8, 3.3x5.3);
      stilbite (10, 4.1x6.2 and 8, 2.7x5.7).
3-D. Chabazite (8, 3.6x3.7); erionite (8, 3.6x5.2); faujasite
      (12, 7.4); offretite (12, 6.4 and 8, 3.2x5.2); zeolite A
      (8, 4.1); zeolite RHO (8-8, 3.9x5.1) zeolite ZK-5
      (8. 3.9).

The underlined figures in the brackets indicate the numbers n of
linked tetrahedra forming the narrowest apertures along the
diffusion pathways and the other figures are the free dimensions
of these apertures. They show, for example for 8-rings, that the
rings may have various configurations according to the framework
in which it appears.

10. Al, Si ORDERING

According to Lowenstein's rule (43) Al-O-Al bonds do not occur
in tectosilicates for which Si/Al > 1. Accordingly where Si/Al = 1
Si and Al must alternate on all tetrahedral sites. Recent magic
angle spinning nuclear magnetic resonance (MASNMR) measurements
appeared to challenge the validity of this rule for zeolite A with
Si/Al = 1 but spectra obtained on zeolites A with values of
Si/Al > 1 led to a re-appraisal which confirmed the rule (44,45).
It is now almost certain that there are no known exceptions to
Lowenstein's rule for tectosilicates with Si/Al ⩾ 1, and hence that
there is Al, Si ordering whenever Si/Al = 1. There can also be
Al, Si ordering in tectosilicates in which Si/Al exceeds unity, an
example referred to in §. 8.4 and Fig. 18 being natrolite where
Si/Al = 3/2.

Faujasites having wide ranges in Si/Al ratios gave MASNMR
spectra which have been interpreted in terms of ordering of Al and
Si at specific ratios corresponding with integral numbers of Al
and Si atoms per sodalite cages (46,47). There is considerable
if not yet total agreement on the ordering schemes. The spectra
themselves serve to give the relative numbers of Si atoms linked
(through O) with 4, 3, 2, 1 and 0 Al atoms, denoted respectively
as Si(4Al), Si(3Al), Si(2Al), Si(1Al) and Si(0Al). Table 12 gives
as a function of the Si/Al ratio the relative percentage populations
of these units for any tectosilicate structure, according to a
binomial distribution and when Lowenstein's rule is valid.

Where Al and Si are distributed on tetrahedral sites involving
more than one sub-lattice there may, as in the felspars (48) be
ordered and disordered forms, subject to the over-riding influence
of the Lowenstein rule. In ordered alkali metal felspars at
equilibrium at low temperatures the Al atoms are found on one sub-
lattice. High temperature equilibrium forms have Al atoms
distributed on more than one sub-lattice. Order-disorder changes
of this kind are sluggish and may best be approachable by synthesis
under high and low temperature conditions provided reaction is also
slow. Rapid hydrothermal syntheses of alkali metal felspar have
yielded disordered forms, presumably metastable (49). This may be
an example of Ostwald's law of successive transformations according
to which less stable phases appear before the more stable ones.

Table 12: Relative percentage populations of Si(nAl) structural units, calculated from the binomial formula for different tectosilicate compositions (46). Lowenstein's rule is obeyed.

$\frac{Si}{Al}$	Si(4Al)	Si(3Al)	Si(2Al)	Si(1Al)	Si(0Al)
1.0	100	0	0	0	0
1.19	49.9	37.9	10.8	1.4	0.0
1.35	30.1	42.1	22.1	5.2	0.5
1.59	15.6	36.9	32.7	12.9	1.9
1.67	12.9	34.5	34.6	15.5	2.5
1.87	8.2	28.5	37.1	21.5	4.7
2.00	6.2	25.0	37.5	25.0	6.3
2.35	3.3	17.7	35.8	32.3	10.9
2.56	2.3	14.5	34.0	35.4	13.8
2.61	2.2	13.9	33.5	35.9	14.5
2.75	1.7	12.2	32.1	37.5	16.5

11. CATIONS AND WATER MOLECULES

Accurate and quantitative location of cations and water molecules has been more difficult than defining the positions of oxygen and T atoms in the anions of zeolites. Reasons for this include:

(i) The total number of cations is rather small compared with the number of O and T atoms in the anions.

(ii) These cations are usually distributed over a number of sub-lattices, and often there is only partial occupancy of sites on a given sub-lattice. Thus the number on a given sub-lattice may be low.

(iii) In a number of investigations powder X-ray photography has of necessity been used, giving more limited information.

(iv) Intracrystalline water is not necessarily all tightly held on specific sites.

A compilation of published work on cation and water locations is available, which however cannot assess accuracy in the locations claimed (50). Such an assessment will not be attempted here, although the role of cations as modifiers of molecular sieve behaviour is of major significance.

78

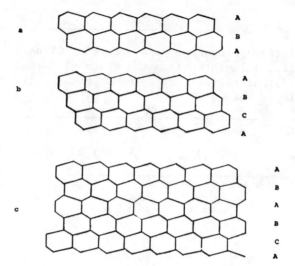

Fig. 24. Stacking of layers of sodalite cages in faujasite and related structures (40). The cages are represented as the line junctions. The view is perpendicular to the c-axis, 110 projection.
(a) Hexagonal AB sequence; (b) cubic ABC sequence; (c) ABABC sequence.

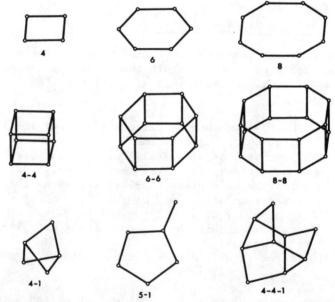

Fig. 25. Secondary building units (SBU) in zeolite building. Their characteristic is that a selected zeolite framework can be built entirely from an appropriately chosen one of these SBU (51).

12. CONDLUDING REMARK

This account of zeolite structures has concentrated upon the anions which are the most accurately defined parts and provide the key to zeolite diversity. When tectosilicate frameworks are constructed by a series of different proceedures one obtains in addition to the many known structures a remarkable number of other frameworks, porous on the scale of molecular dimensions, which represent so far unknown zeolites and point the way to further chemical discovery.

Complementary to the methods of §§. 8.1 to 8.5, one may consider the smallest number of secondary building units (SBU) from which all structures can be made, assuming that the entire framework is made up of one type of SBU only. Nine are reported, as shown in Fig. 25 (51, 52). It cannot however be assumed that these are chemical units actually added to a zeolite crystal during its growth. Also, of the many hypothetical topologies referred to in §. 8, some may be impossible to make because of conformational restrictions (52), for example in T-O-T bond angles.

ACKNOWLEDGEMENTS

I wish to thank Professor J.M. Thomas for Table 12 and especially Professor F. Liebau for permission to use material from his work on silicate classification, and the Academic Press for permission to use considerable material from my two recent books (refs. 22 and 23).

REFERENCES

1. cf. Liebau, F., Reviews in Mineralogy Vol. 5. Orthosilicates, (Editor P.H. Ribbe, Mineralogical Society of America, 1980) Chap. 1.
2. Hoebbel, D., W. Wieker, P. Francke and A. Otto, Z. Anorg. Allg. Chemie 418, (1975) 35.
3. Liebau, F. and I. Pallas, Z. Krist. 155, (1981) 139.
4. Ref. 1, p. 9.
5. Ref. 1, p. 10.
6. Barker, M.G., P.G. Gadd and M.J. Begley, Chem. Comm. (1981) 379.
7. Hefter, J. and M.E. Kenny, Inorg. Chem. 21, (1982) 2810.
8. Hesse, K.-F. and F. Liebau, Z. Krist. 153, (1980) 3.
9. Ref. 1, p. 11.
10. Ref. 1, p. 12.
11. Leibau, F., Structure and Bonding in Crystals (Academic Press Inc., 1981) Chap. 23.
12. Barrer, R.M. and D.E.W. Vaughan, Trans. Faraday Soc., 63, (1967) 2275.
13. Flanigen, E.M., J.M. Bennett, R.W. Grose, J.P. Cohen, R.L. Patton, R.M. Kirchener and J.V. Smith, Nature, 271, (1978) 512.
14. Bibby, D.M., N.B. Milestone and L.P. Aldridge, Nature, 280, (1979) 664.
15. Gies, H., F. Liebau and H. Gerke, Angew. Chem. Int. Edition (English) 21, (1982) 205.
16. Barrer, R.M. and B.M. Munday, J. Chem. Soc. A, (1971) 2904.
17. Gies, H., H. Gerke and F. Liebau, Neues Jahrb. Mineralog. Mh., (1982) 119.
18. Barrer, R.M. Non-stoichiometric Compounds (Editor, L. Mandel-corn, Acadmic Press Inc., 1963) pp. 314-5.
19. Schlenker, J.L., F.G. Dwyer, E.E. Jenkins, W.J. Rohrbaugh, G.T. Kokotailo and W.M. Meier, Nature, 294, (1981) 340.
20. Kokotailo, G.T. and W.M. Meier, Properties and Applications of Zeolites (Editor, R.P. Townsend, Chem. Soc. Special Publication No. 33, 1979) p. 133.
21. Wilson, S.T., B.M. Lok, C.A. Messina, T.R. Cannan and E.M. Flanigen, J. Amer. Chem. Soc., 104, (1982) 1146.
22. Barrer, R.M., Hydrothermal Chemistry of Zeolites (Academic Press Inc., 1982) p. 282.
23. Barrer, R.M., Zeolites and Clay Minerals as Sorbents and Molecular Sieves, (Academic Press Inc., 1978) p. 36.
24. Ref. 23, p. 38.
25. Ref. 23, p. 41.
26. Shoemaker, D.P., H.E. Robson and L. Broussard, Proc. of 3rd Int. Conference on Molecular Sieves (Editor J.B. Uytterhoeven, Zurich, Sept. 3rd-7th, 1973) p. 138.
27. Barrer, R.M. and H. Villiger, Z. Crist., 128, (1969) 352.

28. Smith, J.V. and F. Rinaldi, Mineralog. Mag., _33_, (1962) 202.
29. Solov'eva, L.P., S.V. Borisov and V.V. Bakakin, Sov. Phys.-
    Cryst. _16_, (1972) 1035.
30. Ref. 1, p. 60.
31. Alberti, A., and G. Gottardi, Neues Jahrb. Mineralog, Mh. (1975)
    396.
32. Ref. 23, p. 63.
33. Meier, W.M., Z. Krist., _115_, (1961) 439.
34. Merlino, S., Soc. Ital. Mineralog. Petrolog. Rendicorti, _31_
    (1975) 513.
35. Ref. 23, p. 48.
36. Ref. 23, p. 49
37. Kokotailo, G.T. and S.L. Lawton, Nature, _203_, (1964) 621.
38. Ref. 23, p. 50.
39. Ref. 23, p. 51.
40. Kokotailo, G.T. and J. Ciric, Molecular Sieve Zeolites-I,
    Advances in Chemistry Series No. 101 (Editor R.F. Gould,
    Amer. Chem. Soc., 1971) p. 109.
41. Moore, P.B. and J.V. Smith, Mineralog. Mag. _33_, (1964) 1008.
42. Thomas, J.M., M. Audier and J. Klinowski, Chem. Comm., (1981)
    1221.
43. Lowenstein, W., Amer. Mineralog., _39_, (1954) 92.
44. Thomas, J.M., C.A. Fyfe, S. Ramdas, J. Klinowski and G.C. Gobbi,
    J. Phys. Chem., _86_, (1982), 3061.
45. Melchior, M.T., D.E.W. Vaughan, R.H. Jarman and A.J. Jacobson,
    Nature, _298_, (1982) 455.
46. Klinowski, J., S. Ramdas, J.M. Thomas, C.A. Fyfe and J.S.
    Harman, J. Chem. Soc. Faraday, _78_, (1982), 1025.
47. Melchior, M.T., D.E.W. Vaughan and A.J. Jacobson, J. Amer.
    Chem. Soc., _104_, (1982) 4859.
48. Senderov, E.E., Phys. Chem. Mineral. _6_, (1980), 251.
49. Barrer, R.M. and E.A.D. White, J. Chem. Soc., (1952) 1561.
50. Mortier, W.J. Compilation of Extra-framework Sites in Zeolites
    (Butterworth Scientific Ltd., 1982).
51. Meier, W.M. and D.H. Olson, Atlas of Zeolite Structure Types
    (Structure Commission of the International Zeolite Association,
    1978) p. 9.
52. Gramlich-Meier, R., and W.M. Meier, J. Solid State Chem., _44_
    (1982) 41.

ZEOLITE CRYSTALLOGRAPHY

G. T. Kokotailo

Drexel University
Physics Department
Philadelphia, Pennsylvania

INTRODUCTION

The importance of zeolites in science and technology is well established and the indications are that it will increase. The number of
new zeolites synthesized and structures determined is increasing.
The lack of large crystals suitable for single crystal structure
determination has hindered the progress of structure determination
of the new zeolites synthesized. The knowledge of zeolite structures
is limited; however the topology or the general features of the
framework structure of a fairly large number of zeolites are known.
Good structure information regarding metal framework atom distribution, the position of cations, water and organic molecules is available for only a few zeolites.

The properties of a zeolite are dependent on the topology of its
framework, the size of the free channels, the location, charge and
size of the cations within the framework, the presence of faults and
occluded material, and the ordering of T atoms (framework metal
atoms). Therefore, structural information is important in understanding the absorptive and catalytic properties of zeolites. There
have been a number of reviews of the structure, chemistry and use of
zeolites (1-5).

This paper is directed to the review of the structures of zeolites
ranging from the topology of the frameworks, cation location, T-atom
distribution, faults and imperfections to model building.

CLASSIFICATION OF KNOWN ZEOLITE STRUCTURES

The framework structure of zeolites consists of linked tetrahedra
(metal atoms are tetrahedrally coordinated to four oxygen atoms)

84

Figure 1
Secondary Building Units (2)

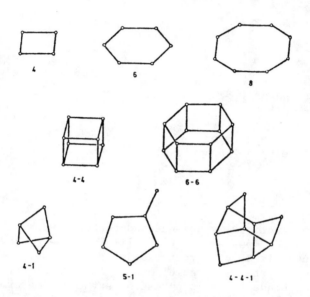

Figure 2
Frameworks of (a) analcite
(b) laumontite (5)

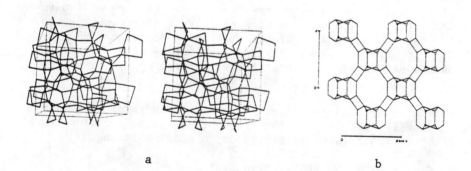

a   b

Figure 3

Chains in (a) natrolite, (b) Brewsterite
(c) ZSM-5

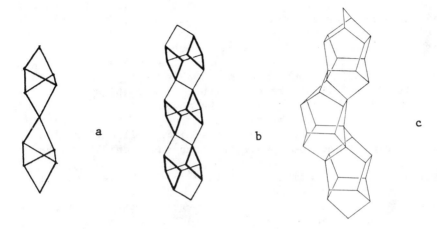

a   b   c

Figure 4

Offretite and Erionite Framework
(a) offretite (b) c-projection of offretite
(c) Erionite (d) c-projection of erionite

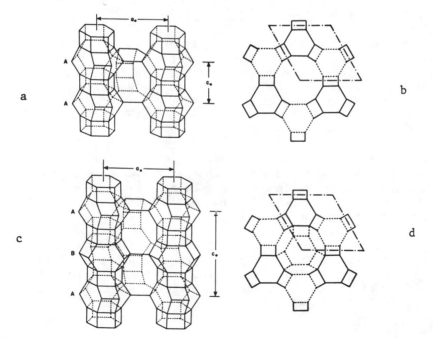

a   b

c   d

which form three dimensional four connected nets. The corner sharing of tetrahedra requires that there are twice as many framework oxygens as T-atoms (metal atoms). Cations are required for T-atom charge balance.

The first classification of zeolites on the basis of common structural units such as parallel 6-rings was made by Smith (6). Meier (2) classified zeolites into seven groups based on secondary building units or polyhedral building blocks which on linking form the framework structures. These units (Figure 1) (single 4-ring, single 6-ring, single 8-ring, double 4-ring, double 6-ring, and 4+1, 5 +1 and 4+4+1 combinations) are sufficient to describe a zeolite framework although 3 and 9 rings should be added. This classification into seven groups should be extended to nine with the addition of the melanophlogite group based on the aluminosilicate analogs of the gas hydrates and the lovdarite group based on 3,5 and 9-rings.

The nine groups of zeolites identified on the basis of their framework structure are given in Table I with idealized cell contents and crystallographic data and channel systems. The isotypes of the species listed in Table I are listed in Table II.

ANALCITE GROUP

The frameworks of the two members of this group, analcite and laumontite, can be derived by interconnecting 4 and 6-rings as shown in Figure 2.

NATROLITE GROUP

The chains (Figure 3a) characteristic of this group consist of linked four 4-ring units. There are three different ways to link these chains resulting in the natrolite (10), edingtonite (12) and thomsonite (10) frameworks (Figure 4). All the structures have a two dimensional 8-ring channel system.

CHABAZITE GROUP

The chabazite group frameworks consist of parallel 6-rings. The offretite-erionite a and c projections are shown in Figures 4a-4d. The stacking sequence involves single 6-rings A, B or C, double 6-rings AA, BB or CC or a combination of both. The stacking sequence of the members of the group are:

Cancrinite	AB	(30)	Levynite	AABCCABBC	(19)
Gmelinite	AABB	(16)	Afghanite	ABABACAC	(25)
Chabazite	AABBCC	(14)	Losod	ABAC	(27)
Offretite	AAB	(18)	Liottite	ABABAC	(28)
Erionite	AABAAC	(17)	TMA-E(AB)	ABBACC	(29)

TABLE I. Classification and Crystallographic Data for Zeolites

	Typical Unit Cell Contents	Crystal Data	Framework Density	Channel System	Reference
Analcite Group					
Analcite	$Na_{16}(AlO_2)_{16}(SiO_2)_{32}\cdot16H_2O$	Cubic Ia3d a=13.7A	18.6[a]		(7,8)
Laumontite	$Ca_4(AlO_2)_8(SiO_2)_{16}\cdot16H_2O$	Monoclinic Am or A$_2$ a=7.6, b=14.8, c=13.1A, γ=112°	17.7	$10^b(4.0\times5.6)1^{c,d}$	(9)
Natrolite Group					
Natrolite	$Na_{16}(AlO_2)_{16}(SiO_2)_{24}\cdot16H_2O$	Orthorhombic Fdd2 a=18.3, b=18.6, c=13.2A	17.8	8(2.6x3.9)3	(10,11)
Thomsonite	$Na_4Ca_8(AlO_2)_{20}(SiO_2)_{20}\cdot24H_2O$	Orthorhombic Pnn2 a=13.1, b=13.1, c=13.2A	17.7	8(2.6x3.9)3	(10)
Edingtonite	$Ba_2(AlO_2)_4(SiO_2)_6\cdot8H_2O$	Orthorhombic P2,2,2 a=9.6, b=9.7, c=6.5A	16.6	8(3.5x3.9)3	(12,13)
Chabazite Group					
Chabazite	$Ca_2(AlO_2)_4(SiO_2)_8\cdot13H_2O$	Trigonal R$\bar{3}$m a=13.2, c=15.1	14.6	8(3.6x3.7)3	(14,15)
Gmelinite	$Na_8(AlO_2)_8(SiO_2)_{16}\cdot24H_2O$	Hexagonal P6$_3$/mmc a=13.8, c=10.0A	14.6	12(7.0)1[e] 8(3.6x3.9)2	(16)
Erionite	$(Ca,MgNa_2K_2)_{4.5}(AlO_2)_9(SiO_2)_{27}\cdot27H_2O$	Hexagonal P6$_3$/mmc a=13.3, c=15.1A	15.6	8(3.6x5.2)3	(17)
Offretite	$(K_2,Ca,Mg)_{2.5}(AlO_2)_5(SiO_2)_{13}\cdot15H_2O$	Hexagonal P$\bar{6}$m2 a=13.3, c=7.6A	15.5	12(6.4)1[e] 8(3.6x5.2)2	(18)
Levynite	$Ca_3(AlO_2)_6(SiO_2)_{12}\cdot18H_2O$	Trigonal R$\bar{3}$m a=13.3, c=23.0A	15.2	8(3.3x5.3)2	(19,20)
Mazzite	$Na_6K_{1.9}Ca_{1.4}Mg_2$ $(AlO_2)_{9.8}(SiO_2)_{26.5}\cdot28H_2O$	Hexagonal P6$_3$/mmc a=18.4, c=7.6A	16.1	12(7.4)1	(21,22,23)
Linde L	$K_6Na_3(AlO_2)_9(SiO_2)_{27}\cdot8H_2O$	Hexagonal P6/mmm a=18.4, c=7.5A	16.4	12(7.1)1	(24)
Afghanite	$(Na_2CaK_2)_{12}(AlO_2)_{24}(SiO_2)_{24}$ $-(Na_2CaK_2)_6(Cl_2,SO_4,CO_3)_6$	Hexagonal P6$_3$/mmc a=12.8, c=10.5A	15.9	6	(25,26)
Losod	$Na_{12}(AlO_2)_{12}(SiO_2)_{12}\cdot19H_2O$	Hexagonal P$\bar{6}$2c a=12.9, c=10.5A	15.8	6	(27)
Liottite	$(CaNa_2K_2)_9(AlO_2)_{18}(SiO_2)_{18}$ $-(CaNa_2K_2)_n(SO_4,CO_3,Cl)_8\cdot2H_2O$	Hexagonal P$\bar{6}$m2 a=12.8, c=5.1A	15.7	6	(28)
TMA-E (AB)	$(Me_4N)_2Na_7(AlO_2)_9(SiO_2)_{27}\cdot26H_2O$	Hexagonal P6$_3$/mmc a=13.3, c=15.2A	15.4	8(3.7x4.8)2	(29)
Cancrinite	$Na_6(AlO_2)_6(SiO_2)_6\cdot CaCO_3\cdot2H_2O$	Hexagonal P6$_3$ a=12.8, c=5.1A	16.7	12(6.2)1	(30,31)

(continued)

	Typical Unit Cell Contents	Crystal Data	Framework Density	Channel System	Reference
Phillipsite Group					
Phillipsite	$(KNa)_5(AlO_2)_5(SiO_2)_{11} \cdot 10H_2O$	Monoclinic P2_1/m $a=9.9, b=14.3, c=8.7A$ $\beta=124°$	15.8	8(4.2x4.4)1[e] 8(2.8x4.8) 8(3)1	(32,33)
Gismondite	$Ca_4(AlO_2)_8(SiO_2)_8 \cdot 16H_2O$	Monoclinic P2_1/a $a=9.8, b=10.0$ $c=10.6A, \gamma=90°$	15.4	8(3.1x4.4)1[e] 8(2.8x4.9)1	(34,35)
Yugawaralite	$Ca_4(AlO_2)_8(SiO_2)_{20} \cdot 16H_2O$	Monoclinic Pc $a=6.7, b=14.0, c=10.0A, \beta=112°$	18.3	8(3.1x3.5)1[e] 8(3.2x3.3)1	(36)
Li A (BW)	$Li_4(AlO_2)_4(SiO_2)_4 \cdot 4H_2O$	Orthorhombic Pna2 $a=10.3, b=8.2, c=5.0A$	19.0	8(3.6x4.0)1	(37)
Heulandite Group					
Heulandite	$Ca_4(AlO_2)_8(SiO_2)_{28} \cdot 24H_2O$	Monoclinic Cm $a=17.7, b=17.9, c=7.4A, \beta=116°$	17.0	8(4.0x5.5)1[e] 10(4.4x7.2)1 8(4.1x4.7)1	(38,39)
Brewsterite	$(Sr,Ba,Ca)_2(AlO_2)_4(SiO_2)_{12} \cdot 10H_2O$	Monoclinic P2_1/m $a=6.8, b=17.5, c=7.7A, \beta=95°$	17.5	8(2.3x5.0)1[e] 8(2.7x4.1)1	(40)
Stilbite	$Na_2Ca_4(AlO_2)_{10}(SiO_2)_{26} \cdot 32H_2O$	Monoclinic F 2/m $a=13.6, b=18.2, c=17.8A, \beta=91°$	16.9	10(4.1x6.2)1[e] 8(2.7x5.7)1	(41,42)
Mordenite Group					
Mordenite	$Na_8(AlO_2)_8(SiO_2)_{40} \cdot 24H_2O$	Orthorhombic Cmcm $a=18.1, b=20.5, c=7.5A$	17.2	12(6.7x7.0)1[e] 8(2.9x5.7)1	(43)
Ferrierite	$Na_{1.5}Mg_2(AlO_2)_{5.5}(SiO_2)_{30.5} \cdot 18H_2O$	Orthorhombic Immm $a=19.2, b=14.1, c=7.5A$	17.7	10(4.3x5.5)1[e] 8(3.4x4.8)1	(44)
Dachiardite	$Na_5(AlO_2)_5(SiO_2)_{19} \cdot 12H_2O$	Monoclinic C2/m $a=18.7, b=7.5, c=10.3A, \beta=108°$	17.3	10[3.7x6.7]1[e] 8[3.6x4.8]	(45)
Bikitaite	$Li_2(AlO_2)_2(SiO_2)_4 \cdot 2H_2O$	Monoclinic P2_1 $a=7.6, b=8.6, c=5.0A, \gamma=114°$	20.2	8(3.2x4.9)1	(46)
Epistilbite	$Ca_3(AlO_2)_6(SiO_2)_{18} \cdot 16H_2O$	Monoclinic C2/m $a=8.9, b=17.7, c=10.2, \beta=124°$	18.0	10(3.5x5.3)1[e] 8(3.7x4.4)1	(47,48)
ZSM-5	$Na_n(AlO_2)_n(SiO_2)_{96-n} \cdot 16H_2O$ $n<27$	Orthorhombic Pnma $a=20.1, b=19.9, c=13.4A$	17.9	10(5.4x5.6)1[e] 10(5.1x5.5)3	(49,50)
ZSM-11	$Na_n(AlO_2)_n(SiO_2)_{96-n} \cdot 16H_2O$ $n<27$	Tetragonal I4̄m2 $a=20.1, c=13.4A$	17.7	10(5.1x5.5)3	(51)

(continued)

	Typical Unit Cell Contents	Crystal Data	Framework Density	Channel System	Reference
Faujasite Group					
Faujasite	$Na_{12}Ca_{12}Mg_{11}(AlO_2)_{59}(SiO_2)_{133} \cdot 260H_2O$	Cubic Fd3m a=24.7A	12.7	12(7.4)3	(52,53,54)
Linde A	$Na_{12}(AlO_2)_{12}(SiO_2)_{12} \cdot 27H_2O$	Cubic Fm3c (Pm3m) a=24.6A (a=12.3A)	12.9	8(4.1)3	(55,56,57)
ZK-5	$Na_{30}(AlO_2)_{30}(SiO_2)_{66} \cdot 98H_2O$	Cubic Im3m a=18.7A	14.7	8(3.9)3[f] 8(3.9)3	(58)
ZSM-3	$[(LiNa)_2AlO_2)_2(SiO_2)_{3.2} \cdot 8H_2O]_m$	Hexagonal a=17.5, c=129A		8(3.9)3	(59)
Zeolite Rho	$(Na,Cs)_{12}(AlO_2)_{12}(SiO_2)_{36} \cdot m \cdot H_2O$	Cubic I43m a=15.1A	14.3	8(3.9×5.1)3	(51)
Paulingite	$(K_2,Na_2,Ca,Ba)_{76}(AlO_2)_{152}(SiO_2)_{525} \cdot 700H_2O$	Cubic Im3m a=35.1A	15.5	8(3.9)3[f] 8(3.9)3	(52)
Merlinoite	$(K_5,Ca_2)(AlO_2)_9(SiO_2)_{23} \cdot 24H_2O$	Orthorhombic Immm a=14.1,b=14.2,c=10.0A	16.0	8(3.1×3.5)1[e] 8(3.5×3.5)1 8(3.4×5.1)1	(53)
Linde N	$Na_{384}(AlO_2)_{384}(SiO_2)_{384} \cdot 518H_2O$	Cubic Fd3 a=36.93		6-rings	(54)
Sodalite	$Na_6(AlO_2)_6(SiO_2)_6 \cdot 2NaCl$	Cubic P43n a=8.9A		6-rings	(30,55)
Melanophlogite Group					
Melanophlogite	$Me_n(AlO_2)_n(SiO_2)_{48-n} \cdot mH_2O$	Cubic P4$_2$32 a=13.4A		6-rings	(56,57,58)
ZSM-39	$(Na,TMA,TEA)_{0.4}(AlO_2)_{0.4}(SiO_2)_{135.6} \cdot mH_2O$	Cubic Fd3m a=19.4A	18.7	6-rings	(59)
Lovdarite Group					
Lovdarite	$K_4Na_{12}(BeO_2)_8(SiO_2)_{28} \cdot 18H_2O$	Tetragonal Poly type P4$_2$/mmc a=7.14,b=7.14,c=21.0A Poly type I4m2 a=7.15,b=7.15,c=42		8-ring 1[e] 9-ring 2 8-ring 1 9-ring 2	(60)

a. Number of atoms per 1000A³.
b. Number of ring members in channel opening.
c. Ring dimension.
d. Dimensions of channel system.
e. Channel system intersect.
f. Separate channel systems.

TABLE II. Zeolite Species Isotypes

Zeolite	Isotype Species
Analcite	CaD (61), Kehoite (62), Leucite, NaB (63), Pollucite, Viseite (64), Wairakite (65)
Laumontite	Leonhardite (66)
Natrolite	Scolecite, Laubanite
Thomsonite	Gonnardite
Edingtonite	K-F (67)
Chabazite	Herschelite, Linde D (68), Linde R (69)
Gmelinite	S (70)
Offretite-Erionite	Linde T (68), O (71)
Levynite	ZK-20 (72)
Mazzite	Omega (73), ZSM-4 (74)
Linde L	BaG (75), P-L (76)
Cancrinite	Cancrinite Hydrate (77)
Phillipsite	Harmatome, Wellsite (78), ZK-19 (79)
Gismondite	Linde B (80), Garronite (81), Pc (82), Pt (82), P (70)
Yugawaralite	Sr-Q (75)
LiA(BW)	$CaAlSiO_4$ (83), $RbAlSiO_4$ (83)
Heulandite	Clinoptilolite (84)
Stilbite	Barrerite (85), Desmine
Mordenite	Na-D (63), Ptilolite, Zeolon
Ferrierite	Sr-D (75)
Dachiardite	Svetlozarite (86)
Faujasite	Linde X (87), Linde Y (88)
Linde A	Alpha (89), ZK-4 (90), ZK-21 (91,92), ZK-22 (91,92)
ZK-5	BaP (93), BaQ (93), P(Cl) (93), Q(Br) (93)
Merlinoite	K-M (67), Linde W (80)
Sodalite	Basic sodalite (63), Danalite, Hydroxysodalite (94), Nosean, Tetracalcium trialuminate (95), Tugputite (96), Ultramarine, Zh (97)

Figure 5

Framework of mazzite
(a) c-projection (b) linking of
gmelinite columns

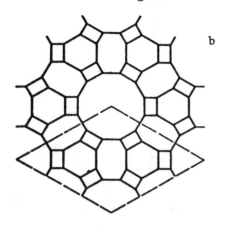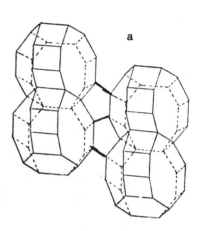

Figure 6

Three Types of 4-ring Chains UUDD,
UDUD, and UDUU (117)

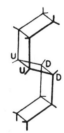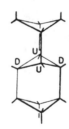

Figure 7

Secondary Units in (a) heulandite group
(b) mordenite group (2)

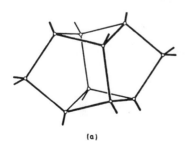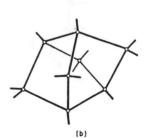

(a) (b)

92

Figure 8

Framework Projections Along Main
Channels of (a) mordenite (b) dachiardite
(c) ferrierite (d) epistilbite (e) bikitaite

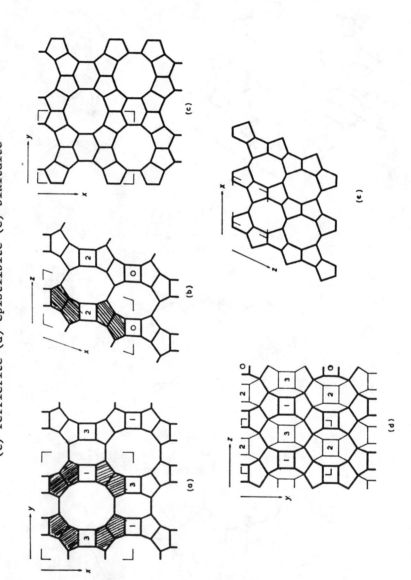

Figure 9

Pentasil Framework with SSII Sequence

a
40 A°
C
S
S
I
I
ZSM-5/ZSM-II INTERMEDIATE (SSII)
FIG. 4

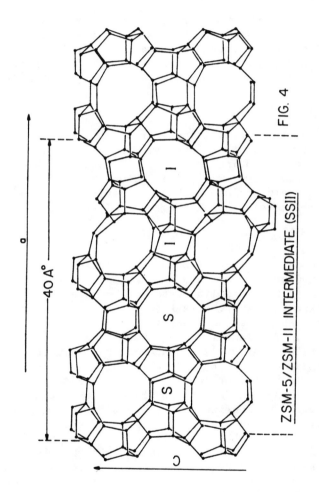

The framework of mazzite (21) consists of columns of gmelinite cages linked through 8-rings with alternate columns staggered by $c/2$ (Figure 5). If cancrinite cages are connected through double 6-rings to form columns and these columns are linked through 6-rings to form 12-ring channels parallel to the C-axis, as in mazzite, the framework of Linde L (24) is formed.

PHILLIPSITE GROUP

The framework of members of this group are based on 4-rings with variations of U (up) and D (down) linkages. Three of the four such variations can be linked to form chains (Figure 6). Phillipsite (32,33) and gismondite (34,35) consist of cross-linked UUDD chains. Li-A(BW) (37) and yugawaralite (36) consist of linked single 4-rings.

HEULANDITE GROUP

A building block containing four 5-rings and two 4-rings is common to the framework of all the members of this group. If linked through a common edge, chains are formed (Figure 3b) which when linked together yield brewsterite (40). Linking these blocks through common vertices yield chains which are constituents of heulandite (38,39) and stilbite (41,42). This group of structures contain some 5-rings.

MORDENITE GROUP

The secondary building block consisting of four 5-rings is common to all members of this group except bikitaite. In mordenite and dachiardite they are linked to form complex chains which in turn are linked in different ways (2). Epistilbite and ferrierite are lamellar structures and they also contain the building block in Figure 7b. The lamellae are normal to a in ferrierite and to b in epistilbite. The projections along the main channels are shown in Figures 8a-8e (2). The projection of bikitaite along b is essentially the same as the projection of epistilbite along a and dachiardite along c. ZSM-5 and ZSM-11 are better described by the chain in Figure 3c. The configurational unit in this chain contains eight 5-rings. If these chains are linked so that alternate pairs are related by a reflection, a layer is formed which is the basic layer in the ZSM-5 and ZSM-11 structures. If this layer is linked so that alternate layers are related by a reflection, S, the ZSM-11 framework is formed (57). If the layers are linked such that alternate layers are related by an inversion I, the ZSM-5 framework results (49). If the stacking sequence is varied, a family of structures results with ZSM-5 and ZSM-11 as the end members. With the lattice parameter doubled there are only two possible stacking sequences, SSII and SISI. The framework for SSII is shown in Figure 9. This variation in stacking sequence has been confirmed by Thomas and Millward (98) with lattice imaging using ultra high resolution electron microscopy.

Figure 10

Arrangement of A Cages in
(a) ZK-5 (b) Rho (c) Paulingite

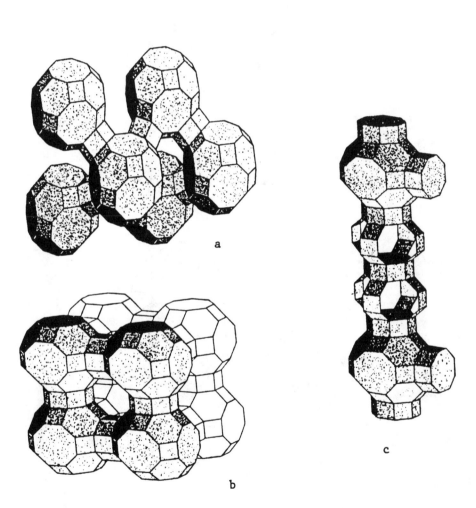

a

b

c

Figure 11

Polyhedral Units in ZSM-39 and
Melanophlogite Structures (a) 12-hedron
(b) 14-hedron (c) 15-hedron (d) 16-hedron

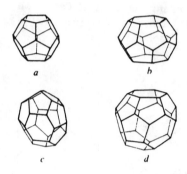

a   *b*

c   *d*

Figure 12

Melanophlogite Framework

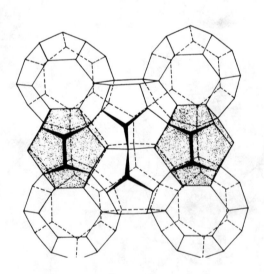

FAUJASITE GROUP

There are three types of cages in this group, sodalite (30), A and
ZK-5. Linde A (55) is formed by linking sodalite cages through
double 4-rings. The large cages in Linde A will be referred to as
the A cage. If the A cages are linked through double 6-rings, the
framework of ZK-5 is formed (58). If they are linked through double
8-rings, the rho (51a) structure results (Figure 10). The cage formed
by the octagonal faces of the A cages in ZK-5 is analogous to the
gmelinite cage and will be referred to as the ZK-5 cage. Columns
of ZK-5 cages linked through double 8-rings when connected together
form the merlinoite (53a) framework. Feldspar and phillipsite are
closely related to merlinoite formed by sliding layers such that the
double 8-rings are broken up. Figure 10a shows part of the framework
of paulingite (52a), the sequence of cages along the cubic axis A,
ZK-5, ZK-5, A each connected to the other through double 8-rings.
No one cage can be considered as the building block, the combination
of cages is required.

The faujasite framework (52) consists of sodalite cages linked
through double 6-rings. If layers of these connected sodalite cages
are linked in an ABC sequence the framework is that of faujasite
with Fd3m symmetry, the same as diamond.

If the stacking sequence is changed to AB, the result is a hexagonal
structure. There are now five 12-ring openings to the large cavity
compared to four in faujasite. The c parameter and the number of
stacking sequences is dependent on the number of layers in the
identity period. ZSM-3 (59) was found to be hexagonal with a=17.5
and c=129 A indicating a nine layer stacking sequence.

Sodalite cages linked through common 6-rings form the sodalite
framework (30).

Falth and Anderrson (54a) found that Linde N, a cubic structure with
a=36.9A, was an intergrowth of ZK-5 and sodalite.

MELANOPHLOGITE GROUP

The relationship between hydrogen bonds linking oxygens in gas hydrates (clathrates) and oxygens linking T-atoms in zeolites is well
known. Hexagonal ice I is isostructural with β-tridymite (58a).
Appleman (57a) found that the rare mineral melanophlogite is isostructural with the 12A cubic gas hydrate. The melanophlogite framework consists of interwoven layers of 12 and 14-hedra (Figures 11 and
12). It is a dense structure with only 5 and 6-ring openings. ZSM-
39 (59a), a high silica synthetic zeolite, was found to be isostructural with the 17A cubic gas hydrate. The ZSM-39 framework consists
of 12 and 16-hedra as shown in Figure 11 , with layers of face
sharing 12-hedra arranged as in Figure 13. These layers are stacked

Figure 13

ZSM-39 Framework

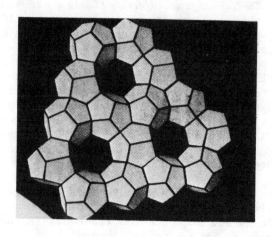

Figure 14

Lovdarite Framework (a) a-axis projection
(b) c-axis projection

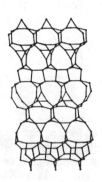

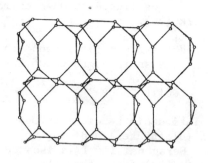

a

b

in ABC sequence. The openings in the framework are limited to 5 and 6-rings. Zeolite analogs of water clathrate structures which typically consist of cages comprising of 5-rings constitute a new family of zeolites.

LOVDARITE GROUP

Lovdarite (60), a unique beryllium zeolite, has a framework consisting of 4 and 8-rings linked via corner sharing 3-rings. It has a two dimensional intersecting channel system bounded by 9-rings (Figure 14). The 9-rings are unique as no other known zeolite has a 9-ring channel system. The 129° equilibrium angle of the Be-O-Si linkage indicates that the three-membered rings are not strained.

T-ATOM DISTRIBUTION

It is desirable to have some knowledge regarding T-atom distribution in zeolites. For several zeolites there is unequivocal evidence of Si,Al ordering (natrolite and gismondite). X-ray and other evidence are consistent with random ordering of T-atoms. The x-ray evidence for the occupancy of tetrahedral sites by Si or Al is based on the Si,Al-O interatomic distances which differ by about .13A (100). This requires accurate atomic coordinates. The available evidence is also in accord with Lowenstein's rule forbidding Al-O-Al bonds. Olson (101) found that hydrated Linde X crystals had Fd3 rather than Fd3m symmetry. The cell content prohibits complete ordering (88 Al, 104 Si). It was determined (102) that Al,Si alternate in the 4-rings in mordenite.

CATION LOCATION

Precise information as to cation positions in zeolites is still rather limited as faults, thermal and positional disorder, partial occupancy act as hindrances. The cation sites and their population in dehydrated mordenite are given in Table III.

TABLE III. Cation Population in Mordenite

Site	H	Na	K	Rb	Cs	Ca	Ba
I or I'	.6Na	3.1				1.7	0.3Ca
II			3.3	3.6	3.8		1.9Ba
III						0.6	0.3Ba
IV		2.6	3.0	3.1	1.9	0.5	1.1Ba
VI		1.5	0.9	0.7	1.7	0.6	
SG	Cmcm	Pbcn	Pbcn	Pbcn	P2$_1$cn?	Cmcm	Pbcn
Ref.	(103)	(104)	(105)	(106)	(107)	(108)	(109)

Site I lies at the end of the side pocket and is too small for large cations. Site II is in the side pocket at the center of the 8-ring. Site IV is in the center of another 8-ring at the junction of the side pocket with the main channel. This site is occupied by all types of cations. Site VI is coordinated to oxygens in walls of the main channel. Lowering of space groups is partially due to displacement of cations.

The number and position of cations is a function of temperature and degree of hydration. The cation site on the threefold axis, outside the 6-rings in ZK-5 is fully occupied in the hydrated state, while at 150°C it is empty (58).

In calcium exchanged offretite and erionite the calcium displaces the K from the center of the cancrinite cages on dehydration (110,111).

In Ce exchanged faujasite the Ce^{+++} ions occupy sites in the center of the 12-ring. On dehydration the cerium ions move into the center of the sodalite cage where each metal ion is coordinated to three framework oxygen ions with Ce-O=2.52, $\pm$.01A and up to three water oxygens with Ce-O=2.44 $\pm$.08 (112). This sodalite cage complex is highly stable.

STACKING FAULTS

The occurrence of stacking faults in a number of zeolites is quite prevalent. Stacking faults can be detected by the presence of broad odd ℓ lines in the diffraction pattern of offretite, by the presence of contrast lines in transmission electron micrographs of erionite (113), by lattice images of ZSM-5 using high resolution electron microscopy (98).

STRUCTURE DETERMINATION

With increasing knowledge of zeolite frameworks there is a considerable understanding of zeolite chemical principles but many zeolites have been synthesized as small or larger but poorer quality crystals. Increased x-ray intensity sources should help in determining the structure of some of these crystals but if it is not possible to use single crystal methods then we have to turn to other methods. Information regarding lattice parameters, symmetry can be obtained from diffraction studies, estimates of channel dimensions from IR and diffusion studies, and ring ellipsisity from diffusion rates.

Physical models using tetrahedral stars to represent T-atoms are connected via tubing to depict zeolite frameworks. These models are built to scale for rapid and accurate estimation of unit cell parameters and atomic coordinates and are useful for determining the symmetry of a structure. The basic building units can be readily identified in these open frameworks.

Trial models may be built using all the available information and satisfying all the known conditions. X-ray patterns can be simulated and compared to experimental ones.

SIMULATION OF PATTERNS

Interatomic distances and bond angles for zeolites of known composition can be predicted within fairly narrow limits. If the lattice parameters are known, the atomic coordinates of individual atoms can be adjusted so that the interatomic distances correspond as closely as possible with predicted distances. The atomic positional parameters can be computed from the prescribed interatomic distances, D_j^o, by a least squares procedure which minimizes the residual function

$$r_w = \sum_{\substack{j \\ m,n}} w_j^2 (D_j^o - D_j^{m,n})^2$$

where w_j is the weight ascribed to the interatomic distance of type j. This DLS (Distance Least Squares) method of refining the positional parameters was described by Meier and Villiger (114). This refinement gives idealized framework models using prescribed interatomic distances and unit cell constants for a given space group. The weight w_j of each error equation is based on bonding considerations or observed bond length variations (115). In zeolites the Si-O bond length relation to the Si-O-Si bond angle is given by the function,

$$d(Si-O) = 1.527 + .068[-\sec(Si-O-Si)]$$

Convergence of the least squares refinement is usually rapid for chemically reasonable structures. A final "R-factor" is provided which can be used to estimate the "goodness" or chemical reasonableness of the structure. If the positional parameters of some of the atoms in a structure are determined by single crystal methods, the missing atoms may be located by model building as in the case of Linde N (54) and the structure further refined.

Simulated structures can be altered in a manner not possible with actual structures. The size of the atoms can be changed by changing the interatomic distances. We can impose restrictions on the lattice parameters. We can put cations in prescribed locations. The DLS refinement will give us the changes in geometry. This should give us valuable insights into chemical principles.

EFFECT OF STACKING FAULTS ON DIFFRACTION PATTERN

Stacking faults can be determined by contrast lines in transmission electron microscopy (113) and by lattice imaging (98) but this is a tedious process and is applicable to individual crystals and not the bulk. In order to determine the effect of stacking faults the

102

Figure 15
Simulated Plots of Ferrierite

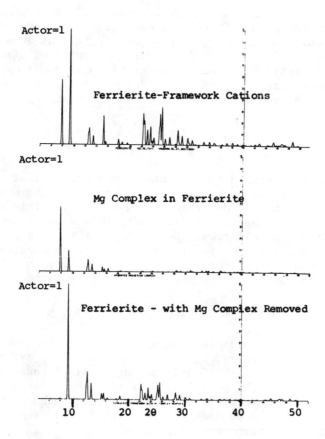

two structures are superimposed and an occupancy of 1 is assigned to all atoms which coincide and an occupancy of ½ to split atoms. The application of this method to offretite-erionite (116) resulted in the determination of the concentration of stacking faults and whether they were random or ordered.

CONTRIBUTION OF CATIONS AND WATER
TO POWDER PATTERN

The structure of ferrierite was determined by Vaughn (44). He found $Mg(H_2O)_6^{++}$ cation complexes in the cavities with 8-rings. Difference in powder pattern intensities for samples from various locations were found to be due to the Mg complexes in the 8-ring cages. Smith plots of the contribution of the cation complexes (116) are shown in Figure 15. By obtaining plots of structures with partial occupancy better agreement of powder data can be obtained.

ACKNOWLEDGEMENTS

I am grateful to W. M. Meier, J. L. Schlenker, S. Sawruk, S. L. Lawton and A. C. Rohrman for valuable discussion, advice and encouragement.

REFERENCES

1. W. L. Bragg and G. F. Claringbull, "Crystal Structure of Minerals", Bell and Sons, London (1965).

2. W. M. Meier in "Molecular Sieves", Soc. Chem. Ind. London (1968), p. 10.

3. D. W. Breck, "Zeolite Molecular Sieves", Wiley Interscience (1974).

4. J. V. Smith, "Zeolite Chemistry and Catalysis", Ed. J.A. Rabo, ACS Monograph 171 (1976), p. 3.

5. R. M. Barrer, "Zeolites and Clay Minerals as Sorbents and Molecular Sieves", Academic Press, London (1978).

6. J. V. Smith, M.S.A., Spec. Pap. 1, 281 (1963).

7. W. H. Taylor, Z. Kristallogr. 74, 1, (1930).

8. C. R. Knowles, F. F. Rinaldi and J. V. Smith, Indian Mineral. 6, 127 (1965).

9. H. Bartle and K.F. Fischer, Nenes Jahrb Mineralog. 33 (1967).

10. W. H. Taylor, C. A. Meek and W. W. Jackson, Z. Kristallogr., 84, 373 (1933).

11. W. M. Meier, Z. Kristallogr. 113, 430 (1960).

12. W. H. Taylor and R. Jackson, Z. Kristallogr. 86, 53 (1933).

104

13. E. Galli, Acta. Cryst. **B32**, 1623 (1976).

14. L. S. Dent and J. V. Smith, Nature, **181**, 1794 (1958).

15. J. V. Smith, R. Rinaldi, L. S. Dent, Glasser
    Acta. Cryst. **16**, 45 (1963).

16. K. Fischer, Neues Jahrb. Mineral. Monatsh. 1 (1966).

17. L. W. Staples and J. A. Gard, Mineral Mag. **32**, 261 (1959).

18. J. M. Bennett and J. A. Gard, Nature **214**, 1005 (1967).

19. R. M. Barrer and I. S. Kerr, Trans. Farad. Soc. **55**, 1915 (1959).

20. S. Merlino, E. Galli and A. Alberti, Tschermaks Min. Petr.
    Mitt. **22**, 117 (1975).

21. E. Galli, Cryst. Struct. Comm. **3**, 339 (1974).

22. E. Galli, E. Passaglia and D. Pongiluppi, Contr. Mineralog.
    and Pet. **45**, 99 (1974).

23. R. Rinaldi, J. J. Pluth and J. V. Smith, Acta. Cryst., **B31**, 1603
    (1975).

24. R. M. Barrer and H. Villiger, Z. Kristallogr. **128**, 352 (1969).

25. P. Bariand, F. Cesbron and R. Geraud, Bull. Soc. Fr. Mineral.
    Crystalogr. **91**, 34 (1968).

26. S. Merlino and M. Mellini, Zeolite 76, Program and Abstracts,
    Tucson (1976).

27. W. Sieber and W. M. Meier, Helv. Chim. Acta., **57**, 1533 (1974).

28. S. Merlino and P. Orlandi, Am. Mineral. **62**, 321 (1977).

29. M. Groner and W. M. Meier, to be published.

30. L. Pauling, Z. Kristallogr. **74**, 213 (1930).

31. O. Jarchow, Z. Kristallogr., **122**, 407 (1965).

32. H. Steinfink, Acta. Cryst. **15**, 644 (1962).

33. R. Rinaldi, J. J. Pluth and J. V. Smith, Acta. Cryst. **B30**,
    2426 (1974).

34. K. Fischer, Am. Mineral **48**, 664 (1963).

35. K. Fischer and V. Schramm, Adv. Chem. Ser. **101**, 250 (1971).

36. I. S. Kerr and D. J. Williams, Z. Kristallogr. **125**, 1 (1967).

37. I. S. Kerr, Z. Kristallogr. **139**, 186 (1974).

38. A. B. Merkle and M. Slaughter, Am. Mineral. **52**, 273 (1967).

39. A. Alberti, Tschermaks. Mineral. Petrogr. Mitt. **18**, 129 (1972).

40. A. J. Perrotta and J. V. Smith, Acta. Cryst. **17**, 857 (1964).

41. E. Galli and G. Gottardi, Mineral. Petrogr. Acta. **12**, 1 (1966).

42. E. Galli, Acta. Cryst. B27, 833 (1971).

43. W. M. Meier, Z. Kristallogr. 115, 439 (1961).

44. P. A. Vaughn, Acta. Cryst. 21, 983 (1966).

45. G. Gottardi and W. M. Meier, Z. Kristallogr. 119, 53 (1963).

46. V. Kocman, R. I. Gait and J. Rucklidge, Am. Mineral. 59, 71 (1974).

47. I. S. Kerr, Nature 202, 589 (1964).

48. A. J. Perrotta, Mineral Mag. 36, 480 (1967).

49. G. T. Kokotailo, S. L. Lawton, D. H. Olson and W. M. Meier, Nature 272, 437 (1978).

50. D. H. Olson, G. T. Kokotailo, S. L. Lawton and W. M. Meier, J. Phys. Chem. 85, 2238 (1981).

51. G. T. Kokotailo, P. Chu, S. L. Lawton and W. M. Meier, Nature 275, 119 (1978).

52. W. Nowacki and G. Bergerhoff, Schweiz. Mineral. Petrogr. Mitt. 36, 621 (1965).

53. G. Bergerhoff, W. H. Baur and W. Nowacki, Neues Jahrb. Mineral. Monatsh. 193 (1958).

54. W. H. Baur, Am. Mineral. 49, 697 (1964).

55. T. B. Reed and D. W. Breck, JACS 78, 5972 (1956).

56. L. Broussard and D. P. Shoemaker, JACS 82, 1041 (1960).

57. V. Gramlich and W. M. Meier, Z. Kristallogr. 133, 134 (1971).

58. W. M. Meier and G. T. Kokotailo, Z. Kristallogr. 121, 211 (1965).

59. G. T. Kokotailo and J. Ciric in "Molecular Sieve Zeolites - 1" Advances in Chemistry Series 101, Am. Chem. Soc. Ed. R.F. Gould (1971), p. 109.

51a. H. E. Robson, D. P. Shoemaker, R. A. Ogilvie and P. C. Manor, Adv. Chem. Ser. 121, 106 (1973).

52a. E. K. Gordon, S. Samson and W. B. Kamb, Science 154, 1004 (1966).

53a. E. Passaglia, D. Pongiluppi and R. Rinaldi, Neues Jahrb. Mineral. Monatsh 355 (1977).

54a. L. Falth and S. Anderrson, Z. Kristallogr. 160, 313 (1982).

55a. J. Loens and H. Schulz, Acta. Cryst. 23, 434 (1967).

56a. B. J. Skinner and D. E. Appleman, Am. Mineral. 48, 855 (1963).

57a. D. E. Appleman, Am. Cryst. Assoc. Mineral. Soc. Joint Meeting, Gatlinberg, Tenn. 80 (1965).

58a. B. Kamb, Science 148, 232 (1965).

59a. J. L. Schlenker, F. G. Dwyer, E. E. Jenkins, W. J. Rohrbaugh, G. T. Kokotailo and W. M. Meier, Nature 294, 340 (1981).

60. S. Merlino, Twelfth Int. Congr. Cryst. (LeDroit and LeClerc Printers, Ottawa, Canada) C189 (1981).

61. L. L. Ames and L. B. Sand, Am. Mineral. 43, 476 (1958).

62. D. McConnell, Mineral. Mag. 33, 799 (1964).

63. R. M. Barrer and E. A. D. White, J. Chem. Soc. 1167 (1952).

64. J. Melon, Ann. Soc. Geol. Belg. 66, 53 (1942).

65. A. Steiner, Mineral. Mag. 30, 691 (1955).

66. C. T. Amirov, V. V. Ilyukin and N. V. Belov, Dokl. Akad. Nauk SSSR 174 667 (1967).

67. R. M. Barrer and J. W. Baynham, J. Chem. Soc. 2882 (1956).

68. D. W. Breck and N. A. Acara, US Patent 2,950,952 (1960).

69. R. M. Milton, Brit. Patent 841,812 (1960).

70. R. M. Barrer, J. W. Baynham, F. W. Bultitude and W. M. Meier, J. Chem. Soc. 195 (1959).

71. R. Aiello and R. M. Barrer, J. Chem. Soc. A, 1470 (1970).

72. G. T. Kerr, US Patent 3,459,676 (1969).

73. E. Flanigen, Neth. Patent 6,710,729 (1968).

74. J. Ciric, Dutch Patent 67-06271 (1967).

75. R. M. Barrer and D. J. Marshall, J. Chem. Soc. 2296 (1964).

76. E. Flanigen and R. W. Grose, Adv. Chem. Ser. 101, 76 (1971).

77. J. Wyart and M. Michel-Levy, Compt. Rend. 229, 131 (1949).

78. P. Cerny, R. Renaldi and R. C. Surdam, Neues Jahrb. Mineral. Monatsh. 312 (1977).

79. G. H. Kuehl, Am. Mineral. 54, 1607 (1969).

80. R. M. Milton, US Patent 3,012,853 (1961).

81. G. P. L. Walker, Mineral. Mag. 33, 173 (1962).

82. A. M. Taylor and R. Roy, Am. Mineral. 49, 656 (1964).

83. S. J. Chung and Th. Hahn, Mat. Res. Bull. 7, 1209 (1972).

84. A. Alberti, Tschmarks. Mineral. Petrogr. Mitt. 22, 25 (1975).

85. E. Passaglia and D. Pongiluppi, Mineral. Mag. 40, 268 (1975).

86. L. R. Gellens, G. D. Price and J. V. Smith, Mineral. Mag. 45, 157 (1982).

87. R. M. Milton, US Patent 2,882,244 (1959).

88. D. W. Breck, US Patent 3,130,007 (1964).

89. C. T. Wadlinger, E. J. Rosinski and C. J. Plank, US Patent 3,375,205 (1968).

90. G. T. Kerr, Inorg. Chem. $\underline{5}$, 1537 (1966).

91. G. H. Kuehl, US Patent 3,355,246 (1967).

92. G. H. Kuehl, Inorg. Chem. $\underline{10}$, 2488 (1971).

93. R. M. Barrer, L. Hinds and E. A. D. White, J. Chem. Soc. 1466 (1953).

94. W. Borchert and J. Heidel, Heidel. Beitr. Mineral. Petrogr. $\underline{1}$, 2 (1947).

95. V. I. Ponomorev, D. M. Kheiker and N. V. Belov, Kristallogr. $\underline{15}$, 981 (1970).

96. H. Sorensen, Am. Mineral. $\underline{48}$, 1178 (1963).

97. S. P. Zhdanov and N. N. Buntar, Dokl. Akad. Nauk. SSSR $\underline{147}$, 1118 (1962).

98. J. M. Thomas and G. R. Millward, J. Chem. Soc. Chem. Commun. 1380 (1982).

99. E. Garnier, P. Gravereau and A. Hardy, Acta. Cryst. $\underline{B38}$, 1401 (1982).

100. J. V. Smith and S. W. Bailey, Acta. Cryst. $\underline{16}$, 801 (1963).

101. D. H. Olson, J. Phys. Chem. $\underline{74}$, 2758 (1970).

102. W. M. Meier, R. Meier and V. Gramlich, Z. Kristallogr. $\underline{147}$, 329 (1978).

103. W. J. Mortier, J. J. Pluth and J. V. Smith, Mater. Res. Bull. $\underline{10}$, 1319 (1975).

104. J. L. Schlenker, J. J. Pluth and J. V. Smith, Mat. Res. Bull. $\underline{14}$, 751 (1979).

105. W. J. Mortier, J. J. Pluth and J. V. Smith, "Mineralogy and Geology of Natural Zeolites", F. A. Mumpton, Ed., Mineral Soc. Am. Short Course Notes Vol. 4, p. 53.

106. J. L. Schlenker, J. J. Pluth and J. V. Smith, Mater. Res. Bull. $\underline{13}$, 77 (1978).

107. J. L. Schlenker, J. J. Pluth and J. V. Smith, Mater. Res. Bull. $\underline{13}$, 901 (1978).

108. W. J. Mortier, J. J. Pluth and J. V. Smith, Mater. Res. Bull. $\underline{10}$, 1037 (1975).

109. J. L. Schlenker, J. J. Pluth and J. V. Smith, Mater. Res. Bull. $\underline{13}$, 169 (1978).

110. G. T. Kokotailo and S. L. Lawton, US Patent 3,640,680 (1972).

108

111. J. L. Schlenker, J. J. Pluth and J. V. Smith, Acta. Cryst. B33, 3265 (1977).

112. D. H. Olson, G. T. Kokotailo, J. F. Charnell, 41st Colloid. Symposium, Buffalo, NY, June 1967.

113. G. T. Kokotailo, S. Sawruk and S. L. Lawton, Am. Mineral. 57, 439 (1972).

114. W. M. Meier and H. Villiger, Z. Kristallogr. 129, 411 (1969).

115. W. H. Baur, Phys. and Chem. Minerals 2, 3 (1977).

116. G. T. Kokotailo and J. L. Schlenker, Advances in X-ray Analysis, Vol. 24 (Plenum Publ. Corp) (1981), p. 49.

117. S. Merlino, Soc. Ital. di Mineral. e Petrogr., Rendecorti 31, 513 (1975).

SYNTHESIS OF ZEOLITES, AN OVERVIEW

L. Deane Rollmann

Mobil Research & Development Corporation
Central Research Division
PO Box 1025
Princeton, NJ 08540

INTRODUCTION

Zeolites have been an important component in petroleum process technology for more than twenty years -- and there is considerable indication that their role will continue and will expand as utilization of our energy resource base broadens worldwide. Despite the large and growing number of distinctive structures and compositions, only a handfull of zeolites, nearly all synthetic, have achieved commercial significance.

This paper is directed at the preparation of zeolites of this special group. The first section develops general principles and techniques in the synthesis of these zeolites, particularly the concepts of templating and compositional control. The second presents examples, to illustrate practical application of these concepts. Composition in zeolites can be controlled during or after synthesis. In the third section, an example of the latter is discussed, the dealuminization of mordenite.

The synthesis of zeolites is an art better chronicled in the patent literature perhaps than in classical scientific publications. Nevertheless, several books and review articles can be cited which will provide useful background information and supplementary detail (1-10). The discussion below summarizes and updates, where appropriate, those reviews. In the interest of clarity, discussion is restricted to three-dimensional, crystalline networks whose framework composition is at least 50 mol percent SiO_2, the remainder being Al_2O_3.

PATTERNS

Most commonly, zeolite crystallization is a nucleation-controlled process occurring from molecularly inhomogeneous, alkaline, aqueous gels, at temperatures between about 80 and 300°C. The particular framework structure which crystallizes can be strongly dependent on the cations present in the gel.

● Reaction mixture composition in a crystallization experiment is best defined by the set of mol ratios given in Table I (10).

Table I. Reaction Mixture Composition

Mol Ratio	Primary Influence
SiO_2/Al_2O_3	Framework composition
H_2O/SiO_2	Rate, crystallization mechanism
OH^-/SiO_2	Silicate molec. wt, OH^- conc
Na^+/SiO_2	Structure, cation distribution
R_4N^+/SiO_2	Framework aluminum content

A large number of silica and alumina source materials can be used in formulating a gel, and the product obtained is often dependent on the sources selected and on their treatment prior to formulation. For example, Na_2SiO_3, silica gel, silica sol, $NaAlO_2$, aluminum sulfate, aluminum turnings or alumina itself are all routinely considered. Clays, either directly or after varying heat treatment, can be used as well (11). In describing a reaction mixture composition in terms of the above mol ratios, Na_2SiO_3 is treated as a mixture of SiO_2 and NaOH; $NaAlO_2$, as Al_2O_3 and NaOH; aluminum sulfate, as Al_2O_3 and H_2SO_4, each with the appropriate amount of accompanying H_2O.

Hydroxide in the OH^-/SiO_2 ratio is calculated by subtracting equivalents of acid added from those of hydroxide. It is not hydroxide ion concentration in the resultant mixture. Organics such as amines, for example, are never included in calculating OH^-/SiO_2 ratios. In careful work, and as OH^-/SiO_2 ratios approach zero, it is important to recognize that Al_2O_3 is incorporated into a zeolite framework as AlO_2^-, i.e., each mol acts as two equivalents of acid and consumes two equivalents of hydroxide:

$$Al_2O_3 + 2OH^- \rightarrow 2AlO_2^- + H_2O$$

To a first approximation, the mol ratios in Table I can be divided according to primary function. SiO_2/Al_2O_3, for example, places a constraint on the framework composition of the zeolite produced, and it may define the structural competitors in a nucleation-controlled experiment. With the exception of the aluminum-rich NaA, zeolites normally incorporate all of the aluminum present in a reaction mixture into their framework structure, leaving varying amounts of silica (silicate) in solution according to hydroxide ion concentration, reaction conditions, etc.

H_2O/SiO_2 and OH^-/SiO_2 strongly influence the "molecular" or "polymeric" species present in a reaction mixture composition and the rate at which those species interconvert, by hydrolysis, to form the ordered, three-dimensional network of the zeolite product. Through their control of hydrolysis rates and polysilicate (or polyaluminosilicate) distribution, they can significantly influence the "winner" among competing, possibly metastable structures in a crystallization experiment.

Cations present in a reaction mixture are often the dominant factor determining which zeolite structure is obtained, as will be discussed below. In addition, their incorporation and presence in a product can be important considerations in subsequent use and handling. Anticipating that discussion, literature data (2) for the synthesis of four different zeolites, Y, Omega, L and TMA-O, are given in Table II. The primary variable, it is asserted, was the cation. (Each preparation was static, 100°C, with SiO_2/Al_2O_3 = 16-20, H_2O/SiO_2 = 14-25, OH^-/SiO_2 = 0.7-0.9.)

Table II. Product Dependence on Cation

Zeolite	Na^+/SiO_2	TMA^+/SiO_2	K^+/SiO_2
Y	0.8	0.0	0.0
Omega	0.6	0.1	0.0
L	0.0	0.0	0.8
TMA-O	0.0	0.1	0.8

As a result of small changes in the cation content of the reaction mixture, completely different zeolite products were obtained.

● Product integrity is best judged by comparison with an "authentic" sample.

Zeolite identification is made largely on the basis of X-ray diffraction, and powder patterns are the common measure of purity in a crystallization product. If a pattern shows no evidence for crystalline or amorphous contaminants, purity is estimated by comparing intensities of reflections (at d-spacings smaller than about 6 A) with those of an authentic sample of the same composition and crystal size. Reflections at higher d-spacing are not used in estimating crystallinity since their intensities are dependent on variables such as moisture content. Except for such large scale commercial products as NaA and NaY, "authentic" samples are normally obtained by repeated and varied crystallization experiments.

Once the X-ray diffraction pattern of a preparation is defined, elemental analysis becomes useful, to corroborate assertions of purity, to detail cation content and to critically explore the question of "templating" in a particular crystallization sequence. In contrast to more conventional chemical synthesis, elemental analysis is generally not a satisfactory or sufficient criterion for purity in zeolites. Almost all zeolite structures have been prepared in a range of framework compositions, as will be shown below. X-ray diffraction, supported by elemental analysis, can often directly probe composition within a particular zeolite framework.

Sorptive and cation exchange properties are a third measure of product integrity among well-characterized zeolites. For example, high purity samples of NaA and NaY will sorb about 25 g H_2O per 100 g of dry zeolite (25°C, 1-4 torr). HZSM-5 and NaY samples should respectively sorb about 11 and 19 percent of their dry weight in n-hexane (25°C, 10-20 torr). An as-synthesized, high purity zeolite sample will conventionally contain at least one cation for every aluminum in the framework.

Microscopy is the fourth essential ingredient in product characterization. Crystal size and crystal morphology are both important, controllable aspects of a synthesis experiment. Moreover, recent electron microbe analysis of aluminum distribution within individual zeolite crystals offers significant insight into actual crystallization processes, as will be discussed below (12).

●   Composition is a variable for any given framework structure.

When composition is cited in recent patents on zeolites, a range
is almost always presented. Although the limits of this range may
not be accurately known, it is now widely recognized that no
elemental composition is unique to any specific framework
structure. Table III presents examples which span the Si:Al range
from 1:1 to infinity.

Table III. Variable Composition in Zeolites

Framework	Name	SiO_2/Al_2O_3	References
A	Linde A	2	2
A	N–A	2.5 to 6	13
A	ZK–4	2.5 to 4	14
Sodalite	Sodalite	2	2
Sodalite	TMA Sodalite	10	15
Faujasite	Linde X	2 to 3	2
Faujasite	Linde Y	3 to 6	2
ZSM–5	ZSM–5	5 to infinity	16

There is abundant evidence in the literature to show that these
ranges indeed reflect differing framework compositions (2, 17–22).

That composition can be controlled by cation type was originally
discovered for the A framework (13,14). It has been emphatically
demonstrated with sodalite (15), where two "model" compositions
are obtained, depending on which cations are present during
crystallization. When sodium ions are used, the sodalite obtained
has a composition $Na_3Al_3Si_3O_{12}$; when tetramethylammonium ions are
used in place of sodium, $(CH_3)_4NAlSi_5O_{12}$ is produced. Both
compositions have the same framework structure, a
three–dimensional network built entirely of truncated octahedra.
Each octahedron in TMA sodalite contains (and can accomodate) only
one tetramethylammonium ion, thereby restricting (by the
requirement of charge balance) the number of negatively charged
AlO_4 tetrahedra. On average, three sodium ions are present in
each octahedron in Na sodalite.

Framework composition can often be varied by simply changing the
SiO_2/Al_2O_3 ratio of the reaction mixture. ZSM–5 and the synthetic
faujasites are examples cited in Table III. With ZSM–5, framework
compositions ranging to over 20,000 can be easily prepared by this
technique. Very high SiO_2/Al_2O_3 ratios require special effort to

exclude adventitious aluminum.

Nucleation and crystal growth may occur with differing facility as the SiO_2/Al_2O_3 ratio of a reaction mixture is changed. Crystals may grow for example in a SiO_2/Al_2O_3 range where nucleation would be difficult – and vice versa. Seeding, low-temperature aging and the use of reactive gels or dispersions are all practical techniques for controlling zeolite purity and/or composition.

The synthesis of zeolite Y illustrates the low-temperature aging approach. Four reaction mixtures were prepared using silica sol, $NaAlO_2$ and NaOH, two at SiO_2/Al_2O_3 = 10 and two at SiO_2/Al_2O_3 = 30. Mol ratios in both sets were OH^-/SiO_2 = 0.6 – 0.7, Na^+/SiO_2 = 0.8 and H_2O/SiO_2 = 16. One from each pair was placed immediately into a steam chest at 90 – 95°C; the second was aged for 24 hours at room temperature before heating. The results are presented in Table IV. Except where noted, the Y samples were all 95 – 100% crystalline.

Table IV. Synthesis of Zeolite Y

Initial SiO_2/Al_2O_3	Aging (hours)	Crystn (days)	Product Zeolite	Product SiO_2/Al_2O_3
10	0	6	P (+ Y)	----
10	24	5	Y (trace P)	5.3
30	0	7	Y	5.9
30	24	6	Y	6.0

It is generally accepted that crystallization of Y will continue until the aluminum in a reaction mixture is exhausted. Supporting that assertion was the fact that, in all the above cases, yield exceeded 90% based on alumina. As a result, there were large differences in residual silica (soluble silicate) in the two cases. At SiO_2/Al_2O_3 = 10, 50 – 60% of the silica was incorporated into the zeolite product; at SiO_2/Al_2O_3 = 30, only about 20%.

Low-temperature aging is commonly used to obtain pure Y from silica sols, and its primary function is probably pre-digestion, "equilibration of the heterogeneous gel with the solution" (2). It is likely to be important in the initial nucleation step as well, however (23-26), which introduces the next assertion.

●   Composition is often variable even within individual zeolite crystals.

NaX seeds have been frequently used to initiate crystallization of zeolite Y (23), and they of course introduce compositional heterogeneity into the crystals which result. Siliceous external shells have been grown onto aluminum-containing ZSM-5 crystals (27). Evidence is increasing however that non-uniform composition may be a common, intrinsic characteristic of synthetic zeolite crystals. Such a result should not be particularly surprising since both solution and gel composition are continuously changing during the course of a crystallization (i.e., as zeolite product is formed, effectively removing constituents from the reaction mixture).

In a large-crystal NaX preparation, for example, there is microprobe evidence that SiO_2/Al_2O_3 ratio increases with increasing distance from the crystal core (28). X-ray photoelectron spectroscopy (XPS) data show aluminum depletion in the surface of NaA, X, Y, and synthetic mordenite crystals (29). Changes in unit cell dimension during crystallization have been cited as evidence for SiO_2/Al_2O_3 gradients in NaY (24), a conclusion however which relies heavily on X-ray diffraction analysis of partially crystalline materials.

Large-crystal ZSM-5 preparations have provided the most striking example of intrinsic compositional heterogeneity to date. Aluminum content increased from core to outer crystal rim, the respective concentrations sometimes differing by a factor of 10 or more (12,30). One note of caution is warranted in generalizing from these observations however. Special techniques have been required to obtain zeolite crystals sufficiently large to permit electron microprobe analysis. These techniques may influence aluminum distribution in the crystals produced. It is nevertheless probable that gradients exist in more conventionally prepared, smaller crystals.

●   Modern analytical techniques will soon define site occupancy within zeolite crystals on an atomic level and may contribute to reaction mixture definition as well.

Until recently there has been no direct probe for aluminum or silicon siting in zeolites, and assertions have been made regarding only average (distributed) siting, largely on the basis of X-ray diffraction data, sorption, ion exchange or catalytic properties. Nuclear magnetic resonance (NMR) promises to provide the desired direct probe.

In very strong magnetic fields, high resolution magic angle spinning NMR spectra have been obtained for both ^{29}Si (I = 1/2, 4.7% abundance) and ^{27}Al (I = 5/2, 100%) in a variety of zeolites (31-38). In structures based on 4-membered rings, ^{29}Si chemical shift differences clearly distinguish SiO_4 tetrahedra according to the number of AlO_4 neighbors (0, 1, 2, 3...) (34,37). In more complex structures, like ZSM-5, numerous crystallographically distint sites exist for Si atoms in the lattice, and numerous different SiO_4 types have been detected (36).

NMR further provides a direct probe for aluminum within zeolite crystals and has confirmed that Al is tetrahedrally coordinated within a multitude of framework structures (35). In the ultrastabilization and dealuminization of synthetic faujasites, both tetrahedral and octahedral (non-framework) aluminums have been detected (37). Only the tetrahedral type was present in the original NaY. In "silicalite" ^{27}Al showed all the aluminum to be tetrahedrally coordinated, with at least two distinct environments, and the authors concluded that, structurally, silicalite and ZSM-5 are essentially indistinguishable (36).

● Organic (and inorganic) cations can "template" particular framework structures.

While the ability of organics to alter the course of a crystallization process is becoming increasingly apparent, the specific function of those organics is sensitively dependent on the details of a given experiment. In general, addition of organics to a reaction mixture can effect changes of four types: (a) A different zeolite structure is obtained; (b) Crystallization rate is strongly enhanced (or inhibited!); (c) The same framework is obtained but with a significantly new chemical composition; and (d) "Microscopic texture", e.g., crystal size, habit, etc. (39), is altered. It is not uncommon in exploratory crystallization experiments for organics to be superfluous, i.e., to simply provide cation balance for varying hydroxide. Only the first and the second of the above (and the second only when the rate is enhanced) represent "templating effects" and then only if it can be shown that the change is not due to the virtually inevitable system perturbations which accompany a new reaction mixture component.

A striking example of the first type was found in crystallization experiments with cationic polyelectrolytes (40), and it demonstrates the detailed analysis which must be performed before "templating" can be suggested. Several relatively low molecular weight polymers of the following type were prepared by reaction of 1,4-diazabicyclo[2.2.2]octane (Dabco) with the compounds $Br(CH_2)_nBr$:

$$-\left[-N^+ \overset{\displaystyle\frown}{\underset{\displaystyle\smile}{}} N^+ - \left(CH_2\right)_n -\right]_x-$$

where n = 3,4,5,6 and 10 and x = 10-60. With 1,4-dibromobutane for example, the polymer was designated "Dab-4 Br". When added to a reaction mixture that produced zeolite Y (and/or P) at 85-90°C, these polymers produced dramatic changes as shown in Table V (SiO_2/Al_2O_3 = 30, H_2O/SiO_2 = 20, OH/SiO_2 = 1.2, Na/SiO_2 = 1.2, 3-13 days).

Table V.   Polymer Effects in Crystallization

Polymer	N^+/SiO_2	Product Zeolite(s)
None	0	Y + P
Dab-4 Br	0.01	Gmelinite (faulted)
Dab-4 Br	0.14	Gmelinite (faulted)
Dab-4 Br	0.23	Pure gmelinite
Dab-4 Br	0.43	Amorphous

The results show clearly that a very small amount of organic can completely alter the course of these nucleation-controlled reactions. Moreover, with these polyelectrolytes, a large excess, which cannot be accommodated within the zeolite product, actually inhibits all crystallization.

If the Dab-4 Br in Table V is indeed "templating" the gmelinite structure, the results should be sensitive to changes in polyelectrolyte molecular structure. Table VI shows that this is indeed the case. A series of polyelectrolytes, together with the monomeric analogs (prepared by the reaction of Dabco with propyl or butyl bromide or iodide), was substituted for Dab-4 Br. Only Dab-4, -5 and -6 polyelectrolytes were effective in producing gmelinite.

Table VI.   Polymer Dependence in Crystallization

Organic	Product Zeolite(s)
Dab-3 Br	P
Dab-4 Br	Pure gmelinite
Dab-5 Br	Gmelinite (faulted)
Dab-6 Br	Gmelinite (faulted)
Dab-10 Br	Y + P
Dab-Pr$_2$	Y + P
Dab-Bu$_2$	Y + P

In theory, gmelinite has a large, 12-ring pore system which should readily admit molecules such as cyclohexane. In fact, both natural and synthetic gmelinites behave like small-pore zeolites, a behavior attributable to chabazite stacking faults. Chabazite fault planes, often observable by x-ray diffraction, effectively block or restrict access to the large gmelinite channels (41).

The "pure gmelinites" described in Tables V and VI are believed to be fault-free. They sorb 7.3% cyclohexane. For comparison, a natural (faulted) sample sorbed only 1.0%. It is proposed that the polyelectrolyte is present in the pore as the gmelinite framework forms, the polymeric nature of the organic preventing formation of stacking faults across that pore.

If the polymer is an integral part of the product structure, located in the pores, certain size and charge balance requirements must be fulfilled. The unit cell in gmelinite is traversed by a single 12-ring channel (7-8 A in diameter) for a distance of 10.0 A.

The Dabco unit is cylindrical with a diameter of about 6.1 A and can thus fit comfortably within the gmelinite pore. In length, the repeating units of Dab-3, -4, -5, -6 and -10 measure 7.5, 8.7, 9.9, 11.0 and 14.5 A. Comparing these measurements with the results in Table VI, all polymers which effected gmelinite synthesis had repeating units 9-11 A in length, matching the unit cell dimension of the pore.

Charge balance is the second constraint. A repeating unit of the Dab-4 polymer contains two equivalents of cation and extends about 8.7 A. The unit cell in gmelinite contains a single 12-ring channel 10 A in length and could therefore hold little more than two quaternary cations. Seven different gmelinite preparations with Dab-4 Br averaged 2.3 N/unit cell. Furthermore, elemental analysis suggested that the polymer was intact. The C/N atomic ratio in the seven samples averaged 5.4, compared with 5.1 in the original polymer.

● Once encapsulated within a zeolite framework, quaternary ammonium cations are protected from hydrolysis.

It is well known that quaternary ammonium cations decompose readily in hot caustic. As the number of examples of their successful use in zeolite synthesis increases however, it is instructive to examine C/N ratios in the synthesis products. Table VII shows remarkably little difference between the C/N expected if the organic remained intact and that found in four as-synthesized zeolites:

Table VII. Average C/N Ratios in Zeolite Products

| | | C/N Atomic Ratio | |
Zeolite	Organic	Expected	Found (ref.)
TMA Offretite	TMA	4.0	4.2 (42,43)
TMA Sodalite	TMA	4.0	4.0 (15)
Dab-4 Gmelinite	Dab-4	5.1	5.4 (40)
ZSM-5	TPA	12.0	13.3 (16,43)

That these organics can indeed be intact within a product zeolite has now been directly demonstrated by ^{13}C NMR on TPA ZSM-5 (44).

EXAMPLES IN SYNTHESIS

Patent literature is the primary information source in zeolite synthesis, but a set of instructional preparations has been assembled which require no specialized equipment and which can serve as an introduction to experimentation in the area. Details of those preparations will be published shortly (43), but an abbreviated version is presented here for easy reference.

Three examples will be given, describing synthesis routes to zeolites A, Y and ZSM-5. They thus include framework compositions known to occur with SiO_2/Al_2O_3 ratios from 1:1 to essentially infinity; they demonstrate such techniques as low-temperature nucleation, templating and variable reactant sources. (A fourth example, TMA Offretite, is given in the above reference but need not be included here.)

- Zeolite A, a preparation for freshman chemistry.

Crystals of NaA can be made in 3-4 hours, a bench-scale preparation requiring only a stirred, heated beaker. A boiling solution of sodium aluminate and NaOH is added to one of sodium meta-silicate, and the resultant mixture is heated with stirring at about 90°C until the suspension will settle quickly when the stirring is stopped. The suspension is then filtered (hot), washed repeatedly with water and dried at about 110°C to yield an 80-90% yield of $Na_2O \cdot Al_2O_3 \cdot 2SiO_2 \cdot {}^-4H_2O$. Product purity is determined by comparing the X-ray diffraction pattern of the solid with that of an authentic sample of NaA. The product, after dehydration at 350-400°C, should sorb about 25% of its weight in water.

In the crystallization, reactants per mole of SiO_2 are one (Al_2O_3), 20 (NaOH) and 550 (H_2O), the water being divided between aluminate and silicate solutions in the ratio 3:2. (Commercial sodium aluminate analyzes about 40% Al_2O_3, 33% Na_2O and 27% water.)

- Zeolite Y, low-temperature aging.

A solution of sodium aluminate and NaOH is added, with vigorous stirring, to 30% silica sol (a colloidal silica suspension). The resultant mixture is aged at room temperature for 2-3 days and then crystallized in a steam chest (about 95°C, no stirring) for 1-2 weeks. Solid is withdrawn, filtered, and analyzed by X-ray diffraction every 2-3 days until NaY purity (diffraction pattern intensity) reaches a limiting value. After hot filtration, washing and drying, a 50-60% yield (based on SiO_2) is obtained with an approximate composition of $Na_2O \cdot Al_2O_3 \cdot 5.3SiO_2 \cdot 5H_2O$. Without the aging, the SiO_2/Al_2O_3 ratio of the product will not exceed 5 and P will be a common contaminant. As with NaA, the purity of the NaY preparation is determined by comparing its X-ray diffraction pattern with that of a authentic sample. The sample should sorb 25% of its weight in water, after dehydration.

Moles of reactants per mole of silica should be 0.1 (Al_2O_3), 0.8 (NaOH) and 16 (water), with the water evenly divided between the silica suspension and the sodium aluminate solution.

- TPA ZSM-5, probable templating.

A solution of sodium aluminate and NaOH is added simultaneously with one of TPA Br and H_2SO_4 to 16% silica sol, and the mixture is immediately mixed, to form a gel. Placed in a steam chest at about 95°C and sampled periodically, the mixture will produce an 80–90% yield of ZSM-5 (based on SiO_2) in 10–14 days. Its molar composition will be approximately as follows: $1.8(TPA)_2O \cdot 1.2Na_2O \cdot 1.3Al_2O_3 \cdot 100SiO_2 \cdot 7H_2O$. Again, the X-ray diffraction pattern should be compared with that of a known sample. A purified, dehydrated sample of ZSM-5 will sorb about 11% n-hexane.

Solutions should be prepared such that moles of reactant per mole of silica are 0.012 (Al_2O_3), 0.54 (NaOH), 0.1 (TPA Br), 0.2 (H_2SO_4) and 45 (H_2O). Sodium-stabilized 30% silica sol is the starting material in above preparation. Additional water is divided among the various solutions in the ratio, one (aluminate): two (TPA Br): one (silica sol). Higher temperatures, for example 140–180°C, will reduce crystallization time. Furthermore, with appropriate adjustment for acid and base, no aluminum need be added to the reaction mixture. In that case, only the aluminum present as a contaminant in the various other reactants will be found in the product ZSM-5.

DEALUMINIZATION

ZSM-5 is one of a number of framework structures which can be crystallized with essentially no alumina. Several other structures can apparently exist in or near that compositional range (45,46), but must first be synthesized in an aluminum-containing form. Although direct synthesis routes will likely be discovered, this section reviews experimental techniques for dealuminizing these structures, with particular emphasis on the very siliceous compositions.

● Acid extraction alone often achieves only partial dealuminization.

The zeolites Y and mordenite are the most commonly targets for dealuminization, and a substantial literature exists on aluminum removal from both. In the case of Y, it is now generally accepted that controlled, direct addition of acid (such as ethylenediamine tetracetic acid (H_4EDTA) can remove at least 50% of the aluminum, to a SiO_2/Al_2O_3 ratio of about 12 (47,48), without significant loss in crystallinity. With mordenite, direct acid leaching can remove up to about 80% of the aluminum ($SiO_2/Al_2O_3 = ~60$) without structure collapse (49).

Beyond this point, i.e., to achieve SiO_2/Al_2O_3 ratios ≥ 100 combined thermal and chemical (acid) treatments are required. Samples of synthetic faujasite with SiO_2/Al_2O_3 = 100-200 have been reportedly prepared by alternate acid leaching and steaming of ultrastable Y's (45). Mordenites in this composition range are prepared by thermal and/or hydrothermal treatment at temperatures above 500°C, followed by varying acid extraction (46,49). Examples are given in Table VIII.

Table VIII. Dealuminization of Y and of Mordenite

Initial Sample (SiO_2/Al_2O_3)	Treatment	Product (SiO_2/Al_2O_3)	Reference
Y (5.3)	H_4EDTA, slow addition	Faujasite (10)	47
Y (5.1)	Na_2H_2EDTA + slow HCl	Faujasite (12)	48
USY[a] (5.2)	2N HCl, 1 h, 90°C	Faujasite (108)	45
Mord (15)	6N HCl, 1 h, reflux	Mordenite (20)	49
Mord (15)	6N HCl, 16 h, reflux	Mordenite (59)	49
Mord (15)	6N HCl, 24 h, reflux	Mordenite (60)	49
Mord (15)	650°C, 3h; 0.5N HCl, 16 h, reflux	Mordenite (73)	49

a – "Ultrastable Y", prepared by treatment under self-steaming conditions at 760-815°C (45)

New techniques for dealuminization, which promise to extend the SiO_2/Al_2O_3 range in the above structures even further, have very recently appeared (50-52). When NaY was treated with $SiCl_4$, for example, a highly crystalline, essentially aluminum-free faujasite reportedly resulted (52).

● In many structures, aluminum removal is site-dependent.

Not all aluminum tetrahedra within a given, siliceous (SiO_2/Al_2O_3 > 5) framework are equivalent. In mordenite, for example, four different crystallographic sites for aluminum potentially exist (49). Aluminum tetrahedra of differing acidity are recognized within the synthetic faujasite framework (53-55). It is therefore very reasonable to expect that aluminum removal, i.e., ease of hydrolysis, will be site dependent.

The strongest evidence to date for such an assertion is the marked non-linearity of the lattice parameter contractions as aluminum (framework charge) is removed from mordenite (49). A very plausible correlation can be developed between the differing a, b and c projections of the four potential AlO_4 sites and the respective non-linearities in the three lattice parameter dependencies on aluminum content in this orthorhombic unit cell. More powerful evidence can be expected as ^{29}Si and ^{27}Al NMR techniques develop. In addition to providing a direct probe in dealuminization, those techniques should clarify a more basic and unanswered question, namely, what is the relationship between an as-synthesized and a dealuminized structure with the same overall composition?

ACKNOWLEDGMENT

This brief summary of current thinking in zeolite synthesis draws extensively on the numerous new zeolites and new preparation techniques discovered in Mobil's Princeton and Paulsboro Laboratories over the past 20-25 years. Thanks are due the many authors whose names appear in the references and whose names will be found on patents describing these discoveries for their valuable input, advice and suggestions as my own experiments progressed.

REFERENCES

1. R. M. Barrer, Zeolites, 1, 130 (1981).

2. D. W. Breck, "Zeolite Molecular Sieves," John Wiley and Sons, New York, 1974.

3. L. B. Sand, Pure Appl. Chem., 52, 2105 (1980).

4. E. M. Flanigen, Pure Appl. Chem., 52, 2191 (1980).

5. M. Mengel, Chem.-Tech., 10, 1135 (1981).

6. G. T. Kerr, Catal. Rev.-Sci. Eng., 23, 281 (1981).

7. H. Robson, ChemTech, 8, 176 (1978).

8. E. M. Flanigen, Adv. Chem. Ser., 121, 119 (1973).

9. S. P. Zhdanov and N. N. Samulevich, Proc. Int. Conf. Zeolites, 5th, ed. L. V. C. Rees, p. 75 (1980).

124

10. L. D. Rollmann, Adv. Chem. Ser., 173, 387 (1979).

11. W. L. Haden, Jr., and F. J. Dzierzanowski, US Patents
    3,663,165 and 3,657,154 (1972), for example.

12. R. von Ballmoos and W. M. Meier, Nature, 289, 782
    (1981).

13. R. M. Barrer and P. J. Denny, J. Chem. Soc., 971 (1961).

14. G. T. Kerr, Inorg. Chem., 5, 1537 (1966).

15. C. Baerlocher and W. M. Meier, Helv. Chim. Acta, 52,
    1853v (1969).

16. R. J. Argauer and G. R. Landolt, US Patent 3,702,886
    (1972).

17. D. M. Bibby, L. P. Aldridge and N. B. Milestone,
    J. Catalysis, 72, 373 (1981).

18. D. H. Olson, W. O. Haag and R. M. Lago, J. Catalysis,
    61, 390 (1980).

19. E. L. Wu, S. L. Lawton, D. H. Olson, A. C. Rohrman, Jr.,
    and G. T. Kokotailo, J. Phys. Chem., 83, 2777 (1979).

20. G. T. Kokotailo, S. L. Lawton, D. H. Olson and
    W. M. Meier, Nature, 272, 437 (1978).

21. D. H. Olson, G. T. Kokotailo, S. L. Lawton and
    W. M. Meier, J. Phys. Chem., 85, 2238 (1981).

22. E. M. Flanigen, J. M. Bennett, R. W. Grose, J. P. Cohen,
    R. L. Patton, R. M. Kirchner and J. V. Smith, Nature,
    271, 512 (1978).

23. E. F. Freund, J. Cryst. Growth, 34, 11 (1976).

24. J.-L. Guth, P. Caullet and R. Wey,
    Bull. Soc. fr. Mineral. Cristallogr., 99, 21 (1976).

25. F. Polak and E. Stobiecka, Bull. Acad. Pol. Sci., Ser.
    Sci. Chim., 26, 899 (1978).

26. H. Kacirek and H. Lechert, J. Phys. Chem., 79, 1589
    (1975).

27. L. D. Rollmann, US Patent 4,203,869 (1980).

28. T. J. Weeks and D. E. Passoja, Clays Clay Miner., 25, 211 (1977).

29. J.-F. Tempere, D. Delafosse and J. P. Contour, "Molecular Sieves-II", ACS Symposium Series, vol. 40, ed. J. R. Katzer, p. 76 (1977).

30. E. G. Derouane, J. P. Gilson, Z. Gabelica, C. Mousty-Desbuquoit and J. Verbist, J. Catalysis, 71, 447 (1981).

31. G. Engelhardt, U. Lohse, A. Samoson, M. Maegi, M. Tarmak and E. Lippmaa, Zeolites, 2, 59 (1982).

32. G. Engelhardt, U. Lohse, E. Lippmaa, M. Tarmak and M. Maegi, Z. Anorg. Allg. Chem., 482, 49 (1981).

33. S. Ramdas, J. M. Thomas, J. Klinowski, C. A. Fyfe and J. S. Hartman, Nature, 292, 228 (1981).

34. C. A. Fyfe, G. C. Gobbi, J. S. Hartman, R. E. Lenkinski, J. H. O'Brien, E. R. Beange and M. A. R. Smith, J. Mag. Resonance, 47, 168 (1982).

35. C. A. Fyfe, G. C. Gobbi, J. S. Hartman, J. Klinowski and J. M. Thomas, J. Phys. Chem., 86, 1247 (1982).

36. C. A. Fyfe, G. C. Gobbi, J. Klinowski, J. M. Thomas and S. Ramdas, Nature, 296, 530 (1982).

37. J. Klinowski, J. M. Thomas, C. A. Fyfe and G. C. Gobbi, Nature, 296, 533 (1982).

38. J. B. Nagy, J. P. Gilson and E. G. Derouane, J. Chem. Soc., Chem. Commun., 1129 (1981).

39. L. D. Rollmann, US Patent 4,205,053 (1980).

40. R. H. Daniels, G. T. Kerr and L. D. Rollmann, J. Am. Chem. Soc., 100, 3097 (1978).

41. K. Fischer, Neues Jahrb. Mineral. Monatsh., 1 (1966).

42. E. E. Jenkins, US Patent 3,578,398 (1971).

43. L. D. Rollmann and E. W. Valyocsik, Inorg. Syntheses, 22, in press (1982).

44. G. Boxhoorn, R. A. van Santen, W. A. van Erp, G. R. Hays, R. Huis and D. Clague, J. Chem. Soc., Chem. Commun., 264 (1982).

45. J. Scherzer, J. Catalysis, 54, 285 (1978).

46. N. Y. Chen and F. A. Smith, Inorg. Chem., 15, 295 (1976).

47. G. T. Kerr, J. Phys. Chem., 72, 2594 (1968).

48. G. T. Kerr, A. W. Chester and D. H. Olson, Acta Phys. Chem. (Hung.), 24, 169 (1978).

49. R. W. Olsson and L. D. Rollmann, Inorg. Chem., 16, 651 (1977).

50. C. D. Chang, US Patent 4,273,753 (1981).

51. P. Fejes, I. Kiricsi, I. Hannus, A. Kiss and G. Schobel, React. Kinet. Catal. Lett., 14, 481 (1980).

52. J. Klinowski, J. M. Thomas, M. Audier, S. Vasudevan, C. A. Fyfe and J. S. Hartman, J. Chem. Soc., Chem. Commun., 570 (1981).

53. R. Beaumont and D. Barthomeuf, J. Catalysis, 27, 45 (1972).

54. R. J. Mikovsky and J. F. Marshall, J. Catalysis, 44, 170 (1976).

55. E. Dempsey, J. Catalysis, 39, 155 (1975).

STUDY ON THE MECHANISM OF CRYSTALLIZATION OF ZEOLITES A, X AND Y

Fred Roozeboom [a,*], Harry E. Robson [a] and Shirley S. Chan [b]

a) Exxon Research and Development Laboratories, P.O. Box 2226,
Baton Rouge, La. 70821, U.S.A. and
b) Exxon Research and Engineering Company, P.O. Box 45, Linden,
N.J. 07036, U.S.A.

ABSTRACT

Samples of A, X, and Y synthesis gels at 98 $^{\circ}$C were withdrawn after
crystallization for various times, then centrifuged while hot. The
solid and liquid phases were examined by Laser Raman spectroscopy
(LRS), X-ray diffraction (XRD) and chemical analysis (Al, Si, Na).

LRS and XRD on solid samples detect zeolite structures at about the
same time, i.e., when crystallites are about 500 $\overset{\circ}{A}$ in size. LRS
on the liquids gives additional information. For instance, the
$Al(OH)_4^-$ in the liquids (band at 621 cm^{-1}) disappeared before any
zeolite structure was detected. Some solute species appeared in
the course of X and Y crystallization, having broad, weak Raman
bands (around 448, 600, 777, and 936 cm^{-1}), due to free, unreacted
monomeric and dimeric silicate ions.

The above and chemical analysis results indicate the dissolution of
some silicon containing (monomeric and polymeric) ions from the
amorphous aluminosilicate gels. The hydroxylated ions condense
with $Al(OH)_4^-$ to form a large variety of complex aggregates.
These complexes may be soluble polymeric aluminosilicates related
to zeolitic fragments and thus form the nuclei of crystal growth.
A mechanism for the formation of 4 and 6 ring vs. 5 ring systems is
proposed.

---

* Present address:
Esso Chemie B.V., P.O. Box 7225, 3000 HE Rotterdam, The Netherlands.

INTRODUCTION

During the past two decades a number of papers has been published
on the kinetics and mechanism of zeolite formation. In most cases
the synthesis of zeolite A and faujasite zeolites was studied
(1-17) since these are the easiest and fastest to synthesize.
Recently also highly siliceous zeolites, such as the ZSM-5 (10, 17)
and mordenite systems (11) have received attention.

From the beginning results have been controversial. Zhdanov argued
in a detailed review (1) in favor of a solution transport mechanism,
first put forward by Barrer et al. (2) and later by many other
investigators (3-11). Barrer et al. proposed the nucleation to
be the result of the polymerization of aluminate, silicate and
possibly more complex ions in the liquid phase, the ions being
continuously supplied by the dissolution of the solid gel material.

McNicol et al. studying A and faujasite type zeolites (12, 13)
proposed a solid phase transformation mechanism, involving zeolite
crystallization in the solid gel phase via condensation between
hydroxylated Si-Al tetrahedra. This mechanism was also favored
by Polak et al (14, 15) who studied the mechanism of formation
of X and Y zeolites. Flanigen (16) proposed a similar mechanism,
involving a reordering of the hydrogel to an ordered crystalline
state via surface diffusion in the absence of liquid phase transport. Derouane et al. (17, 18), studying the synthesis of zeolites
with ZSM-5 topology, concluded that both the liquid phase ion transportation mechanism and solid hydrogel phase transformation
mechanism are important, depending on the silica source and the gel
formulation used. In the former mechanism only a few nuclei are
formed, yielding large crystallites, whereas the latter involves
numerous nuclei yielding polycrystalline aggregates. These workers
also performed infrared spectroscopic tests showing the existence of
ZSM-5 which was not X-ray detectable in intermediate phases (17, 19).

Most investigators at the present time tend to agree with Barrer's
concept (2), especially since some recent Raman spectroscopic
(20, 21) and ^{29}Si NMR spectroscopic (22-25) data suggest the
existence of solute aluminosilicate species. Sofar, Raman investigations have been conflicting. McNicol et al. (12, 13) observed no
spectral changes in the liquid phase during crystallization and
thus proposed a solid phase transformation. However, similar
experiments by Angell and Flank (9) showed spectral changes in the
liquid, giving evidence for a solution transport mechanism. Guth et
al. (20) reported indirect evidence for the occurrence of aluminosilicate complexes in solution by comparing the spectra of silicate
and aluminate ions separately in NaOH solution with the spectrum of
silicate and aluminate ions present together in solution.

The objective of the present work was to seek evidence for the existence of (soluble or solid) zeolitic precursor species in the earlier stages of zeolite formation and to add to our understanding of the mechanism of zeolite formation. Thus, we analyzed both solid and liquid components of zeolite A, X and Y synthesis gels by laser Raman Spectroscopy (LRS) for identification of molecular species and chemical analysis for Al, Si and Na content. Solids were also tested by X-ray diffraction (XRD) for crystallinity. The main reason for using Raman spectroscopy is its unique ability to examine solid samples as well as solution samples, and its ability to measure their lower frequency modes as opposed to infrared spectroscopy. Unlike the other Raman investigators (9, 12, 20) we had a Raman spectrometer equipped with an optical multi-channel analyzer for signal averaging, and thus had the possibility of resolving intermediate species which often exist at relatively low concentrations species.

EXPERIMENTAL

Zeolite synthesis. Three simple zeolite synthesis cases were selected i.e. type A, and type X and Y, because of their fast formation at relatively low temperatures. The gels studied were formed by mixing aqueous solutions of sodium aluminate, sodium hydroxide and colloidal silica sol (Ludox HS-40).
Zeolite A crystallization was from a gel with composition $2.1\ Na_2O.Al_2O_3.2SiO_2.80H_2O$. No aging was carried out since it has been reported to have no effect (9) and no seed was added. For zeolites X and Y the gel compositions were $3.5\ Na_2O.Al_2O_3.$ $5SiO_2.80H_2O$ and $3.3\ Na_2O.Al_2O_3.9SiO_2.140H_2O$ respectively. The zeolites were synthesized using published methods of seeding (26, 27), the seed slurry composition being $13.3\ Na_2O.Al_2O_3.12.5SiO_2.267H_2O$. The amount of seed was 5 mol % of the total Al. By seeding the crystallization time was reduced to 8 hours for zeolite X and 12 hours for zeolite Y.

Raman Spectroscopy. Polyethylene bottles of the master synthesis gels were withdrawn from a constant temperature oven (98°C) after various times of static crystallization, centrifuged while still hot and filtered to separate the liquid phase and solid phase. The solid samples were further washed with water, dried at 110°C and calcined at 500°C for 2 hr in air then cooled to room temperature.

An argon ion laser was tuned to the 514.5 nm line for excitation. Liquid samples at room temperature were placed in 10mm x 10mm quartz cuvettes. The Raman signals were collected using 90° scattering geometry with a F1.2 lens. The laser power at the sample location was set at 70-90 mW. The solid samples were pelletized to 13 mm diameter wafers for mounting in a sample holder.

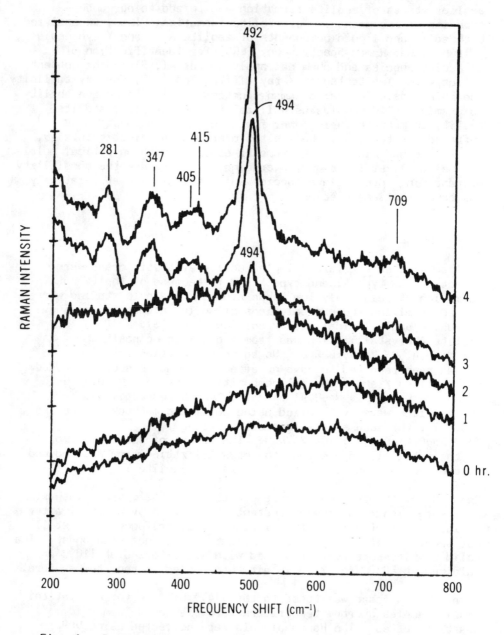

Fig. 1   Raman spectra of solid phase in zeolite A synthesis

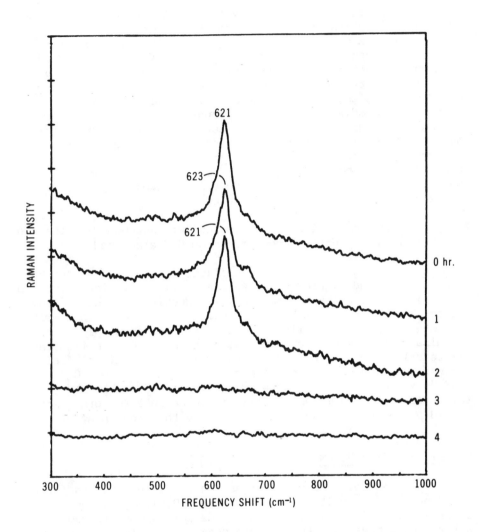

Fig. 2 Raman spectra of liquid phase in zeolite A synthesis

The collection optics in this case was a back scattering geometry and the laser power was lowered to 10-30 mW.

The Raman spectra were analyzed by a triplet monochromator, model DL203 with F4 optics made by Instrument SA, Metuchen, NJ, and an optical multichannel analyzer system, model OMA2 equipped with an intensified photodiode array detector, made by Princeton Applied Research, Princeton, NJ. This system made it possible to collect a complete spectrum over a range of thousand cm^{-1} simultaneously in the matter of seconds or minutes. Thus each sample could be examined immediately after its preparation procedure without any chance of aging which might affect its structural integrity. The total accumulation time needed for each spectrum reported here was in the range 20-100 sec. The digital display of the spectrum was calibrated to give 1.7 cm^{-1}/channel whereas the overall spectral resolution was about 8 cm^{-1} which was adequate for vibrational band widths over 20 cm^{-1}.

X-Ray diffraction. Solid phase samples were scanned with the aid of a Philips X-ray diffractometer, using CuK-alpha radiation.

Chemical analysis. Chemical compositions (Al, Si and Na content) of both solid and liquid phases were determined by plasma spectroscopy using a Jarrell-Ash Atomcomp III Inductively Coupled Plasma/ Atomic Emission Spectrometer which performs a simultaneous multi-element measurement. Solid samples were fused in a salt mixture by a Claisse Fluxer, which automatically pours the molten flux into a dilute acid solution for final dissolution. The initial fusion mixture comprises 0.1 g. of sample plus 1.5 g. of a $Li_2CO_3/Li_2B_4O_7$ mixture. This melt was dissolved in 100 ml 5% HNO_3, then diluted to 250 ml and finally injected into the argon plasma. Liquid phases were injected directly into the argon plasma.

RESULTS AND DISCUSSION

Raman Spectroscopy. Figure 1 gives the resulting spectra for the solid phases in zeolite A synthesis after 0, 1, 2, 3, and 4 hours of crystallization. The spectra of the initial gel (0 hours) and after 1 hour have no interpretable peaks. The peaks indicated in the other spectra are all due to zeolite A formation and agree reasonably well with those reported by Angell (28). He reported one strong band at 490 cm^{-1} and four weak bands at 700, 410, 340, and 280 cm^{-1}. The corresponding band frequencies of our measurement centre around 492, 714, 405, 347 and 281 cm^{-1}. We did not observe significant shifts of these band positions, which might suggest the growth from a solid precursor.

Figure 2 shows the spectra of the corresponding liquid phases. It is seen that in the early stages, i.e., during induction and even after 2 hours when crystallization has set in, a strong peak was observed at 621 cm^{-1} which is assigned to the AlO_4 symmetric stretching mode of the monomeric $Al(OH)_4^-$ aluminate anion (29). Other authors found this peak at frequencies ranging from 618 cm^{-1} (12) to 625 cm^{-1} (29). No evidence for solute anionic aluminosilicate precursor species was noted in the liquid phase during the course of the crystallization. It is not likely that Raman inactive precursor species are formed suggesting the absence of well defined precursor species in solution other than $Al(OH)_4^-$. This means that numerous ill-defined precursor species may be formed, each having insufficient concentration for a Raman peak to be resolved. The observations also indicate that no significant net dissolution of the solid gel phase occurs throughout the synthesis of zeolite A: only aluminate is being consumed from the liquid phase (disappearing Raman peak) on incorporation in the solid phase and no silicate peaks appear.

Figures 3 and 4 show the corresponding data for zeolite X and Figures 5 and 6 for zeolite Y. The peaks indicated in Figure 3 all arise from solid zeolite X (511, 376, and 291 cm^{-1}) and those in Figure 5 from zeolite Y (511, 369, and 298 cm^{-1}). Literature reports the peaks to be at 505, 375, and 282 cm^{-1} (plus very weak at 1075 and 990 cm^{-1}) for zeolite X and at 503, 350, and 300 cm^{-1} (and 1110 cm^{-1}) for zeolite Y (28). Thus, these measurements agree reasonably well with the literature data to within experimental limits.

The peaks or bands in Figures 4 and 6 clearly show that the $Al(OH)_4^-$ ion (peak at 621 cm^{-1}) is present in the initial gel of both zeolite X and zeolite Y. As in the case of zeolite A, this peak disappears after the first 2-3 hours of crystallization. But unlike zeolite A, some soluble species are observed in the course of the crystallization. The identification of these species, having bands around 448, 600, 777, and 936 cm^{-1} for zeolite X and 600 and 777 cm^{-1} for zeolite Y, is difficult because of the broadness and weakness of the bands. In general, a broad Raman band points to a variety of structures rather than to a very well defined structure of one distinct ion (like $Al(OH)_4^-$).
This variety of structures is common for silicate ions, which are known to be present in different degrees of hydroxylation and polymerization depending on concentration, pH, temperature and pressure (30).

134

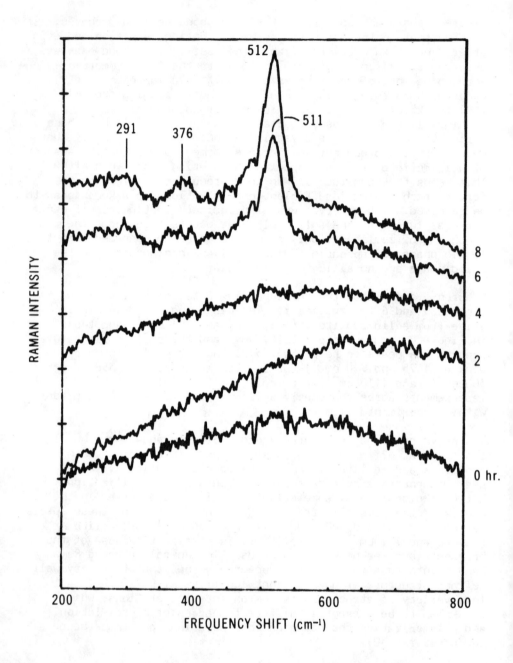

Fig. 3   Raman spectra of solid phase in zeolite X synthesis

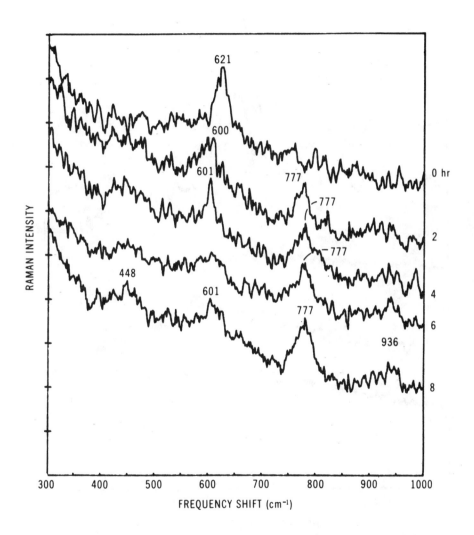

Fig. 4　Raman spectra of liquid phase in zeolite X synthesis

136

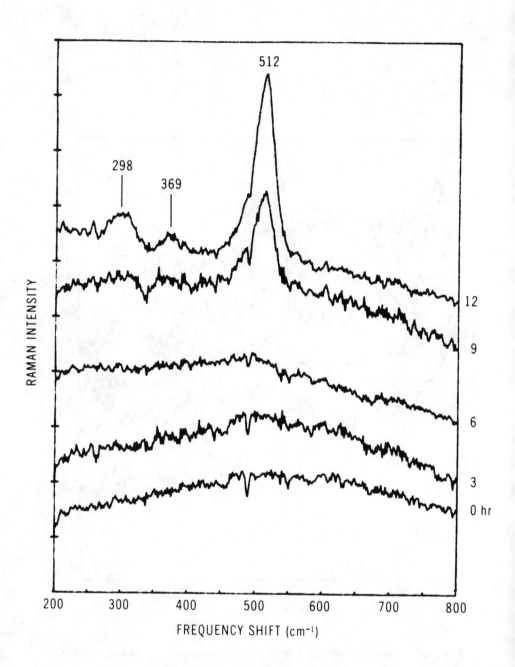

Fig. 5   Raman spectra of solid phase in zeolite Y synthesis

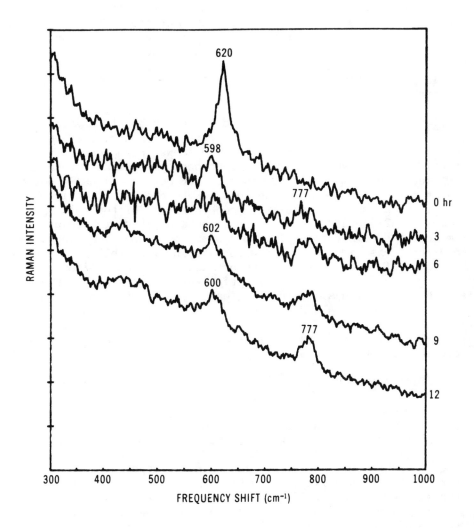

Fig. 6   Raman spectra of liquid phase in zeolite Y synthesis

Chemical Analysis Tables I and II summarize the analysis results, along with the other results.

Zeolite A. In accordance with what was observed with Raman spectroscopy, the alumina concentration decreased after 3 hours of synthesis time. From the results on the solid phase, we can conclude that aluminate disappeared from solution and is incorporated in the solid phase. As to the silicate concentration of the liquid phase, this concentration decreased less drastically to comparably low levels.

Zeolites X and Y. From Tables I and II we can see that in case of both zeolite X and zeolite Y in the first 2-3 hours, i.e., before any crystallites were detected, aluminate was disappearing from solution as observed with Raman spectroscopy and incorporated in the amorphous aluminosilicate phase. For both zeolites we observed in the same period an increase in silicate concentration in the liquid and a decrease in silicon concentration of the solid phase. After this 2-3 hour period, the silicon content decreased or remained the same for the liquid phases and roughly remained the same for the solids. From these data it is clear that in the first 2-3 hours of the crystallization or nucleation some silicon containing ions (monomeric or polymeric) were dissolved from the amorphous (alumino)silicate gel. These hydroxylated ions apparently condensed with the $Al(OH)_4^-$ ions present to form a large variety of aggregates, which may be the nuclei for crystal growth. Thus, the whole mechanism of zeolite A, X and Y formation is a solution transport mechanism.

X-Ray Diffraction

The X-Ray diffraction patterns for the various solid phases of zeolites A, X, and Y are summarized in Table II. In comparing XRD and Raman results, we can conclude that Raman spectroscopy detects zeolite formation in the solid phases at about the same stage as XRD: the characteristic XRD peaks appear at the same point in the crystallization as the Raman peaks, i.e., when crystallites are about 500 Å in size as observed with Scanning Electron Microscopy (SEM). As in the Raman spectra, no other XRD peaks appeared than those of the zeolite we intended to synthesize. Thus, no intermediate phases were detected by the two techniques.

Background Fluorescence

Raman spectroscopy of the solid samples does not give additional information; by the time that zeolite crystals are large enough to be Raman detectable, these crystals give strong XRD patterns (> 500 Å).

TABLE 1

SUMMARY OF LASER RAMAN SPECTROSCOPY AND CHEMICAL ANALYSIS DATA AS A FUNCTION
OF CRYSTALLIZATION TIME IN ZEOLITE SYNTHESIS FOR LIQUID SAMPLES

Zeolite Synthesized	Cryst. Time/h	Band Position cm⁻¹	Band Intensity and Shape a)	Probable Species	Chemical Analysis (wt %)		
					Al_2O_3	SiO_2	Na_2O
A	0	621	I, S	$Al(OH)_4^-$	8.3	1.2	10.3
	1	623	I, S	,,	4.5	1.2	6.4
	2	621	I, S	,,	3.1	0.3	7.1
	3	---	---	---	0.2	0.3	3.6
	4	---	---	---	0.3	0.4	3.4
X	0	621 / 777	I, S / W, B	$Al(OH)_4^-$ / $SiO_2(OH)_2^{2-}$	1.8	0.8	8.6
	2	936, 448 / 600 / 777	W, B / W, B / W, B	,, / Disilicate / $SiO_2(OH)_2^{2-}$	0.2	10.6	7.6
	4	936, 448 / 601	W, B / W, B	,, / Disilicate	816 ppm	9.9	7.6
	6	777 / 936, 448 / 601	M, B / M, B / W, B	$SiO_2(OH)_2^{2-}$ / ,, / Disilicate	867 ppm	8.5	7.8
	8	777 / 936, 448 / 601	I, B / W, B / M, B	$SiO_2(OH)_2^{2-}$ / ,, / Disilicate	478 ppm	8.6	7.8
Y	0	620	I, S	$Al(OH)_4^-$	0.8	0.2	6.3
	3	777 / 598	W, B / M, B	$SiO_2(OH)_2^{2-}$ / Disilicate	365 ppm	8.7	4.7
	6	777 / 602	W, B / W, B	$SiO_2(OH)_2^{2-}$ / Disilicate	395 ppm	9.3	4.6
	9	777 / 602	M, B / W, B	$SiO_2(OH)_2^{2-}$ / Disilicate	536 ppm	7.8	4.8
	12	777 / 600	M, B / W, B	$SiO_2(OH)_2^{2-}$ / Disilicate	0 ppm	7.7	5.2

a) I = Intense, M = Medium, W = Weak, S = Sharp, B = Broad

TABLE II

SUMMARY OF LASER RAMAN SPECTROSCOPY, XRD AND CHEMICAL ANALYSIS DATA AS A FUNCTION OF CRYSTALLIZATION TIME IN ZEOLITE SYNTHESIS FOR SOLID SAMPLES

Zeolite Synthesized	Cryst. Time/h	Band Position/cm^{-1}	Band Intensity and Shape a)	Probable Species	XRD Spectrum b)	Chemical Analysis (wt%) Al$_2$O$_3$	SiO$_2$	Na$_2$O
A	0	--	--	--	Amorphous	12.5	70.8	11.1
	1	--	--	--	??	14.3	62.7	15.1
	2	494	W, B	Zeolite A	Zeolite A, W	18.9	59.2	15.5
	3	494	S, S	Zeolite A	Zeolite A, S	31.0	38.3	19.3
		281,347	W, B	Zeolite A				
		405–415	W, B	Zeolite A				
	4	492	I, S	Zeolite A	Zeolite A,VS	29.3	35.3	19.9
		281,347	W, B	Zeolite A				
		405–415	W, B	Zeolite A				
X	0	--	--	--	Amorphous	15.4	49.4	20.3
	2	--	--	--	??	27.6	46.6	22.1
	4	--	--	--	??	28.5	46.6	19.0
	6	511	I, S	Zeolite X	Zeolite X, VS	26.8	47.3	16.8
		376,291	W, B					
	8	512	I, S	Zeolite X	Zeolite X, VS	26.4	46.0	16.0
		376,291	W, B	Zeolite X				
Y	0	--	--	--	Amorphous	12.0	55.6	15.5
	3	--	--	--	??	19.8	47.1	20.6
	6	--	--	--	??	26.4	49.6	17.1
	9	512	M, S	Zeolite Y	Zeolite Y, S	23.6	51.8	16.0
		369,298	W, B	Zeolite Y				
	12	512	I, S	Zeolite Y	Zeolite Y, VS	21.5	61.8	15.8
		369, 298	W, B	Zeolite Y				

a) I = Intense, M = Medium, W = Weak, S = Sharp, B = Broad
b) VS = very strong, S = Strong, W = Weak

Literature data have recently been reported for various silicate
ions (19): 1005 (shoulder), 930 (intense), 830 (shoulder), 778
(intense), and 448 cm^{-1} (medium) for the monomeric species,
$SiO_2(OH)_2^{2-}$, 605 cm^{-1} (medium) for dimeric silicate, and 632,
555, 542, and 520 cm^{-1} (all weak) for polymeric, especially
cyclic, trimeric anions. Thus one may ascribe the band
positions at 936, 777, 600, and 448 cm^{-1} to monomeric and
dimeric silicate anions. As discussed later in this paper (Tables
I and II), the chemical analysis results show that after some 2-3
hours of synthesis the silicon concentration in the liquid phase of
X and Y remains the same or decreases, whereas the Raman bands
become more intense. Thus, it is clear that the bands are due
to the depolymerization of polymeric silicate species into monomeric
and dimeric silicate ions.

The SiO_2/Al_2O_3 ratio in both the initial gel formulation
and the final product of zeolite A synthesis was 2.0. Thus
all silicate species present were consumed by condensation with
aluminate species. In the case of the faujasites the synthesis
gels were rich in silica: the SiO_2/Al_2O_3 ratios in the
initial gel and final product were 5.0 and 3.0 for zeolite X and
9.0 and 4.8 for zeolite Y. Thus, in contrast to zeolite A, the
appearance of free silicate ions was observed during the crystallization reaction.

Angell and Flank (9) observed some silicate species with bands
around 450 and 800 cm^{-1} (probably monosilicate) in the separated
liquid phase of their initial zeolite A synthesis gel.
This is presumably due to the sodium silicate, which they used as a
silica source. We used colloidal silica sol which does not give
rise to any signal in the gel.

Apparently the condensation of $Al(OH)_4^-$ and the depolymerized,
hydroxylated silicate ions requires the same 2-3 hours for all three
zeolites. After this induction period the aluminate disappears and,
in case of X and Y, free silicate ions appear in solution.
Schwochow and Heinze (31) argued that the zeolite species crystallized from separate liquid phases depends on the composition of the
liquid phase and, at constant composition, also on the size of the
polymeric silicate ions. They concluded that polymeric ions
promote the formation of a phillipsite type phase over a faujasite
type and that at the time of nucleation the dissolved silica must
be predominantly in a monomeric state to crystallize as faujasite
type structures.

As to the mechanism of zeolite formation, the Raman spectra
support the solution transport mechanism (see later).

Part of this "late" Raman detectability may be due to the large fluorescent background that all solid zeolites samples give rise to. This background has also been observed by other authors (28, 32, 33) on the examination of metal oxide materials. They also reported, as we found in our present work, that prolonged exposure to the laser beam decreased the fluorescence, in general. However, even this reduced the fluorescence hardly sufficiently for the observation of Raman spectra in several samples, e.g., AO-A2,XO-X4, and YO-Y6. It has been suggested that the cause of the excessive background could be the fluorescence due to transition metal impurities, especially Fe^{3+}, Cr^{3+} and Mn^{2+} (13). Angell (28) obtained a high purity zeolite Y which contained less than 17 ppm Fe and showed much less fluorescence. Other high purity materials (zeolite A and X) gave rise to a relatively strong Raman spectrum.

Thus, it may very well be that in our case some relevant information was obscured by fluorescence. Egerton et al (34, 35) reported that in the case of silica and silica-alumina materials the background fluorescence can be minimized by heating the samples in oxygen at higher temperatures (500°C). It is reported for zeolites, however, that activation of zeolites under similar conditions always results in higher backgrounds (28).

ON THE RAMAN DETECTABILITY AND IDENTIFICATION OF COMPLEX ALUMINO-SILICATE SPECIES IN SOLUTION

Recently, Guth et al.(20, 21) published some papers on a Raman investigation of strongly basic (NaOH) solutions of diluted sodium aluminate and silicate. They recorded spectra in the 580-680 cm^{-1} range only. In aluminate solutions, an intense polarized peak at 623-625 cm^{-1} was observed and they attributed this peak to the free $Al(OH)_4^-$ ion. Only this band had strong enough Raman cross-section to allow exploitation whereas those of the silicate and aluminate or complex anions were much too weak. In the presence of silicate, the aluminate band at 623 cm^{-1} (Si/Al=1) decreased to varying extents as we found, thus denoting indirectly the formation of aluminosilicate species, as in our case with zeolite A. For silicate/aluminate mixtures with Si/Al=6, Guth et al. (19) observed, besides the disappearance of the $Al(OH)_4^-$ peak at 623 cm^{-1} and a decrease of the $SiO_2(OH)_2^{2-}$ peaks, the appearance of small peaks or shoulders. Their most "intense" peak was at 577 cm^{-1}. Yet, in all these cases the spectra are too poor to conduct a structural study of these complexes (20). Again, this might indicate that a large variety of polymeric aluminosilicate complexes is formed, none of them having enough concentration or existing for long enough time to be reasonably detectable.

ON THE MECHANISM OF RING FORMATION

Meier (36, 37) classified zeolite structures according to the secondary building units (SBU). These units are basically single or
double 4, 6 or 8 ring systems and more complex 4-1, 4-4-1 and 5-1
units. For simplification we will subdivide them into even-number
membered ring systems [4, 6, 8] and 5 ring systems. The latter group
may be extended to odd-number membered ring systems [5 and 9], if we
include the 9 ring systems, recently reported for lovdarite by Merlino
(38). For solutions with Si/Al > 5, Guth et al. (20) suggested that
a relatively stable species exists as an intermediate, i.e. the
$Al(OSiO_3)_4^{13-}$ species. Derouane et al. (17) also postulated this
species to be the species transported through the liquid phase of
silicon-rich gels.
We assume this species will be formed already at Si/Al = 4 in
solution. At Si/Al > 4 the solute Al-species are completely
saturated with Al-O-Si bonds formed by the condensation reaction:

$$4SiO_2(OH)_2^{2-} + 4OH^- + Al(OH)_4^- \longrightarrow \left[\begin{array}{c} SiO_3 \\ | \\ O \\ \downarrow \\ O_3Si-O-Al\!\blacktriangleright O-SiO_3 \\ | \\ O \\ | \\ SiO_3 \end{array} \right]^{13-} + 8H_2O \qquad (Ia)$$

We postulate the remaining silicate anions in the silicon-rich
mixture (Si/Al > 4), which have not been used for the condensation
Ia, to condense to polysilicate structures which can eventually
react with the $Al(OSiO_3)_4^{13-}$ species to 5 ring systems:

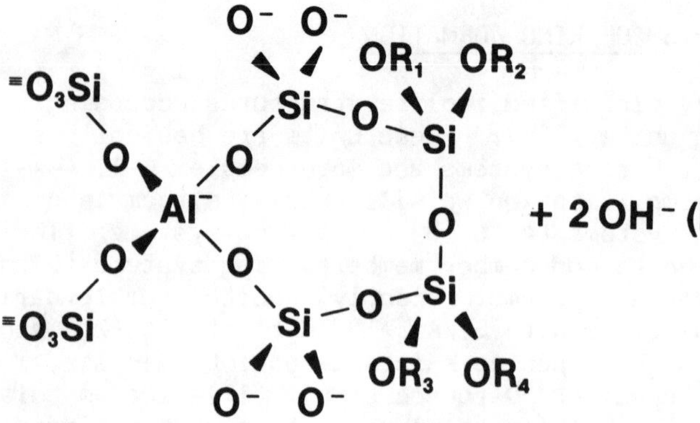

R1 to R4 represent groups ranging from hydrogen to more complicated Si-O-Si and Si-O-Al networks.

At Si/Al < 4 the solute $Al(OH)_4^-$ species are not completely saturated with Al-O-Si bonds and the condensation reaction may for example be:

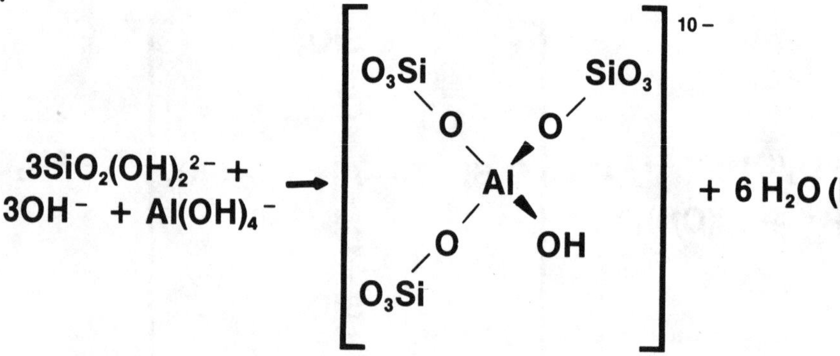

Since the mixture is now relatively rich in Al we may have other (solid or solute) polymeric aluminosilicate species reacting with the solute unsaturated aluminosilicate species from reaction IIa, yielding even number ring systems. For the 4 ring system an example may be:

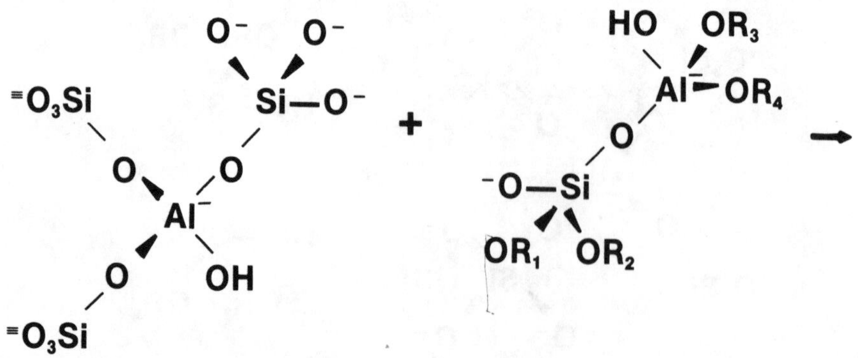

$$\begin{array}{c}
 O^- \quad O^- \\
\text{=}O_3Si \qquad\qquad \underset{Si}{\diagdown} \\
\diagdown \quad O \quad \diagup O \diagdown O \\
O \quad \overset{}{\diagup} Al \quad \diagdown O \quad \diagup Al \blacktriangleright OR_3 \\
\end{array}$$

For the 6 and 8 ring systems one can design similar reactions. The Si/Al ratios in the liquid phase of the initial gel (see Table I) were measured to be 0.44, 1.44 and 1.05 for the A, X and Y synthesis gels respectively, thus all < 4 and allowing type IIa condensations.

R1, R2, R3, and R4 indicated in reaction IIb can range from hydrogen to more complicated Si-O-Si and Si-O-Al networks. This means that a large variety of complex ions is present in the solution, each species having too low concentration due to its growth to larger species (e.g. D-4-Rings for zeolite A and D-6-Ring: for zeolite X/Y and possibly larger species).

In our opinion this variety of species in solution prevents the observation of specific Raman peaks during nucleation. Eventually the growing solute species become viable crystallization centers, precipitating from solution. The precipitate consists of crystals large enough to be Raman and X-ray detectable, the signals showing up simultaneously in the course of crystallization.

ACKNOWLEDGEMENT

Thanks are due to Prof. P.J. Gellings (Twente University of Technology, The Netherlands) and Prof. D.P. Shoemaker (Oregon State University, Corvallis, Oregon) for their comments on the manuscript and to Exxon Research & Engineering Co. for permission to publish this paper. Mrs. K. Gerth is acknowledged for typing the manuscript.

146

REFERENCES

1. Zhdanov, S.P., Advan. Chem. Ser. 101 (1971) 20
2. Barrer, R.M., J.W. Baynham, F.W. Bultitude and W.M. Meier,
    J. Chem. Soc. 195 (1959)
3. Kerr, G.T., J. phys. Chem. 70 (1966) 1047
4. Kerr, G.T., J. phys. Chem. 72 (1968) 1385
5. Culfaz, A., and L.B. Sand, Advan. Chem. Ser. 121 (1973) 140
6. Cournoyer, R.A., W.L. Kranick and L.B. Sand, J. phys. Chem. 79
    (1975) 1578
7. Kacirek, H., and H. Lechert, J. phys. Chem. 79 (1975) 1589
8. Freund, E.F., J. Crystal Growth 34 (1976) 11
9. Angell, C.L., and W.H. Flank, in "Molecular Sieves II" (Katzer,
    J.R., ed.), ACS Symposium series 40 (1977) 194
10. Chao, K.J., T.C. Tasi, M.S. Chen, and I.J. Wang, J. Chem. Soc.
    Faraday Trans. I 77 (1981) 547
11. Hawkins, D.B., Clays and Clay Minerals 29 (1981) 331
12. McNicol, B.D., G.T. Pott, and K.R. Loos, J. phys. Chem. 76 (1972)
    3388
13. McNicol, B.D., G.T. Pott, K.R. Loos, and N. Mulder, Advan. Chem.
    Ser. 121 (1972)
14 Polak, F., and A. Cichocki, Advan. Chem. Ser. 121 (1973) 209
15 Polak, F., and E. Stobiecka, Bull. Acad. Pol. Sci. 26 (1978)
    899
16 Flanigen, E.M., Adv. Chem. Ser. 121 (1973) 119
17 Derouane, E.G., S. Detremmerie, Z. Gabelica, and N. Blom, Appl.
    Catal. 1 (1981) 201
18 Gabelica, Z., N. Blom, and E.G. Derouane, Appl. Catal. 5 (1983)
    227
19 Jacobs, P.A., E.G. Derouane, and J. Weitkamp, J.C.S. Chem.
    Comm., 591 (1981)
20. Guth, J.L., P. Caullet, P. Jacques, and R. Wey, Bull. Soc. Chim.
    France 3-4 (1980) 121
21. Guth, J.L., P. Caullet, and R. Wey, Proc. Fifth Int. Conf. on
    Zeolites (Rees, L.V.C., editor), Heyden, Londen, 1980, p.30.
22. Dibble, W.E., B.H.W.S. de Jong, and L.W. Cary, Proc. Int. Symp.
    Water-Rock Interact., Edmonton, Canada, July 14-24, 1980, p.47
23. Engelhardt, G., D. Hoebbel, M. Tarmak, A. Samoson, and E.
    Lippmaa, Z. anorg. allg. Chem. 484 (1982) 22
24. Ueda, S., N. Kageyama, and M. Koizumi, Proc. Sixth Int. Conf.
    Zeolites, Reno, Nevada, July 10-15, 1983
25. Dent Glasser, L.S., and G. Harvey, ibid
26. McDaniel, C.V., and H.C. Duecker, US Patent 3,574,538 (1971)
27. Vaughan, D.E.W., G.C. Edwards, and M.G. Barrett, US Patent
    4,178,352 (1979)
28. Angell, C.L., J. phys. Chem. 77 (1973) 222
29. Moolenaar, R.J., J.C. Evans, and L.D. McKeever, J. phys. Chem.,
    74 (1979) 3629

30. Barby, D., I. Griffiths, A.R. Jacques, and D. Pawson, in "The modern inorganic chemicals industry" (Thompson, R., ed.), Chemical Society, London, 1977, p. 320.
31. Schwochow, F.E., and G.W. Heinze, Advan. Chem. Ser. 101 (1971) 102
32. Hendra, P.J., and E.J. Loader, Trans. Farad. Soc. 67 (1971) 828
33. Hendra, P.J., J.R. Horder, and E.J. Loader, J. Chem. Soc. A 1766 (1971)
34. Egerton, T.A., A.H. Hardin, Y. Kosirowski, and N. Sheppard, Chem. Comm. 888 (1971)
35. Egerton, T.A., and A.H. Hardin, Catal. Rev. 11 (1975) 71
36. Meier, W.M., Molecular Sieves, Society of Chemical Industry, London, 1968, p.10
37. Meier, W.M., and D.H. Olson, Atlas of Zeolite Structure types, Juris Druck Verlag, Zürich, Switzerland, 1978
38. Merlino, S., Twelfth Int. Congr. Cryst., Le Droit and Le Clerc Printers, Ottawa, Canada, 1981, p. C 189

PART II

PHYSICAL CHARACTERIZATION AND SORPTION FUNDAMENTALS

THE PHYSICAL CHARACTERIZATION OF ZEOLITES

H. Lechert

Institute of Physical Chemistry, University of Hamburg
Laufgraben 24, D-2000 Hamburg 13, Germany

1. INTRODUCTION

The physical characterization of a material, such as a zeolite begins in principle by visual inspection. This inspection can be refined by the use of instruments which enlarge small particles. Such instruments may be: an ordinary microscope, an electron microscope, or a raster scan electron microscope. Crystalline materials may often be identified by the shape of a crystal which also may give information concerning the presence of crystalline impurities. Furthermore, from the shape of the crystals and their size distribution conclusions may even be drawn which render information concerning the conditions of crystallization or later transformations which will take place, and the colour of the crystals reveals the presence of ionic impurities, such as iron which occurs in natural or industrial zeolites.

In most cases, the structure of the crystalline lattice of the zeolites may be obtained by X-ray diffraction methods. These methods range from simple powder patterns up to sophisticated single crystal methods which allow the localization of all atoms in a unit cell and the calculation of the electron distribution between these atoms.

The literature, especially the patent literature, shows that powder patterns are used as a means of identifying the structure of a specific zeolite.

The X-ray methods may be supplemented by neutron diffraction experiments for the localization of light atoms, especially protons. The neutron scattering experiments may also be used to study the mobility of water inside zeolite cavities. For samples which are made up of small crystals and for the investigation of special problems, such as the distribution of Si- or Al-atoms in the alumo-silicate lattice, electron diffraction may be applied.

In addition to the diffraction methods spectroscopic methods are used to study structure problems of zeolites. In this field of study the IR-spectroscopy and recently the ^{29}Si-NMR-spectroscopy has been applied.

Strictly speaking, the term 'physical characterization' is applied in zeolite chemistry mostly in connection with the description of special structural elements, responsible for the properties needed in the various applications, e.g. the acid OH-groups in catalysis.

In this field the different methods of spectroscopy are unique.

With the IR-spectroscopy a large number of studies has been carried out on the characterization of the strength and the accessibility of the acid OH-groups in different zeolites. The Lewis sites have been characterized by studying the IR-spectra of complexes with sorbed organic bases. Another application of the IR-spectroscopy is the sorption of molecules with different electronic structures on transition metal ions. These types of sorption complexes may also be studied by UV-, visible, or ESR-spectroscopy. Transition metals which are contained in zeolites, incidentally - they are used as catalysts in various reactions with hydrogen - have been studied extensively by means of Moessbauer-, ESR-, and XPS-spectroscopy.

The possibility of investigating acidic groups and their interaction with sorbed molecules is also described by different NMR-spectroscopy techniques, which have also found wide application in the study of mechanisms concerning molecular mobility in the cavities of zeolite structures.

Furthermore, under the heading of 'physical characterization' sorption measurements of any kind have to be summarized, giving information on the heat and entropies of sorption and thus on the interaction of molecules with various shapes and polarity with special structures of the walls of zeolite cavities. The accessibility of the cavities may be studied by diffusion measurements.

Last but not least the technique of thermal analysis shall now be mentioned: by this technique the force of interaction and the amount of sorbed molecules may be studied. The different mechanisms of thermal degradation may also be investigated conveniently using this method which yields information on the intermediate states which are often identical with the active catalytic species. Furthermore, information regarding the condition under which the zeolite is stable and where its crystalline structure is destroyed, can be obtained. Investigations of this kind are of great importance for the description of the limits of stability of zeolite catalysts in technical processes.

In the following sections the problems of the methods mentioned shall be described in more detail with examples of application for different problems of the chemistry and technology of zeolites. Some further supplementary methods shall be discussed in short:

2. SOME PROBLEMS OF THE INSPECTION OF CRYSTAL SIZE AND CRYSTAL SHAPE

A visual inspection of zeolite crystals and the inspection using a weakly enlarging instrument, such as a magnifying glass is important for geologists out in the field. In most cases, especially when inspecting synthetic zeolites, a microscope is needed. A microscope is usually used as a first means of identification by looking at the crystal shape of some specific zeolites. The shape of the crystals may sometimes be misleading because it depends on the conditions of growth. More detailed information on the crystal classification may be obtained using a polarization microscope. A very impressive example of the use of the optical microscope may be found in the kinetic studies of an X-type zeolite published in a paper by ZHDANOV and SAMULEVICH (1). These authors collected about 20 of the largest crystals, out of a batch of growing NaX-crystals, and observed their further growth under the microscope. From the increase of the crystal size they obtained the linear growth rate. At the end of the crystallization a representative sample of the product was taken and a histogram of the crystal size distribution was obtained from microscopic measurements. Extrapolating this distribution function with the growth rate which has previously been determined, the kinetics of the nucleation and the kinetics of the growth of the zeolite can be obtained.

In the field of electron optical instrumentation a wide range of techniques is available giving information concerning crystal habit and crystal size and other more specific characteristics. The recent literature in this field has been summarized by BAIRD (2). The main techniques used in zeolite research are connected to transmission electron microscopy (TEM), whereby the technique yielding most information is connected with the use of the scanning electron microscope (SEM) accompanied by a scanning micro-probe analysis.

With the transmission electron technique magnification up to about 10^6 can be obtained corresponding to a point to point resolution of about 3 Å. Depending on the sample, this resolution can be improved by using special imaging techniques known as bright-field, dark-field, or lattice imaging which shall not be discussed in detail here.

In the scanning electron technique a fine beam of electrons is scanned over the surface of the sample using a system of deflection coils. The various signals produced by the interaction of the electron with the surface, such as secondary electrons, back-scattered electrons, or X-rays can be used to form an image. Magnifications in the range of 20 - 50,000 are available with a resolution of about 100 Å. Non-metallic samples are usually covered with a thin film of carbon and gold to ensure a sufficient electric conductivity to prevent a surface charge which leads to distorted pictures. Another effect is the protection of heat sensitive material.

The output of the secondary electron varies according to the accelerating voltage of the beam (5 - 50 kV) and the structural characteristics of the sample as well as the particular angle of the incident beam with respect to the surface features. The changes in the secondary electron current induced by these features exhibit therefore a 3-dimensional character of the image.

The back-scattered electrons give a signal varying with the respective atomic number. Measuring the wave length of the induced characteristic X-ray radiation with special detectives an elementary analysis of the area hit by the beam can be carried out. This technique is known as electron micro-probe analysis.

In principle, this scanning technique and the analysis of the characteristic X-ray radiation can also be applied in transmissions. This method is usually called 'Scanning Transmission Electron Microscopy' or STEM. This method has the advantage of better resolution as compared with the TEM, but has the disadvantage of difficult sample preparation.

In the case of zeolites, high resolution transmission studies are very rare. The reason for this is that the preparation of the specimen is very difficult. After the first attempts made by MENTER (3) and, much later, by BURSILL et al. (4) it was recognized that structural information could be obtained by the use of electron microscopy. To demonstrate the possibilities of this method, a schematic picture of high-resolution image of a (001) plane of a NaA single crystal is shown schematically compared to a model of the structure drawn in the respective scale (fig. 1).

It can be seen that the tunnel structure of the zeolite is excellently visible in these pictures. Studies of this kind have been extended to NaA samples which were essentially amorphous showing in

Fig. 1. Left schematic drawing of a high resolution image of a
        (001) plane of a NaA single crystal (4).
        Right model of a (001) plane of the A zeolite structure

the TEM image the same sub-unit fragments as in the crystal (5).

With the TEM technique especially SCHMIDT (6) and also
EXNER et.al. (7) have studied the distribution of metals in
zeolite catalysts and their behaviour during the catalytic process.

As has been indicated above the transmission technique generally suffers from the relatively great difficulties in the preparation of the specimen.

The most powerful method in the investigation of zeolite problems is the scanning electron microscopy (SEM). This method has
above all the advantage of a simple preparation of the samples and
gives quick information on the shape and the distribution of the
size of the crystals and also of the presence of amorphous material.

In fig. 2 an example for the use of the SEM in zeolite synthesis is given. The picture shows two examples of A- and X-type zeolites which were grown from batches with very similar composition,
varying only in the excess of alkali. A higher excess of alkali
leads to a higher rate of nucleation and finally to smaller particles. The A-zeolite crystallizes in well-defined cubes which are
sometimes truncated by octahedral planes, depending on the condition of growth. The X-zeolite crystallizes in octahedra. Fig. 2d
shows the X-crystal balls of agglomerated smaller crystals which

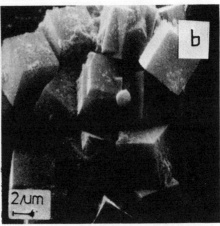
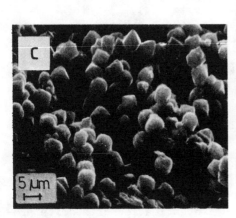
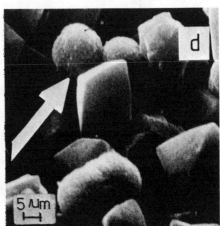

Fig. 2. Raster scan micrographs of A and X zeolites grown from
batches with different alkalinity causing a different
rate of nucleation and different particle size.

a. $NaAlO_2$ 0.8 Na_2O SiO_2 40 H_2O A zeolite
b. $NaAlO_2$ 0.8 Na_2O SiO_2 200 H_2O A zeolite

c. $NaAlO_2$ 1.5 SiO_2 1.37 $NaOH/SiO_2$ 100 H_2O X zeolite
d. $NaAlO_2$ 1.5 SiO_2 1.37 $NaOH/SiO_2$ 150 H_2O X zeolite
                                               with 25 % of P

The P zeolite can be seen in the picture d. by the balls
marked by the arrow.

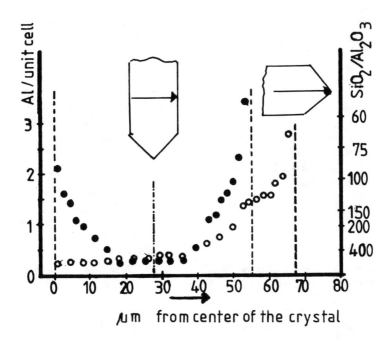

Fig. 3. Dependence of the Al-content in different directions in
ZSM-5 crystals on the distance from the center of the
crystal (9).

are typical for the P-type zeolite which often occurs as an impurity during the crystallization of faujasite-type zeolites especially
with higher Si/Al-ratios.

Extensive kinetic studies of the growth of faujasite based
mainly on the analysis of the particle size distribution from
scanning electron micrographs have been carried out by KACIREK et
al. (8).

A very interesting study done with the micro-probe analysis
has been published by v. BALLMOOS and MEIER (9). The authors have
studied the distribution of silicon and aluminium in ZSM-5 crystals.
These results are very important in explaining the mechanism of
crystallization of the zeolites, due to the rate of nucleation as
well as the growth increase with the Si-content. Furthermore, there
is some importance in explaining the catalytic activity of the substances because the number of active centers within the channel

system is a function of the Si/Al-ratio. v. BALLMOOS has embedded large ZSM-5 crystals and polished them in different crystallo-graphic directions. From scanning experiments it could be seen that the aluminium content in the center of the crystals was almost ze-ro so that the conclusion could be drawn that obviously a nuclea-tion of a species with a very high Si/Al-ratio occurs. The conse-quences for the catalytic activity are not yet quite clear. Fig. 3 shows the dependence of the Al-content determined in different di-rections.

In summary it may be said that by inspecting a zeolite sample with the appropriate instrument, the shape of the crystal, the type of zeolite, and the impurities which are present in the sample may be identified. By observing the time dependency of the crystal size kinetic studies are also possible. Furthermore, special techniques such as electron-microscopy yield detailed information on the di-stribution of single components in the sample. By using high-reso-lution TEM-techniques structural details of the zeolite lattice may be resolved.

3. PRINCIPLES OF THE CHARACTERIZATION OF ZEOLITES BY DIFFRACTION TECHNIQUES

The most frequently used method for identifying and describ-ing a special zeolite structure, primarily used for patent purpo-ses, is the X-ray powder diffraction method. The reason for this is that this technique is very simple and that zeolites occur mainly as small crystals where single-crystal techniques cannot be applied.

An extended collection of structures described in this way can be found in the book of BRECK (10).

In practice, the pattern obtained from Debye-Scherrer-, a Guinier-, or a diffractometer experiment are indexed and the spac-ings of the obtained (hkl) values are calculated from the angle of the respective peak.

The intensities of the individual X-ray deflections are given in a relative scale using the abbreviations vs, s, ms, m, w, vw, and vvw meaning 'very strong', 'strong', 'medium-strong', 'medium', 'weak', 'very weak', and 'very very weak'. This is preferentially done when evaluating photographs. When inspecting diffractometer patterns usually the most intense peak is set to be 100 and the others are set relative to this. By this procedure in most cases an identification of the respective zeolite and of the most important impurities can be achieved. If the impurity has at least one strong reflection located at an angle sufficiently far away from the re-flections of the main component, impurities occurring only at a few percent can be detected.

The unambiguos identification of a specific zeolite by such a set of powder reflections can sometimes be very difficult, if the structures of several zeolites in question show only slight variations. This has been very impressively shown by BRECK (10) comparing the powder diagrams of gismondine, phillipsite, P-zeolite, and some others.

In fig. 4 another example is shown which has some importance in the identification of preparations in the ZSM-series for patent applications.

Fig. 4 depicts the X-ray powder-diffractograms of the two pentasils ZSM-5 and ZSM-8. The difference claimed by the patent applications (11, 12) are connected in the group of reflections which are marked. In the ZSM-5 sample this group consists of reflections which are not very well resolved, whereas for the ZSM-8 this group is very well resolved. A third member of this family shows a systematic extinction of the h+k+l = 2n+1 reflections and a merging of doublets into singlets which give rise to the interpretation that ZSM-11 has a body-centred tetragonal and ZSM-5 an orthorhombic unit cell.

The channel structures which have been determined by KOKOTAILO et al. (13, 14) are shown schematically in fig. 5. Whether the ZSM-5 and the ZSM-8 pattern actually belong to different structures is not yet quite clear.

Generally, by inspecting a powder pattern one can try and analyze the given structure. This has been done, particularly for synthetic zeolites with some success. The intensities obtained by the powder patterns may be lost in the background or by an overlapping of reflections so that details concerning the structures, e.g. the exact positions of the cations cannot be resolved. There are several methods reported in the literature for analyzing the peak profiles and to avoid possible disadvantages. One advantage of the powder data is the reduced secondary extinction which disturbs, for single crystals, especially the strong reflexes so that in some cases powder data may be used as a supplement to single crystal data.

For an exact determination of the crystal structure single crystals of at least 20 - 50 µm are necessary and single crystal techniques need to be applied using as many X-ray intensities as possible.

Due to the fact that in most diffractional processes the phase is lost and crystal structure determination is, in principle, a trial-and-error procedure, whereby the observed intensities are to

160

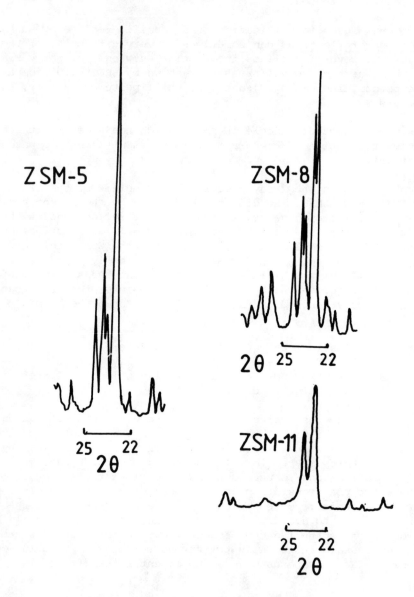

Fig. 4. X-ray powder diagrams of ZSM 5, ZSM 8 and ZSM 11 around
the region of $2\theta = 22°-25°$.
ZSM 8 shows a splitting in the highest peak which is not
observed for ZSM 5 and ZSM 11.
The pattern of ZSM 11 has distinctly fewer lines than the
others as has been explained in the text.

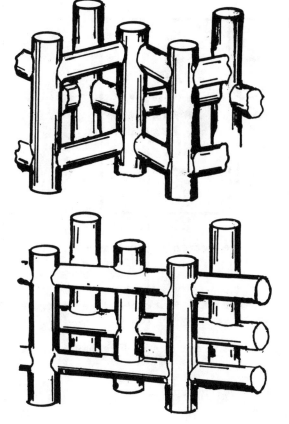

Fig. 5. Schematic draw-
ing of the chan-
nel system of

a.   ZSM-5

b.   ZSM-11

be matched to the intensities calculated from structural models.
The procedures necessary for a detailed structure determination
are beyond the scope of this article. A summary of the modern
methods of structure refinements by X-ray methods may be found in
the book of LUGER (15).

The problems which occur especially in the case of zeolite
have been summarized by FISCHER (16).

One of these difficulties is given by the determination of the
exact space group symmetry. This difficulty has its origin mainly
in the scattering factors of the Si- and Al-atoms which are very
similar due to the similar electron density surrounding these a-
toms. Therefore, the distribution of these atoms in the alumo-silicate framework seems to be random in most cases and the space group
appears to have a higher symmetry than is actually the case. If the
refinement is done using a space group with an average Al-O- and
Si-O-distance which is too symmetric, then ordering effects may be
overlooked.

162

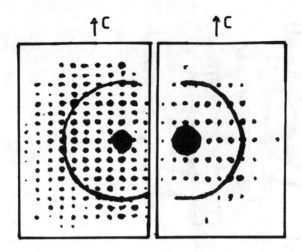

Fig. 6. Schematic drawing of the electron diffraction pattern of
the h0l zones of erionite (left) and offretite (right)

Some problems of this ordering are discussed in an article of
SMITH (17), where further literature may be found.

A special method of finding appropriate models has been developed by MEIER (18).

For difficulties of this kind the application of electron
diffraction is advantageous.

A further problem is in the stacking faults possible in the
various families of zeolites.

In the erionite-offretite system very careful electron diffraction studies have been carried out by GARD (19).

Fig. 6 shows a schematic drawing of the h0l zones of erionite
and offretite. It can be distinctly seen that every second row of
the l-reflexes is missing in the offretite system. In synthetic
samples called zeolite T streaks parallel to c can be observed indicating disorder in the layers of sixmembered rings from which
both structures can be constructed.

By the localization of cations inside the cavity systems of
various zeolite structures further problems occur. The literature
which has been published in this field up to now has been summarized with the most recent results in the atlas by MORTIER (20).

4. SPECTROSCOPIC METHODS FOR THE CHARACTERIZATION OF ZEOLITES

According to the two points of view of a characterization of zeolite samples mentioned in the introduction the discussion of the methods in this chapter shall be separated into procedures giving information on the alumosilicate framework and structural problems concerning the cations within the channel systems of the different zeolites and the description of the effects of modifications of the zeolite samples in order to obtain suitable properties for the application, especially in catalysis.

4.1 The Infrared Spectroscopy and the Framework Structures of Zeolites

Beside the diffraction methods the infrared spectroscopy has gained some importance for the structural characterization of zeolites, especially in situations where no unambiguous X-ray patterns can be obtained.

The most important arguments for the application of this method are the straightforward sample preparation and an instrumentation which is readily available in many laboratories.

Typical for the spectra of the alumosilicate lattice is the socalled mid-infrared region of $400 - 1300$ cm^{-1}.

After a series of papers by various authors on zeolites and other tectosilicates FLANIGEN, KHATAMI and SZYMANSKI (21) have done pioneering work in measuring spectra of a great variety of zeolites and collecting the arguments for their interpretation. An excellent review of the most important literature has been given by FLANIGEN (22).

According to FLANIGEN, SZYMANSKI and KHATAMI (21) the mid-infrared vibrations can be classified into two types, named internal and external vibrations.

The internal vibrations belong to the Si/AlO$_4$- or - as they are commonly called - the TO$_4$-tetrahedra. These vibrations are present in the spectra of all zeolites with small changes regardless of the type of the framework of the zeolite.

The external vibrations depend on the structure and are assigned to linkages of the TO$_4$-tetrahedra typical for a special framework topology. Groups of this kind are the rings and double rings, present in a great variety of zeolites and also the various polyhedral units.

Thus, the different zeolites have typical spectra, which are e.g. suited for an identification at least of the zeolite family in

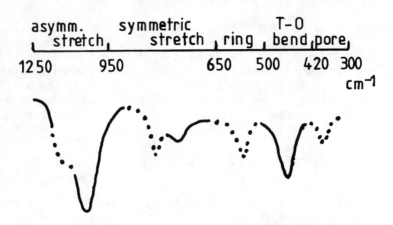

Fig. 7. Assignments of the different infrared wave number regions
with a spectrum of zeolite Y.
Solid lines: internal tetrahedra, structure insensitive
Dotted lines: external linkages, structure sensitive

many cases.

The exact assignment of the bands is rather difficult. An assignment can be only done by a careful empirical comparison of the spectra from zeolites with different structural elements known from X-ray diffraction studies. In a series of papers comparisons with Raman spectra were made to check special assignments.

Fig. 7 shows a spectrum of a NaY zeolite containing the different assignments adopted until now.

The frequency regions where the different kinds of vibrations are located are summarized in the Table 1.

Table 1. Zeolite Infrared Assignments, cm^{-1}

Internal Tetrahedra		External Linkages	
Asym. stretch	1250-950	Double ring	650-500
Sym. stretch	720-650	Pore opening	300-420
T-O bend	420-500	Sym. stretch	750-820
		Asym. stretch	1050-1150 sh

From the bands summarized in Table 1 the two most intense bands are at 950-1250 cm^{-1} and at 420-500 cm^{-1}. These are common for all zeolites. The first is assigned to an asymmetric stretching mode and the second to a bending mode of the T-O bond. The bands in the region of 650-720 cm^{-1} are assigned to a symmetric stretching within the TO$_4$-tetrahedra.

All other bands are more or less distinctly dependent on the crystal structure. This can be seen from the following figure 8 where samples of the A and the faujasite structure with different Si/Al-ratios are compared.

For all zeolites a nearly linear decrease has been observed for the asymmetric stretch stretch band near 980 to 1100 cm^{-1} with increasing fraction of Al in the tetrahedral sites, as can be seen from Fig. 8.

In the region of the external bands, a medium strong band near 500-650 cm^{-1} is present in all structures with double rings as e.g. zeolite A, the faujasites, chabazite, gmelinite and the offretite-erionite system. Zeolites without double rings show only weak intensities in this region. Inconsistencies are observed with the structures of P and Omega, which contain no double rings.

For the band of double sixmembered rings an influence of the Si/Al-ratio of the zeolite could be observed.

Further, distinct dependencies of the lattice vibrations on the cations in the zeolite could be established in the symmetric stretching region as well as in the bending vibrations of the TO$_4$-tetrahedra.

Especially deKANTER et al. (24) were able to demonstrate linear relationships of several types of bands from the reciprocal sum of the radii of the O^{-2}-ion and the respective cation. From investigations of this kind informations on the cation distribution within the zeolite lattice can be obtained.

An important application of the lattice vibrations in the mid-infrared region is the investigation of nucleation mechanisms during the crystallization of zeolites. In the paper of FLANIGEN (22) a series of spectra of growing NaX is demonstrated, showing typical bands of the faujasite from the initial gel to the well crystallized sample. Especially mentioned is a band near 575 cm^{-1} which may be assigned to double sixmembered rings.

Furthermore, two papers of JACOBS et al. (25) and COUDOURIER et al. (26) shall be mentioned using IR-spectroscopy for the identification of small particles of ZSM-zeolites, showing no X-ray pattern but the typical effects of the catalytic activity of these substances. This activity could be closely related to the presence of a band near 550 cm^{-1} from which also an analysis of the quantity

166

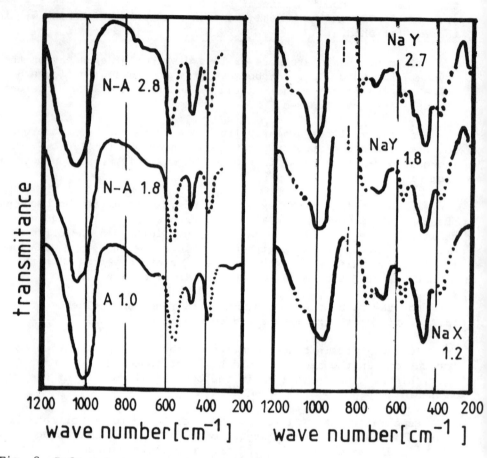

Fig. 8. Infrared spectra of zeolite samples with different Si/Al-
ratio.
Left: A and NA zeolites
Right: X and Y zeolites

of the zeolite could be carried out.

4.2 IR-Studies of Surface Structures Important for Sorption and Catalysis

The structures which are involved in the most applications in catalytical reactions are OH-groups in different situations with respect to the local structure in the zeolite framework. These groups act as more or less strong Brønstedt sites and are situated at the Si or Al atoms of the framework or at the cations. Further, in dehydroxylated zeolites at the aluminium atoms Lewis sites may be formed. Special effects of transition metals present in the structure shall not be discussed in this connection. Typical wave numbers of the bands of the OH groups lie between 3000 and 4000 cm^{-1} and demand some care in the preparation of the samples because of the radiation loss from scattering. The preferred technique is here the use of thin self-supporting wafers made of small particles.

Because of the strong bands of water in the frequency region mentioned, typical investigations are carried out on the dehydrated samples or with samples with very low water contents. The presence of water is easily detected by the presence of its bending vibration near 1650 cm^{-1}.

The history of the investigations in this field and the various arguments of an interpretation of the observed bands in different zeolites have been summarized by WARD (27). A more recent survey has been given by KUSTOV et.al. (28).

Dehydrated samples of A, X, and Y zeolites with alkali cations show no structural OH groups if hydrolysis is avoided. The only band which is observed in these samples has a wave number near 3740 cm^{-1} and belongs to terminating silanol groups at the edges of the crystals or at amorphous regions within the crystals. Ba containing samples show sometimes the same behaviour.

For zeolites containing divalent cations, again the band at 3740 cm^{-1} and additional bands near 3650 cm^{-1} and 3540 cm^{-1} are observed which also appear in decationized zeolites and can be assigned to different kinds of SiOH groups.

A common interpretation of these effects is a hydrolysis mechanism

$$Me^{2+}(H_2O) + \overline{O}\text{-Zeolite} \quad = \quad Me^{+}OH \ + \ HO\text{-Zeolite}$$

About an assignment of a band at 3570-3600 cm^{-1} to the $Me^{+}OH$ groups no agreement could be obtained.

Careful studies have been carried out with rare earth exchanged Y zeolites. In general, after removal of the physical adsorbed water at 450°C hydroxyl groups near 3740, 3640 and 3540 cm^{-1} remain.

An additional band at 3470-3520 cm^{-1} is sensitive against the radius of the RE-cation and has to be assigned to $RE^{2+}(OH)$ groups originating from

$$RE^{3+}(H_2O) + \overline{O}\text{- Zeolite} = RE^{2+}(OH) + HO\text{-Zeolite}$$

but this band vanishes upon heating to 200°C.

More detailed assignments of the bands can be obtained from studies of the interaction of the acidic hydroxyl groups with organic bases like pyridine or piperidine. In this way the band at 3640 cm^{-1} has been assigned to OH groups in the supercages of the Y zeolites because it is changed under the influence of a sorption of the mentioned bases.

The 3470-3520 cm^{-1} band is not influenced under the sorption of piperidine which is consistent with X-ray results, showing that the RE ions are in the S_1' positions inside the cubooctahedra as long as the band is observed and go into the hexagonal prisms at higher temperatures, which is accompanied by a vanishing of the band.

The deammoniation and the formation of the hydrogen forms of NH$_4$X and NH$_4$Y has been carefully studied by UYTTERHOEVEN et al. (30). The evolution of the ammonia could be observed by the disappearance of the bands of the ammonium ion near 1450 cm^{-1} and 1670 cm^{-1}.

Summarizing this and a series of other papers it could be stated, that the deammoniation starts at 100°C and is finished at about 400°C depending on the conditions of the heat treatment. The decationized Y zeolite has three major bands at 3740, 3640 and 3540 cm^{-1}. X-ray studies by OLSON and DEMPSEY (31) of the T-O distances in a HY sample lead to the result that the band at 3640 cm^{-1} belongs to an OH groups at the oxygen atoms of hexagonal prisms connecting the cubooctahedra, usually denoted by O$_1$. The band at 3540 cm^{-1} was assigned to the O$_4$-ions and is directed into the hexagonal prisms, which agrees with the most sorption experiments. Deviations are explained by a mobility of the protons. Treatment at higher temperatures leads to the formation of Lewis sites by dehydroxylation, which starts at 400-500°C. During this procedure an increase of the 3740 cm^{-1} band can be observed indicating the formation of amorphous silica. As long as the structure is intact the bands can be restored by a readdition of water.

A great variety of papers has been published on decationized zeolites containing defined amounts of different cations, ultra-stabilized zeolites and zeolites steamed under different conditions which are adopted to the treatment in technical reactors. A good deal of the literature up to 1976 has been summarized in the paper of WARD (27) already mentioned. Generally, the results in these fields are difficult to survey on a common basis, because of the varying conditions of the sample preparation and treatment.

JACOBS (32) has pointed out in a recent summarizing paper that

the wave numbers of the OH stretching vibrations can be correlated
with the acidity of the respective groups. Looking at the spectra
of a variety of zeolites the wave numbers of a special OH vibration decreases almost linearily with increasing Si/Al ratio of the
zeolite. This has been related to an increased acid strength because of a decreasing interaction of the proton with the framework.

These arguments have been checked by quantum chemical CNDO
calculations of OH groups at O-bridges with a varying Si/Al ratio
of the neighbouring region (33). Another correlation has been
found to the average electronegativity defined by SANDERSON (34).
Looking at an arrangement of atoms with the composition $X_x Y_y Z_z$ and
and electronegativities of the single components S_X, S_Y and S_Z the
average electronegativity S of this arrangement is given according
to SANDERSON (34) by

$$S = (S_X{}^x S_Y{}^y S_Z{}^z)^{1/(x+y+z)} \tag{1}$$

Plotting the exact wave number e.g. of the 3650 cm^{-1} vibration a-
gainst this parameter a linear relationship is obtained. Measuring
now, the spectrum of an unknown zeolite, at least indications of
the acidity of its OH groups can be obtained.

The changes of the wave numbers under the influence of sorbed
molecules obey also a linear relationship with the parameter S. The
same is true for the reactivities of a number of acid catalyzed reactions.

Finally, it shall be mentioned that IR spectroscopy allows to
distinguish Brønstedt and Lewis acidity in a zeolite sample which
may be a partly dehydroxylated HY or H mordenite. The pyridine-
proton complex shows a very strong band at 1545 cm^{-1} and strong
bands at 1655, and 1627 cm^{-1}. The respective complex of a pyridine
molecule with a Lewis site has a very strong band at 1455-1442 cm^{-1}
and a strong band at 1490 cm^{-1} which is, however, common to both
complexes.

Similar bands occur with 2,6-dimethylpyridine. The advantage
of this molecule is that because of a steric hindrance of the formation of the complex with the Lewis site, the molecules react first
with the Brønstedt sites and after the saturation of these sites
with the Lewis sites (35). A more detailed paper on Lewis sites
has been published by JACOBS and BEYER (29)

4.3 Nuclear Magnetic Resonance Studies for the Characterization of Zeolites

Up to about five years ago in NMR investigations on zeolites
almost only protons were used, looking at the adsorbed molecules
and its mechanisms of mobility and in some cases on the OH groups

present inside the cavities of acid zeolites.

Only a few papers dealt with the resonance of nuclei of the cations. For these investigations mostly quadrupolar nuclei were studied giving information on the field distribution around the cation.

One of the most important events for zeolite research is the development of high resolution spectroscopy of solids. This development has two important reasons. One is the possibility to create high magnetic fields by the use of superconducting magnets. High fields increase both the sensitivity of the detection and the chemical shift of nuclei in different chemical surroundings. The chemical shift is the most important information obtainable from a high resolution spectrum. All other interactions of the nuclei, especially the dipole-dipole interaction, are unaffected by an increase of the magnetic field. The other important technique developed for routine use in the last year is the magic-angle-spinning (MAS) by which the dipole-dipole interaction of the nuclei can be minimized below the resolution necessary to detect chemical shifts. Further importance have the fourier transform and computer averaging techniques, which allow the accumulation of a large number of spectra and an improvement of the signal-to-noise ratio. Finally, the double resonance techniques shall be mentioned which are in the most cases used for the decoupling of the dipole-dipole interaction of the nuclei in question with the protons of the surroundings.

The most valuable informations which could be obtained from these techniques lie in the ^{29}Si resonance of the alumosilicate framework of the zeolites. For the investigation of sorbed molecules ^{13}C- and ^{15}N-spectra have got some importance.

4.4 The ^{29}Si Resonance and the Distribution of Silicon and Aluminium in the Alumosilicate Framework of Zeolites

The technique of the ^{29}Si resonance in zeolite chemistry was first introduced by LIPPMAA et al. (36, 37), who showed that well resolved high resolution spectra can be obtained with a chemical shift characteristic for the local environment of the framework silicon. For a sample of zeolite NaX with Si/Al = 1.8 a spectrum with five peaks could be obtained belonging to the five possible local structures Si(nAl), where n can adopt the numbers 0 to 4. The chemical shifts could be assigned to the configurations of n Al surrounding one Si given in the Table 2.

A summary of a great variety of structures investigated by ^{29}Si NMR has been given in the book of BARRER (38).

Table 2. Chemical shift ranges vs. TMS for Four Coordinated Si environments

	Chemical shift/ppm
Si(4Al)	-83 to -87
Si(3Al Si)	-88 to -94
Si(2Al 2Si)	-93 to -99
Si(1Al 3Si)	-97 to -107
Si(4Si)	-103 to -114

From the data of the Table 2 follows a nearly linear relationship between the number of Al atoms around a Si and the chemical
shift value, taking into account that the shift is slightly dependent on the structure of the individual zeolite.

The most detailed structural investigations have been published by LIPPMAA et al. (36, 37) and by MELCHIOR et al. (39, 40)
on the distribution of silicon and aluminium in the alumosilicate
lattices of X,Y,A and NA zeolites.

The first problem which has already been mentioned in the
discussion of the diffraction methods, is the validity of the
Loewenstein rule. This rule states that two four coordinated Al
atoms can never be nearest neighbours in alumosilicate structures.
The validity of this rule demands for the A zeolite which has a
Si/Al ratio of 1 a strictly alternating order of the Si and Al.
Actually the spectrum shows only a single line which is located,
however, at -89 ppm belonging to the Si(3Al,Si) coordination in
many alumosilicates (41). Therefore, the authors suggested an ordered structure consisting of pairs of Al-O-Al and Si-O-Si. This
structure could be finally ruled out by studying a series of Na
zeolites with varying Si/Al ratio showing that the observed value
for the A structure can be very well understood as perfectly ordered according to the Loewenstein rule.

For the faujasites MELCHIOR et al. (39) have carried out careful investigations. The authors compared the obtained spectra with
models of different stages of refinement. At first it could be
shown that the Loewenstein rule is strictly valid.

From a calculation of the average number of the nearest
neighbours from a random distribution and a distribution according
to the Loewenstein rule compared with the respective values obtained from the spectra, it can be seen from fig. 9 that the random distribution can be ruled out very clearly.

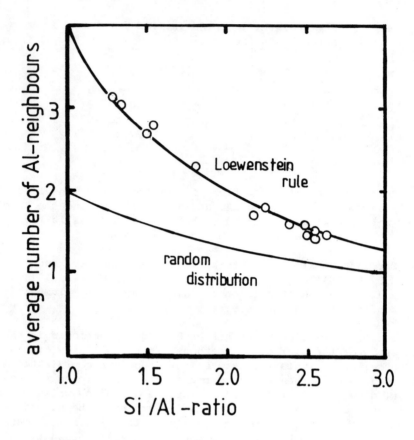

Fig. 9. Average number of Al neighbours for a Si atom in faujasite
structures with different Si/Al ratio, calculated for a
random distribution and according to the Loewenstein rule
compared with the numbers evaluated from the ^{29}Si resonance
(39), which are represented by the circles.

Further it could be shown that an improved fit to the experimental spectra is possible if the number of the next nearest Al-Al
neighbours is minimized in a cubooctahedron for the respective
Si/Al ratio. For the cubooctahedra in which one, two and four Al
are replaced by Si the distributions given in fig. 10 should be
favourable. From a suitable mixing of these configurations optimal
fits to the experimental spectra can be obtained.

Some of the spectra obtained for faujasite at 11.9 Mc are demonstrated in fig. 11. As a final result it can be stated, that
the structure with the most sixmembered rings with two Al in p-po-

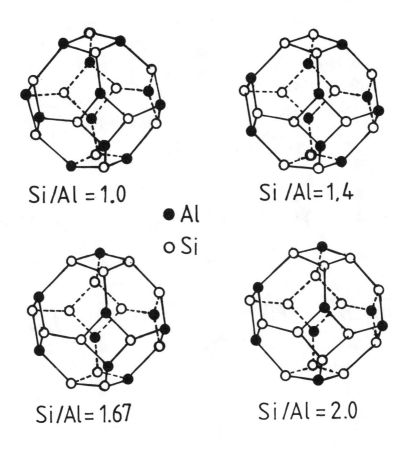

Si/Al = 1.0   Si/Al = 1.4

● Al
o Si

Si/Al = 1.67   Si/Al = 2.0

Fig. 10. Cubooctahedra with minimal numbers of Al-O-Si-O-Al
         linkages for different Si/Al ratios (39).

sition are the most favourable compositions as far as they fulfill
the topological constraints demanded by the overall structure.

4.5 The Use of Quadrupole Nuclei for the Characterization of
Structures in Zeolites

Nuclei with a spin I > 1/2 have an electric quadrupole moment
which experiences a torque in electric field gradients caused by
the crystal fields and bond electrons within a solid structure.
Whereas from single crystals very detailed informations can be obtained about the field gradient tensor and the field distribution
around the nucleus, these informations are for powders and especi-

174

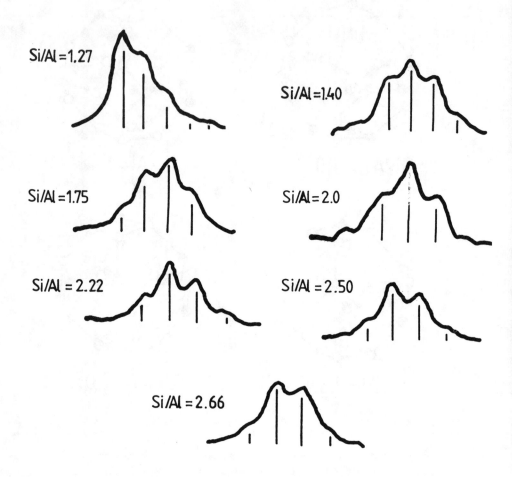

Fig. 11. ^{29}Si spectra of faujasites with different Si/Al ratios.
The measured spectra are simulated by constructing structures from different amounts of the cubooctahedra shown
in Fig. 10 (39).

ally for complicated structures like zeolites rather poor.

The electric field gradient tensor can be described by its
value in the direction of the highest field inhomogeneity, called

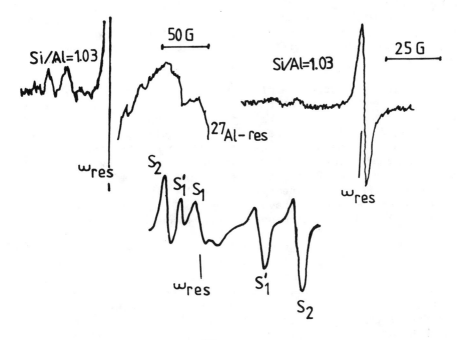

Fig. 12. Comparisons of two ^{23}Na spectra of a dehydrated NaX
sample with Si/Al = 1.03 at different amplification and
span with a computer simulated spectrum, using the data
of model calculations of the electric field gradients at
the different sites of the sodium ions.

"field gradient" and by the deviation from rotational symmetry
perpendicular to this direction, which is described by a dimensionless parameter between 0 and 1 called "asymmetry parameter".

A very careful single crystal experiment of natrolite has
been published by PETCH and PENNINGTON (42). The authors showed
that the Al as well as the Na ions in the natrolite structure are
in chemical equivalent positions and obtained very exact values
for the field gradient tensors.

The absolute values for the field gradient at the sites of
the Al show that the AlO$_4$ tetrahedron is only weakly distorted.
The field gradient at the Na sites is rather large compared with
other sodium containing crystals.

In a series of papers published by the author (43-46) the
^{23}Na resonance and the resonances of other nuclei of the alkali
group in faujasite and A zeolites have been measured. The results
have been compared with model calculations of the electric field

gradient tensor. The aim of these investigations was to find a model suitable to describe the electric field distribution within the zeolite cavities, which is interesting for the explanation of the phenomena connected with the catalytic activity of these substances.

In Fig. 12 two spectra of an X zeolite are compared with a calculated spectrum. It can be seen that in the calculated and in the measured spectra two peaks appear at lower fields the position of which agrees well with the position calculated for the S_2 and the S_1' sites. The strong intensity in the center of the line cannot be assigned exactly, but it must belong to the extra framework ions for which several different sites are reported (see e.g. the atlas of MORTIER (20)). The spectra can be explained by the assumption that these sites have varying field gradients and high asymmetry parameters, which seems to be plausible.

For the calculation of the field gradients a point multipole model has been used. In this model at first the point charge contribution to the fields and field gradients at the sites of all ions of a cubooctahedron was evaluated. With the obtained values, the strengths of the induced dipoles and quadrupoles were calculated for different polarizabilities of the oxygen ions and the contributions of these multipoles added to the point charge contribution. A similar model has been used by DEMPSEY (47) who has carried out the calculation self-consistent without the induced quadrupoles.

The calculations have been done for different positions of the mentioned extra framework cations in the dehydrated zeolite and for different models with adsorbed molecules. In the fully hydrated zeolite NaX a single slightly asymmetric line can be observed which is appreciably narrower than the central line belonging to the transitions $m = -1/2 \longleftrightarrow m = +1/2$ in the dehydrated zeolite.

By measurements of pulsed experiments on NaY samples BASLER (48) has found that this resonance must be attributed to S_2 sites with an adsorbed water molecule which changes its site rather quickly. The field gradients which can be obtained from this experiment for the S_2 sites are in good agreement with the calculated values for the described configuration of a S_2 Na ion with an attached water molecule.

BOSACEK et al. (49) have studied the ^{27}Al resonance in decationated samples of Y zeolite in comparison to the respective NaY, and samples with different degrees of cation exchange.

The spectrum obtained in dehydrated NaY can be explained by the presence of sites with a distribution of field gradients and

and asymmetry parameters.

When the Na^+ ions are exchanged against NH_4^+ ions and the samples are treated at 400°C OH groups are formed near the Al ions and a large field gradient is created causing a very broad and weak line which cannot be observed any more. The formation of ultrastabilized zeolite in which a certain amount of Al ions is in extralattice positions is studied by complexing these ions by ace-tylacetone. The complex provides a rather symmetric surrounding of the Al nucleus and shows a narrow line from which the number of these ions can be easily evaluated.

4.6 Some Special Questions Solved by Proton NMR Spectroscopy

In this section some problems of special structural investi-gations by proton resonance measurements shall be discussed. The problems of sorption and mobility investigations by proton NMR spectroscopy are beyond the scope of this article.

In a paper of BASLER and MAIWALD (50) the OH groups in the cavities of A zeolite has been studied on a variety of samples of different origin. These samples have been carefully checked for structural integrity by X-ray and sorption measurements. As has been observed also for samples of faujasite zeolites, the water molecules in the supercages and the cubooctahedra can be quanti-tatively separated by NMR pulse experiments. The respective decay functions after a 90°-pulse are shown for a number of zeolites in fig. 13. The short decay belongs to the water molecules inside the cubooctahedra and the long one to the molecules in the supercages.

By a careful analysis of the temperature behaviour of both decay functions it could be shown that in the short decay a third decay function should be hidden with almost the same decay time, which could be assigned to $Al(OH)_3$ or similar entities within the cubooctahedra. The amount of these entities is distinctly depend-ent on the conditions of the growth of the samples. It could be demonstrated that out of 31 samples grown under different condi-tions or in different laboratories no sample was without these aluminate inclusions.

Similar investigations have been carried out with the zeolite ZK 5, containing only large cavities (51). This structure is ex-cellently suited for studies of hydrolysis processes within these large cavities.

Directly after the removal of the organic cation about two OH groups could be observed in one cavity, causing an exchange of protons with the water molecules.

This exchange expresses itself in a minimum of the tempera-

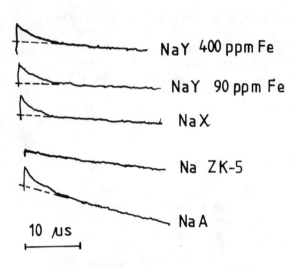

Fig. 13.

Magnetization decays after a 90°-pulse of a NMR pulse experiment on protons in different zeolites.

The short decay belongs to water molecules inside the cubooctahedra, the long decay to the molecules inside the large cavities.

ture function of the transverse relaxation time as it is shown in fig. 14a. Treating the sample with 0.1 n NaOH the signal of the OH groups falls below the limit of detection and only one kind of protons can be observed in the magnetization decays as well as in the temperature functions of the relaxation times (fig. 14b). The absolute values of the relaxation time in the short component can be calculated assuming an interaction of the protons with the aluminium of the lattice taking the usual distances of an AlOH bond. Finally, the problem of the mobility of the protons of the acid OH groups in the cavities of zeolites shall be discussed in short.

This problem has been studied with NMR methods by MESTDAG et al. (52), by FREUDE et al. (53) and by FREUDE and PFEIFER (54). FREUDE and PFEIFER have shown from the temperature dependence of the transverse relaxation times in decationized Y and mordenite that the correlation times of the motion of the protons are dependent on the temperature of the pretreatment. A strong increase of the mobility with residual ammonia could be observed. A systematic study showed that the pyridinium ions have an increased mobility, too. The number of the mobile protons could be obtained by measuring the proton NMR relaxation after the sorption of deuterated pyridine. The absolute number of pyridinium ions could be also determined by ^{13}C resonance measurements, where characteristic shifts can be observed in the spectra of the protonated and the nonprotonated species.

The ratio of the number of the acid sites and the correlation

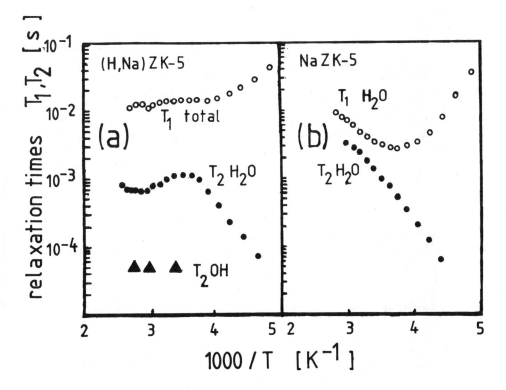

Fig. 14. Temperature functions of the longitudinal relaxation
times T_1 and the temperature functions of the nuclear
relaxation times T_1 and T_2.

a. T_2 of water (black circles), T_2 of OH groups (tri-
angles)
T_1 of all protons (open circles) of (H,Na)ZK 5 with
298 mg of water/g;

b. T_2 (black circles) and T_1 (open circles) of water in
NaZK 5 with 317 mg of water/g.

time of the mobility of its protons which can be regarded as the
rate of creation of free or protons attached to a sorbed molecule
is a good measure for the acidity of the respective OH group.

180

4.7 The Application of Moessbauer Spectroscopy in Metal Containing Zeolites

In zeolite research Moessbauer spectroscopy is applied mostly for the investigations of iron in its different oxidation states and local environments. The oxidation state may be seen from the shift of the observed lines and the environment expresses itself in a characteristic structure of the spectrum because the ^{57}Fe nucleus usually applied for these investigations has an electric quadrupole moment which may interact with gradients of the electric field of its surroundings as it has been discussed in connection with the NMR experiments with quadrupolar nuclei. Furthermore, magnetic ordering in a sample can be detected because the ^{57}Fe nucleus has the spin 3/2.

Investigations of this kind have received considerable attention because of their importance in adsorption and catalysis. The different oxidation states of the iron and especially the conditions of the existence of the zerovalent iron inside the cavities has become important for the preparation of catalysts for hydrogenation reactions.

The characteristic argumentation shall be shown from results obtained by SCHMIDT (55).

Fig. 15 shows a Moessbauer spectrum of a Y zeolite which has been exchanged under very careful pH conditions under nitrogen with ferrous sulphate to prevent the formation of ferric hydroxide. The samples were partly reduced, by exposing them to Na vapour at 673 K. The central doublet is due to iron clusters, which can be shown to be superparamagnetic by magnetic measurements. This superparamagnetism expresses itself in the weak six peak component in the spectrum. The results obtained from these investigations have been confirmed by electron microscopic experiments.

Numerous investigations by Moessbauer spectroscopy have been reported by REES et al. (56, 57) and by DELGASS et al. (58).

4.8 Further Application of Spectroscopic Methods

In this section some further spectroscopic methods shall be mentioned in short which have been applied to solve special problems in zeolite research which are not typical characterization problems.

A special field of research in zeolite chemistry is the investigation of transition metals and transition metal complexes. The transition metal complexes have been proposed since a long time as reactive intermediates in heterogeneous catalysis. Inspite

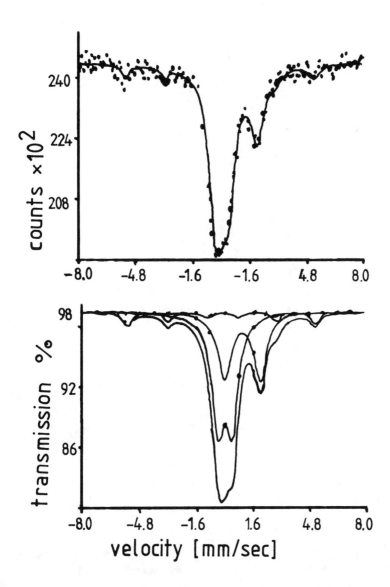

Fig. 15. Moessbauer spectrum of a Fe^0-Y/Fe^{2+}-Y zeolite (a) at
298 K and its computer fit (b).

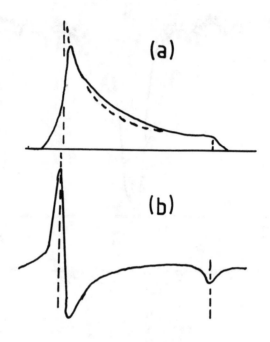

Fig. 16. ESR powder absorption spectrum for an axially symmetric
g-tensor (a) and its derivative (b).

of the great importance comparatively few work has been done in
this field. Some of the most important papers have been published
by LUNSFORD (59). The theoretical background has been summarized
by KASAI and BISHOP (60). The shape of an electron spin resonance
(ESR) spectrum is mainly determined by the g-tensor and the hyperfine coupling tensor. The g-tensor measures the deviation of the
g-value from the value g_e of the free electron caused by a spin-
orbit coupling which is dependent on the direction in the crystal.
Because zeolites are usually not available in single crystals
which are large enough to measure this tensor only the socalled
powder patterns are usually observed. Fig. 16 shows a powder
pattern of an axially symmetric g-tensor, and its derivative which
is measured in the ESR technique. The hyperfine splitting is caused by an interaction with the nuclear spins in the neighbourhood
of the electronic spin. In the usual practice the parameters of
the g-tensor and the hyperfine splitting are taken approximately

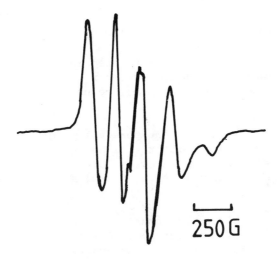

250 G

Fig. 17. ESR spectrum of NO adsorbed on Cu^+-Y zeolite.
The splitting is due to the hyperfine coupling with the
^{14}N spin. The disturbance in the central part is possibly
due to the presence of a small amount of Cu^{2+} ions.

from the measured spectrum and then a computer fit is made to find
the exact values, and to decide whether additional interactions
have to be taken into account.

A typical spectrum is shown in fig. 17 with NO sorbed on
Cu(I)Y. The spectrum is similar to that shown by KASAI and BISHOP
(60). The splitting is due to the interaction with the ^{14}N nucleus.

Generally from the symmetry of the g-tensor conclusions on
the symmetry of the crystal field at the site of the cation or of
the sorption complex can be drawn, which can be refined by the use
of the data from the hyperfine interaction.

The transition metal cations and also the sorption complexes
have been studied also by optical spectroscopy in the reflectance
technique giving also information on the crystalline or the ligand
field at the site of the ion. The basic problems of this field
have been summarized by KLIER and KELLERMANN (61). A recent comparison of all methods on the rhodium complexes in zeolites has

been made by LUNSFORD (62).

Some papers on the valence state of transition metals in zeolites can be found by the XPS method (62).

5. THE CHARACTERIZATION BY ADSORPTION OF SPECIFIC PROBE MOLECULES

Sorption processes will be treated in a separate lecture of
this course. Therefore, in this chapter only some sorption experiments shall be mentioned, which may be used as a means of characterization of a specific zeolite.

A first information on the pore volume can be obtained from
the sorption capacity for water. The amount of larger pores may be
determined by the sorption capacity for larger molecules which cannot enter e.g. the cubooctahedra of faujasite. A molecule often
used for these purposes is cyclohexane.

Sorption capacities are usually determined by balance techniques or for gaseous components volumetric. Another possibility is
the gas chromatographic method. The accuracy of such a determination is mostly better than one percent. Thus, the sorption capacities are often used to characterize the purity e.g. of a synthetic
zeolite from inclusions of silica or other disturbances of the
structure.

In earlier times the zeolites have been classified according
to their ability to adsorb or to exclude molecules of a particular
size. Details are summarized in the book of BRECK (10).

The sorption isotherms characterizing the sorption behaviour
for any specific substance can be measured in more detail by the
same methods as the sorption capacities. The balance technique can
be regarded as having the highest accuracies, where especially
also diffusion data can be obtained observing the establishment of
the equilibrium.

Very conveniently the isotherms can be measured by gas chromatographic step or pulse methods, where the concentration of the
sorbate in a gas stream over the zeolite is changed in a step or a
pulse of the substance is injected into the stream of a carrier
gas. From the response function the isotherm can be obtained. For
a pulse the isotherm may be calculated e.g. by the relations given
by HUBER and GERRITSE (63).

The gas chromatographic method is preferable if isotherms
over a wide range of equilibrium pressures are needed. Especially
for liquids low partial pressures are difficult to maintain exactly over a long time which is necessary if the isotherms are meas-

ured by the balance.

From the isotherms at different temperatures the heats and entropies of sorption may be evaluated giving information on the strength of the interaction of the molecules with the cavity system and among themselves.

Special effects have to be expected, if reactions inside the cavities take place, as this is the case e.g. at sorption processes of bases in zeolite cavities with acidic groups. An example will be discussed in connection with the thermoanalytical methods.

6. CHARACTERIZATION OF ZEOLITES BY THERMAL ANALYSIS

Thermoanalytical methods are among the most important tools of the characterization of zeolites.

Generally, thermal analysis describes a group of methods whereby the dependence of the parameters of any physical property of a substance on temperature is measured.

The two techniques measuring the change of heat and the change of weight are the methods used preferably for the characterization of zeolite properties.

These methods are called differential thermal analysis (DTA) and thermogravimetric analysis (TGA).

In both methods the sample and possibly a reference sample are heated or cooled at a controlled rate. In the DTA technique the difference in temperature between a substance and a reference material against either time or temperature is recorded. If any heat releasing or heat consuming process takes place in the sample, the temperature of the sample increases or remains behind the temperature of the reference. If the process is finished the temperatures of both specimen become equal again. The peak, obtained in the recorded curve can be evaluated to get the kinetics and the amound of the heat transfer.

In the thermogravimetric analysis the weight of the sample is recorded in dependence on the temperature.

In modern devices both principles of measurement are often realized in one apparatus.

The DTA method has a sensitivity of about 10^{-4} Joule. With the TGA method weight changes of about 10^{-8} g can be detected. A summary of the most important effects observed in zeolites has been recently published by DIMITROV et.al. (64).

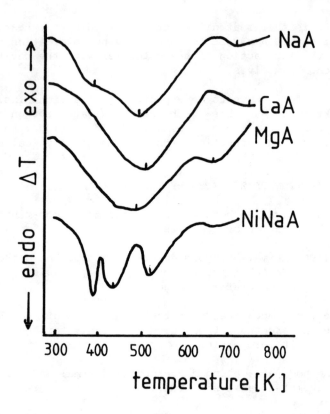

Fig. 18. DTA curves in the region of the dehydration for A zeolite
with different cations.

In the typical behaviour of a zeolite being heated and subjected to differential thermal analysis three typical regions can
be distinguished. The first region begins slightly above room temperature, has its maximum mostly near 500 K and is finished at
about 750 K. This region expresses itself as an endotherm in the
DTA curves and is caused by the evolution of water and possibly
other volatile substances in the zeolite cavities.

Between about 900 and 1500 K often two exotherms can be observed which are associated with the collapse of the zeolite
lattice and sometimes at much higher temperatures recrystalliza-
tion to a new phase.

In the first region often a stepwise evolution of water can
be observed. Fig. 18 shows DTA curves for A zeolites containing
different cations.

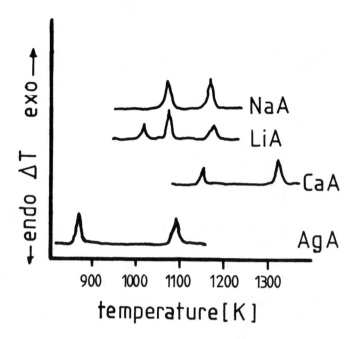

Fig. 19. High temperature effects in the DTA curves of different
A zeolites.

A detailed thermoanalytical study has been carried out by
DYER and WILSON (65) on NaA, who measured beside the DTA curves
thermogravimetric data which were compared with X-ray experiments
looking at the sites of the cations and water molecules. According to these studies the first endothermic effect at about 395 K
is connected with an evolution of 10 water molecules per unit cell
which are bonded very loosely. A second peak at 438 K is connected
with a loss of 8 water molecules, which have been sorbed at the
sodium ions in the S_1 positions. Of the remaining 10 molecules
six are released at temperatures between 445 and 625 K. The four
molecules inside the sodalite cavities leave the structure at
higher temperatures.

Changing the cations characteristic changes in the DTA curves
are observed which depend on the degree of the exchange and may
often be explained by the typical complexes with these ions.

The high temperature effects have been extensively studied
by BERGER et al. (66). Some typical exothermic effects of A zeolites with different cations are shown in Fig. 19.

The process of the dehydration of faujasite zeolites is more complicated. This is especially also due to the fact that at higher temperature dehydroxylation occurs. Special attention has been devoted to the deammoniation and the formation of the hydrogen form. In dependence on the composition of the sample for the end of the deammoniation of various zeolites temperatures between 570 and 770 K are reported, so that the deammoniation and the dehydroxylation often cannot be resolved. The process can only be studied exactly by analyzing the released substances simultaneously to the DTA measurement.

Furthermore, in the high temperature effects influences of ultrastabilization can be detected. A summarizing article on this topic has been published by McDANIEL and MAHER (67).

For the two methods a large number of variations have been reported giving answer to questions different properties of zeolites and zeolite catalysts.

Finally, a special method shall be mentioned with which centers of different acid strength can be characterized. The acid form of the zeolite in question is exposed to ammonia at temperatures near 500 K, where no more physical adsorption occurs and the sorbed ammonia is then desorbed by programmed heating at higher temperatures near 850 K detecting the amount in a TGA experiment. The temperature at which an evolution of ammonia is observed, is then a good measure of the acidity of the respective site.

Although thermal analysis gives valuable information on a series of properties of zeolites and zeolite catalysts, it has not successfully provided a reproducible and standard method for measuring thermal properties because too many factors of the particular instrument and the conditions of the experiment influence the measured parameters. Although it is often quite useful for direct comparisons, it is difficult to compare results reported by different authors.

References

1. Zhdanov, S.P. and N.N. Samulevich, Nucleation and Crystal Growth of Zeolites in Crystallizing Alumosilicate Gels. Proceedings of the 5th International Conference on Zeolites, L.V.C. Rees, ed. (Heyden, London 1980) p. 75
2. Baird, T., Characterization of Catalysts by Electron Microscopy, in Specialist Periodical Report, Catalysis Vol. 5, G.C. Bond and G. Webb, eds. (The Royal Society of Chemistry Burlington House London W1V OBN 1982) p. 172
3. Menter, J.W., Advances in Physics 7 (1959) 299
4. Bursill, L.A., E.A. Lodge and J.M. Thomas, Nature (London) 286 (1980) 111
5. Thomas, J.M. and L.A. Bursill, Angewandte Chemie 19 (1980) 745
6. Schmidt, F., Characterization of Metal Zeolites, in Studies in Surface Science and Catalysis, Vol. 12, Metal Microstructures in Zeolites, P.A. Jacobs, N.I. Jaeger, P. Jíru and G. Schulz-Ekloff, eds. (Elsevier Publ. Comp. Amsterdam-Oxford-New York 1982) p. 191
7. Exner, D., N.I. Jaeger, K. Möller, R. Nowak, H. Schrübbers, G. Schulz-Ekloff and P. Ryder, Electron Microscopical Analysis of Monodispersed Ni and Pd in a Faujasite X Matrix, in Studies in Surface Science and Catalysis, Vol. 12, Metal Microstructures in Zeolites, P.A. Jacobs, N.I. Jaeger, P. Jíru and G. Schulz-Ekloff, eds. (Elsevier Publ. Comp., Amsterdam-Oxford-New York 1982) p. 205
8. Kacirek, H. and H. Lechert, Journal of Physical Chemistry 79 (1975) 649
9. v. Ballmoos, R. and W. Meier, Nature (London) 289 (1981) 782
10. Breck, D.W., Zeolite Molecular Sieves (John Wiley and Sons, New York 1974)
11. Argauer, R.J. and G.R. Landolt, 1972, U.S. Patent 3,702,886
12. Mobil Oil Corp., 1971, Dutch Patent 7,014,807
13. Kokotailo, G.T., S.L. Lawton, D.H. Olson and W.M. Meier, Nature (London) 272 (1978) 437
14. Kokotailo, G.T., P. Chu, S.L. Lawton and W.M. Meier, Nature (London) 275 (1978) 119
15. Luger, P., Modern X-Ray Analysis on Single Crystals (de Gruyter, Berlin-New York 1980)
16. Fischer, K.F., Zeolite Structure Refinement, in Molecular Sieves, Advances in Chemistry Series 121. W.M. Meier and J.B. Uytterhoeven, eds. (Amer. Chem. Soc. Washington D.C. 1973) p. 31
17. Smith, J.V., Origin and Structure of Zeolites, in J.A. Rabo, Zeolite Chemistry and Catalysis, ACS Monograph 171 (Amer. Chem. Soc. Washington D.C. 1976) p. 3
18. Meier, W.M., Symmetry Aspects of Zeolite Frameworks in Molecular Sieves, Advances in Chemistry Series 121, W.M. Meier and J.B. Uytterhoeven, eds. (Amer.Chem.Soc.Washington D.C.1973)p.39

19. Gard, J.A. and J.M. Tait, Structural Studies in Erionite and Offretite, in Molecular Sieve Zeolites I, Advances in Chemistry Series 101, R.F. Gould, ed. (Amer. Chem. Soc. Washington D.C. 1971) p. 230
20. Mortier, W.J., Compilation of Extra Framework Sites in Zeolites (Butterworth Scientific Ltd. Guilford 1982)
21. Flanigen, E.M., H.A. Szymanski and H. Khatami, Infrared Structural Studies of Zeolite Frameworks, in Molecular Sieve Zeolites I, Advances in Chemistry Series 101, R.F. Gould, ed. (Amer. Chem. Soc. Washington D.C. 1971)
22. Flanigen, E.M., Structural Analysis by Infrared Spectroscopy in J.A. Rabo, Zeolite Chemistry and Catalysis, ACS Monograph 171 (Amer. Chem. Soc. Washington D.C. 1976) p. 80
23. Wright, A.C., J.P. Rupert and W.T. Granquist, Amer. Mineralogist 53 (1968) 1293
24. de Kanter, J.J.P.M., I.E. Maxwell and R.J. Trotter, Journ.Chem. Soc. Chem. Commun. (1972) 733
25. Jacobs, P.A., E.G. Derouane and J. Weitkamp, Journ. Chem. Soc. Chem. Commun. (1981) 591
26. Coudurier, G., C. Naccache and J.C. Vedrine, Journ. Chem. Soc. Chem. Commun. (1982) 1413
27. Ward, J.W., Infrared Studies of Zeolite Surfaces and Surface Reactions, in J.A. Rabo, Zeolite Chemistry and Catalysis, ACS Monograph 171 (Amer. Chem. Soc. Washington 1976) p. 118
28. Kustov, L.M., V.Yu. Borovkov and V.B. Kazansky, Journal of Catalysis 72 (1981) 149
29. Jacobs, P.A. and H.K. Beyer, Journal of Physical Chemistry 83 (1979) 1174
30. Uytterhoeven, J.B., L.G. Christner and W.K. Hall, Journal of Physical Chemistry 69 (1965) 2117
31. Olson, D.H. and E. Dempsey, Journal of Catalysis 13 (1969) 221
32. Jacobs, P.A., Catalysis Reviews Sci. Eng. 24(3) (1982) 415
33. v. Ballmoos, R. and W.M. Meier, Nature (London) 289 (1981) 782
34. Sanderson, R.T., Chemical Bonds and Bond Energy, 2nd ed. (Academic Press, New York 1976)
35. Jacobs, P.A. and C.F. Heylen, Journal of Catalysis 34 (1974) 267
36. Lippmaa, E.T., M.A. Alla, T.J. Pehk and G. Engelhardt, Journ. Amer. Chem. Soc. 100 (1978) 1929
37. Lippmaa, E.T., M. Magi, A. Samoson, G. Engelhardt and A.R. Grimmier, Journ. Amer. Chem. Soc. 102 (1980) 4889
38. Barrer, R.M., Hydrothermal Chemistry of Zeolites (Academic Press, London-New York-Paris-San Diego-San Francisco-Sao Paulo-Sydney-Tokyo-Toronto 1982)
39. Melchior, M.T., D.E.W. Vaughan and A.J. Jacobson, private communication to appear in Journ. Amer. Chem. Soc.
40. Melchior, M.T. private communication, ACS Symposium Series Intrazeolite Chemistry to be published 1982
41. Klinowski, J., S. Ramdas, J.M. Thomas, C.A. Fyfe and J.S. Hartman, Journ.Chem.Soc., Faraday Transactions 78 (1982) 1025

42. Petch, H.E. and K.S. Pennington, Journal of Chemical Physics 36 (1962) 1216
43. Lechert, H., Ber. d. Bunsenges. physik. Chemie 77 (1973) 697
44. Lechert, H. and H. Henneke, Ber. d. Bunsenges. physik. Chemie 78 (1974) 347
45. Lechert, H., Catalysis Reviews Sci.-Eng. 14(1) (1976) 1
46. Lechert, H. and H. Henneke, ^{23}Na-Resonance in Zeolites of the A-Type and its Interpretation by Computer Simulation of the Measured Spectra, in Molecular Sieves -II ACS Symposium, Series 40, J.R. Katzer, ed. (Amer. Chem. Soc. Washington D.C. 1977) p. 53
47. Dempsey, E., Journal of Physical Chemistry 73 (1968) 3660
48. Basler, W.D., Zeitschr. f. Naturforschung 35a (1980) 645
49. Bosácek, V., D. Freude, T. Fröhlich, H. Pfeifer and H. Schmiedel, Journal of Colloid and Interface Science 85 (1982) 502
50. Basler, W.D. and W. Maiwald, Journal of Physical Chemistry 83 (1979) 2148
51. Basler, W.D., Journal of Physical Chemistry 81 (1977) 2102
52. Mestdagh, H.H., W.E.Stone and J.J.Fripiat, Journal of Physical Chemistry 76 (1972) 1226
53. Freude, D., W. Oehme, H. Schmiedel and B. Staudte, Journal of Catalysis 49 (1977) 123
54. Freude, D. and H. Pfeifer, NMR Studies Concerning Brønstedt Acidity of Zeolites, in Proceedings of the 5th International Conference on Zeolites, L.V.C. Rees, ed. (Heyden, London 1980) p. 732
55. Schmidt, F., W. Gunsser and J. Adolph, Formation of Iron Clusters in Zeolites with Different Supercage Sizes, in Molecular Sieves II, ASC Symposium Series 40, J.R. Katzer, ed. (Amer. Chem. Soc. Washington D.C. 1977) p.
56. Rees, L.V.C., Moessbauer Spectroscopic Studies of Ferrous Ion Exchange in Zeolite A, in Studies in Surface Science and Catalysis, Vol. 12, Metal Microstructures in Zeolites, P.A. Jacobs, N.I. Jaeger, P. Jíru and G. Schulz-Ekloff, eds. (Elsevier Publ. Comp., Amsterdam, Oxford, New York 1982) p. 33
57. Dickson, B.L. and L.V.C. Rees, Journ. Chem. Soc. Faraday Transactions I 70 (1974) 2038
58. Delgass, W.N., R.L. Garten and M. Boudart, Journal of Chemical Physics 50 (1969) 4603
59. Lunsford, J.H., Transition Metal Complexes in Zeolites, in Molecular Sieves II, ACS Symposium Series 40, J.R. Katzer, ed. (Amer. Chem. Soc., Washington D.C. 1977) p. 473
60. Kasai, P.H. and R.J. Bishop, Electron Spin Resonance Studies of Zeolites, in J.A. Rabo, Zeolite Chemistry and Catalysis, ACS Monograph 171 (Amer. Chem. Soc., Washington D.C. 1976) p. 350
61. Kellermann, R. and K. Klier, Intrazeolitic Transition-metal Ion Complexes, in Specialist Periodical Reports, Surface and Defect Properties of Solids - Vol. 4 (The Chemical Soc., Burlington House, London, W1V 0BN 1975) p. 1

62. Lunsford, J.H., The Chemistry of Ruthenium in Zeolites, in Studies in Surface Science and Catalysis, Vol. 12, Metal Microstructures in Zeolites, P.A. Jacobs, N.I. Jaeger, P. Jíru and G. Schulz-Ekloff, eds. (Elsevier Publ. Comp., Amsterdam, Oxford, New York 1982) p. 1
63. Huber, J.F.K. and G. Gerritse, Journal of Chromatography 58 (1971) 137
64. Dimitrov, Ch., Z. Popova, S. Mladenov, K.-H. Steinberg and H. Siegel, Zeitschrift für Chemie 21 (1981) 387
65. Dyer, A. and M.J. Wilson, Thermochimica Acta 5 (1973) 91
66. Berger, A.S., T.I. Samsonova and L.K. Jakovlev, Izvest. Akad. Nauk USSR, Ser. Chim. (1971) 2129
67. McDaniel, C.V. and P.K. Maher, Zeolite Stability and Ultrastable Zeolites, in J.A. Rabo, Zeolite Chemistry and Catalysis, ACS Monograph 171 (Amer. Chem. Soc. Washington D.C. 1976) p. 285

STRUCTURAL CHARACTERIZATION OF ZEOLITES BY HIGH RESOLUTION MAGIC-ANGLE-SPINNING SOLID STATE ^{29}Si-NMR SPECTROSCOPY

Zelimir Gabelica [a], Janos B.Nagy [a], Philippe Bodart [a], Guy Debras [a], Eric G. Derouane [a,c] and Peter A. Jacobs [b]

a-Facultés Universitaires de Namur
   Laboratoire de Catalyse, Rue de Bruxelles, 61
   B-5000 Namur, Belgium

b-Centrum voor Oppervlaktescheikunde en Colloïdale
   Scheikunde, Katholieke Universiteit Leuven,
   De Croylaan, 42 , B-3030 Leuven (Heverlee) Belgium

c-Present address :
   Mobil Technical Center, Central Research Division,
   P.O. Box 1025, Princeton, NJ 08540, U.S.A.

ABSTRACT
     The structure of various natural and synthetic zeolites was investigated by high resolution magic-angle-spinning (HRMAS) solid state ^{29}Si-NMR spectroscopy. The NMR analysis is based on the great sensitivity of the ^{29}Si-chemical shifts either on the chemical environment (i.e. the number of aluminium atoms in the second coordination shell of silicon), on the crystal symmetry (i.e. the number of crystallographically different sites) or on local strains in the crystal.

     Silicon-aluminium orderings could be described in zeolites such as faujasites, mordenite, ferrierite, ZSM-5, ZSM-8, ZSM-11, ZSM-39 and ZSM-48 by examining systematically either progressively dealuminated samples or synthetic zeolites with different Si/Al ratios. The Si/Al ratios were determined from the relative line intensities. It is concluded that aluminium atoms occupy preferentially specific sites in ZSM-5, ZSM-11, mordenite, ferrierite, ZSM-35 and probably also in ZSM-39. ZSM-8 and ZSM-48 zeolites are shown to contain combined 4- and 5- membered rings forming layer sequences similar to those characterizing pentasil structures.

     Finally, HRMAS ^{29}Si-NMR was also used to follow the different silicon-aluminium arrangements that occur during progressive transformations of amorphous gels into crystalline ZSM-5 phases.

194

1. INTRODUCTION

The high resolution magic-angle-spinning (HRMAS) solid state NMR has become a powerful tool for detailed structural investigations of solid zeolitic materials. Short-range silicon-aluminium framework orderings within crystallographically non-equivalent sites of various natural and synthetic Al-rich zeolites have been resolved (1-9). The position of the ^{29}Si-NMR resonance essentially depends on the number of tetrahedral Al atoms in the second coordination sphere of Si (10). An interval of about 5 ppm usually separates two neighbouring NMR resonances belonging to each Si(nAl) configuration. The assignment of the lines to the various Si-Al orderings is usually achieved by following their intensity variations for different dealuminated materials. Such a procedure was successfully applied in the case of faujasite (11-13), mordenite (9) and pentasil materials (14). The percentage of each Si(nAl) unit in the structure can be evaluated by measuring the corresponding line intensity.

Quantitative determination of Si/Al ratios in the zeolite framework can also be derived from NMR data (4,6).

Recent studies have shown that the chemical shifts which characterize the ^{29}Si-NMR lines belonging to defined Si(nAl) configurations were also influenced by the actual geometry of the T-O-T linkages (T = Si or Al) or by longer range structural features such as the (T-O)p ring sizes (15,16). This was turned to account advantageously for the effect of the actual ring structure and strains that occur in high siliceous zeolites (ZSM-5, ZSM-11, silicalite, ...) (14,17,18), where Si(OAl) configurations are predominant.

In addition, ^{29}Si-NMR lines belonging to silanol groups ($\equiv$Si-O-H) in dealuminated faujasites (12) and in ZSM-5 materials (19) were unambiguously identified, using the cross-polarization technique.

The purpose of the present paper is first to show how the informations obtained from the HRMAS solid state ^{29}Si-NMR can be used in resolving framework distribution of Si-Al atoms in various zeolitic structures. Secondly, this technique is also used to follow the progressive structural changes of the Si-Al orderings during the transformation of amorphous gels into crystalline ZSM-5 phases.

2 EXPERIMENTAL

2.1 Materials

Na-X and Na-Y zeolites were synthesized by Dr D. Barthomeuf (Lyon, France). Mordenite, a Na-zeolon sample, was supplied by the Norton Company. ZSM-5 and ZSM-11 were synthesized according to published methods (20-23), using different Si/Al ratios. The highly pure ZSM-5 zeolite, parent material for various chemical modifications, was obtained using diluted conditions, as described earlier (24). Progressive crystallization of ZSM-5 zeolites was conducted following two different procedures as reported previously (21,25,26). From each procedure, intermediate phases with increasing degrees of crystallinity were isolated and analyzed. ZSM-5 was obtained from both patent (27) and own-modified (22) recepies. ZSM-39 was synthesized as described by Dwyer et al.(28). The other zeolites : ferrierite, ZSM-35 and ZSM-48 were obtained by own syntheses,according to procedures published in the current or patent literature (29).

The chemical composition of each sample was determined by Proton Induced γ-Ray Emission (PIGE) (35), by Energy Dispersive X-Ray Analysis (EDX) (23) or by Atomic Absorption Spectroscopy (AAS). Water and organic contents of the pentasil materials were measured by thermal methods (21-23,31,32). The structural identity of all the zeolites was ascertained by the classical powder X-ray diffraction method.

2.2 ^{29}Si-NMR measurements

The HRMAS ^{29}Si-NMR spectra were obtained at room temperature on a Bruker CXP-200 spectrometer operating in the Fourier Transform mode, using a "one cycle" type measurement. An r.f.-field of 49.3 Oe was used for the $\pi/2$ pulses of ^{29}Si (ν_o = 39.7 MHz). Magic-angle-spinning was at a rate of 3.1 kHz, using Delrin conical rotors. Time intervals between pulse sequences were 3.0 s. 2,000 to 20,000 free induction decays were accumulated per sample. Chemical shifts (δ), in ppm, were measured with respect to tetramethylsilane (TMS), used as external reference.

3 RESULTS AND DISCUSSION

3.1 Zeolites with crystallographically equivalent tetrahedral sites : faujasite

The structure of faujasite is built up of corner sharing TO_4 tetrahedra (T = Si or Al). 24 tetrahedra are joined to form a cuboctahedron or sodalite cage. Cages are stroked tetrahedrally

to form a cubic diamond lattice. All the T sites are crystallographically (but not chemically) equivalent [33]. Five types of
groupings can be distinguished with Si linked through oxygen to
0,1,2,3 or 4 Al atoms. In the case of faujasite, there is no
overlap in the ^{29}Si chemical shifts for the above groupings and a
structural model could be developped that successfully explains the
Si,Al framework distribution, in relation with the NMR data,over a
wide range of compositions.

Provided the Lowenstein rule is obeyed in the zeolite, i.e.
no Al-O-Al linkages are present, the Si/Al ratio of the faujasite
lattice can be estimated from the relative signal intensities of
^{29}Si [4] :

$$(Si/Al)_{NMR} = \frac{\sum_{n=0}^{4} I_{Si(nAl)}}{\sum_{n=0}^{4} 0.25 \, n \, I \, Si(nAl)} \qquad [1]$$

It can be used as an excellent quantitative method of measuring
zeolite composition independently of the elemental chemical analyses. A non-coincidence between NMR Si/Al ratios and those obtained by chemical methods indicates that non-framework Al or Si
are present (34).

Detailed structural models and Si,Al orderings in faujasite
have been developped and discussed in many recent papers (1-8,
10-13,16,35-37). Figure 1 shows examples of HRMAS spectra of
three faujasites, a NaX zeolite with Si/Al = 1.25 and two NaY zeolites with Si/Al = 1.82 and 2.4. The intensities of the NMR lines
characterizing the five Si(nAl) (n = 0,1,2,3,4) configurations are
computed in Table 1. Good agreement is obtained between the
(Si/Al) ratios as calculated using equation [1] and those measured
by the EDX technique.

TABLE 1
^{29}Si-NMR characteristics of faujasites

Sample	Si/Al (EDX)	Relative populations of various Si(nAl) configurations,in % (δ in ppm from TMS)					Si/Al (NMR)
		Si(4Al) (-84.6)	Si(3Al) (-89.1)	Si(2Al) (-94.5)	Si(1Al) (-99.5)	Si(0Al) (-103.4)	
Na-X	1.25	53.8	26.5	12.2	5.0	2.5	1.23
Na-Y	1.82	15.9	31.4	32.7	15.5	4.5	1.7
Na-Y	2.4	2.9	15.0	39.4	37.3	5.4	2.3

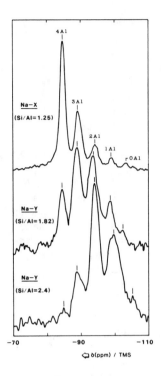

Figure 1.   HRMAS ^{29}Si-NMR spectra of three faujasite zeolites.

3.2   Zeolites with crystallographically non-equivalent tetrahedral sites

In zeolites whose structures exhibit different framework topologies, some overlap in the ^{29}Si-NMR lines characterizing various Si(nAl) configurations can occur because the chemical shifts of an NMR resonance belonging to a given configuration does also depend on the actual symmetry of that Si(nAl) site and/or on the size of the (T-O)p ring to which that configuration belongs. Typical examples are encountered in mordenite (9,37), pentasil zeolites (15,16,19) and ZSM-39 (38) zeolite. In such cases, the unambiguous attribution of an NMR line within the overlap region becomes highly uncertain. Therefore, various experimental procedures are developped to derive the actual Si, Al orderings in zeolitic frameworks from ^{29}Si-NMR spectra : progressive dealumination of the sample or synthesis of the zeolite with various Si/Al contents.

3.2.1. Mordenite group.   A common feature of the zeolite frameworks belonging to the mordenite group is the existence of 6-ring sheets. In mordenite, these are linked to each other through single

4-membered rings and in ferrierite,   through single 6-membered rings.

Single crystal X-ray refinement of mordenite (39) has suggested that the Al atoms must be located on the 4-membered rings.
Using that proposal, the theoretical distribution of the Al atoms
within the mordenite structure can be calculated for various Si/Al
ratios. The so obtained total number of Si(nAl) configurations
distributed within one unit cell can be computed and the corresponding theoretical NMR line intensities compared to the experimental
data.

The HRMAS ^{29}Si-NMR spectra of Na-mordenite (Si/Al = 5.5) shows
three lines at -95, -105 and -110 ppm (Fig.2), which are attributed
respectively to tetrahedral arrangements of Si(2Al) , Si(1Al)   and
Si(0Al).   The parallel decrease of the first two lines and the
increase of that at -110 ppm,with progressive dealumination of the
parent sample (Fig. 2), corresponds to the theoretical prediction
and confirms the assignment. Quantitative data are given in
Table 2. A more complete NMR study of mordenite and its dealuminated forms is presented elsewhere (9).

TABLE 2

HRMAS ^{29}Si-NMR characterization of various mordenite samples

Sample	Si/Al (AAS)	Relative line intensities (%) [a]			Si/Al (NMR)
		Configuration: Si(2Al) δ/ppm: -95	Si(1Al) -105	Si(0Al) -110	
Na-Mor	5.5	13.2	44.7	42.1	5.4
H-Mor$_1$ [b]	20.5	0	29.8	70.2	21.6
H-Mor$_2$ [c]	31.2	0	22.4	77.6	30.7

(a) Normalized to 100 for Na-Mor
(b) Na-Mor, leached with 4M HNO$_3$ at 90°C for 24h
(c) Na-Mor, leached with 14M HNO$_3$ at 90°C for 24h.

The HRMAS ^{29}Si NMR spectrum of ferrierite and of its Si-richer
analogue H-ZSM-35 shows three lines located at -101, -108 and
-113 ppm. They are attributed to the Si(2Al) , Si(1Al)   and
Si(0Al) configurations respectively.
The ferrierite unit cell contains two types of tetrahedrally coordinated T atoms. The relative intensities of the three lines,
as experimentally measured on both ferrierite and H-ZSM-35, correspond well to those evaluated theoretically for similar Si/Al ratios. For the latter, it was assumed that Al atoms were specifically located in 6-membered rings linking one dimensional sheets

in the structure (sites T_1) (40), and that a 6-membered ring with one single Al atom is more stable than one containing two Al atoms. A similar theoretical computation, assuming the Al atoms preferentially located in 6-membered rings of the sheets, at a maximum distance from each other (sites T_2), does not fit the experimental data (Table 3).

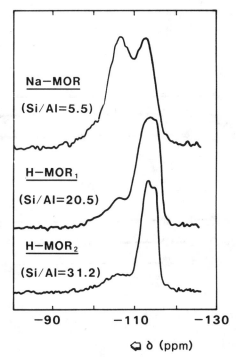

Figure 2. HRMAS ^{29}Si-NMR of mordenites : effect of dealumination.

TABLE 3

HRMAS ^{29}Si-NMR of ferrierite and H-ZSM-35 zeolite : comparison of theoretical and experimental line intensities

Sample	Si/Al	Relative line intensities (%)		
		configuration : Si(2Al) δ/ppm : -101	Si(1Al) -108	Si(0Al) -113
fictitious (Al on T_2)	8.2	0	48.6	51.4
fictitious (Al on T_1)	8.2	5.9	36.8	57.3
Ferrierite	8.2	8	39	53
fictitious (Al on T_2)	10.4	0	39	61
fictitious (Al on T_1)	10.4	3.7	31.7	64.6
H-ZSM-35	10.4	6	34	60

3.2.2. Pentasil zeolites. The HRMAS ^{29}Si-NMR spectra of H-ZSM-5 and
H-ZSM-11 zeolites having different Si/Al ratios (from 30 to 1000)
have been studied in detail (14,16). At low resolution, essentially three resonances were observed at -105, -113 and -115 ppm.
The decreasing intensity of the -115 ppm line with the increasing
Si/Al ratio is directly linked to the parallel intensity increase
of the -113 ppm line, while the intensity of the -115 ppm line
remains constant in both H-ZSM-5 and H-ZSM-11 (Fig. 3).

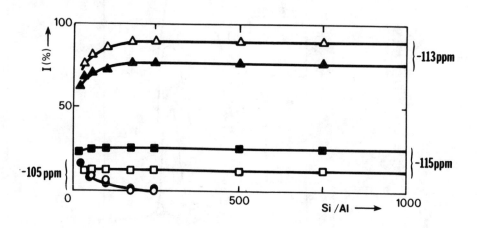

Figure 3. Variation of the ^{29}Si-NMR line intensities, as a function
of Si/Al ratios in H-ZSM-5 (white spots) and in H-ZSM-11
(black spots) zeolites.

These variations are easily explained if the -105 and -113 ppm
lines are attributed to Si (1Al) and Si(0Al) configurations respectively. The Si/Al ratios computed using equation [1] , I_{total}/
/0.25 I (-105) (every Al is surrounded by 4 Si), match exactly the
experimental values. In addition, the intensity ratio I (-115)
ZSM-11/I (-115), ZSM-5, equal to 2, is directly proportional to
the amount of 4-membered rings present is one unit cell of each
zeolite: four in ZSM-5 and eight in ZSM-11. The -115 ppm line can
therefore be attributed to Si(0Al) configurations located in 4-
-membered rings. This also implies that Al atoms are exclusively
located in the 5-membered rings of both zeolites and thus not statistically distributed throughout the pentasil lattice.
In the ^{29}Si-NMR spectrum of highly siliceous H-ZSM-5 (Si/Al=1000),
up to 8 line components can be distinguished (Fig.4).

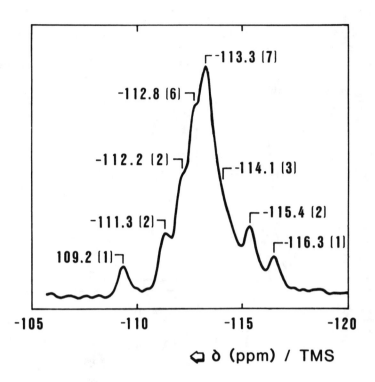

Figure 4.   High resolution ^{29}Si-NMR spectrum of H–ZSM-5
        (Si/Al = 1000).

        This multiplicity arises from crystallographically non equi-
valent Si(OAl)   arrangements. Using the intensities of the well
resolved signals located at -109.2 ppm and at -116.3 ppm as a
base unit of one, the total intensity is found to be approximately 24, in agreement with previous findings (16). The number of
Si atoms in the 24 non-equivalent sites in the repeat-unit of the
ZSM-5 structure is given in parentheses in Fig. 4. The unambiguous assignment of all these lines still needs further investigations. Nevertheless, the following inferences can be made :
(i)   three magnetically different Si atoms in the 4-membered rings
        contribute to the three line components observed in the
        -115 ppm region (14). This reveals that strain must exist
        in these rings, in both ZSM-5 and ZSM-11 ;
(ii)   the substitution of Si by Al only takes place in the 5-mem-
        bered rings, where the energy is probably lower than in the
        strained 4-membered rings.

        The chemical shifts and relative line intensities in ^{29}Si-
NMR spectra of samples which by XRD were found to be ZSM-8 and
ZSM-48,appear to be very similar to those of ZSM-5 and ZSM-11
(Fig. 5). This suggests that both zeolites could also belong to

the pentasil family. In particular, the resonance line at -115 ppm shows that ZSM-8 and ZSM-48 frameworks also contain 4-membered rings. For ZSM-8, the relative intensity of this line falls between those measured for ZSM-5 and ZSM-11. This suggests that ZSM-8 could be a mixture of ZSM-5 and ZSM-11 or an intergrowth of these two stuctures as suggested by Kokotailo (41).

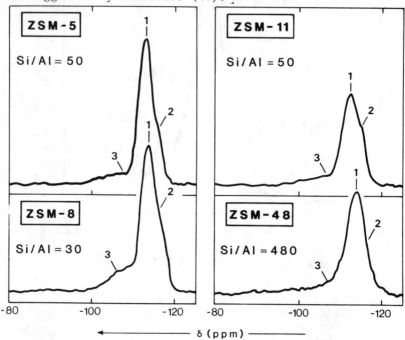

Figure 5. HRMAS ^{29}Si-NMR spectra of four pentasil zeolites.

3.2.3. ZSM-39 zeolite.

This Si-rich zeolite belongs to the "Clathrate group". Its framework consists of an arrangement of TO_4 12-and 16-hedra in which three types of T atoms (T_1, T_2 and T_3) were identified (42). Their nature and their distribution in the unit cell are detailed in Table 4. The HRMAS ^{29}Si-NMR spectrum of ZSM-39 shows three well resolved resonances at δ = -109,-115 and -120 ppm (Fig.6).

Although the number of relevant Si(nAl) configurations is restricted to Si(JAl) and Si(OAl) due to the high Si/Al ratios, the intensity ratio of the three NMR lines cannot be explained without computing all the possible Al potential sitings in the lattice. This has been envisaged in detail elsewhere (38).

TABLE 4

Repartition of T atoms in the unit cell of ZSM-39

T atom	Ring structure	Number/u.c.	Number of neighbouring T atoms of type		
			T_1	T_2	T_3
T_1	shared between ⬠	8	0	4	0
T_2	in ⬠	32	1	0	3
T_3	shared between ⬠ and ⬡	96	0	1	3

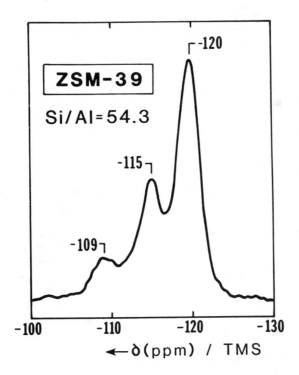

Figure 6.   HRMAS ^{29}Si-NMR spectrum of ZSM-39 zeolite.

Table 5 compares the experimental and the various theoretical line intensities obtained assuming either a specific location of the Al atoms on T_1, T_2 or T_3 sites or their statistical distribution between all the T sites in the lattice.

204

TABLE 5

Comparison between the relative theoretical intensities calculated
from various Al sitings in the lattice and the experimental ones
measured for H-ZSM-39 (Si/Al = 54.3)

δ (ppm)	Relative theoretical intensities (%) assuming Al sited				Exper. values (NMR)
	in T_1	in T_2	in T_3	statistically	
-104	0	1.8	0	0.5	0
-109	11.5	4.2	7.8	7.1	9
-115	16.6	27.6	27.7	27.0	29
-120	71.9	66.4	64.5	65.4	62

The best agreement between experimental and theoretical values
is obtained when Al cations are specifically located on sites T_3
i.e. in the 6-membered rings. However, because of the high proportion of the T_3 sites, a statistical distribution also gives a close
agreement.
^{29}Si-NMR spectra of ZSM-39 samples with a lower Si/Al ratio are
needed to differenciate unequivocally between the two distributions.

3.3 Crystallization of ZSM-5 from amorphous gels

Informations on structural evolution of various constituent
species formed within gel mixtures or solutions, precursors to crystallization of mordenite (43) or ZSM-5 (25,26), have been obtained
recently using HRMAS multinuclear (^{13}C, ^{27}Al, ^{29}Si) NMR.

The hydrothermal synthesis of ZSM-5 can involve at least two
different mechanisms whose various aspects have been developped
in detail elsewhere (21,23). Procedure A yields large single crystals of ZSM-5 which grow slowly in an Al-rich gel through a liquid
phase ion transportation mechanism. Procedure B gives rapidly small
ZSM-5 polycrystalline aggregates which appear very early within the
hydrogel, where they remain stabilized as very small sized "X-ray
amorphous" zeolites (21,25,44).

Several intermediate phases formed during the crystallization
of ZSM-5 conducted using both procedures A and B were isolated and
investigated by HRMAS ^{29}Si-NMR.

The spectra of gels exhibit very broad resonances. Their maxima
are located between -100 and -111 ppm, depending on their actual
Si/Al ratio. The lines become narrower and are shifted towards
higher fields, as ZSM-5 is progressively formed within the gels.
This evolution is shown in Fig. 7 for 3 phases from synthesis A.

The lines appearing below -111 ppm must essentially belong to Si atoms which have only Si as neighbours, while those located above - 100 ppm should reflect various Si(nAl) configurations, where n > 0. In that case, Si atoms are surrounded either by Al, randomly arranged in amorphous phase, or by silanol groups which should appear in that region in amorphous or crystalline ZSM-5 (19). The variation of the corresponding relative intensities, respectively noted by I (δ < -111) and I (δ > -100), as well as that of the total ^{29}Si-NMR intensity, as a function of synthesis time, leads to interesting conclusions.

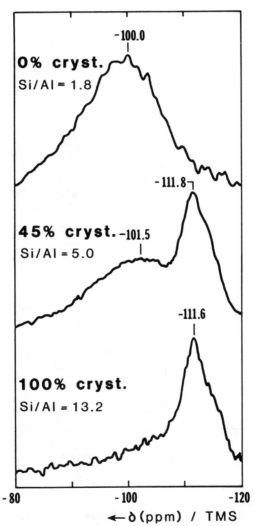

Figure 7. HRMAS ^{29}Si-NMR spectra of some intermediate phases obtained using procedure A.

3.3.1. ZSM-5 synthesized according to procedure A.

$I(\delta < -111)$ and $I(\delta > -100)$ respectively increase and decrease as the crystallization proceeds, confirming that the number of Si(nAl) (n > 0) configurations decreases in a parallel way with the decreasing global Al concentration in the solid phases (Si/Al progressively increases from 1.8 (0 % crystallinity) to 13.2 (100 % crystallinity)), while more Si(OAl) configurations appear ordered in a crystalline zeolite phase. (Fig. 8,A)

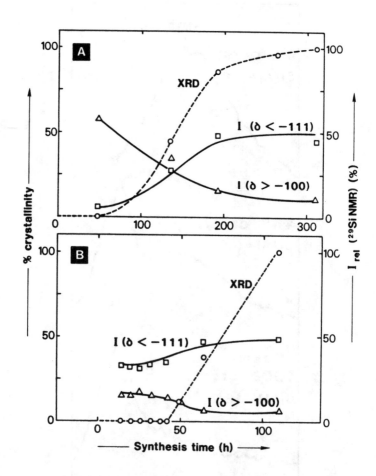

Figure 8. Variation of the relative ^{29}Si-NMR line intensities of the peaks located below -111 ppm ($\delta < -111$) and above -100 ppm ($\delta > -100$) and of the % crystallinity (XRD) for various A and B-type intermediate phases, as a function of crystallization time.

The progressive Al-depletion of the solid phases is also characterized by the sygmoidal increase of the total intensity, which follows
the XRD crystallinity (Fig. 9A).

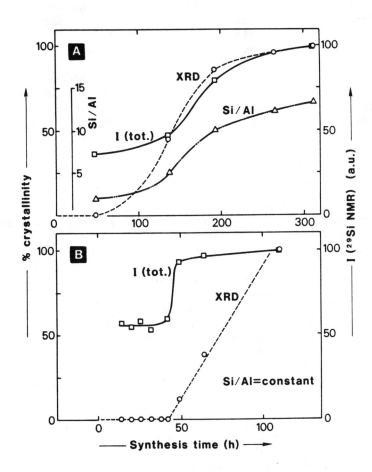

Figure 9 Variation of the total ^{29}Si-NMR intensities and of the %
of crystallinity of A and B-type intermediate phases, as
a function of synthesis time.

3.3.2. ZSM-5 synthesized according to procedure B. Oppositely to
the synthesis A, in the beginning of the process B, I (δ > -100) is
low, while I (δ < -111) is already high (Fig. 8B). Both intensities
show little variations as the crystallization proceeds. This only
reflects the low amount of Si(nAl) configurations within the Si-rich

(gel + zeolite) phases in which Si/Al ≃ 35 and remains constant
during the synthesis course.
By contrast, I(total) remains constant only during the time interval in which the X-ray amporphous ZSM-5 crystallites are detected
(Fig. 9B). It shows, however, a sharp increase when the ZSM-5 crystallites begin to grow. Because Si/Al remains constant during the
crystallization, the ordering of Si atoms within a ZSM-5 lattice is
the only way to explain that increase. This latter is probably due
to shorter ^{29}Si-NMR relaxation times for Si located within an ordered lattice than for Si surrounded by a random environment in the
gel. This phenomenon is presently being investigated by determining
the T_1 relaxation times in the gel and crystalline phases. This increase however appears to be a remarkable qualitative illustration
of the ordering process which occurs during the growth of zeolite
crystals.

4. CONCLUSIONS

HRMAS ^{29}Si-NMR porved to be a powerful technique to determine
the distribution of the Al atoms in various zeolitic frameworks.

The chemical shifts on the ^{29}Si-NMR lines essentially depend
on the chemical environment of the Si atoms. Moreover, for a given
Si(nAl) configuration, they are also sensitive to the actual symmetry of the Si-O-T linkages as well as to local framework strains.
In such cases, the silicon-aluminium ordering can be resolved by
following the relative intensities of each line component characterizing a given configuration, as a function of the Si/Al ratio in
the zeolite. It is concluded that most of the investigated materials
(zeolites belonging to the mordenite or pentasil family and ZSM-39)
are characterized by preferential distribution of Al atoms on specific sites.

For zeolites whose structure and Si-Al orderings are known, the
measure of the intensity of each line component leads to a straightforward determination of the Si/Al ratios. However, for intermediate
alumino-silicate phases obtained during the synthesis, the line intensities were found to be proportional to the degree of ordering
of the phases. This property was therefore exploited to follow
qualitatively the progressive crystallization and growth of (ordered)
zeolitic frameworks from amorphous (random)alumino-silicate gels.

5. ACKNOWLEDGEMENTS

The authors wish to thank Mr. G. Daelen for his skilful help in
taking the NMR spectra. P.A. Jacobs acknowledges a research position
as "Senior Research Associate" from NFWO-FNRS (Belgium) and P. Bodart
thanks IRSIA (Belgium) for financial support.

6 REFERENCES

1. J. Klinowski, J.M. Thomas, C.A. Fyfe and J.S. Hartman, J. Phys. Chem. 85, 2590 (1981).
2. E. Lippmaa, M. Mägi, A. Samoson, M. Tarmak and G. Engelhardt, J. Am. Chem. Soc. 103, 4992 (1981).
3. G. Engelhardt, E. Lippmaa and M. Mägi, J. Chem. Soc., Chem. Commun. 1981, 712.
4. G. Engelhardt, U. Lohse, E. Lippmaa, M. Tarmak and M. Mägi, Z. Anorg. Allg. Chem. 482, 49 (1981).
5. S. Ramdas, J.M. Thomas, J. Klinowski, C.A. Fyfe and J.S. Hartman, Nature, 292, 228 (1981).
6. J. Klinowski, S. Ramdas, J.M. Thomas, C.A. Fyfe and J.S. Hartman, J. Chem. Soc. Faraday Trans. II, 78, 1025 (1982).
7. J.M. Thomas, C.A. Fyfe, J. Klinowski and G.C. Gobbi, J. Phys. Chem. 86, 3061 (1982).
8. M.T. Melchior, D.E.W. Vaughan, R.H. Jarman and A.J. Jacobson, Nature, 298, 455 (1982).
9. G. Debras, J. B.Nagy, Z. Gabelica, P. Bodart and P.A. Jacobs, Chem. Lett. 1983, 199.
10. E. Lippmaa, M. Mägi, A. Samoson, G. Engelhardt and A.-R. Grimmer, J. Amer. Chem. Soc. 102, 4889 (1980).
11. J. Klinowski, J.M. Thomas, M. Audier, S. Vasudevan, C.A. Fyfe, and J.S. Hartman, J. Chem. Soc. Chem. Comun. 1981, 570.
12. G. Engelhardt, U. Lohse, A. Samoson, M. Mägi, M. Tarmak and E. Lippmaa, Zeolites, 2, 59 (1982).
13. I.E. Maxwell, W.A. van Erp, G.R. Hays, T. Couperus, R. Huis and A.D.M. Clague, J. Chem. Soc. Chem. Commun. 1982, 523.
14. J. B.Nagy, Z. Gabelica, E.G. Derouane and P.A. Jacobs, Chem. Lett. 1982, 2003.
15. J. B.Nagy, J.P. Gilson and E.G. Derouane, J. Chem. Soc. Chem. Commun. 1981, 1129.
16. C.A. Fyfe, G.C. Gobbi, J. Klinowski, J.M. Thomas and S. Ramdas Nature, 296, 530 (1982).
17. P.A. Jacobs, M. Tielen, J. B.Nagy, G. Debras, E.G. Derouane and Z. Gabelica, in "Proc. Sixth Intern. Conf. Zeolites", Reno, 1983 (submitted).
18. J. B.Nagy, Z. Gabelica, G. Debras, P. Bodart, E.G. Derouane and P.A. Jacobs, J. Molec. Catal., in press.
19. J. B.Nagy, Z. Gabelica.and E.G. Derouane, Chem. Lett. 1982,1105.
20. P.A. Jacobs, J.A. Martens, J. Weitkamp,and H.K. Beyer, Faraday Discuss. Chem. Soc. 72, 353 (1981).
21. E.G. Derouane, S. Detremmerie, Z. Gabelica and N .Blom, Appl. Catal. 1, 201 (1981).
22. Z. Gabelica, E.G. Derouane and N. Blom, Appl. Catal. 5, 109 (1983).
23. Z. Gabelica, N. Blom and E.G. Derouane, Appl. Catal. 5, 227 (1983).
24. P.A. Jacobs,and R. Von Ballmoos, J. Phys. Chem. 86, 3050 (1982).
25. Z. Gabelica, J. B.Nagy and G. Debras, J. Catal., submitted.

26. Z. Gabelica, J. B.Nagy, G. Debras,and E.G. Derouane, in "Proc. Sixth Intern. Conf. Zeolites" Reno, 1983 (submitted).
27. N.Y. Chen, U.S. Pat. 3,700,585 (1972).
28. F.G. Dwyer and E.E. Jenkins, U.S. Pat. 4,287,166 (1981).
29. R.M. Barrer,"Hydrothermal Chemistry of Zeolites", Academic Press, London and New York, 1982.
30. G. Debras, E.G. Derouane, J.P. Gilson, Z. Gabelica and G. Demortier, Zeolites, 3, 37 (1983).
31. Z. Gabelica, J.P. Gilson, G. Debras and E.G. Derouane, in "Thermal Analysis, Proc. Seventh Intern. Conf. Thermal Anal.", (B. Miller, ed.) vol. II, Wiley-Heyden, New York, 1982; pp. 1203.
32. J. B.Nagy, Z. Gabelica and E.G. Derouane, Zeolites, 3, 43 (1983).
33. W.H. Meier and H.J. Moeck, J. Sol. State Chem. 27, 349 (1979).
34. J. Klinowski, J.M. Thomas, C.A. Fyfe and G.C. Gobbi, Nature, 296, 533 (1982).
35. M.T. Melchior, D.E.W. Vaughan and A.J. Jacobson, J. Amer. Chem. Soc. 104, 4859 (1982).
36. M. Mägi, A. Samoson, M. Tarmak, G. Engelhardt and E. Lippmaa, Dokl. Akad. Nauk SSSR (Engl. Transl) 261, 1159 (1981).
37. J. Klinowski, J.M. Thomas, M.W. Anderson, C.A. Fyfe and G.C. Gobbi, Zeolites, 3, 5 (1983).
38. P. Bodart, J. B.Nagy, G. Debras, Z. Gabelica, E.G. Derouane and P.A. Jacobs, Submitted for publication.
39. V. Gramlich, Ph.D. Diss. ETH n°4633, Zürich (1971).
40. R.M. Barrer "Zeolites and Clay Minerals as Solvents and Molecular Sieves" Academic Press, London and New York, 1978.
41. G.T. Kokotailo, Eur. Pat. 18,090 (1980) and U.S. Pat. 4,289,607 (1981).
42. J.L. Schlenker, F.G. Dwyer, E.E. Jenkins, W.J. Rohrbaugh, G.T. Kokotailo and W.M. Meier, Nature, 294, 340 (1981).
43. P. Bodart, Z. Gabelica, J. B.Nagy and G. Debras, in "Zeolites Science and Technology", Lisbon, 1983,(this meeting).
44. P.A. Jacobs, E.G. Derouane and J. Weitkamp, J. Chem. Soc. Chem. Commun. 1981, 591.

MULTINUCLEAR SOLID-STATE NMR STUDY OF MORDENITE CRYSTALLIZATION

Philippe Bodart, Zelimir Gabelica, János B.Nagy and Guy Debras [a]

Facultés Universitaires de Namur
Laboratoire de Catalyse, Rue de Bruxelles, 61
B-5000 Namur, Belgium

a- Present address :
   Labofina S.A., Zoning Industriel,
   B-6520 Feluy, Belgium.

ABSTRACT

Solid intermediate phases obtained from amorphous aluminosilicate gels during mordenite crystallization have been characterized by multinuclear solid-state NMR spectroscopy. The progressive reorganization of the amorphous gel into the crystalline ordered zeolitic phase has been detected by high resolution magic-angle-spinning (HRMAS) solid state ^{29}Si-NMR. Broad band solid state ^{27}Al and ^{23}Na-NMR were used to follow the incorporation of Al- and Na-atoms into the mordenite lattice and channels. An evidence of aluminium gradient in the crystallites is obtained from the comparison between surface and bulk analysis methods.

1 INTRODUCTION

During the last thirty years, the synthesis of zeolites has been extensively developed to provide new or improved materials for catalysis and/or molecular sieving (1,2). However, the synthesis processes, i.e. nucleation and crystal growth from amorphous aqueous aluminosilicate gels are still poorly understood (3). Recently, multinuclear solid state NMR has proven to be a powerful tool in the investigation of zeolite structures (4-14). This method was also successfully used to characterize intermediate solid phases obtained during crystallization of ZSM-5 materials (13-15).

Since the first synthesis of the silica-rich zeolite mordenite by Barrer (16), considerable literature data have been published on its preparation (17-23), the mechanism of this crystallization process remains however quasi unknown. The aim of this paper is to characterize by solid state multinuclear NMR the structural rearrangements that occur during the progressive transformation of amorphous gels into crystalline mordenite.

2 EXPERIMENTAL

2.1 Mordenite synthesis

The synthesis of mordenite was based on published methods of preparation (19,22). An aqueous aluminosilicate gel having the molar composition $2 \cdot 4$ $Na_2O \cdot Al_2O_3 \cdot 11 \cdot 1$ $SiO_2 \cdot 220$ H_2O was prepared from Na-silicate (Merk, art. 5621), silicagel (Davison grade 950) and Na-aluminate (Riedel de Häen, art. 13404). It was divided into several portions and sealed in identical 20 ml pyrex tubes. The latter were heated at 165°C under autogeneous pressure for given periods of time and progressively removed from the oven, in order to isolate materials with increasing degrees of crystallinity. After cooling, each sample (gel + zeolite) was filtered, washed with cold water and dried at 120°C overnight, before characterization. The percentage of mordenite in the as-synthesized phases was evaluated by the conventional X-ray diffraction technique (XRD) (Philips PW 1349/30 diffractometer, Cu K_α radiation), using reflections occuring at 2θ angles of 22.3, 25.6, 26.3 and 27.9 degrees. 100% crystallinity was assigned to the most crystalline phase of the series, which proved to be 115% crystalline with respect to a commercial H-Zeolon from the Norton Company, generally used as standard reference.

2.2 NMR spectra

Solid state NMR spectra were obtained at room temperature, using a Bruker CXP-200 spectrometer operating in the Fourier transform mode. An r.f. field of 49.3 Oe was used for the $\pi/2$ pulses of ^{29}Si (39.7 MHz). The Delrin conical rotor was spun at a rate of 3.1 kHz. Time intervals between pulse sequences were 3.0 s and over 15000 free induction decays were accumulated per sample. Chemical shifts (δ in ppm) were measured from tetramethylsilane (TMS). ^{27}Al high power NMR spectra were recorded at 52.1 MHz. Chemical shifts were measured with respect to $Al(H_2O)_6^{3+}$, used as external reference. Time intervals of 0.1 s were used between pulse sequences and 5000 free induction decays were accumulated per sample. ^{23}Na-NMR spectra were recorded at 52.9 MHz, in static conditions. Waiting times between pulse sequences were 0.2 s and 2000 free induction decays were accumulated per sample.

3   RESULTS AND DISCUSSION

3.1   XRD crystallinity of the intermediate phases

Figure 1 shows the variation of crystallinity for some intermediate solid phases as a function of synthesis time. A classical
sigmoïd curve is obtained as usually observed for various zeolites
crystallizing from non-seeded systems (21-23).

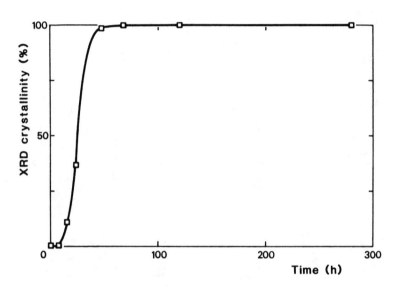

Figure 1.   Variation of relative crystallinity of the intermediate
solid phases   as a function of synthesis time.

3.2   Solid state HRMAS ^{29}Si-NMR study

The solid state HRMAS ^{29}Si-NMR spectrum of 100% crystalline
mordenite is reported in fig. 2. The three resonance lines that
appear at δ = -110, -105 and -96 ppm, correspond to Si-atoms having respectively 0, 1 and 2 Al-atoms in their second coordination
shells (11).

214

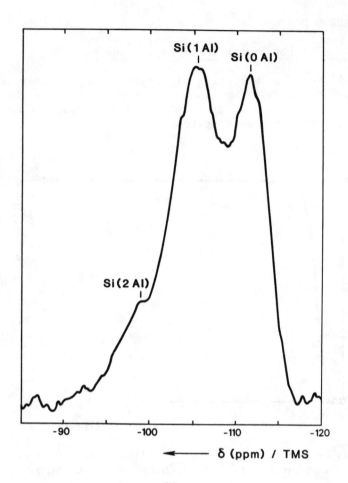

Figure 2. Solid state HRMAS ^{29}Si-NMR spectrum of mordenite.

Figure 3 shows HRMAS ^{29}Si-NMR spectra of intermediate phases obtained at various crystallization times. The first spectrum (A) consists of a broad resonance line centered at $\delta = -95$ ppm, characterizing Al-rich amorphous gel phases. With increasing crystallization times,the amorphous phase becomes richer in silicon and as a result, its still broad resonance line shifts to higher fields (Table 1) (fig. 3, A,B,C).

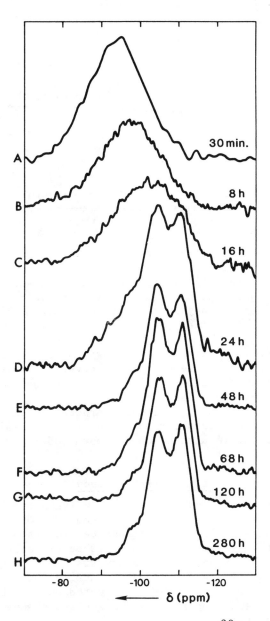

Figure 3. Evolution of the solid state HRMAS ^{29}Si NMR spectrum
of intermediate phases formed during mordenite crystallization.

216

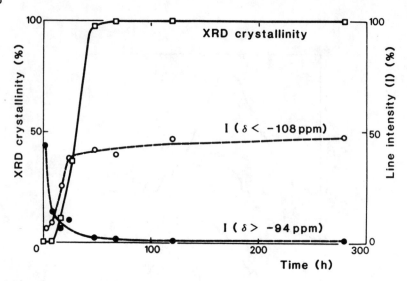

Figure 4.   Variation of ^{29}Si-NMR line intensities at $\delta > -94$ ppm
(amorphous phase) and $\delta < -108$ ppm (mordenite Si(0 Al)
resonance line) as a function of synthesis time (NMR
line intensities relative to the total intensity).

The line corresponding to the amorphous phase decreases rapidly and disappears at the end of the crystallization step. Oppositely, the mordenite Si(0 Al) line intensity increases with increasing crystallinity. These results indicate that ^{29}Si-NMR line
intensities can be validly used to characterize the cristallinity
of the intermediate phases formed during the synthesis course of
mordenite. Similar results have been obtained in the study of
ZSM-5 synthesis (15).

3.3 Broad band solid state ^{27}Al-NMR study

Figure 5 shows ^{27}Al-NMR spectra of a synthetic Na-mordenite
and of the same sample treated with aqueous HCl 0.2N. The Na-
mordenite is characterized by a resonance line at $\delta \approx 53$ ppm,
corresponding to Al-atoms sited in tetrahedral position in the
framework. Another resonance line appears near $\delta \approx 0$ ppm after
the HCl leaching. It is due to extra-lattice octahedrally coordinated Al-atoms, which have been extracted out of the aluminosilicate framework and deposited in the zeolite channels or on its
external surface (11).

TABLE 1

Variations of Si/Al ratios and ^{29}Si chemical shifts (δ), of some amorphous intermediate phases formed during the first step of mordenite crystallization.

Crystallization time (h)	Si/Al (a)	δ(ppm) (b)
0.5	3.2	-95.4
8	4.8	-97.8
16	5.2	-102.3

(a) determined by energy dispersive X-ray analysis (EDX) (24)
(b) from TMS

The line intensity measured above -94 ppm (δ > -94 ppm) corresponds essentially to the amorphous solid, while the one at δ < -108 ppm contributes essentially to the Si(0 Al) resonance line of the crystalline mordenite present in the intermediate (zeolite + gel) phase. In between the -94 and -108 ppm region, the resonance lines from the amorphous phase as well as those pertaining to the Si(2 Al) and Si(1 Al) configurations in crystalline mordenite, will contribute to the spectrum. Table 2 and figure 4 illustrate the variation of the ^{29}Si-NMR intensities at δ > -94 ppm and δ < -108 ppm with the XRD crystallinity.

TABLE 2

Evolution of the ^{29}Si-NMR line intensities as a function of crystallization time

Crystallization time (h)	X-Ray diffraction crystallinity (%)	^{29}Si NMR intensities(%) (a)	
		δ > -94 ppm	δ < -108 ppm
0.5	0	43.6	6.9
8	0	14.1	8.5
16	11	6.9	25.8
24	37	10.1	38.2
48	99	1.9	41.6
68	100	1.8	39.8
120	100	0.7	47.0
280	100	0.7	47.9

(a) relative to the total ^{29}Si-NMR line intensity

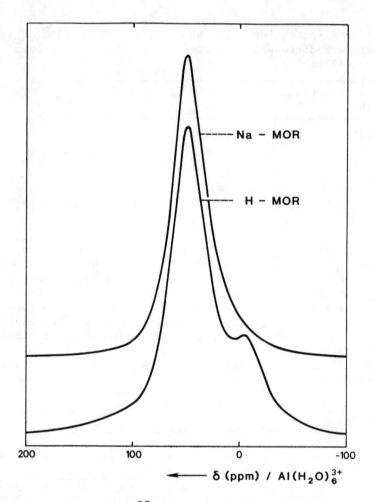

Figure 5. Solid state ^{27}Al-NMR spectra of mordenite :
a. Na-mordenite (Si/Al = 6.5)
b. the same, leached with HCl 0.2 N (Si/Al = 6.5)

Table 3 sums up the ^{27}Al chemical shifts (δ) and linewidths for different solid phases that appear during mordenite crystallization. The value of ^{27}Al chemical shifts suggests that Al-atoms in the intermediate phases basically occupy tetrahedral positions during all the crystallization process. Furthermore, the corresponding linewidths decrease as the crystallization proceeds and remain stabilized at about 1.9 kHz as soon as 100% crystalline mordenite is obtained (Table 3 and Figure 6).

TABLE 3

^{27}Al-NMR characterization of the intermediate solid phases formed during mordenite crystallization.

Crystallization time (h)	^{27}Al-NMR	
	δ(ppm) [a]	Linewidths (kHz)
0.5	55	5.3
8	53	3.6
16	53	3.5
24	53	2.6
48	53	1.9
68	53	1.9
120	53	1.8
280	53	1.9

(a) vs $Al(H_2O)_6^{3+}$

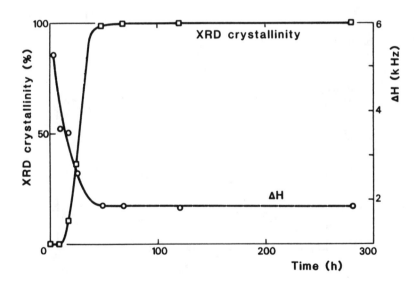

Figure 6. Variation of the ^{27}Al-NMR linewidths (ΔH) and of the XRD crystallinity of the intermediate phases as a function of synthesis time.

This decrease can be related to the progressive ordering of the gel phase during the growth process. Indeed, an ^{27}Al-NMR linewidth can be considered as directly related to the homogeneity of Al-sites distribution within the solid phase. In amorphous phases, the distribution is random, while in ordered crystalline phases all tetrahedral sites are regularly arranged.

TABLE 4

Evolution of ^{27}Al-NMR line intensities with XRD crystallinity and the Al-concentration

XRD crystallinity (%)	$C_{Al} \times 10^4$ (a) (mol g^{-1})	I_{Al} (a.u.)
0	33.0	18.5
0	23.0	13.6
11	23.4	14.0
37	22.0	13.4
99	19.4	12.8
100	18.9	12.2
100	17.6	12.3
100	14.4	10.8

(a) determined by EDX (24)

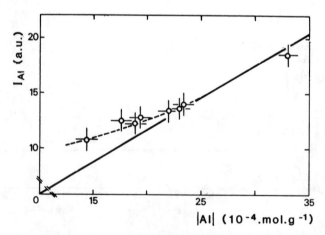

Figure 7. Correlation between ^{27}Al-NMR line intensities and Al concentration, measured in some intermediate phases obtained during mordenite crystallization.
───── = small sized particles entirely probed by EDX
- - - - = larger particles not entirely probed by EDX

The straight line (calibration curve), shown in figure 7, was established from analyses of phases consisting of small sized (less
than 2 μm) particles which are entirely probed by the EDX technique
(Table 4) (24). The more crystalline and Si-richer phases are composed of particles averaging 30 μm in diameter, for which the EDX
technique detects only Al- and Si-atoms located in the outer shell.
As the amount of aluminium detected by ^{27}Al-NMR (a bulk analysis
method) is different from that determined by a "surface" analysis
(EDX) (figure 7, dotted line), it is concluded that an Al-concentration gradient must exist in the 100 % crystalline mordenite particles : their inner core is richer in aluminium than their outer
rim. Si/Al ratios of a 100 % crystalline mordenite, as measured by
different analytical methods, are compared in table 5. Their regular decrease with the depth of the analysis confirms the existence of an Al-gradient.

TABLE 5

Variation of the Si/Al ratio of mordenite (crystallization time
68 hours, 100 % XRD crystallinity) as a function of the depth
of analysis.

Analytical method	Depth of analysis	Si/Al
XPS (ESCA) (a)	30 Å	8.0
EDX	2 μm	6.5
PIGE (b)	10 μm	5.2
^{27}Al-NMR (c)	bulk analysis	5.7

(a) X-ray photoelectron surface analysis (25)
(b) Proton induced γ-ray emission (26)
(c) Determined using the calibration curve of fig. 7.

3.4 Broad-band ^{23}Na-NMR study

The ^{23}Na-NMR spectra of intermediate phases obtained during
mordenite crystallization mainly consist of a broad resonance
line centered at δ = -17 ppm (vs aqueous $NaClO_4$) (figure 8).

222

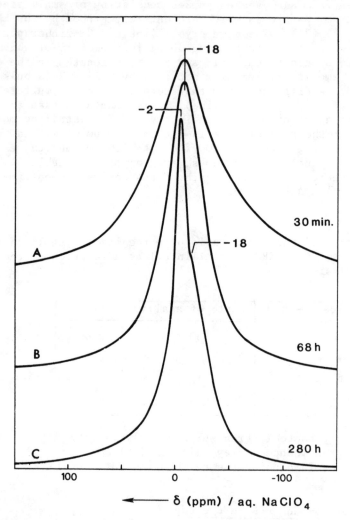

Figure 8. ^{23}Na-NMR spectra of some intermediate phases obtained during mordenite crystallization.

During the synthesis process, the ^{23}Na-NMR linewidth (ΔH) rapidly decreases and remains nearly constant as the crystallization is complete (fig. 8 and 9).

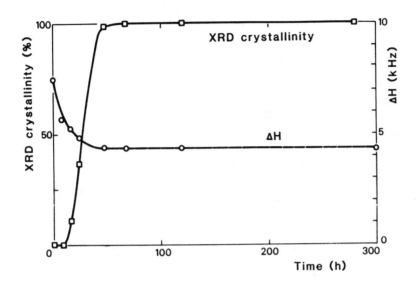

Figure 9.   Variation of ^{23}Na-NMR linewidths and XRD crystallinity
as a function of synthesis time.

This illustrates the progressive incorporation of sodium ions
into the zeolite lattice.   In the intermediate phases, the distribution of the chemical shifts is broad.   Moreover, the electric
field gradient on the ^{23}Na-nucleus must also be different in an
amorphous intermediate phase with respect to the 100 % crystalline
phase,hence a broad NMR line is observed. Oppositely, in the ordered
crystalline phases, the chemical shift distribution in narrower
and the linewidth of the corresponding NMR resonance decreases.

When the 100 % crystalline mordenite is left in the autoclave
for a long time, a new resonance line appears at $\delta \approx -2$ ppm.   It
can be tentatively attributed to sodium ions incorporated into a
parasite species (such as analcime) (fig. 8).

Figure 10 and Table 6 compare the ratio of the normalized
intensities of ^{27}Al- and ^{23}Na-NMR lines with the Al/Na atomic ratio
measured by EDX.   A relatively good correlation is observed.   This
demonstrates that the combination of ^{27}Al- and ^{23}Na-NMR techniques
can be valuably used both to show the incorporation of Al- and Na-
atoms in the zeolite and to estimate the Al/Na ratio in the sample,
provided a calibration curve.

TABLE 6

Variation of the NMR intensity ratio $\frac{I_{Al-NMR}}{I_{Na-NMR}}$, with the $\frac{Al}{Na}$ atomic ratio measured by EDX in the intermediate phases.

$\frac{Al}{Na}$ (EDX)	$\frac{I_{Al}}{I_{Na}}$
0.83	0.58
0.91	0.48
0.98	0.60
1.04	0.59
1.13	0.74
1.14	0.68
1.13	0.77
1.13	0.66

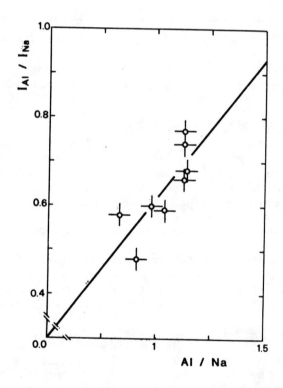

Figure 10. Evolution of the ^{27}Al- and ^{23}Na-NMR intensity ratio with Al/Na atomic ratio as determined by EDX

4   CONCLUSIONS

Multinuclear solid state NMR reveals to be a powerful tool for the characterization of intermediate solid phases occuring during crystallization of zeolites. HRMAS ^{29}Si-NMR spectra can be validly related to the ordering of the initial amorphous solid. The progressive incorporation of Al-atoms in an ordered zeolitic lattice is characterized by ^{27}Al-NMR. This technique is also able to discriminate between tetra- and octa-coordinated Al-atoms. It can be used to determine the Al-content of a solid phase, provided a calibration curve. From the comparison of the Si/Al ratios determined by different surface and bulk analysis methods, evidence can be obtained for the aluminium gradients in the small crystallites. In the particular case of mordenite, the outer shell of the 100 % crystalline particles contains less aluminium than their inner core. The evolution of ^{23}Na-NMR lines shows the incorporation of sodium ions into the zeolite lattice. This technique can be used together with ^{27}Al-NMR to determine the Al/Na ratio of the studied sample.

5   ACKNOWLEDGMENTS

The authors wish to thank Mr. G. Daelen for his appreciated help in taking the NMR spectra and Mr. F. Vallette for his technical assistance. P. Bodart thanks IRSIA (Belgium) for financial support.

6 REFERENCES

1. D.W. Breck "Zeolite Molecular Sieves", J. Wiley & Sons, New York (1974).
2. R.M. Barrer "Hydrothermal Chemistry of Zeolites", Academic Press,London (1982).
3. L.B. Sand in "Proceedings of the 5th International Conference on Zeolites", (L.V. Rees, ed.), Heyden, London, p. 1 (1980).
4. G. Engelhardt,U. Lohse, E. Lippmaa, M. Tarmak and M. Mägi, Z. Anorg. Allg. Chem., 482, 49 (1981).
5. J. Klinowski, S. Ramdas, J.M. Thomas, C.A. Fyfe and J.S. Hartman, J. Chem. Soc., Faraday Trans II, 78, 1025 (1982).
6. J.M. Thomas, C.A. Fyfe, J. Klinowski and G.C. Gobbi, J. Phys. Chem., 86, 3061 (1982).
7. C.A. Fyfe, G.C. Gobbi, J.S. Hartman, R.E. Lenkinski and J.H. O'Brien, J. Magn. Reson., 47, 168 (1982).
8. C.A. Fyfe, G.C. Gobbi, J. Klinowski, J.M. Thomas and S. Ramdas, Nature, 296, 530 (1982).
9. J. B.Nagy, J.-P. Gilson and E.G. Derouane, J. Chem. Soc.,Chem. Commun., 1981, 1129.
10. J. B.Nagy, Z. Gabelica, E.G. Derouane and P.A. Jacobs, Chem. Lett., 1982, 2003.
11. G. Debras, J. B.Nagy, Z. Gabelica, P. Bodart and P.A. Jacobs, Chem. Lett., 1983, 199.
12. J. B.Nagy, Z. Gabelica, G. Debras, P. Bodart, E.G. Derouane and P.A. Jacobs, J. Mol. Catal., in press.
13. Z. Gabelica, J. B.Nagy, P. Bodart, G. Debras, E.G. Derouane and P.A. Jacobs, in "Zeolites Science and Technology", Lisbon, 1983 (this meeting).
14. J. B.Nagy, Z. Gabelica and E.G. Derouane, Zeolites, 3, 43 (1983).
15. Z. Gabelica, J. B.Nagy, G. Debras and E.G. Derouane, Proc. 6th Int. Conf. Zeolites, Reno, 1983 (submitted).
16. R.M. Barrer, J. Chem. Soc., 1948, 2518.
17. L.L. Ames and L.B. Sand, Amer. Miner., 43, 476 (1958).
18. L.B. Sand in "Molecular Sieves", Society of Chemical Industry, London, p. 71 (1968).
19. L.B. Sand, U.S. Pat. 3,436,174 (1969).
20. O.J. Whittemore Jr., Amer. Miner., 57, 1146 (1972).
21. A. Culfaz and L.B. Sand, Adv. Chem. Ser., 121, 140 (1973).
22. P.K. Bajpai, M.S. Rao and K.V.G.K. Gokhale, Ind. Eng. Chem. Prod. Res. Dev., 17, 223 (1978).
23. S. Ueda, H. Murata, M. Koizumi and H. Nishimura, Amer. Miner., 65, 1012 (1980).
24. Z. Gabelica, N. Blom and E.G. Derouane, Appl. Catal. 5, 227 (1983).
25. E.G. Derouane, J.-P. Gilson, Z. Gabelica, C. Mousty-Desbuquoit and J. Verbist, J. Catal., 71, 447 (1981).
26. G. Debras, E.G. Derouane, J.-P. Gilson, Z. Gabelica and G. Demortier, Zeolites, 3, 37 (1983).

SORPTION BY ZEOLITES
PART I. EQUILIBRIA AND ENERGETICS

R.M. Barrer

Chemistry Department
Imperial College of Science and Technology
London SW7 2AY

ABSTRACT

Steric factors governing penetration of the host zeolite by guest species have been considered. After equilibrium is established quantification of the results can be made in several ways. Selectivity of sorption requires interpretation of heats and entropies of uptake in terms respectively of universal (or non-specific) and of specific components of heat, and in terms of the physical state of the sorbed species. These aspects have also been considered. Isotherm modelling has been discussed in terms of the virial isotherm equation, and in terms of a site model in which a site is identified with a cavity and so is able to hold a small cluster of molecules.

1. INTRODUCTION

Zeolites and porous crystalline silicas provide stable, high capacity, micropore sorbents with diverse molecule sieving properties. Each framework topology provides its own unique system of channels and cavities. For example, in zeolites in which wide straight channels are found the detail of these channels is quite different, as illustrated for zeolites LTL, MAZ and MER in Fig. 1 (1). One expects and finds differences in sorption behaviour for each topology. The behaviour is further modified for zeolites by the number, location and size of the intracrystalline cations which neutralise the negative charge on the framework. Thus, since cations are exchangeable, modified sorbents can be made from a given framework topology by cation exchange. The uptake of guest molecules by zeolite host crystals has equilibrium and energetic properties,

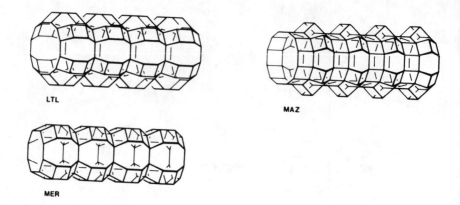

Fig. 1.   The channels in zeolite L, LTL, mazzite, MAZ, and
merlinoite, MER (1).

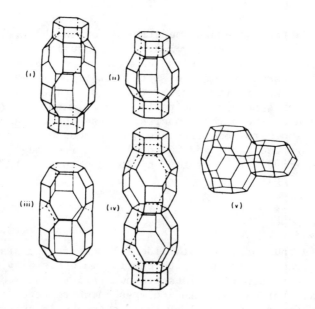

Fig. 2.   Some polyhedral voids found in zeolites.
(i)   The chabazite 20-hedron, capped with hexagonal
prisms.
(ii)   The gmelinite 14-hedron of type II.
(iii)   The erionite 23-hedron
(iv)   The levynite 17-hedron of type I.
(v)   The losod 17-hedron of type II with associated
11-hedral cancrinite cage.

considered in this Part, and kinetic aspects, discussed in Part II.
In terms of the defined structures of channels and cavities zeolites
provide model systems for quantitative studies of sorption.

2. CONDITIONS FOR PENETRATION OF ZEOLITE LATTICES

Provided a zeolite crystal is open enough to admit the guest
species under consideration sorption complexes form in which the
amount of guest sorbed in a given weight of zeolite depends only
upon the pressure of the vapour of the guest and the temperature.
However, the ease of penetration of the outgassed zeolite by guest
molecules depends upon the following factors:

1. The size and shape of the windows controlling entry to
   the channels and cavities in the zeolite.
2. The size and shape of the guest molecules.
3. The number, locations and size of the exchangeable cations.
4. The presence or absence of defects such as stacking faults
   which may narrow diffusion pathways at planes where such
   faults occur.
5. The presence or absence of detrital material left in the
   channels during synthesis or introduced subsequently,
   for example, by chemical means such as silanation (2).
6. The presence or absence of other strongly held guest
   molecules like water, ammonia (3) and salts (4,5), introduced intentionally in metered amounts.

The sixth factor is referred to in Part II. Here we will
refer primarily to the first four factors.

2.1. Free Dimensions of Windows

The windows or openings which control entry to the intracrystalline pores and channels are rings of linked $(Al,Si)O_4$ tetrahedra circumscribing channels like those shown in Fig. 1, or
allowing access to polyhedral cavities like those illustrated in
Fig. 2. The important openings from the present viewpoint are
those composed of 8, 10 or 12 linked tetrahedra. These 8-, 10-
and 12-rings are lined on their inner peripheries by oxygen atoms,
and in the frameworks of zeolites having different topologies they
may be variously elongated, or puckered to different conformations,
such as crown, boat or chair configurations. Therefore a given
n-ring may have different free dimensions and so may impose diverse
molecule sieving behaviour for the same value of n, according to
the zeolite in which it occurs. The variations in free dimensions
are shown in Table 1 for typical zeolites with 8-, 10- and 12-ring
windows. By "free dimensions" one means that the space to which
the dimensions refer is not impinged upon by the peripheral oxygens
lining the inside of each ring.

Table 1: Free dimensions in Å of 8-, 10- and 12-ring windows in some zeolites, taking 2.75 Å as diameter of framework oxygens (9)

Zeolite	8-ring	Zeolite	10-ring	Zeolite	12-ring
Brewsterite	2.3x5.0	Dachiardite(b)	3.7x6.7	Cancrinite hydrate	6.2
	2.7x4.1	Laumontite	4.0x5.6	Offretite(b)	6.4
Bikitaite	3.2x4.9	Stilbite(b)	4.1x6.2	Mordenite(b)	6.7x7.0
Levynite	3.3x5.3	Ferrierite(b)	4.3x5.5	Gmelinite(b)	7.0
Chabazite	3.6x3.7	Heulandite(b)	4.4x7.2	Linde L	7.1
Li-ABW	3.6x4.0	ZSM-11	5.1x5.5	Mazzite	7.4
Dachiardite(a)	3.6x4.8	ZSM-5	5.1x5.5	Faujasite	7.4
Erionite	3.6x5.2		5.4x5.6		
TMA-EAB	3.7x4.8				
Paulingite	3.9				
Linde A	4.1				

(a) There are also 10-ring windows in this zeolite.
(b) There are also 8-ring windows in this zeolite.

2.2. Molecular Dimensions

The idea of using simple molecules of known dimensions to characterise molecular sieve zeolites was introduced some time ago (6,7) and has proved most useful in grouping zeolites into different categories of molecular sieve (6,7,8). Based on Pauling's values of bond lengths and van der Waals radii of atoms and molecules the dimensions of some yardstick molecules in $\overset{o}{A}$ are as follows (8):

Ne	3.2_0	CCl_4	$6.8_8(7.1_2)$
Ar	3.8_3	CBr_4	$7.4_6(7.7_2)$
Kr	3.9_4	CI_4	$8.2_2(8.5_0)$
Xe	4.3_7	$C(CH_3)_4$	$6.6_8(7.0_8)$
CH_4	$4.0(4.4_4)$	SF_6	$6.0_6(6.0_6)$
CF_4	$5.3(5.4_4)$		

For CH_4, 4.0 $\overset{o}{A}$ is a value assuming it to be a smooth sphere. The value in brackets takes into account the tetrahedral structure. For the other molecules the figures in brackets are the diameters of the circumscribing spheres and those not in brackets are the heights of the molecules sitting on a triangular base. One or other of these dimensions should be critical in determining whether the molecule will enter the zeolite lattice.

For several dumbell-shaped molecules the cross-sectional diameter and (in brackets) the lengths are as follows (8):

H_2	$2.4(3.1_4)$	C_2H_6	$4.4_4(4.6_0)$
O_2	$2.8(3.9_0)$	C_2F_6	$5.3_3(5.1_1)$
N_2	$3.0(4.0_8)$	C_2Cl_6	$6.8_8(6.2_3)$

Where the length exceeds the cross-sectional diameter it will be the latter which is the critical dimension for penetrating the zeolite.

Situations have been abundantly demonstrated in which one molecular species is totally excluded while another is readily sorbed by a given zeolite (6,7,8). The separation can be quant-itative in a single step and can be most important for large scale separations and in shape selective catalysis. Table 2 illustrates sieve characteristics for three zeolites, Ca-A, ZSM-5 and faujasite (Na-X) for which the windows are respectively 8-, 10- and 12-rings. For fully stretched n-paraffins the cross-sectional dimension critical for penetration is that in the plane of the zig-zag (4.9 $\overset{o}{A}$); for benzene this dimension is the distance across in the plane of the molecule. For non-symmetrical molecules the best orientation for penetration can be seen by presenting a scale model of the guest to a scale model of the window.

The critical dimensions for entry are seen to be greater than the free dimensions of the openings (Table 2). This happens because the guest molecules and the lattice atoms (here oxygens) are not hard spheres but are deformable and also because included in lattice vibrations there are breathing frequencies for the openings as a whole. Accordingly n-paraffins diffuse rather easily into Ca-A, although significant energy barriers are involved. The energy barriers increase rapidly as the critical dimensions increase, and those for total exclusion are soon reached, as indicated in Table 2.

Temperatures at which the uptake occurs can play an important part in molecule sieving. Because of the energy barriers involved diffusivities, D, follow an Arrhenius relation: $D = D_0 \exp\text{-}E/RT$. Thus, the lower the temperature the more sensitive the molecule-sieving process becomes. At low temperatures ($-183^{\circ}C$) it is possible to separate O_2 and Ar quantitatively using Na-A or levynite (their critical dimensions being 2.8 Å and 3.8 Å respectively), and also to obtain large rate differences between O_2 and N_2 (3.0 Å). Such diffusional aspects will be referred to in Part II.

2.3. Blocking of Windows by Cations

In Ca-A, ZSM-5 and faujasite the windows are not blocked by cations and the sieve character could therefore be assessed from the free dimensions of the windows deduced from the structures. However after exchanges $Ca^{2+} \rightarrow 2Na^{+}$ or $2K^{+}$ the number of cations in zeolite A is doubled and Na^{+} and K^{+} ions occupy positions in 8-ring windows. Na-A no longer admits n-paraffins at room temperature, but still sorbs O_2, N_2, Ar and small polar molecules. Likewise Ca-chabazite will sorb n-paraffins but Na-chabazite does not sorb even oxygen (11). Zeolite RHO sorbs only water and ammonia until converted into its H-form when it sorbs permanent gases and n-paraffins copiously. The windows are octagonal prisms of free diameter 3.9 x 5.1 Å, but these windows appear to be selective cation traps (12).

When in zeolite Na-A one exchanges $2Na^{+}$ by Ca^{2+} in stages n-paraffins begin to be sorbed freely when about 30% of the Na^{+} is replaced (13). This sensitive range occurs when enough 8-ring windows have been freed of cations to give a fraction of clear pathways through the three-dimensional network of channels. Similar behaviour is observed in other exchanges of mono- by divalent cations (11,14).

2.4. Control of Access by Stacking Faults

Stacking faults can occur in cancrinite and gmelinite and sometimes in offretite. In cancrinite an ab... sequence of layers

Table 2: Sieve behaviour of a narrow, an intermediate and a wide port zeolite (10).

	Zeolite		
	Ca-A	ZSM-5	Faujasite (Na-X or Na-Y)
Access of Guest via:	8-rings	10-rings	12-rings
Free dimensions of rings ($\mathring{A}$)	4.1	5.6x5.4 (straight channels) 5.5x5.1 (sinusoidal channels)	7.4
Critical dimension of guest for entry ($\mathring{A}$)	~4.9	~6.9	~8.8
Critical dimension for exclusion ($\mathring{A}$)	~5.6	~7.8	~10.0
Sorbed	n-paraffins	n- and simple iso-paraffins; benzene, toluene, xylenes, 1,2,4-trimethyl benzene, napthalene	n-, iso-, neo- and cyclo-paraffins; many aromatics including 1,3,5-trisopropyl benzene
Excluded	iso-, neo- and cyclo-paraffins and aromatics	pentamethyl- and 1,3,5-trimethyl benzene	tertiary-perfluoro-propylamine

containing single hexagons (6-rings) can be interrupted by an
abc··· sequence typical of sodalite. In gmelinite an AB···
sequence of layers containing hexagonal prisms is interrupted by
an ABC··· sequence, and in offretite an Ab··· sequence is interrupted by AbAc··· sequence of erionite. In all these situations
wide channels curcumscribed by 12-rings are blocked, by windows no
wider than the 6-rings of sodalite or the 8-rings of chabazite and
erionite for cancrinite and for gmelinite and offretite respectively.
It is not difficult however to make offretites free of such intergrowths and stacking faults.

In the case of cancrinite in particular stacking faults may
not be the only cause of blocking. Both cancrinite and sodalite
are notable for trapping salts and caustic soda along with zeolitic
water during synthesis, and there is some analytical evidence that
detrital silicate may be a further blocking factor in cancrinite
(5).

3. DISTRIBUTION PATTERNS OF GUEST MOLECULES

The pore space in zeolites is parcelled up into cavities and/
or channels of molecular dimensions. The resultant patterns of
pathways through which guest molecules of the right size and shape
can move can be placed in three categories:

1. All pathways are parallel and non-interconnected (1-dimens-
    ional (1-D) channel systems as in mordenite, mazzite,
    laumontite or zeolite L.
2. The pathways whether parallel or not are interconnected to
    give 2-dimensional (2-D) channel systems. Guest molecules
    may migrate between layers but cannot move from one layer
    to a parallel layer in the crystal (heulandite, levynite,
    stilbite and ferrierite).
3. The pathways may be so interconnected as to allow guest
    molecules to migrate in 3-dimensions (3-D channel systems
    as in chabazite, erionite, zeolite A and zeolites ZSM-5,
    RHO and ZK-5).

Detailed channel geometries are however different for each
framework topology, as was seen in Fig. 1 for the 1-D channels of
zeolites L (LTL), mazzite (MAZ) and merlinoite (MER). The distribution of the guest molecules in the host zeolites is determined
by the 1-D, 2-D or 3-D nature of the channels in which they are
located. Thus in 1-D systems they are present as parallel filaments
supported by the channel walls. For example in zeolite L there are
restrictions along each channel of 7.1 Å free diameter alternating
with wider parts of $\sim$ 12 Å free diameter. When the channel is full
of small molecules like water or oxygen these form liquid-like beads
or clusters in the wide part connected into filaments through the

7.1 Å openings to other beads. In faujasite each 26-hedron of type II of free diameter about 12 Å is connected through 7.4 Å openings to four more such tetrahedra. Liquid-like clusters in each 26-hedron are connected to those in its four neighbours to give a 3-D pattern of connected clusters arranged like the bond pattern in diamond.

The smaller the free diameter of the connecting windows the more isolated each molecular cluster becomes from its neighbours while the smaller the cavity and/or the larger the guest molecule

Table 3: Cluster sizes at saturation of cavities in chabazite and faujasite (zeolite X).

Zeolite	Cavities	Guest molecules per cavity
Chabazite	20-hedra (6 x 8-rings 2 x 6-rings 12 x 4-rings)	12-14 H_2O ~7.7 NH_3 ~6 Ar, N_2, O_2 ~4.9 CH_3NH_2 ~4.3 CH_3Cl ~3.1 CH_2Cl_2 ~2.0 I_2
Faujasite	26-hedra (4 x 12-rings 4 x 6-rings 18 x 4-rings)	~32 H_2O (28 + 4)[*] 17-19 Ar, N_2, O_2 ~7.5 I_2 ~7.8 CF_4 ~6.5 SF_6 ~5.8 C_2F_6 ~5.6 cyclopentane ~5.4 benzene ~4.6 toluene ~4.5 $n-C_5H_{12}$ ~4.1 cyclohexane ~4.1 perfluorocyclobutane ~4.1 $C_2F_4Cl_2$ ~3.5 $n-C_7H_{16}$ ~3.4 C_3F_8 ~2.9 $n-C_4F_{10}$ ~2.8 iso-C_8H_{18} ~2.3 perfluoromethylcyclohexane ~2.1 perfluorodimethylcyclohexane

[*] Four of the water molecules are thought to be in the sodalite-type 14-hedra also present in faujasite.

the fewer are the molecules per cluster (Table 3). Thus in the 14-hedral cavities $_o$ of sodalite hydrate each cavity has a free diameter of ~ 6.6 Å and is connected by 6-ring windows of ~ 2.1 Å free diameter to eight nearest neighbour 14-hedra. It can $_o$ accommodate four water molecules (van der Waals diameter ~ 2.8 Å), as a nearly isolated cluster due to the small free diameter of the windows. At high pressure and temperature it $_o$ can accommodate one only Ar or Kr (15) of diameter 3.8_3 and 3.9_4 Å respectively. This corresponds with interstitial solution as the limit to the larger clusters illustrated in Table 3 for the 20-hedra present in chabazite and the 26-hedra in faujasite.

4. TYPES OF ISOTHERM AND GUEST-ZEOLITE COMPLEX

Under appropriate conditions zeolites can sorb non-polar and polar molecules, salts or metals. In addition metals may be introduced by reducing cationic forms of the zeolite or by sorbing volatile metallic compounds such as carbonyls and then decomposing these. Salts may be introduced, in competition with water, during hydrothermal synthesis, or from aqueous solution into the already formed zeolite, as well as from salt melts or vapours.

Isotherms of non-polar guest molecules are as a rule of type I in Brunauer's classification (16) as shown in Fig. 3 (17). The more condensable the guest molecule or the lower the temperature the more rectangular the isotherms become. On the other hand the less condensable the sorbed molecule or the higher the temperature the more nearly does the isotherm approach the Henry's law limit (uptake proportional to pressure).

There are however sorption complexes characterised by very strong molecule-molecule interactions between pairs of guest molecules which dramatically change the isotherm contours to types IV or V in Brunauer's classification. These will be illustrated for an electronegative element, a salt and a metal. Thus Fig. 4 (18) shows isotherms nearly of type V for the reversible uptake of phosphorus in zeolite Na-X. The strong upward inflexion may arise from the onset of polymerisation of smaller phosphorus species such as P_4 within the zeolite.

When salts are incorporated into sodalite during hydrothermal synthesis keeping water, caustic soda and metakaolinite constant in the reaction mixture the isotherms of salt uptake are again of type I as shown in Fig. 5 (19). However when salts were taken up from aqueous solutions into pre-formed zeolites the isotherms in Fig. 6 (20) were of type III, with curvature in the opposite sense to those in Fig. 5. Their shape is determined by a Donnan equilibrium whereas the shape in Fig. 5 may arise because the salts are acting as templates during actual crystal nucleation and growth (21). When

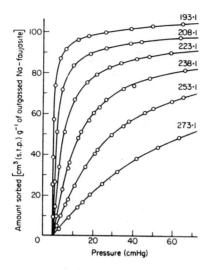

Fig. 3.   Type I isotherms of CF₄ in Na-faujasite at various
          absolute temperatures (17).

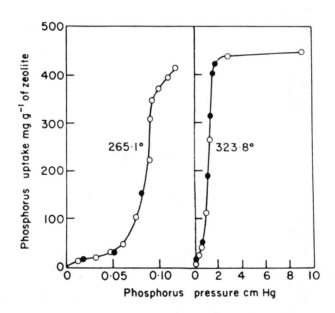

Fig. 4.   Isotherms for uptake of phosphorus in Na-X (18).
          Temperatures are in °C. 0 = adsorption points;
          ● = desorption points.

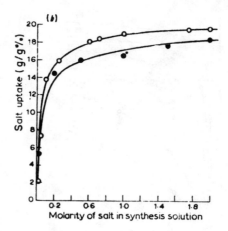

Fig. 5.   Isotherms for uptake of NaClO₄ (O) and NaClO₃ (●) from
aqueous solutions during synthesis of sodalite (19).
The conditions of formation were: 32g NaOH; 2g meta-
kaolin; 200 ml distilled water, to which the desired
amounts of salt were added.   Reaction in polypropylene
bottles rotated at 80°C for 6 days.

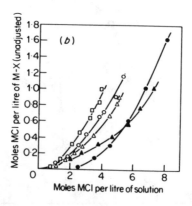

Fig. 6.   Isotherms for salts at 25°C in pre-formed zeolite X (20)
    □ = KCl   ● = LiCl
    O = NaCl   ▲ = CsCl
    △ = CaCl$_2$

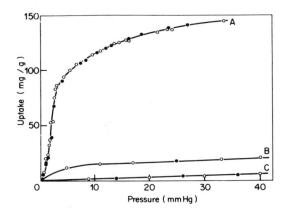

Fig. 7.   Isotherms for uptakes in zeolite K-L at 245°C   of
           A Vapourised NH$_4$Cl (i.e. NH$_3$ + HCl in 1:1 ratio)
           B HCl gas alone
           C NH$_3$ alone (22).

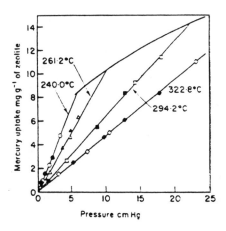

Fig. 8.   Isotherms for uptake of Hg in Na-X (23). The upper
           curve represents the limit to the uptake set by the
           saturation vapour pressure of liquid Hg at each
           experimental temperature. O, Δ, □ and ◇ are adsorption
           points; ●, ▲, ■ and ◆ are desorption points.

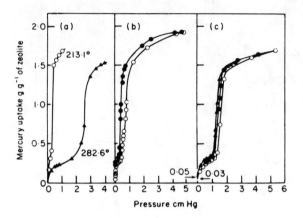

Fig. 9. Sorption of Hg in Ag-X (23)
(a) Uptakes at two temperatures ($^{\circ}$C)
(b) A sorption-desorption cycle at 235.2 $^{\circ}$C
(c) Two successive sorption-desorption cycles at 270°C

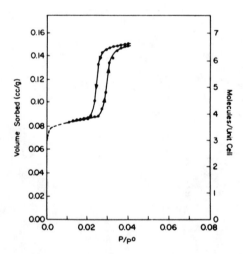

Fig. 10. A sorption-desorption cycle for uptake of p-xylene in
zeolite H-ZSM-5 (SiO_2/Al_2O_3 = 226) at 70°C (24).

NH$_4$Cl was sorbed from its vapour into zeolites a third isotherm contour was found, of type V , and rather similar to that in Fig. 4. This is illustrated in Fig. 7 (22). When ammonium chloride is vapourised it dissociates virtually completely into a 1:1 mixture of HCl + NH$_3$. Therefore from its vapour it is this mixture which is sorbed into the zeolite. In Fig. 7 curve C is the reversible isotherm of NH$_3$ alone in zeolite K-L at 245°C; curve B is this isotherm for pure HCl; and curve A shows what happens when the 1:1 mixture of NH$_3$ + HCl (i.e. NH$_4$Cl vapour) is sorbed. There is clearly a strong interaction between NH$_3$ and HCl within the zeolite.

When metals are initially atomically dispersed in zeolites, for example by reductions such as

$$\text{Ni-zeolite} + \tfrac{1}{2}H_2 \rightarrow \text{H-zeolite} + Ni,$$

and the system is subjected to further heating, the metal atoms tend to nucleate into clusters within the zeolite or outside it as small crystallites. This tendency can be studied for intra-crystalline nucleation using mercury as the guest species. When the concentration of mercury atoms is low nucleation does not occur and intrazeolite sorption follows Henry's law, as shown in Fig. 8 for uptake in Na- and Pb-X (23). Na$^+$ and Pb^{2+} are derived from elements higher in the electrochemical series than Hg. In contrast with this behaviour when sorption occurred in Hg- or Ag- zeolites the isotherms became once more of type IV in the Brunauer classi-fication, as seen in Fig. 9 (23). The ions Hg^{2+} or Ag$^+$ originally in the zeolite are derived from elements as low as or lower in the electrochemical series than mercury, so that reductions can occur:

$$
\left.
\begin{aligned}
Hg^{2+} + Hg &\rightarrow Hg_2^{2+} \\
2Ag^+ + 2Hg &\rightarrow Hg_2^{2+} \\
2Ag^+ + Hg &\rightarrow Hg^{2+}
\end{aligned}
\right\} \tag{1}
$$

Such reductions, which may represent the first step in the isotherms of Fig. 9, appear to trigger off clustering processes:

$$
\left.
\begin{aligned}
Ag + xHg &\rightarrow AgxHg \\
Hg_2^{2+} + xHg &\rightarrow Hg_{x+2}^{2+}
\end{aligned}
\right\} \tag{2}
$$

Isotherms with contours like those of P, NH$_4$Cl or Hg are rare compared with those shown for CF$_4$ in Fig. 3. It is thus of interest that a type IV isotherm has been found for p-xylene in ZSM-5 in which a clear step, this time with some hysteresis, occurs (24). The explanation offered was not in terms of strong molecule-molecule interaction, but that at a certain intracrystalline concentration they pack in a new configuration, more economical of space (Fig. 10).

5. SELECTIVITY IN MIXTURE SEPARATION AND HEATS OF SORPTION

That mixtures can be separated, often quantitatively and in a single step, by molecule sieving has been very fully established. This is illustrated in Table 4 in which the crystalline zeolite was powdered natural chabazite (7,8). Ca-chabazite is one of a class of zeolites able to separate n-paraffins from iso- neo- and cyclo-paraffins and aromatics. Others in this class are Ca-A, erionite, zeolite ZK-5 and the hydrogen form of zeolite RHO.

Strong selectivities are also possible when both components of a binary mixture can enter the zeolite and equilibrate with it. To understand these selectivities one may consider the components of the physical bond between guest molecules and host crystals. These include

Dispersion energy, ϕ_D
Close-range repulsion energy, ϕ_R
Polarisation energy, ϕ_P
Field-dipole energy, $\phi_{F\mu}$
Field gradient-quadrupole energy, $\phi_{\dot{F}Q}$
Guest-guest self-energy, ϕ_{SP}

Electric moments in the guest molecule may be permanent, or they may be induced by the local electrostatic field of strength F. Thus polarisation energy can have components of field-induced dipole or dipole-induced dipole.

The components ϕ_D, ϕ_R and ϕ_P are termed "non-specific" because they are always involved in the host-guest bond. With the advent of high silica zeolites, hydrogen zeolites and porous crystalline silicas local fields F and field gradients $\dot{F}$ can be much reduced so that ϕ_P can become small. The components $\phi_{F\mu}$ and $\phi_{\dot{F}Q}$ for molecules with permanent dipole moments μ or molecular quadrupole moments Q are termed "specific" components of the host-guest bond because they do not arise with non-polar molecules like the rare gases. They can be very large in aluminous zeolites and can lead to unusually high heats of sorption and selectivities (e.g. H_2O or NH_3).

The guest-guest interactions giving ϕ_{SP} may in extreme cases be chemical as well as physical in nature, as for P and for (NH_3 + HCl) or in clustering of Hg atoms (§. 4). Usually however only physical interactions are involved. These are universally dispersion and close-range repulsion, and for molecules with permanent electric moments there may be terms in ϕ_{SP} such as

Table 4

Resolutions of Mixtures Using Chabazite (8)

Mixture	Component(s) Sorbed	Conditions and Comments
$CH_3OH + (CH_3)_2CO$	CH_3OH	As liquid at $\sim 20°C$. Rapid and quantitative.
$CH_3OH + (CH_3)_2 \cdot O$	CH_3OH	As above
$CH_3OH + CS_2 + CH_3CN + C_6H_6$	CH_3OH, CS_2 and CH_3CN	As above
$CH_3OH + H_2O + CH_3OCOCH_3$	CH_3OH, H_2O	As above
$CH_3OH + COCH_3 \mid COCH_3$	CH_3OH	Sorption via vapour in equilibrium with solution. Complete separation.
$C_2H_5OH + C_6H_5CH_3$	C_2H_5OH	As liquid at $\sim 20°C$. Separation slow but complete.
$C_2H_5OH + CHCl_3 \; CCl_4$	C_2H_5OH	As above
$C_2H_5OH + CH(CH_3)_2OH$	C_2H_5OH	As above
$C_2H_5OH + C(CH_3)_3OH$	C_2H_5OH	As above
$C_2H_5OH + n\text{-}C_7H_{16}$	C_2H_5OH	As above
$C_2H_5OH + H_2O + (C_2H_5)_2O$	C_2H_5OH, H_2O	As above
$C_2H_5OH + CH_3COC_2H_5$	C_2H_5OH	As liquid at $112°C$. Separation rapid and complete.
$C_2H_5OH + CH_2Br_2$	C_2H_5OH	As liquid at $\sim 20°C$. Slow but complete separation.
$CH_2O + H_2O + CH(CH_3)_2OH$	CH_2O, H_2O	As liquid at $\sim 20°C$. Rapid and quantitative.
$CH_2O + H_2O + CH_3I$	CH_2O, H_2O	As above
$CO_2 + CH(CH_3)_2OH$	CO_2	As above. CO_2 initially in solution.
$SO_2 + CHCl_3$	SO_2	As above. SO_2 initially in solution.
$N_2O_3 + C_6H_6$	N_2O_3	As above. N_2O_3 in solution.
$H_2S + C_6H_6$	H_2S	As above. H_2S in solution.
$CS_2 + CH_3COCH_3$	CS_2	As liquid at $\sim 20°C$. Rapid and quantitative.
$CS_2 + CHCl_3$	CS_2	As above
$CS_2 + C_6H_6$	CS_2	As above
$CS_2 + N(CH)_4CH$	CS_2	As above
$C_2H_5SH + C_6H_6$	C_2H_5SH*	As liquid at $50°$. Partial removal in a week.
$CH_3NH_2 + C_2H_5OH + N(CH_3)_3$	CH_3NH_2, C_2H_5OH	As liquid at $\sim 20°C$. Complete removal of both constituents within 16 hours.
$C_2H_5NH_2 + (C_2H_5)_2NH$	$C_2H_5NH_2*$	As liquid at $\sim 20°$. Slow but complete separation.
$(C_2H_5)_2NH + iso\text{-}C_8H_{18}$	$(C_2H_5)_2NH*$	As liquid at $180°C$. Separation only partial in five days.
$CH_3CN + Thiophen$	CH_3CN	As liquid at $\sim 20°C$. Rapid and quantitative.
$CH_3CN + CH_2Br_2$	CH_3CN	As above
$CH_3CN + CH_2O + H_2O + n\text{-}C_7H_{16}$	CH_3CN, CH_2O, H_2O	As above
$C_2H_5CN + CH(CH_3)_2OH$	C_2H_5CN*	As liquid at $100°C$. Separation complete within two days.
$HCl + CHCl_3$	HCl	As liquid at $\sim 20°C$. Dissolved HCl removed quickly and completely.
$Cl_2 + C_6H_6$	Cl_2	As above
$Br_2 + CCl_4$	Br_2	As above. Equilibrium separation nearly complete within 16 hours.
$I_2 + CHCl:CCl_2$	I_2*	As liquid at $\sim 20°C$. Partial removal within four days.
$CH_3I + C_6H_6$	CH_3I*	As liquid at $\sim 20°C$. Nearly complete in 12 days.
$CH_2Cl_2 + CH(CH_3)_2OH$	CH_2Cl_2	As liquid at $\sim 20°C$. Rapid and complete.
$CH_2Cl_2 + CH_3COC_2H_5$	CH_2Cl_2	As liquid at $112°C$. Rapid and complete.
$CH_2Cl_2 + Dioxane$	CH_2Cl_2*	As liquid at $112°C$. Partial separation in three days.
$CH_2Cl_2 + sym\text{-}C_2H_4Cl_2$	CH_2Cl_2	As liquid at $\sim 20°C$. Separation complete.
$C_2H_5Cl + CH(CH_3)_2OH$	C_2H_5Cl*	As above
$CH_3Br_2 + iso\text{-}C_8H_{18}$	CH_3Br_2*	As liquid at $\sim 20°C$. Separation complete within 12 days.
$CH_3Br_2 + C_6H_6$	CH_3Br_2*	As above
$CH_3Br_2 + CH(CH_3)_2OH$	CH_3Br_2*	As liquid at $112°C$. Separation nearly complete within 24 hours.
$C_2H_5Br + C_6H_6$	C_2H_5Br*	As above. Separation complete.
$C_2H_5Br + CH(CH_3)Cl_2$	C_2H_5Br*	As liquid at $97°C$. Separation complete within two days.

* Slowly sorbed at room temperature.

$\phi_{\mu\mu}$, $\phi_{\mu Q}$ and ϕ_{QQ}, which represent dipole-dipole, dipole-quadropole and quadropole-quadropole terms. When the amount sorbed is small, as in the Henry's law range, ϕ_{SP} can be neglected.

It is possible to estimate the specific components, $\phi_{F\mu}$ and ϕ_{FQ}, separately from the non-specific ones, ϕ_D, ϕ_R and ϕ_P. ϕ_D and ϕ_P are functions of the polarisability, α, and therefore for a series of non-polar guest molecules one may plot the initial value (when ϕ_{SP} = 0) of the differential molar heat of sorption, $\Delta\overline{H}$, against the polarisability. The curves obtained are illustrated in Fig. 11 (25). Within the range of polarisabilities covered and for simple non-polar species they calibrate the host crystal. When the initial heat for a molecule with a permanent electric moment is plotted on the same diagram the difference between the point on the reference curve corresponding with the polarisability of the polar molecule and the actual heat measures the contribution of $\phi_{F\mu}$ and/or ϕ_{FQ} to this heat. Such a separation of specific and non-specific components is illustrated in Table 5 (25). The specific components for N_2, N_2O, CO_2, NH_3 and H_2O are always significant and for NH_3 and H_2O are dominant. The analysis makes it clear why zeolites are excellent desiccants of industrial gases and liquids; and why N_2 (which has a molecular quadrupole moment) is selectively sorbed compared with O_2 (which has an almost zero moment).

To a first approximation the dispersion energy can be considered additive for each pair of interacting atoms, one being in the guest molecule and one in the host zeolite. This means that for n-paraffins in zeolites such as A, L, or faujasite $-\Delta\overline{H}$ can increase continually with carbon number. In (Ca,Na)-A at 50°C the heat is already about 24 kcal mol^{-1} for n-$C_{12}H_{26}$ (26). Heats of physical sorption can indeed exceed the heats of many chemical reactions.

Zeolite sorbents are usually energetically heterogeneous. This is shown by a decrease in $-\Delta\overline{H}$ as the amount sorbed increases, and is exemplified for CH_4 in zeolite H-L in Fig. 12 (27). When polar molecules are sorbed (NH_3, H_2O) the decrease in $-\Delta\overline{H}$ with uptake is sometimes very sharp, usually more so than for non-polar guest molecules. However guest-guest interactions resulting in the component ϕ_{SP} can offset the decrease in $-\Delta\overline{H}$ as the amount sorbed increases. This effect can be observed in Fig. 12. It becomes more important the more condensable the paraffin, so that for n-C_4H_{10} for example instead of declining with uptake increasing $-\Delta\overline{H}$ = q_{st} actually increases.

When saturation of the intracrystalline sorption volume is approached $-\Delta\overline{H}$ = q_{st} can through the contribution of ϕ_{SP} pass through a maximum and then, as sorption on external surfaces culminating in capillary condensation between crystallites becomes the dominating process, $-\Delta\overline{H}$ declines towards the heat of condensation.

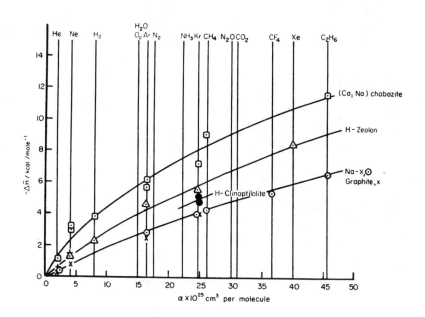

Fig. 11. Initial values of $-\Delta \overline{H}$ plotted against polarisability of sorbate for small relatively symmetrical molecules in several sorbents (25).

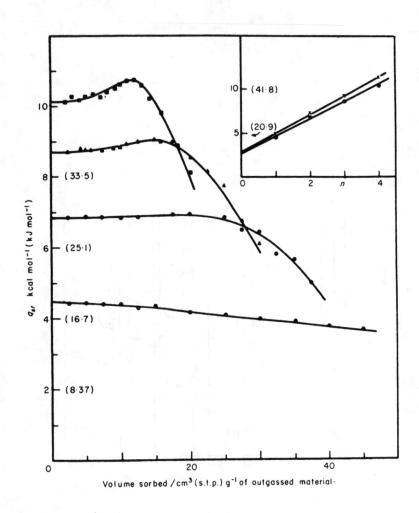

Fig. 12. Plots of $q_{st} = -\Delta\overline{H}$ against uptake for CH_4 (bottom) to $n\text{-}C_4H_{10}$ (top). Inset: plots of initial value of q_{st} against carbon number for K-L (top) and H-L (bottom). Figures in brackets are heats in kJ mol^{-1} (27).

Table 5: Division of components of initial heats, $\Delta\overline{H}$ (cal mol^{-1}), of guests in several zeolites (25).

Zeolite	Outgassed at $^\circ$C	Guest	$-\Delta\overline{H}$		
			Total	Dispersion + Repulsion + Polarisation	Dipole + Quadrupole
Chabazite	480	N_2	9000	6,450	2,550
	450	N_2O	15,000	9,100	6,200
	480	NH_3	31,500	7,500	24,000
H-mordenite	350	H_2	6,200	4,500	1,700
		CO	11,100	6,750	4,350
Na-mordenite	350	N_2	7,000	4,500[a]	2,500
		CO	15,700	6,750[a]	8,950
Faujasite(Na-X)	350	N_2	6,500	3,100	3,400
		CO	12,200	4,200	8,000
		NH_3	18,000	3,750	14,250
		H_2O	∿34,000	2,650	∿31,350
Faujasite(Na-Y)	350	CO	8,200	4,850[b]	3,350

(a) Assuming $\phi_D + \phi_R + \phi_P$ does not differ between H- and Na-mordenite.

(b) Assuming $\phi_D + \phi_R + \phi_P$ does not differ between Na-X and Na-Y.

6. THERMODYNAMIC CONSIDERATIONS

As with any distribution equilibrium thermodynamic analysis can be very informative, and indeed has been anticipated in §5 in discussing the differential heat of sorption.

6.1. Heat and Entropy of Sorption

An equilibrium condition for the distribution of a molecular species between the gas phase in which its chemical potential is μ_g and the zeolite in which this potential is μ_S is

$$\Delta\mu = \Delta\overline{H} - T\Delta\overline{S} = 0 \qquad (3)$$

where

$$\Delta\mu = \mu_s - \mu_g$$
$$\Delta\overline{H} = \overline{H}_s - \overline{H}_g$$
$$\Delta\overline{S} = \overline{S}_s - \overline{S}_g$$

$\Delta\overline{S}$ is the differential entropy of sorption per mole and $\overline{H}$ and $\overline{S}$ denote differential enthalpy and entropy respectively. For a pure molecular species $\overline{H}_g = H_g$ and $\overline{S}_g = S_g$ where H_g and S_g are enthalpy and energy per mole.

$\Delta\overline{H}$ may be determined calorimetrically, or from the Clapeyron-Clausius equation

$$\left(\frac{\partial \ell np}{\partial T}\right)_{n_s} = -\frac{\Delta\overline{H}}{RT^2} = \frac{q_{st}}{RT^2} \tag{4}$$

The term on the l.h.s. is the slope of a plot of ℓnp against T for a constant uptake, n_s, of guest species in a fixed weight of zeolite.

6.2. Equilibrium

The dimensionless equilibrium constant, K, for partition of the guest between the zeolite and the external phase is

$$K = a_s/a_g = C_s\gamma_s/C_g\gamma_g \tag{5}$$

where a denotes activity, C is concentration and γ is the activity coefficient. For a perfect gas and in the Henry's law dilute range of uptake where each $\gamma \to 1$ eqn. 5 becomes

$$K = C_s/C_g \tag{5a}$$

Thus in this dilute range of uptake eqn. 5a serves to give the dimensionless thermodynamic equilibrium constant. The standard energy and entropy of sorption are then

$$\Delta E^{\ominus} = RT^2 d\ell nK/dT \tag{6}$$

$$\Delta S^{\ominus} = R \ell nK + RT d\ell nK/dT \tag{7}$$

In experimental studies the Henry's law limit to the isotherm is often expressed as

$$K_p = C_s/p \tag{8}$$

Since for a perfect gas $p = C_g RT$ one has for this case

$$K = K_p RT \tag{9}$$

If sorption occurs at constant pressure, p, and the change in volume per mole sorbed is ΔV the enthalpy and energy of sorption are related by $\Delta H = \Delta E + p\Delta V$. However, $p\Delta V \sim -RT$ and so $\Delta H = \Delta E - RT$. It follows from this result and eqn. 9 that

$$\Delta H^{\Theta} = RT^2 d\ln K_p/dT \tag{10}$$

Plots of $\ln K_p$ against $(T/K)^{-1}$ are shown for some gases in H-chabazite in Fig. 13 (28). From the slopes the standard heats ΔH^{Θ} may be found, and are illustrated in Table 6. ΔH^{Θ} is in the order of the polarisabilities of molecules with negligible electric moments but is augmented for N_2 and CO_2 by the quadrupole moments of these two molecules.

6.3. Entropy of Sorbed Molecules

From $\Delta \overline{S} = \Delta \overline{H}/T$ and $\Delta \overline{S} = \overline{S}_s - S_g$ one may, when the phase external to the zeolite is a perfect gas, obtain the differential entropy per mole of sorbed guest for each amount sorbed. This entropy is

$$\overline{S}_s = S_g^{\Theta} + R \ln p^{\Theta}/p + \Delta \overline{H}/T \tag{13}$$

The entropy S_g^{Θ} of many gases at the standard pressure $p^{\Theta}(= 1 \text{ atm})$ may be interpolated from tables for the experiment at temperature T. $\overline{S}_s$ is finite and has the following characteristics:-

1. In energetically nearly uniform sorbents $\overline{S}_s$ decreases continually with increasing intracrystalline uptakes. However as the relative vapour pressure approaches unity and sorption upon

Table 6: Standard heats of sorption in kJ mol^{-1} for some gases in H-chabazite (28).

Gas	Temperature interval (T/K)	$-\Delta H^{\Theta}$	Gas	Temperature interval (T/K)	$-\Delta H^{\Theta}$
H_2	78.1–90.2	8.1_2	Ne	77.9–90.0	5.40
O_2	135.2–196.7	14.3_5	Ar	137.8–183.2	14.2_0
N_2	126.0–209.7	20.0_5	Kr	144.5–220.2	18.5_5
CO_2	229.6–373.2	37.9_6	Xe	211.4–296.0	24.7_0

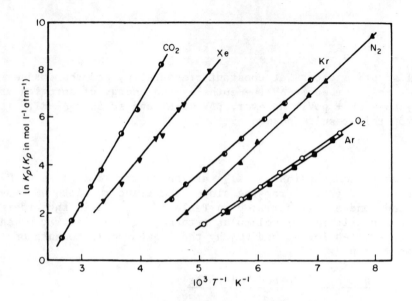

Fig. 13. ℓn K_p plotted against reciprocal absolute temperature for some gases in H-chabazite (28).

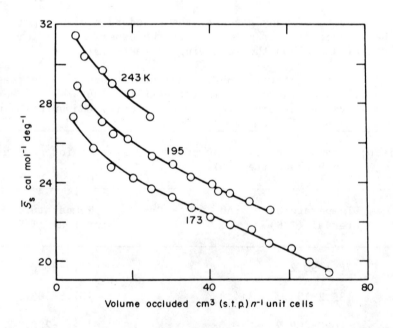

Fig. 14. Differential entropy per mole for sorption of N_2 in K-faujasite (K-X) at several absolute temperatures (29).

external surfaces takes over from intracrystalline sorption $\overline{S}_S$ rises again.

2. In energetically non-uniform sorbents (true of most zeolites) $\overline{S}_S$ may initially decline only a little as uptake increases, or may initially increase with uptake. It then passes through a maximum and finally declines again, as in case 1 above.

3. At different temperatures the curves $\overline{S}_S$ against uptake all tend to run parallel with each other. $\overline{S}_S$ has a well defined positive temperature coefficient (Fig. 14 (29)).

$\overline{S}_S$ may be considered in terms of a thermal part, $\overline{S}_{Th}$, and a configurational part, $\overline{S}_C$. For the ideal Langmuir isotherm ($K = \Theta/p(1 - \Theta)$) the thermal part should be independent of the degree of filling, Θ, of intracrystalline pore space, while the configurational part is given by

$$\overline{S}_C = R \ln [(1 - \Theta)/\Theta] \tag{12}$$

Thus as $\Theta \to 0$, $\overline{S}_C \to + \infty$ and as $\Theta \to 1$, $\overline{S}_C \to - \infty$. At $\Theta = 0.5$ $\overline{S}_C = 0$ and thus $\overline{S}_S = \overline{S}_{Th}$. However, the Langmuir isotherm is an idealisation from which real systems differ to a greater or lesser degree.

6.4. Heat Capacity of Intracrystalline Guest Molecules

The differential heat capacity per mole of sorbed guest, $\overline{C}_S$, is, for a given uptake related to $\overline{S}_S$ by

$$\overline{C}_S = T \left(\frac{\partial \overline{S}_S}{\partial T} \right)_{n_S} \tag{13}$$

Over a finite temperature interval, δT, in which $\overline{S}_S$ changes by $\delta \overline{S}_S$ eqn 13 may be replaced by

$$\overline{C}_m = T_m \left(\frac{\delta \overline{S}_S}{\delta T} \right)_{n_S} \tag{13a}$$

where the subscript m denotes the mean value over the interval δT. When the guest molecules were the rare gases the values of $\overline{C}_m$ are exemplified in Table 7. The value anticipated for an ideal Einstein oscillator is about 25 JK^{-1} mol^{-1}. For hydrocarbons in Na-X $\overline{C}_m$ increases with carbon number, but was always less than the value for corresponding bulk liquid (30).

Table 7: $\overline{C}_m(JK^{-1}\ mol^{-1})$ for Kr and Xe in some zeolites

Zeolite	T_m/K	Uptake (cm^3 at s.t.p. g^{-1})	$\overline{C}_m$ Kr	Xe
Na-Y	473	15	22	24
Ca-X	473	15	20	21
Ca-A	473	15	20	25
H-mordenite	473	15	18	22
Na-mordenite	473	15	20	19
Chabazite	473	15	21	19
H-offretite	149.2	2[a]	22.6	–
H-erionite	164.7	4[a]	22.	–
H-L	166.2	3[a]	22.2	–

[a] molecules per unit cell

7. ISOTHERM FORMULATION

A full treatment of the isotherm equation would need to take account of the following experimental features.

1. At a given relative pressure apparent saturation capacities for a given guest species decrease as the temperature rises.

2. At constant sorption potential the thermal expansion co-efficients of guest species within zeolites are not very different from those of corresponding liquids (31).

3. Fewer large molecules than small ones saturate the intra-cystalline pore space (Table 3). Thus classical site models with one molecule per site are inadequate in that the so called site and the number of sites would need to be different according to the molecular volume of each guest.

4. In intracrystalline pores and channels the molecules of a given guest species are not all bound with the same energy. The binding energy varies with position of a molecule relative to the walls of the pores and channels and to the cations present.

5. Molecule-molecule interaction in the clusters or filaments of guest molecules within the zeolite cannot normally be ignored (§. 5).

The first two of the above properties, and the mobility of

sorbed molecules, suggest liquid-like properties of clusters and
filaments of guest species in the zeolite. One may then assume
that there is a mean hydrostatic stress intensity, P, in this fluid
which is related to C_s by a virial equation:

$$P = C_s RT(1 + A_1 C_s + A_2 C_s^2 + A_3 C_s^3 + \ldots) \tag{14}$$

If the gas phase behaves ideally ($p = C_g RT$) then a thermodynamic
argument gives the virial isotherm equation as

$$K = \frac{C_s}{C_g} \quad \exp[2A_1 C_s + (3/2) A_1 C_s^2 + (4/3)A_3 C_s^3 + \ldots] \tag{15}$$

or the corresponding expression for K_p = K RT (eqn. 9) in which C_g
is replaced by p. The A_i (i = 1, 2, 3 ...) correspond with virial
coefficients in the virial equation of bulk guest species in sense,
but because of the restricted environment not in numerical values.
The coefficients A_i will therefore take care of molecule-molecule
interactions. The A_i may be functions of temperature but not of
C_s. The exponential term is the activity coefficient, γ_s, of the
sorbed molecules. As C_s approaches its saturation value the number
of coefficients A_i needed in eqns. 14 and 15 increases. However
Fig. 15 (28) shows as an example that with not more than three
coefficients A_i isotherms of some gases in H-chabazite are well
represented. When C_s becomes small the exponential term in eqn. 15
declines to unity so that Henry's law is then obtained. Thus the
isotherm equation lends itself to thermodynamic analysis. Its great
generality does not however give much insight about events at
molecular level.

The second approach which will be considered is a site model in
which each cavity is regarded as a site-capable of accommodating a
cluster of up to m guest molecules. The isotherm has been developed
using detailed balancing (32) and by statistical mechanics (33)
with the expected equivalent results. In terms of detailed balancing
the isotherm equation is written as

$$\theta = \frac{R_1 + (m-1)R_1 R_2 + \dfrac{(m-1)(m-2)}{1.2} R_1 R_2 R_3 + \ldots + (R_1 R_2 \ldots R_m)}{1 + m R_1 + \dfrac{m(m-1)}{1.2} R_1 R_2 + \ldots\ldots + (R_1 R_2 \ldots R_m)} \tag{16}$$

In eqn. 16

$$R_{i+1} = \left[\vec{k}_i / \overset{\leftarrow}{k}_{(i+1)} \right] p = K_{i+1} p \tag{17}$$

254

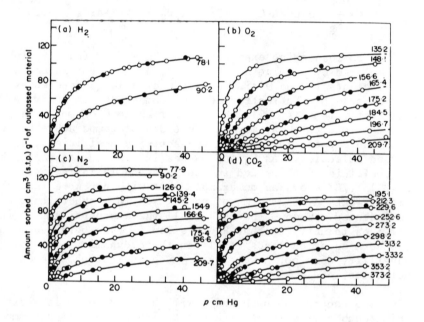

Fig. 15. Isotherms for gases in H-chabazite at the absolute
temperatures indicated (28).
O = experimental points
● = calculated using the virial isotherm equation.

$\vec{k}_i$ is the rate constant for condensation into the cavity (or site) already holding i molecules and $\overleftarrow{k}_{(i+1)}$ is the rate constant for evaporation of a molecule from a cavity carrying (i + 1) guest molecules. K_{i+1} is the equilibrium constant between i and (i + 1) molecules per cavity. In the ideal case when all $\vec{k}_i$ are equal and so are all $\overleftarrow{k}_{(i+1)}$ one has

$$R_1 = R_2 = \ldots = R_m = Kp \tag{18}$$

and eqn. 18 reduces to Langmuir's equation

$$\Theta = \frac{Kp}{1 + Kp} \; ; \quad K = \frac{\Theta}{p(1 - \Theta)} \tag{19}$$

Langmuir's equation is also recovered from eqn. 16 when m = 1, so that it becomes in either of these two ways an important limiting case. The validity of eqn. 18 would mean negligible molecule-molecule interaction within the zeolite; conversely such inter-actions will result in values of R_i which change with i. Reference to Table 3 suggests values for m for a number of guest molecules in chabazite and faujasite (zeolite X). m will be the nearest whole number for the cluster size.

In measurements of isotherms at 673 K for C_{10}, C_{12}, C_{14}, C_{16}, and C_{18} n-paraffins in (Mg, Na)-A and (Ca, Na)-A the value of m was near to unity for each hydrocarbon, independently of carbon number (34). This suggests that each chain is coiled so that, except in the act of diffusing, each molecule is confined to one cavity and there is no room for more than one coil per 26-hedral cavity (35). Here, where m = 1, it is of interest that the strongly curved iso-therms obeyed Langmuir's isotherm reasonably well. The semi-isolation of one coil in each cage would reduce molecule-molecule interaction, which in turn would favour the Langmuir isotherm.

When, in the above study, the value of K given by eqn. 19 was plotted against carbon number K at first increased but then passed through a maximum. A reason for this could be seen from

$$-RT \ln K = \Delta G^\Theta = \Delta H^\Theta - T\Delta S^\Theta \tag{20}$$

ΔH^Θ and ΔS^Θ are each negative, so that $-T\Delta S^\Theta$ is positive and ΔH^Θ and $-T\Delta S^\Theta$ oppose each other in determining K. The values of $-\Delta H^\Theta$ and $-\Delta S^\Theta$ both increase with carbon number, but the coiled n-paraffins of highest carbon number fit more tightly within the cavity and so increments in $-T\Delta S^\Theta$ as carbon number increases finally outweigh increments in ΔH^Θ in influencing ΔG^Θ or K. The more restricted the coiled hydrocarbon is within its cavity the greater the loss in entropy when sorption occurs.

256

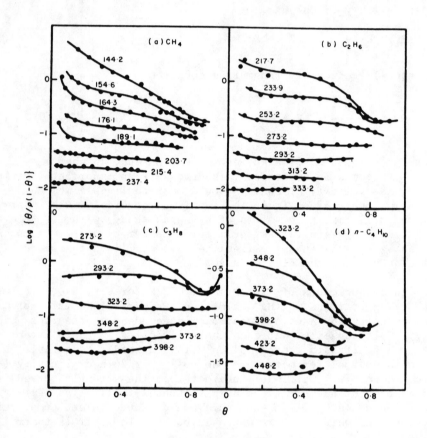

Fig. 16.   Log $[\theta/p(1 - \theta)]$ plotted against θ for some hydrocarbons in H-chabazite, at various absolute temperatures (36).

A sensitive way to check how nearly actual isotherms approach the ideal case is to plot $\Theta/p(1 - \Theta)$, or its logarithm or reciprocal against Θ. Langmuir's isotherm gives a straight line parallel to the axis of Θ. Examples of such plots are given in Fig. 16 (36) for C_1, C_2, C_3 and n-C_4 paraffins in H-chabazite at each of a series of temperatures. The behaviour is often characteristic in that

1. at low temperatures negative slopes are obtained;
2. as the temperature rises the slopes become less negative; and
3. at still higher temperatures horizontal regions may be obtained, or regions with positive slopes.

The slopes of the lines will be the nett result of the interplay of the usual energetic heterogeneity (which on its own would make $-\Delta\overline{H}$ decrease with increasing Θ and would thus result in negative slopes); of molecule-molecule interaction (which is usually exothermic and if so tends to give positive slopes); and of values of $R_{(i+1)}$ which as a result of energetic heterogeneity and molecule-molecule interaction would vary with i (so that the unabridged eqn. 18 would be required rather than the Langmuir limiting case). Values of m in chabazite would be about 6 for CH_4, 4 for C_2H_6, 3 for C_2H_8 and 2 for n-C_4H_{10}.

3. CONCLUDING REMARK

It is hoped that the necessarily limited examples given have shown the importance of combining accurate thermodynamic data with modelling for isotherms and energetics in interpreting the physical bond in host-guest complexes, selectivity in sorption and the character of the intracrystalline fluid. Considerable progress has been made in this important area but much remains to be done.

REFERENCES

1. Gramlich-Meier, R. and W.M. Meier, J. of Solid State Chemistry, 44, (1982), 41.
2. Barrer, R.M., Zeolites and Clay Minerals as Sorbents and Molecular Sieves, (London, Academic Press, 1978) pp. 353 *et seq.*
3. Ref. 2, pp. 307 *et seq.*
4. Ref. 2, pp. 398 *et seq.*
5. Barrer, R.M., Hydrothermal Chemistry of Zeolites (London, Academic Press, 1982), Chap. 7.
6. Barrer, R.M., J. Soc. Chem. Ind., 44, (1945) 130 and 133.
7. Barrer, R.M. and L. Belchetz, J. Soc. Chem. Ind., 44, (1945), 131.
8. Barrer, R.M., British Chem. Eng., 4, (1959) 267.
9. Meier, W.M. and D.H. Olson, Atlas of Zeolite Structure Types (Structure Commission of the International Zeolite Ass., 1978).
10. Barrer, R.M., J. of Inclusion Phenomena, 1, (1983), in press.
11. Ref. 2, p. 295 *et seq.*
12. Barrer, R.M. and M.A. Rosemblat, Zeolites 2, (1982) 231.
13. Ref. 2, p. 282.
14. Takaishi, T., Y. Yatsurugi, A. Yusa and T. Kuratomi, J. Chem. Soc. Faraday Trans. I, 70, (1974) 97.
15. Barrer, R.M. and D.E.W. Vaughan, J. Phys. Chem. Solids, 32, (1971) 731.
16. Brunauer, S., The Adsorption of Gases and Vapours (Oxford Univ. Press, 1944) p. 150.
17. Barrer, R.M. and P.J. Reucroft, Proc. Roy. Soc., 258A, (1960) 431.
18. Ref. 2, p. 391.
19. Barrer, R.M., E.A. Daniels and G.A. Madigan, J. Chem. Soc. Dalton Trans., (1976) 1805.
20. Ref. 5, p. 336.
21. Ref. 5, pp. 316-7.
22. Ref. 5, p. 325.
23. Ref. 2, p. 394-5.
24. Olson, D.H., G.T. Kokotailo, S.L. Lawton and W.M. Meier, J. Phys. Chem., 85, (1981) 2238.
25. Barrer, R.M., J. Coll. and Interface Sci., 21, (1966) 415.
26. Fiedler, K., H.-J. Spangenberg and W. Schirmer, Monatsber., 9, (1971) 1.
27. Ref. 2, p. 119.
28. Barrer, R.M., and J.A. Davies, Proc. Roy. Soc., 320A, (1970) 289.
29. Ref. 2, p. 245.
30. Ref. 2, pp. 346-249.
31. Ref. 2, p. 128.
32. Ref. 2, pp. 145-147.
33. Brauer, P., A.A. Lopatkin and G.Ph. Stepanez, in Molecular Sieve Zeolites (Amer. Chem. Soc. Advances in Chemistry Series

No. 102, 1971) p. 97.
34. Fiedler, K., A. Roethe, K.P. Roethe and D. Gelbin, Z. Phys. Chem. Leipzig, 257, (1978) 979.
35. Barrer, R.M., J. Chem. Techn. Biotechnol., 31, (1981) 71.
36. Ref. 2, p. 117.

SORPTION BY ZEOLITES
PART II.   KINETICS AND DIFFUSIVITIES

R.M. Barrer

Chemistry Department
Imperial College of Science and Technology
London SW7 2AY
England

SUMMARY

An objective in the study of sorption-desorption kinetics is
the accurate determination of differential intrinsic and self-
diffusivities and their concentration dependance. Attention has
been drawn to experimental problems in such measurements, to
situations where intracrystalline diffusion is not rate-controlling
and to ways in which these alternative controls may be recognized
and minimised. Some properties of self- and intrinsic diffusivities
have been reviewed, including their dependance on molecular
dimensions and shape in relation to window apertures and the types
of cation present in the zeolite.

1.   INTRODUCTION

Sorption rates in beds of zeolite crystals play an important
part in separations of mixtures dependent on partial and total
molecule sieving. These rates are equally significant in catalysis
where reactants must reach intracrystalline catalytic centres
against a counter flow of resultants out of the crystals.
Accordingly increasing study is being devoted to this area. The
present account will be limited to rates of sorption and desorption
of single species.

2.   SORPTION KINETICS

If the rates of uptake are sufficiently slow, giving long

half-life times, these rates can reasonably be ascribed to intracrystalline diffusive flow. While most studies have involved beds
of zeolite powders or of bonded pellets containing zeolite crystals,
a limited number have been made using one large crystal (1,2) or
a small number of pieces of such a crystal (3). In these large
pieces there is no difficulty in studying intracrystalline diffusion, but in fine powders often in the size range ~0.1µ to ~10µ
or in bonded pellets containing these small crystals complications
may arise. Possible rate-controlling steps are:

1. Intracrystalline diffusion.
2. Intercrystalline diffusion and flow.
3. Transmission through surface skins.
4. Evolution of heat on sorption and cooling on desorption
   with resultant time-dependent drifts of sorption
   equilibria.
5. Combinations of two or more of the above possibilities.

Accordingly interpretation of sorption kinetics needs care.

2.1. Evaluation of Intracrystalline Diffusivities

For long half-life times such that intracrystalline diffusion
is normally rate-controlling the sorption rate curves are available
for interpretation in terms of intracrystalline diffusivities.
Those of interest are differential intrinsic and self-diffusivities,
$\overline{D}$ and $\overline{D}^*$, and, where these are functions of concentration, the
corresponding integral diffusivities $D = (1/C) \int_o^C \overline{D}dC$ and
$D^* = (1/C) \int_o^C D^* dC. \overline{D}$ and $\overline{D}^*$ are related approximately by the
Darken relation $\overline{D} \simeq \overline{D}^* d\ell na/d\ell nC$ (see §.5). In these relations C
is the concentration and a the activity of sorbed guest.

In the simplest situation when $\overline{D}$ is constant $\overline{D} = D$ and the
diffusion equation is, for concentration C and time t,

$$\partial C/\partial t = D \text{ div grad } C \tag{1}$$

If sorption occurs into a powder of isotropic crystals approximated
as spheres all of radius r_o, and if the boundary conditions are

$$C = C_\infty \text{ at } r = r_o \text{ for } t > 0$$

$$C = C_o \text{ for } 0 < r < r_o \text{ at } t = 0 \tag{2}$$

the solution is

$$\frac{M_t}{M_\infty} = \frac{Q_t - Q_o}{Q_\infty - Q_o} = 1 - \frac{6}{\pi^2} \sum_{n=1}^{\infty} \frac{1}{n^2} \exp\left(-\frac{Dn^2\pi^2 t}{r_o^2}\right) \tag{3}$$

or alternatively

$$\frac{M_t}{M_\infty} = \frac{6}{r_o} \left(\frac{Dt}{\pi}\right)^{\frac{1}{2}} \left\{1 + 2\pi^{\frac{1}{2}} \sum_{n=1}^{\infty} \text{ierfc} \frac{n\,r_o}{\sqrt{Dt}}\right\} - \frac{3Dt}{r_o^2} \qquad (4)$$

In these expressions C_∞ is the equilibrium concentration of guest within the crystals and C_o is the initial concentration. Q_t, Q_∞ and Q_o are amounts of guest in the crystals at $t = t$, ∞ and 0 respectively, while ierfc x denotes $[(1/\sqrt{\pi})\exp\text{-}x^2\text{-}x(1\text{-erf } x)]$. The series in eqn. 3 converges only slowly for small t; that in eqn. 4 converges only slowly for large t. For small t eqn. 4 gives

$$\frac{M_t}{M_\infty} = \frac{6}{r_o} \left(\frac{Dt}{\pi}\right)^{\frac{1}{2}} \qquad (5)$$

so that M_t/M_∞ is proportional to $t^{\frac{1}{2}}$, with slope $\frac{6}{r_o}\left(\frac{D}{\pi}\right)^{\frac{1}{2}}$. D is then obtainable. For large t eqn. 3 reduces to

$$\frac{M_t}{M_\infty} = 1 - \frac{6}{\pi^2} \exp -\left(\frac{D\pi^2 t}{r_o^2}\right) \qquad (6)$$

so that a plot of $\ln(1\text{-}M_t/M_\infty)$ against t approaches a straight line of slope $-\frac{D\pi^2}{r_o^2}$ and this serves also to give D.

The boundary conditions in eqn. 2 require that uptake should occur at constant pressure of guest in the gas phase just outside the surfaces $r = r_o$ of the crystals. Alternatively sorption can occur at constant volume and variable pressure. At $t = 0$ a dose of gas enters the sorption volume and is taken up by the host. Provided Henry's law governs the equilibrium isotherms ($C_\infty = kp$), new equations corresponding with 3 and 4 describe the uptake. For large t that corresponding with eqn. 3 gives

$$\ln\left(\frac{Q_\infty - Q_t}{Q_\infty - Q_o}\right) = \ln\left[\frac{6K(K + 1)}{9(K + 1) + \alpha_1^2 K^2}\right] - \frac{\alpha_1 Dt}{r_o^2} \qquad (7)$$

where K = amount of guest in the gas phase/amount in the crystals, at equilibrium, and α_1 is the first root of

$$\text{Tan}\alpha = 3\alpha/(3 + K\alpha^2) \qquad (8)$$

A plot of $\ln\left(\frac{Q_\infty - Q_t}{Q_\infty - Q_o}\right)$ against t again serves to find D. For

264

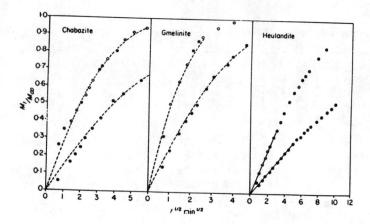

Fig. 1.   Sorption kinetics of water at constant pressure into crystals of chabazite, gmelinite and heulandite (3). The dashed lines are calculated curves. The temperatures of the runs in $^\circ C$ are: chabazite, ⊕, 75.4; ●, 30.8; gmelinite, ⊙, 62.5; ●, 31.7; heulandite, ⊙. 77.8; ●, 37.4.

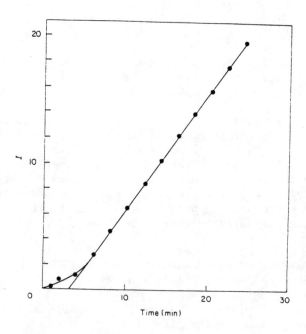

Fig. 2.   Plot of $I = \int_{o}^{t} (Q_t/Q_\infty)\,dt$ against t for $n\text{-}C_6H_{14}$ at $33.4^\circ C$ in zeolite H-RHO (4).

small t the equivalent of eqn. 4 gives

$$\frac{M_t}{M_\infty} = \frac{6}{r_o} \cdot \frac{(1 + K)}{K} \left(\frac{Dt}{\pi}\right)^{\frac{1}{2}} \tag{9}$$

and a plot of M_t/M_∞ against $\sqrt{t}$ again serves to give D.

Examples of rate curves for the uptake of H_2O by sizeable pieces of single crystals of chabazite, gmelinite and heulandite at each of two temperatures are shown in Fig. 1 (3) for the constant pressure boundary conditions of eqn. 2. When t is small they show the validity of the $\sqrt{t}$ eqn.

Diffusion coefficients can be determined in other ways. One such involves plotting $\int_0^t (M_t/M_\infty) dt$ against t, as shown for $n-C_6H_{14}$ at 33.4°C in zeolite RHO in Fig. 2 (4). This plot approaches an asymptote of unit slope which makes an intercept L_a on the axis of t. For several geometries and for constant pressure, or for constant volume combined with Henry's law, the results are (5):

	L_a Constant pressure	Constant volume, Henry's law
Spheres, radius r_o	$r_o^2/15D$	$r_o^2 K/15D(K+1)$
Long cylinders, radius r_o	$r_o^2/8D$	$r_o^2 K/8D(K+1)$
Sheet, thickness 2ℓ	$\ell^2/3D$	$\ell^2 K/3D(K+1)$

These results allow ready evaluations of D for appropriate geometries and constant diffusivities.

2.2. Self-diffusivities

Differential self-diffusivities can be measured through exchange diffusion between the zeolite having a given loading of sorbate and the gas phase of the sorbate. Either the zeolite or the gas phase contains labelled molecules for time t > 0. Labelling may for example be with radio-carbon, or with deuterium. The boundary conditions of §. 2.1 are convenient and with them the mathematical relations already given are applicable.

Another effective way of measuring self-diffusion is by pulsed field gradient NMR. For a bed of zeolite powder an over-all or effective diffusivity is given by

$$D_{eff} = \left\{ \overline{D}^* + \frac{p \, \overline{D}^*_g}{\gamma^2 \sigma^2 g^2 \tau p \overline{D}^*_g + 1} \right\} \simeq \frac{\langle w^2 \rangle}{6\Delta} \qquad (10)$$

In this expression Δ is the time interval between pulses of width σ and amplitude g. $\langle w^2 \rangle$ is the mean square displacement of a molecule over the interval Δ; τ is the lifetime of guest molecules within the crystals (all of radius r_o). γ is the gyromagnetic ratio, p is the fraction of the molecules in the bed which are in the gas phase and $\overline{D}^*_g$ is the differential self-diffusivity of the guest in this gas phase. There are two extremes in the above expression:

1. $\tau \gg \Delta$ and so $\langle w^2 \rangle^{\frac{1}{2}} \ll r_o$. In this limit $D_{eff} \to \overline{D}^*$.
2. $\tau \ll \Delta$ and so $\langle w^2 \rangle^{\frac{1}{2}} \gg r_o$. Here $D_{eff} \to p\overline{D}^*_g$.

These two extremes can be illustrated by results for cyclohexane in Na-X, when $\ln D_{eff}$ is plotted against K/T (Fig. 3 (6)). Below -65^oC $D_{eff} \to \overline{D}^*$ while above $+10^o$ $D_{eff} \to p\overline{D}^*_g$, the temperature coefficient being primarily that of p. The rather flat intermediate region is one where translational diffusion is restricted mainly to the intracrystalline pore space of Na-X because the thermal energy is insufficient for most molecules to evaporate from the crystals. Besides giving $\overline{D}^*$ and $p\overline{D}^*_g$ the NMR method, from plots of D_{eff} against K/T, allows one to find the temperature range where gas phase and intracrystalline diffusion are both influencing D_{eff}, as seen in Fig. 3.

2.3. Determining the Energy of Activation for Diffusion

Intracrystalline diffusion involves migration of molecules from one site to another often with a squeeze past an obstruction such as a narrow window. Thus unit diffusion steps require an activation energy, E, and D can be expressed in terms of the Arrhenius relation

$$D = D_o \exp{-E/RT} \qquad (11)$$

In plots of M_t/M_∞ against t at two temperatures T_1 and T_2 one measures the times t_1 and t_2 for M_t/M_∞ to reach a chosen value. Then, from eqn. 3 for example, one must have for constant pressure

$$D_1 t_1 = D_2 t_2 \qquad (12)$$

Accordingly, from eqns. 11 and 12

$$-\ln t_2/t_1 = \ln D_2/D_1 = \frac{E}{R} \left[\frac{1}{T_1} - \frac{1}{T_2} \right] \qquad (13)$$

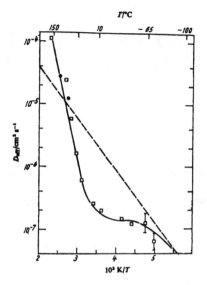

Fig. 3. D_{eff} for cyclohexane in Na-X (10) with $\sim$ 2 molecules per large cavity ($\square$), and comparison with $\overline{D}^*$ (dashed line) and with corresponding values from sorption rates using the method of moments ($\bullet$).

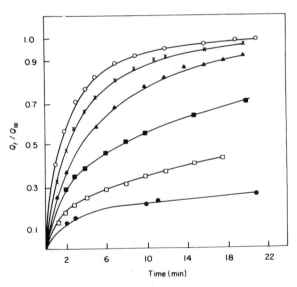

Fig. 4. Sorption-desorption kinetics of n-hexane in zeolite H-RHO (4). Q_t is the amount sorbed or desorbed at time t and Q_∞ is the final amount sorbed or desorbed. Temperatures of runs in °C are:

Sorption		Desorption	
O,	33.4	O,	33.4
x,	95.6	☐,	95.6
▲,	141.4	■,	14.4

so that E is obtained. This method is applicable for any distribution of crystal shapes and sizes because it involves only eqn. 10 and __experimental__ curves of M_t/M_∞ against t.

3. COMPLICATING FACTORS

The expressions in § 2.1 relating M_t/M_∞ and time refer to strict boundary conditions, constant diffusivity, crystals all of one size and shape, isothermal conditions throughout each bed within each crystal and intracrystalline diffusion as the rate-controlling step. However eqns. 5 and 9, in which M_t/M_∞ is proportional to $\sqrt{t}$, apply to small t for any distribution of crystal size and shape. It is only necessary that t is such that each crystallite is behaving as a semi-infinite medium (i.e. the deep interiors are free of diffusant).

When the above requirements of the relationships in § 2.1 are compared with real situations the following disturbing factors may arise (cf. § 2):

1. Boundary conditions as given in § 2.1 are not maintained, and in the case of variable pressure constant volume runs sorption isotherms do not obey Henry's law.

2. Where intracrystalline diffusion is fast the rate-determining step may cease to be governed only by intracrystalline diffusion but by surface barriers or by inter-particle flow (§ 2.2., eqn. 10).

3. Intra-crystalline diffusivities may be functions of the concentration of diffusant.

4. Because of liberation of the heat of sorption, the uptake of guests does not occur under isothermal conditions. Heating and then cooling of the bed and of each crystallite superposes drifts in equilibria and rates expected for isothermal uptakes. These effects are further discussed in order, below.

3.1. Maintenance of Boundary Conditions

The problem of maintaining rigorous boundary conditions is illustrated by a recent study of n-hexane uptake in the hydrogen form of zeolite RHO (Fig. 4 (4)). The boundary conditions sought for sorption were those of eqn. 2, and the corresponding ones for desorption were

(i) $C = C_o$ in each crystal at time t = 0

(ii) $C = 0$ just within each crystal at r_o for t > 0.

The uptakes were measured gravimetrically using a silica spring balance with a bed of 0.3 g of zeolite powder in a glass bucket. Fig. 4 shows that

(a) sorption rates measured as Q_t/Q_∞ vs t were much larger than desorption rates; and

(b) for sorption Q_t/Q_∞ at given t had a negative temperature coefficient whereas in desorption Q_t/Q_∞ had a positive temperature coefficient. The higher the temperature therefore the more nearly did Q_t/Q_∞ vs t for sorption and desorption approach one another.

The above behaviour was considered in the following terms. Although the constant pressure boundary conditions of eqn. 2 were sought, when the valve from the liquid n-hexane reservoir to the sorption volume was opened a considerable period elapsed before the equilibrium vapour pressure was established in the sorption volume. At low temperature, say T_1, all pressures in the sorption volume correspond with points along the flat top of the very rectangular sorption isotherm, and therefore with saturation uptake. The boundary conditions of eqn. 2 were accordingly met:

(i) C = 0 at t = 0 in each crystal

(ii) C = C_∞ at t > 0 just within each crystal at r_0.

As T increases the isotherm becomes less and less rectangular so that the changing pressure means that C just within a crystal varies with time and only slowly approaches C_∞. Thus Q_t/Q_∞ is less than it would have been at time t if the second of the above boundary conditions had been valid at once.

A second factor also operates. If the rate of uptake by the crystals is comparable with the rate of supply of vapour by inter-crystal flow there will be a pressure gradient into the bed. At the low temperature, T_1, the rectangular isotherm still ensures the approximate validity of the second of the above boundary conditions. However, for the less rectangular isotherms at higher T this boundary condition will be less and less correct and the deviation will increase with depth in the bed. This behaviour augments that of the previous paragraph in making Q_t/Q_∞ even less than it would have been had the boundary conditions of eqn. 2 been met. If the two effects influence the temperature dependance of Q_t/Q_∞ more than does the diffusivity, with its positive temperature dependance, then the observed negative temperature coefficient in curves of Q_t/Q_∞ vs t for sorption will result.

We next consider why desorption is so much slower than sorption and has a positive temperature coefficient (Fig. 4). On reducing the pressure of sorbate outside the bed to the lowest possible vacuum level, within the bed, as the crystals shed n-hexane, there will remain residual pressures which are greatest at depth. At the low temperature, T_1, the rectangular isotherm ensures large departures from the second boundary condition for desorption (C = 0 for t > 0 just within each crystal surface). Therefore Q_t/Q_∞ is much less at given t than it would be had this second boundary condition been met. It will therefore be much less also than Q_t/Q_∞ for sorption at time t and temperature T_1 where the boundary conditions of eqn. 2 are met.

As temperature increases and the isotherms become less and less rectangular the residual pressures at depth in the bed will result in smaller departures from the condition C = 0 for t > 0 just within each crystal surface, and Q_t/Q_∞ for given t will increase as compared with Q_t/Q_∞ at the low temperature T_1. When this effect is combined with the positive temperature coefficient of intracrystalline diffusivity the positive temperature coefficient in curves of Q_t/Q_∞ vs t for desorption (Fig. 4) is to be expected.

It can be concluded that for rectangular isotherms, provided sorption is sufficiently slow to minimise heating of the bed and to avoid rate control by interparticle flow, sorption kinetics can serve to give diffusivities, D, but desorption may not. Also, as isotherm curvature becomes less and less, in the Henry's law limit curves of Q_t/Q_∞ for sorption and desorption should coincide and either could, under constant pressure or constant volume variable pressure conditions, serve to give D.

The major importance of boundary conditions is thus clear, through their influence on the interpretation of the kinetics. Equations such as those in § 2.1 are valid for this purpose only when the boundary conditions for which they were derived are strictly maintained. This can experimentally be difficult to achieve.

3.2. The Rate-determining Step

For intra-crystalline diffusion to be rate-determining the process must be much slower than the rate of supply of guest molecules to the surfaces of the crystals comprising the bed. In some zeolites, such as Ca-A or faujasite (zeolite X and Y), this is far from the case for many guest molecules. Attempts to evaluate self-diffusivities for intra-crystalline diffusion from sorption kinetics have often led to gross discrepancies when compared with those determined from pulsed field gradient NMR, (§ 2.2) as illustrated by the figures in Table 1 (7). The NMR procedure can give $\bar{D}*$ for intracrystalline diffusion for high values of $\bar{D}*$ and is therefore in these situations a suitable standard for comparisons (8, 9, 10). The table shows that the

Table 1: Comparison of $\bar{D}^*$ (cm^2 s^{-1}) from pulsed NMR and from sorption Kinetics (7).

System	Temp. (°C)	Molecules per cavity (NMR)	$\bar{D}^*$ (NMR)	$\bar{D}^*$ (sorption rates)
CH$_4$/Ca-rich A	23	5	2×10^{-5}	5×10^{-10}
C$_2$H$_6$/Ca-rich A	23	3	2×10^{-6}	10^{-10}
C$_3$H$_8$/Ca-rich A	23	4	5×10^{-8}	3×10^{-11}
n-C$_7$H$_{16}$/Na-X	164	0.6	5×10^{-5}	3×10^{-9}
cyclo-C$_6$H$_{12}$/Na-X	164	1.3	4.5×10^{-5}	4×10^{-9}
C$_6$H$_6$/Na-X	164	1	2×10^{-6}	10^{-10}
C$_2$H$_4$/(Na,Ag)-X	23	3	2×10^{-5} (0% Ag)	5×10^{-7} (0% Ag)
	23	3	7×10^{-6} (10% Ag)	10^{-7} (10% Ag)
	23	3	2×10^{-6} (20% Ag)	5×10^{-8} (20% Ag)
	23	3	2×10^{-7} (50% Ag)	2×10^{-8} (50% Ag)
	23	3	10^{-7} (100% Ag)	10^{-8} (100% Ag)

discrepancies can be very great indeed; they indicate that in sorption kinetics the rate-controlling step is not just intra-crystalline diffusion but must involve other factors such as inter-crystal flow and heat effects. The kinetic data have in many cases been wrongly interpreted in terms of intra-crystalline diffusion alone.

If intra-crystalline diffusion is rate-controlling then the values of the diffusivity obtained from sorption kinetics must

(a) be independent of the size of the crystallites used in the measurements, for beds of the same form and depth; and
(b) be independent of the thickness of the bed for crystal-lites all of the same size.

These criteria have not often been applied. As a third criterion, of course, kinetics and pulsed NMR must give the same values of self-diffusivity for equal loadings of the crystals.

Attempts which have been reasonably successful have been made to reconcile discrepant values of $\overline{D}*$ from pulsed field gradient NMR and from sorption kinetics by making the crystallites larger (Fig. 5) and making the bed of extreme thinness (6, 11), optimally no more than the thickness of a single crystal. Extreme care in the kinetic measurements is required even so, as exemplified by the half-life times, $t_{\frac{1}{2}} \simeq 2.1 \times 10^{-2} \; r_o^2/D$, involved for synthetic zeolites in the usual size range 0.1 to 10 μ (i.e. radius r_o from 10^{-5} cm to 10^{-3} cm). $t_{\frac{1}{2}}$ is given in Table 2 (12) for various values of r_o and D. The dashed line in the table divides the half-lives from sorption rates into those too small for accurate measurements with normal equipment from those which are long enough for accurate determination. In kinetic measurements crystal dimensions and diffusivities should be such as to give values of $t_{\frac{1}{2}}$ below the dashed line.

Table 2: Half-life times, $t_{\frac{1}{2}}(s)$ for:

r_o(cm); D(cm² s⁻¹) →	10^{-6}	10^{-8}	10^{-10}	10^{-12}	10^{-14}
10^{-5}	2.1×10^{-6}	2.1×10^{-4}	2.1×10^{-2}	2.1	2.1×10^{2}
10^{-4}	2.1×10^{-4}	2.1×10^{-2}	2.1	2.1×10^{2}	2.1×10^{4}
10^{-3}	2.1×10^{-2}	2.1	2.1×10^{2}	2.1×10^{4}	2.1×10^{6}
10^{-2}	2.1	2.1×10^{2}	2.1×10^{4}	2.1×10^{6}	2.1×10^{8}
10^{-1}	2.1×10^{2}	2.1×10^{4}	2.1×10^{6}	2.1×10^{8}	2.1×10^{10}

Measurements of $\overline{D}*$ by pulsed NMR on the other hand are normally suitable only for self-diffusivities above about 10^{-8} cm^2 s^{-1}. This method would therefore be unsuitable for small diffusivities for example those of n-paraffins in some zeolites such as (Na,Ca)-A, Ca-rich chabazite or erionite. In (Na,Ca)-A tracer diffusion using hydrocarbons labelled with radiocarbon has been successful in measurements of $\overline{D}*$ (13). Table 2 also shows that by suitably increasing the crystal dimensions values of diffusivities of any magnitude, from very small ($< 10^{-14}$ cm^2 s^{-1}) to very large ($> 10^{-6}$ cm^2 s^{-1}) are possible, using sorption kinetics.

Under conditions where intracrystalline diffusivities do not wholly or even partially govern sorption kinetics it may still be possible to describe the system in terms of an effective diffusivity, D_{eff}. Conditions can be made such that in the intercrystal spaces molecule-crystal collisions greatly exceed molecule-molecule collisions, so that molecular streaming (Knudsen flow) occurs in the inter-crystal spaces. We consider flow through a cylindrical bed bounded by planes $x = 0$ (the entry face) and $x = \ell$ (the exit face), and with no flow through the curved surface (i.e. the bed is in an open cylindrical container). The total flow, J, entering the bed across the plane $x = 0$ per unit area of this plane can be written as

$$J = -D_{eff} \left(\frac{dC'}{dx}\right)_{x=0} = -D_{eff} \left[\varepsilon \frac{dC'_g}{dx} + (1 - \varepsilon) \frac{dC'_i}{dx} \right]_{x=0} \tag{14}$$

D_{eff} is an effective diffusivity, ε is the porosity of the bed (exluding intracrystalline porosity), C is the total number of molecules per unit volume of porous medium at $x = 0$, and C'_g and C'_i are the numbers of molecules per unit volume of gas phase and of crystals, averaged across the plane at $x = 0$. Each gradient is an average of all local gradients across the plane. Also we may write

$$J = J_g + J_i \tag{15}$$

where J_g and J_i are respectively gas phase and intracrystalline components of J and

$$J_g = -D_g \varepsilon \left(\frac{dC'_g}{dx}\right)_{x=0} \tag{15a}$$

$$J_i = -D_i (1 - \varepsilon) \left(\frac{dC'_i}{dx}\right)_{x=0} \tag{15b}$$

where the gradients are again averages over the plane $x = 0$, D_g is the Knudsen flow diffusivity and D_i the intracrystalline diffusivity. Combination of eqns. 14 to 15b gives

$$D_{eff} = \left[D_g \varepsilon \frac{dC'_g}{dx} + D_i (1-\varepsilon) \frac{dC'_i}{dx} \right] \Big/ \left[\varepsilon \frac{dC'_g}{dx} + (1-\varepsilon) \frac{dC'_i}{dx} \right] \qquad (17)$$

Because in the transient state the gradients in eqn. 17 are functions of time D_{eff} may have significant time dependance when $D_g \neq D_i$. In the steady state of flow, however, such time dependances will have disappeared, for the boundary conditions

(i) $C = 0$ for $0 < x < \ell$ at $t = 0$
(ii) $C = C_o$ at $x = 0$ for $t > 0$
(iii) $C = C_\ell \ll C_o$ at $x = \ell$ for $t > 0$

A particular interest is to show in a given system whether inter- or intracrystalline flow is the dominant component of J through $x = 0$. This is possible to find when Henry's law governs sorption equilibrium $(C'_i = kC'_g)$, D_g is independent of C'_g and in this range D_i is independent of C'_i. Eqn. 17 now reduces to

$$D_{eff} = \left[D_g \varepsilon + k D_i (1-\varepsilon) \right] \Big/ \left[\varepsilon + k(1-\varepsilon) \right] \qquad (18)$$

so that D_{eff} is constant. For uniform packing of the bed dC/dx in the steady state is the same at all planes $x = X$ and equals $-(C_o-C_\ell)/\ell$ so that, from $J = D_{eff}(C_o-C_\ell)/\ell$, D_{eff} is found. ε and k are available independently. D_g is best assessed using helium as a vitually non-sorbed reference gas to evaluate D_g^{He} from $J^{He} = J_g^{He}$. If M is the molecular weight of the diffusant and M_{He} that of helium, then

$$D_g \qquad D_g^{He} \sqrt{\frac{M_{He}}{M}} \qquad (19)$$

Thus D_i in eqn. 18 can be found and thence the ratio

$$\frac{J_i}{J_g} = \frac{k D_i (1 - \varepsilon)}{D_g \varepsilon} \qquad (20)$$

The steady-state flow method has been applied not to zeolite compacts alone, but to graphite-zeolite A compacts (14). Some results are given for nitrogen in Table 3, in which an extended analysis served to give not only D_g and D_i but also a surface diffusivity D_s, associated largely with the graphite. The intra-crystalline diffusivities of N_2 in zeolite A at the experimental temperatures are reasonable in value. The method merits further attention for large intracrystalline diffusivities.

Table 3: Analysis of mixed flows J_g, J_s and J_i for N_2 in Graphite–zeolite A compacts (14).

Compact:	Graphite + Na-A				Graphite + Ca-A			
Temp (°C):	0	23	50	75	0	23	50	75
J_i/J_g	0.32	0.29	0.19	0.20	0.48	0.37	0.26	0.20
J_s/J_g	0.36	0.30	0.24	–	0.27	0.22	0.19	–
$10^3 D_g$ (cm^2 s^{-1})	0.75	0.72	0.69	0.69	1.1_2	1.1_0	1.0_6	1.0_4
$10^4 D_s$ (cm^2 s^{-1})	0.51	0.75	1.11	–	0.5_4	0.79	1.2_2	–
$10^6 D_i$ (cm^2 s^{-1})	0.9_5	3.22	3.75	6.5	1.4_6	2.3_8	3.50	5.1_0
E_s (kcal mol^{-1})		2.4				2.8		
E_i (kcal mol^{-1})		3.6				3.1		

276

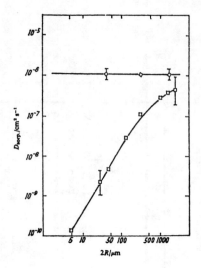

Fig. 5.   Intracrystalline self-diffusivity ($\Diamond$) from pulsed
           field gradient NMR and apparent diffusivity from
           sorption rates of CH_4 in chabazites at $0^\circ C$, as functions
           of crystallite radii (6).

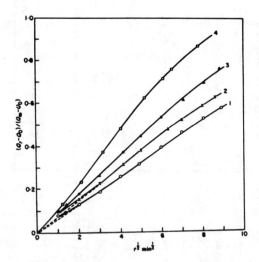

Fig. 6.   Plots of $M_t/M_\infty = (Q_t-Q_O)/(Q_\infty-Q_O)$ against $\sqrt{t}$ for
           $n\text{-}C_6H_{14}$ in 30.55% Ca-exchanged (Ca,Na)-A, at constant
           pressure and at 348K (15). $Q_\infty = 120.05$ mg g^{-1}.

           Curve 1: $Q_O = 0$ Curve 3: $Q_O = 63.37$ mg g^{-1}.
           Curve 2: $Q_O = 48.08$ mg g^{-1} Curve 4: $Q_O = 96.99$ mg g^{-1}.

3.3. Concentration-Dependent Diffusivities

When the differential intrinsic diffusivity in the zolite is a function of concentration C within the crystals* one may add diffusant in small increments, so that under constant pressure boundary conditions (eqn. 2) $(C_\infty - C_0)$ is small and, in this interval in C, $\overline{D}$ may be considered constant. Eqns. 3 to 6 are then still valid and heat evolution as a complicating factor is minimised.

When isotherms are rectangular in shape the equilibrium pressure is often so small that constant pressure sorption of increments is impossible. Constant volume variable pressure addition is also not suitable because even though $(C_\infty - C_0)$ is small the pressure surge may initially raise intracrystalline concentrations well above C_∞ near the surface of each crystal, and the assumption that $\overline{D}$ is constant because $(C_\infty - C_0)$ is small is no longer tenable. The method of small increments has therefore been modified as follows (15). The crystals were equilibrated to an initial uptake, Q_0, and then were brought to saturation uptake at Q_∞ under nearly constant pressure of sorbate vapour. The value of Q_0 was progressively increased, while Q_∞ was of course constant. Initially linear plots of $M_t/M_\infty = (Q_t-Q_0)/(Q_\infty-Q_0)$ against $\sqrt{t}$ were still obtained (Fig. 6) but, in the expression $M_t/M_\infty = \dfrac{6}{r_0}\sqrt{\dfrac{Dt}{\pi}}$, D must be an average or integral diffusivity over the interval $(Q_\infty-Q_0)$. Assuming therefore that

$$D(Q_\infty-Q_0) = \int_{Q_0}^{Q_\infty} \overline{D}dQ = -\int_{Q_\infty}^{Q_0} \overline{D}dQ \qquad (22)$$

where $\overline{D}$ is the differential diffusivity corresponding to a particular $Q_t (Q_\infty > Q_t > Q_0)$ one may evaluate integral diffusivities from $M_t/M_\infty = \dfrac{6}{r_0}\sqrt{\dfrac{Dt}{\pi}}$ and $\overline{D}$ from plots of D against Q_0 for any value of Q_0, because eqn. 22 can alternatively be written as

$$\overline{D} = D - (Q_\infty-Q_0) \left(\frac{\partial D}{\partial Q_0}\right)_{Q_\infty} \qquad (23)$$

This method was used to evaluate diffusivities of $n\text{-}C_4$, $n\text{-}C_6$ and $n\text{-}C_9$ paraffins in $(Na,Ca)\text{-}A$. The extents of exchange were regulated so that uptake was slow, to ensure that intracrystalline diffusion was rate controlling and perturbation of the kinetics by heat

---

*The subscript i of § 3.2 is omitted here where only intracrystalline flow need be considered.

evolution was minimised. Illustrative values of $\bar{D}$ for $n-C_4H_{10}$ in (Na,Ca)-A are given in Table 4. Isotherms of the paraffins in zeolite A are of type 1 in Brunauer's classification (16) and often approximate to Langmuir's isotherm (17). For the ideal isotherm based on his postulates the intrinsic differential diffusivity should be proportional to $(1-\theta)[d\ln p/d\ln\theta]$ since $(1-\theta)$ is the chance that a vacant site occurs into which a molecule on an adjacent site may jump and $d\ln p/d\ln\theta$ is the Darken activity correction. This correction for Langmuir's isotherm equation $(kp = \theta/(1-\theta))$, is $(1-\theta)^{-1}$, so that $\bar{D}$ should be constant, and therefore $\bar{D} = D$. The trends in $\bar{D}$ in Table 4 can be considered as one measure of deviations of the actual isotherms from Langmuir's idealisation.

Analysis of the temperature coefficients of $\bar{D}$ in terms of the Arrhenius equation $D = D_0 \exp-E/RT$ extrapolated to $Q_0 = 0$ are given in Table 5 (15). E tends to decrease as the extent of exchange

Table 4: $\bar{D}$ for $n-C_4H_{10}$ in 30.55, 32,54 and 34.10% Ca-exchanged Zeolite A (15).

% Ca^{2+}	T/K	Q_∞ (mg g^{-1})	Q_0 (mg g^{-1})	$10^{17}\,\bar{D}$ (m^2 s^{-1})
30.55	348	87.2_2	0	0.74
			5	0.74
			20	0.65
			40	0.55
			60	0.44
			70	0.40
32.54	323	106.5_3	0	1.8_2
			5	1.7_5
			20	1.5_2
			40	1.42
			60	1.29
			70	1.30
34.10	273	116.0_3	0	1.9_9
			5	1.9_9
			20	1.9_9
			40	1.9_4
			60	1.7_9
			70	1.5_8

Table 5

(a) $E(kJ\ mol^-)$ in $\bar{D} = D_o\ exp\text{-}E/RT$ for $Q_o = 0$

Diffusant	% Ca exchange in (Ca,Na)-A			
	30.55	32.5	34.10	35.48
$n\text{-}C_4H_{10}$	40.7	47.3	36.4	$-$
$n\text{-}C_6H_{14}$	61.1	58.2	48.5	38.9
$n\text{-}C_9H_{20}$	74.5	67.4	60.7	53.1

(b) $D_o\ (m^2\ s^{-1})$ in $\bar{D} = D_o\ exp\text{-}E/RT$ for $Q_o = 0$

$n\text{-}C_4H_{10}$	1.8×10^{-9}	7.3×10^{-10}	3.3×10^{-10}	$-$
$n\text{-}C_6H_{14}$	6.1×10^{-9}	1.3×10^{-8}	1.5×10^{-9}	2.1×10^{-11}
$n\text{-}C_9H_{20}$	3.8×10^{-8}	2.3×10^{-8}	4.8×10^{-8}	9.2×10^{-10}

of Na^+ by Ca^{2+} increases, and D_o tends to decrease as E decreases. The energy barriers are considerable for the rate-controlling unit diffusion steps because if there is a significant Na^+ content many of these ions partially block the 8-ring windows through which diffusant must pass. Even in absence of cations the windows have free diameters of about 4.1 A compared with a critical dimension of a stretched hydrocarbon chain of about 4.9 Å.

3.4. Evolution of the Heat of Sorption

The exothermal heat of sorption produces both local and extended temperature changes in the bed of sorbent. That is, in each crystallite transient temperature gradients appear, the heat being produced initially at the surfaces, with the temperature wave then spreading both into the body of the crystals as well as heating the ambient gas phase. After an interval, the temperature of the whole bed becomes more uniform but is above that of the thermostat in which it may be immersed. At still longer times the temperature falls asymptotically to that of the thermostat. Attempts have been made to allow for the heat effects in the second and third stages (18) but any comprehensive treatment allowing for heat evolution and heat and matter flows and their interaction, together with time-dependent drifts in equilibria during heating and cooling of beds must be very difficult. It is therefore preferable to aim at conditions as near isothermal as possible by:

1. Making the beds as thin as possible, optimally a layer one crystal thick.
2. Circulating the vapour or liquid guest through the bed to remove heat as quickly as possible.

3. Avoiding very rapid sorption, by having suitably large crystals (see Table 2).
4. Adding only small increments of diffusant at a time.

4. SOME ADDITIONAL EFFECTS

Diffusion inside zeolite crystals may be influenced by factors other than those described above. It has been suggested that surface skins may sometimes be present which form an added resistance to flow. Such skins may exist in patches only, so that the area through which diffusants enter the zeolite crystals is reduced or the skins could cover the whole external surface. In the latter circumstance the crystals could become non-sorbents if the skin was glassy and impermeable. For total control of sorption rates by surface skins $(1-M_t/M_\infty)$ should decline exponentially with time (19) rather than following initially the eqn. 5 $(M_t/M_\infty = \frac{6}{r_o} \sqrt{\frac{Dt}{\pi}})$. However eqn. 5 has been well verified in many rate studies which suggests that surface skins are not normally important. Hydrothermal treatments which diminish sorption rates have on this evidence been considered to result in skins (19 a and b).

Pre-treatments may be such that the crystals suffer chemical damage (e.g. grinding, hydrothermal treatments and even the outgassing procedures). With bonded pellets the bond may block some areas of the crystallite surfaces. Within the crystallites there may be variable amounts of detrital material occluded during synthesis; and lattice defects such as stacking faults may occur periodically or at random to an extent varying between preparations. Also the Si/Al ratios of different preparations of the same zeolite may vary according to the methods of preparation, and therefore the cation concentrations and locations in the intracrystalline channels and cavities. All these factors can result in different values of diffusivities among different samples of the same zeolite. When such results, obtained in different laboratories, are to be compared pre-treatment and synthesis conditions should be standardised and/or samples from each laboratory should be exchanged and the measurements made on both samples in each laboratory.

A purpose of §§ 2 to 4 has been to draw attention to ways in which the study of sorption kinetics can be improved for determinations of intracrystalline diffusivities, by indicating some complications and ways of avoiding or minimising these.

5. SELF-DIFFUSIVITY AND DARKEN'S RELATION

Exact relations have been derived between differential intrinsic

and differential self-diffusivities (20,21,22). Here we will
consider only the simpler approximately valid Darken relation
which is

$$\overline{D} = \overline{D}*(C) \; d\ln a/d\ln C \qquad (24)$$

The activity correction $d\ln a/d\ln C$ for an ideal gas phase is re-
placeable by $d\ln p/d\ln C$, which can be found from the slopes of
isotherms (plotted as $\ln C$ against $\ln p$). The correction so determined assumes that everywhere there is local equilibrium between
gaseous and intrazeolite sorbate.

In the Henry's law range of sorption ($C \propto p$) $d\ln p/d\ln C$ is
unity. Outside this range for type I isotherms $d\ln p/d\ln C$ becomes
greater than unity and for rectangular isotherms may become very
large as the uptake approaches its saturation value. Darken's
relation has been tested for water diffusing in natural chabazite,
heulandite and gmelinite (3). Both intrinsic and self-diffusivities
were measured, the latter using H_2O to displace D_2O. The crystals
were large pieces of the zeolite, and an inert carrier gas was
charged with water vapour at constant relative pressures and cir-
culated continually through the samples suspended from a silica
spring balance in an open mesh glass-fibre bucket. Conditions were
such as to ensure vitually complete saturation of zeolite by either
D_2O or H_2O.

The results in Table 6 show that Darken's relation is, as
expected from the more refined theory (20,21,22), only approximately
valid. However the dominant correction is certainly that for the
activity ($d\ln p/d\ln C$).

Pulsed field gradient NMR has been used to study self-diffusion
of $n-C_4$ to $n-C_{18}$ alkanes in zeolite X over the range -100 to $+200^\circ C$
(23). This very complete study showed the following behaviour:

1. For comparable degrees of filling of the intracrystalline
   pore space $\overline{D}*$ decreased with increasing carbon number.
2. At constant temperature $\overline{D}*$ decreased monotonically for each
   paraffin as the amount sorbed increased. It has already
   been indicated that for an ideal Langmuir sorption one
   would expect $\overline{D}* = \overline{D}*_{\theta=0} (1-\theta)$ where θ is the degree of
   filling. For n-paraffins this is normally an over-
   idealisation, but a decrease in $\overline{D}*$ with θ is as expected.
3. For a given temperature and paraffin $\overline{D}*$ was within an order
   of magnitude of $D*$ for the corresponding liquid paraffin.
4. For a constant amount sorbed $\overline{D}*$ depended exponentially on
   temperature ($\overline{D}* = D_o \exp{-E/RT}$).
5. The heat of sorption for a given degree of filling, θ, of
   intracrystalline pore space increased continuously with
   increasing carbon number, but activation energies, E, for

self-diffusion approached an asymptotic limit and were never more than a rather small fraction of the heat of sorption (Fig. 7 (23)).

It has been suggested (24) that the behaviour of E means that as the n-paraffin chains increase in length each unit diffusion step tends increasingly to involve segmental rotations of part only of the molecule round a C-C bond. This results in changes of position of the centre of mass of the molecule, but not necessarily in translation of all parts of the molecule simultaneously. The longer the chains the more probable the segmental mechanism. If for longer chains this mechanism involves segments of similar size, eventually nearly independent of chain length, then, as observed, E would become almost constant. Segmental unit diffusion steps can also account for the decrease in $\bar{D}^*$ with increasing carbon number, because the number of unit steps required

Table 6: Relation between $\bar{D}$ and $\bar{D}^*$ for water in several zeolites near saturation of the zeolite (3).

Zeolite	$T(^\circ C)$	$10^8 x\bar{D}^*$ $(cm^2\ s^{-1})$	$\dfrac{d\ln p}{d\ln C}$ (a)	$10^6 x\bar{D}^* \dfrac{d\ln p}{d\ln C}$ $(cm^2\ s^{-1})$	$10^6\ \bar{D}$ $cm^2\ s^{-1}$
Chabazite B	75	46.2	23.0 (b)	10.6_5	11.7
	65	31.8	24.0 (b)	7.6_0	8.9
	55	21.4	25.5 (b)	5.4_5	6.6
	45	14.1	27.0 (b)	3.80	4.8
	35	9.0	28.0 (b)	2.50	3.4
Heulandite	75	9.7_7	30.0	3.00	4.9
	65	6.0_7	32.0	1.95	3.5
	55	3.70	34.0	1.26	2.4
	45	2.19	35.5	0.78_0	1.6
	35	1.24	37.0	0.46_5	1.05
Gmelinite	55	7.33	26.5	1.95	2.3
	45	4.99	28.0	1.40	1.5
	35	3.31	29.5	0.97	1.0

(a) Determined from the nearly flat top of the isotherm.
(b) Determined from isotherms measured in sample A of chabazite.

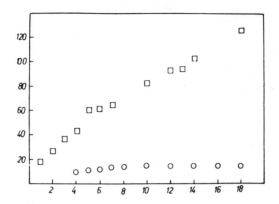

Fig. 7.   Heats of sorption and energies of activation for self-
diffusivities, for n-alkanes in Na-X zeolite, as
functions of carbon number. The upper curve gives the
heat of sorption, the lower curve the activation energy
(23). The ordinate is in kJ mol^{-1} and the abscissa is
the carbon number.

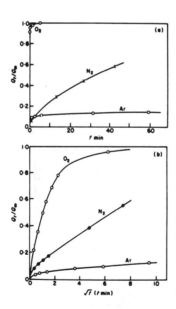

Fig. 8.   (a) Sorption rates of O_2, N_2 and Ar in levynite at
$-184^{\circ}C$ (25). For O_2, $Q_{\infty} = 10.02$, for N_2 it is 9.77
and for Ar, 10.01, all in cm^3 at s.t.p. g^{-1}.
(b) Sorption rates of O_2, N_2 and Ar in Ca-mordenite at
-78°(26).

284

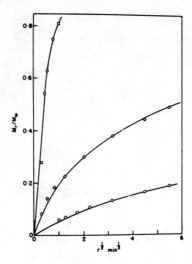

Fig. 9.   Sorption rates at 30°C in Na-Y of liquid 1,3,5-
            trimethyl-(▣), 1,3,5-triethyl-(◇) and 1,3,5-triiso-
            propyl-benzene (O) (27).

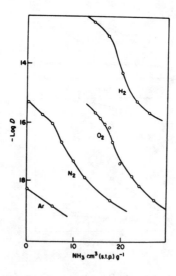

Fig. 10.   Diffusivities in cm² s⁻¹ of H_2, O_2, N_2 and Ar in
            mordenite at -183°C, as functions of the amount of
            pre-sorbed NH_3 (30).

for migration of an entire molecule from one cavity to the next in zeolite X will increase with chain length.

6. MOLECULE SIEVING

According to the free dimensions of the meshes through which guest molecules must migrate intracrystalline diffusivities are found to change very greatly with the molecular dimensions of the guest. Windows or meshes controlling molecule sieving are 6-, 8-, 10- and 12-rings of varied conformations, and octagonal prisms. With some zeolites there is additional obstruction due to location of cations in or near the windows. Crystals with restricted windows differentiate between groups of small molecules and exclude larger ones altogether; crystals with larger windows differentiate between groups of bigger molecules, and so on. Even quite small differences in molecular dimensions can result in dramatically different rates of sorption. This is illustrated in Fig. 8 for uptakes of O_2, N_2 and Ar in levynite (25) and Ca-mordenite (26) which function as fine-mesh sieves. Fig. 9 shows a similar behaviour in a wide-mesh sieve, Na-Y, for 1,3,5-trimethyl-1,3,5-triethyl- and 1,3,5-triisopropyl-benzene, all as liquids (27).

The relation between molecular dimensions and diffusivities is illustrated for fine mesh sieves K-A and K-mordenite in Table 7 (28). There are five orders of magnitude between diffusivities of Ne (3.2 Å diameter) and Kr (3.9 Å) in K-A at 20°C; and also between H_2 (2.4 x 3.1 Å) and Kr (3.9_4 A) in K-mordenite at −78°C. There is a parallel increase in the energy barrier, E, involved in each unit diffusion step.

The strong influence of the exchange cation upon the uptake of Ar by various ion-exchanged forms of mordenite at −78°C is shown in changes of $\bar{D}/r_0^2$ (26) with cation:

Ca-mordenite	$\bar{D}/r_0^2$:	1.51×10^{-8} s^{-1}
K-mordenite		2.8×10^{-6} s^{-1}
Ba-mordenite		5.5×10^{-6} s^{-1}
Na-mordenite		3.8×10^{-6} s^{-1}
Li-mordenite		4.0×10^{-4} s^{-1}
NH_4(H)-mordenite		1.35×10^{-3} s^{-1}

From the sequence above differing cation positions as well as cation size may play a part. Again the extreme range is five orders of magnitude.

One may also alter the diffusivities of guest molecules by pre-sorbing metered amounts of a strongly held guest (e.g. H_2O or NH_3) which is virtually immobile at the temperature at which less polar guest molecules are to be sorbed. This effect can be very large as

Table 7: Relation between diffusivities and molecular dimensions (27).

Zeolite	Molecule	Equilibrium Dimensions ($\overset{\circ}{A}$)	$\bar{D}$ or $\bar{D}*(cm^2\ s^{-1})$ (and $T^{o}C$)	E (kcal mol^{-1})
K-A	Ne	3.2	2.9×10^{-13} (20)	7.0
	H_2	2.4x3.1	2.0×10^{-12} (20)	9.9
	Ar	3.8_3	1.6×10^{-17} (20)	12.6
	N_2	3.0x4.1	9.8×10^{-18} (20)	16.2
	Kr	3.9_4	1.8×10^{-18} (20)	16.4
K-mordenite	H_2	2.4x3.1	2.7×10^{-13} (−78)	2.5
	O_2	2.8x3.9	2.0×10^{-15} (−78)	4.4
	N_2	3.0x4.1	9.2×10^{-16} (−78)	4.8
	Ar	3.8_3	2.4×10^{-16} (−78)	8.4
	Kr	3.9_4	1.8×10^{-18} (−78)	10.0

in Fig. 10 when H_2, O_2, N_2 and Ar were sorbed at −183°C in mordenite containing metered pre-sorbed amounts of NH_3 (30).

7. CONCLUDING REMARK

This account has emphasised some experimental problems and means of overcoming these, because much previous work on rates of uptake of simple gases in zeolite A and of these and hydrocarbons in zeolites X and Y has, as shown by pulsed field gradient NMR, been misinterpreted.

Some properties of intracrystalline diffusivities have been described, emphasising in particular their great sensitivity to molecular dimensions of diffusing species. As indicated in § 1 diffusion can be of first importance not only in separation processes, but particularly in catalysis where reactants must reach active centres within crystals and resultants must migrate out of the crystals. Such counter-current flows are much more difficult to interpret because of cross-coefficients between flows. Formulation of transport is now best done in terms of so-called irreversible thermodynamics and phenomenological straight and cross-coefficients. Not much has been done so far in the areas of co- and counter-diffusion of mixtures in zeolites (31,32).

REFERENCES

1. Tiselius, A., Z. Phys. Chem., A169, (1934), 425.
2. Tiselius, A., Z. Phys. Chem., A174, (1935), 401.
3. Barrer, R.M. and B.E.F. Fender, J. Phys. Chem. Solids, 21, (1961), 12.
4. Barrer, R.M. and M.A. Rosemblat, Zeolites, 2, (1982), 231.
5. Ash, R., R.M. Barrer and R.J.B. Craven, J. Chem. Soc. Faraday Trans. II, 74, (1978), 40.
6. Karger, J., and J. Caro, J. Chem. Soc. Faraday Trans. I, 73, (1977), 1363.
7. Barrer, R.M., Zeolites and Clay Minerals and Sorbents and Molecular Sieves (London, Academic Press, 1978), p. 317.
8. Karger, J., Z. Phys. Chem., Leipzig, 248, (1971), 27.
9. Karger, J., S.P. Shdanov and A. Walter, Z. Phys. Chem. Leipzig, 256, (1975), 319.
10. Caro, J., J. Karger, H. Pfeifer and R. Schollner, Z. Phys. Chem. Leipzig, 256, (1975), 698.
11. Karger, J., and D.M. Ruthven, J. Chem. Soc. Faraday Trans. I, 77, (1981), 1485.
12. Ref. 7, p. 318.
13. Quig, A., and L.V.C. Rees, Proc. 3rd Internat. Conference on Molecular Sieves (Zurich, Sept. 3-7, 1973) p. 277.
14. Barrer, R.M., and J.H. Petropoulos, Surface Sci., 3, (1965), 126 and 143.
15. Barrer, R.M., and D.J. Clarke, J. Chem. Soc. Faraday Trans. I, 70, (1974), 535.
16. Brunauer, S., The Adsorption of Gases and Vapours, (New York, Oxford Univ. Press, 1944) p. 150.
17. Ref. 7 pp. 112-114.
18. Ruthven, D., in Properties and Applications of Zeolites, Editor, R.P. Townsend (London, The Chemical Society, 1980), p. 43.
19. Karger, J., and P. Hermann, Ann. Phys. Leipzig, 31, (1974), 277.
19a. Bulow, M., P. Struve and S. Pikus, Zeolites, 2, (1982), 267.
19b. Karger, J., W. Heink, H. Pfeifer, M. Rauseher and J. Hoffmann, Zeolites, 2, (1982), 275.
20. Ash, R., and R.M. Barrer, Surface Sci., 8, (1967), 461.
21. Karger, J., Surface Sci., 36, (1973), 797.
22. Karger, J., Surface Sci., 57, (1976), 749.
23. Karger, J., H. Pfeifer, M. Rauscher and A. Walter, Z. Phys. Chem., Leipzig, 259, (1978), 784.
24. Barrer, R.M., Symposium on the Characterisation of Porous Solids (Neuchatel, 9-13th July, 1978).
25. Barrer, R.M., Nature, 159, (1947), 508.
26. Barrer, R.M., Trans. Faraday Soc., 45, (1949), 358.
27. Satterfield, C.N., and C.S. Cheng, Amer. Inst. Chem. Eng., 68th National Mtg., Houston, Feb. 28-Mar. 4th (1971) Adsorption Pt. 1, Paper 16f.

28. Ref. 7, p. 292.
29. Walker, P.L. Jr., L.G. Austin and S.P. Nandi, in Chemistry and Physics of Carbon, Vol. 2, (Dekker, 1966) pp. 257-371.
30. Barrer, R.M., and L.V.C. Rees, Trans. Faraday Soc., 50, (1954), 852 and 989.
31. Karger, J., M. Bulow and W. Schirmer, Z. Phys. Chem. Leipzig, 256, (1975), 144.
32. Moore, R.M. and J.R. Katzer, J. Amer. Inst. Chem. Eng., 18, (1972), 816.

PART III

CATALYSIS

UNIFYING PRINCIPLES IN ZEOLITE CHEMISTRY AND CATALYSIS

JULE A. RABO

Union Carbide Corporation
Tarrytown, New York 10591

INTRODUCTION

In recent years molecular sieve catalysts have
assumed an increasingly important role in industrial
catalysis. Applications of zeolite catalysts are ex-
panding from the traditional petroleum refining to new
and improved fuel processing applications, and to new
roles in both the petrochemical and chemical industries.
Up to the present, all commercial applications of zeolite catalysts have been carried out with acidic zeolites.
Recent investigations of zeolite chemistry revealed
several important features which appear common to both
alkali and acidic zeolites. This new chemical evidence
raises the possibility that the underlying physicochemical
features of both types of zeolites play a role in catalysis.

Chemistry and Catalysis With Alkali Zeolites

Chemistry

The intracrystalline pore-cavity system in zeolites,
often called the zeolitic surface, is surrounded by the
zeolite crystal lattice, and consequently it is strongly
influenced by the zeolite crystal field. This crystal
field, pervading the intracrystalline pore-cavity system, renders zeolites solid electrolytes. Depending on
the ionic character of the crystal which is mainly controlled by the alumina content, zeolites show properties
of weak to very strong electrolytes. A unique characteristic of these solid electrolytes is that by admitting

various adsorbates or reactants, one can change the chemic
composition of the electrolyte. Moreover, by evaluating
the effect of the zeolite upon appropriately chosen
occluded molecules, one can monitor the strength of zeolite electrolyte.

It seems reasonable to expect that the effect of the
zeolite as electrolyte upon occluded molecules, and vice
versa, is mainly influenced by thermodynamics, namely by
the tendency to minimize the free energy of the zeolite-
guest material system. Unfortunately, the lattice
energies of zeolite crystals are not known. Their calculation with any degree of reliability is impossible
without knowing the covalent and ionic character of the
linkages between lattice atoms. Several observations,
expecially the preference of zeolite cations for water
over framework oxygen, indicate that the bond between
framework cations and oxygen is strongly covalent.
Nevertheless, a certain degree of ionic character must
exist, and a crystal field corresponding to the ionic
character must pervade the whole porous crystal. Occluded
molecules, especially the polar or strongly polarizable
ones, will be polarized by the zeolite, and they themselves will exert a similar effect on the zeolite crystal
as well. The overall interaction between zeolite and
adsorbate must tend to minimize the free energy of the
zeolite-adsorbate system. Therefore, the simplest way
to forecast thermodynamically favored reactions in zeolites is to evaluate their effect on the stability of
the zeolite crystal.

Inspection of the ionic lattice energy equation
shows that with ordinary ionic crystals the free energy
is reduced by placing more ion pairs or ions of higher
valence within the same crystal volume. According to
this simple relationship, the occlusion of added ionic
material in the porous zeolite lattice should generally
lower the lattice energy, and consequently increase the
thermodynamic stability of the overall system. Of course,
the occlusion of added ionic material may affect the
zeolite structure and symmetry, ultimately affecting the
Madelung constant of the zeolite crystal. This may render
the interaction between zeolite and the occluded ionic
species complicated and difficult to analyze. In spite
of these limitations, the study of interactions between
zeolites and occluded ionic material is a convenient
technique to assess the role of the zeolite electrolyte
upon zeolite chemistry, and ultimately on zeolite
catalysis.

In order to evaluate the influence of zeolites as electrolytes on catalytic phenomena, some of the relevant data on salt occlusion and redox chemistry in zeolites will be reviewed.

It has been suggested that the porous zeolite lattice has strong affinity for the occlusion of added ionic species (1). This occluded ionic material may consist of ionic compounds such as salts, or it may consist of atoms or molecules which are not ionic outside the zeolite, but become ionized upon occlusion into the zeolite crystal (2). There are several examples of both phenomena.

The occlusion of salts in zeolites has been described earlier (1). The experimental evidence with zeolites A (3) and Y clearly shows that ionic compounds, such as salts, readily penetrate the zeolite crystal, filling up the available space in large intracrystalline cavities. Salts, especially salts of univalent anions, can even penetrate the sodalite cages in Y zeolite(1). This is surprising because the size of the O_6-ring port of the sodalite cage is only about ~2.4 Å, much smaller than an anion. It was found that the occlusion of salts into the sodalite cages requires high temperatures in order to provide the high activation energy needed to enlarge the O_6-ring port, probably by the temporary cleavage of O-Al or O-Si bonds. Interestingly, both halide salts as well as salts of large complex anions such as ClO_3^- and NO_3^- can be occluded in the sodalite cage.

From the point of view of this discussion, the most important aspect of salt occlusion is the enhanced stability of the zeolite-salt occlusion compound. It has been reported that in every successful salt occlusion experiment both the Y zeolite and the occluded salt were rendered thermally more stable, showing decomposition temperatures well beyond the decomposition temperature of the zeolite or of the guest salt itself. It was also found that the guest salt occluded in the sodalite cages cannot be removed from the host Y zeolite crystal even by steaming at 700°C (1). Thus the zeolite lattice is stabilized by the occluded salt and vice versa.

The affinity of zeolites for ionic species is best shown by the ionization of occluded atoms and molecules. The most frequently observed ionization reaction in zeolites is the ionization of water by cation hydrolysis. This is a particularly important reaction in zeolites because this reaction generates acidic OH groups attached to framework cations which are directly responsible for acidic behavior (4,11). These acidic OH groups in zeolites have been well characterized by IR spectroscopy (5) as well as by other techniques (see next section).

Other ionization effects in zeolite lattices have been observed by the occlusion of alkali metals and certain gas molecules. It was reported that alkali metals readily reduce both alkali X and alkali Y zeolites (6). The sodium ions of the host crystal capture the electron from the occluded alkali atoms forming symmetrical Na_4^{3+} or Na_6^{5+} centers in the large cavities:

$$NaY \xrightarrow{\quad Na^o \quad} Na_4^{3+}Y$$

$$NaX \xrightarrow{\quad Na^o \quad} Na_6^{5+}X$$

It has also been reported that certain transition metal zeolites readily ionize adsorbed NO radicals, forming electron transfer complexes with zeolite cations (2):

$$Cu^{2+}Y \xrightarrow{\quad NO \quad} \{Cu^+ - NO^+\}Y$$

$$Ni^{2+}Y \xrightarrow{\quad NO \quad} \{Ni^+ - NO^+\}Y$$

Another interesting ionization phenomenon is the interaction of NO and NO_2 radicals within zeolites. Here, these radicals become ionized, presumably forming a salt-like complex in the zeolite (2):

$$NaY + \xrightarrow{\quad NO + NO_2 \quad} \{NO^+ - NO_2^-\}NaY$$

The electrolytic strength of X and Y zeolites in these redox reactions is best demonstrated by the fact that these redox reactions are all highly endothermic in the gas phase outside the zeolite crystal. For example, the formation of the NO^+ and NO_2^- ions from NO and NO_2 radicals is endothermic by about 5.5 eV ($\sim$126 kcal/mol), not considering coulombic interaction between the products. In spite of this, this reaction readily occurs on NaY at moderate temperatures. Similarly, the interaction between sodium metal and NaY, the highly endothermic ionization of sodium metal (I.P. = 5.14 eV, $\sim$118 kcal/mol), is readily accomplished at 300°C with a Na_4^{3+} center formed in each large cavity.

These examples demonstrate well the high affinity of zeolites for ionic species. They also provide information on the large contribution of zeolites to the ionization of occluded species, displaying the characteristics of a very strong electrolyte.

It should also be noted that several of the ionization processes described above readily occur both with Na-zeolite and with a variety of cation exchanged zeolites.

Catalysis with Alkali Zeolites

The important industrial applications of zeolite catalysts are based on carbenium ion chemistry. This chemistry is induced by strong-acid hydroxyl groups in zeolites (see next section). At present, no acid-free, alkali cation zeolite catalyst is used in industrially important processes. Nevertheless, it seems worthwhile to review here the influence of alkali zeolites as electrolytes upon hydrocarbon reactions because of the large chemical influence they exert upon occluded molecules (see previous section). The important effects discussed above include strong adsorption and a very strong affinity for ionic materials, even including the ionization of occluded molecules. Such a strong display of chemical activity, particularly in the ionization reactions, suggests that these phenomena are probably also relevant in some way to the carbenium ion type catalysis carried out on acidic zeolites. In order to evaluate the effect of the zeolite as electrolyte, without contribution from acidic hydroxyl groups, the effect of alkali zeolites in hydrocarbon cracking will be briefly reviewed.

The influence of alkali zeolites (K-Y) free of acidic hydroxyls on the cracking of hydrocarbons was investigated with hexanes as reactants {7,8}. At about 500°C the K-Y catalyzed cracking gave conversion levels up to 5 times higher relative to the same reactor filled with quartz chips at the same temperature. The product compositions obtained from n-hexane and its isomers over K-Y were markedly different from those obtained over acidic zeolites. Significantly, they were also quite different from the product obtained in a reactor filled with quartz chips at the same reaction conditions (see Table 1).

TABLE 1

Products from Pyrolysis and K-Y Catalyzed[a] Cracking at 500°C {7}, (mol/100 mol converted)

	n-Hexane		2-Methylpentane		3-Methylpentane	
	Thermal	K-Y	Thermal	K-Y	Thermal	K-Y
Hydrogen	2.8	3.1	24.4	4.3	5.6	3.9
Methane	50.3	23.5	27.9	21.7	63.5	57.3
Ethylene	56.5	26.0	11.5	6.5	23.7	23.0
Ethane	37.7	37.4	42.2	43.5	44.2	44.3
Propylene	55.7	58.5	57.8	44.2	20.5	15.7
Propane	10.7	38.2	19.2	36.4	–	0.5
1-Butene	31.3	12.0	–	–	7.6	11.2
2-Butene	–	19.6	–	–	34.2	34.8
Isobutene	–	–	40.5	40.1	–	–
Isobutane	–	–	0.2	1.6	–	–
1-Pentene	7.3	1.9	1.0	2.3	–	–
2-Pentene	–	4.7	8.5	8.4	13.3	6.4
2-Methyl-1-butene	–	–	–	–	20.9	–
3-Methyl-1-butene	–	–	4.5	1.6	–	1.0
2-Methyl-2-butene	–	–	–	1.7	–	19.5

[a]Rate enhancements: ∠5-fold.

The analysis of the reaction mechanism based on product distributions obtained from all hexane isomers showed
that over K-Y the free radical type mechanism prevailed,
but with specific changes in the rates of certain reaction
steps relative to the "hot tube." A mechanistic study
revealed that the K-Y does not change the selectivity of
alkyl radicals either in the choice between different
β-scission options or between different H atom abstraction steps when positionally different choices are available. However, the K-Y has a very large effect upon the
ratio of the rates of H atom abstraction steps over the
β-scission steps. Mechanistic analysis of the cracked
products obtained from the hexane isomers shows that
over K-Y the ratio of rate of H-abstraction/rate of
β-scission is increased for free alkyl radicals by a
factor of 6 to 9. This effect accounts for all significant differences in product formation between the non-
catalyzed and the K-Y "catalyzed" reactions {7-9}.

Speculating on the reason why zeolites enhance the
rate of H-abstraction over β-scission, it was considered
that H-abstraction is a bimolecular reaction step whose
rate is strongly influenced by the concentration of the
reactant, whereas β-scission, being a unimolecular process, is unaffected by changes in reactant concentration.
Accordingly, the "catalytic" effect of K-Y is the result
of increased reactant concentration within the zeolite
cavities relative to the surrounding gas phase {9,10}.

The concentration or adsorption of hydrocarbon within
zeolite crystals is, of course, well documented in adsorption studies carried out at low temperatures. For the
temperature range used in the hexane cracking study, the
hexane loading on K-Y at $500^{o}C$ was estimated from the
reported Arrhenium plot of n-hexane loading on NaX at
temperatures up to $300^{o}C$. The estimated hexane loading
at $500^{o}C$ is substantial, and it is consistent with the
reactant concentration suggested by the mechanistic
analysis.

It is concluded from the hexane cracking study that
zeolites concentrate hydrocarbon reactants within the
zeolite crystal, that this concentration is substantial,
and that this effect is responsible for the substantial
changes in product formation. Methanistically, the
specific result of this effect is a substantial enhancement of the rates of bimolecular reaction steps {9}.
The high hydrocarbon concentration within Y zeolite
crystals is the direct result of the high affinity of
this strong electrolyte for ionic or polar matter.

Adsorbed hydrocarbons become polarized upon adsorption, and they are consequently strongly held in the zeolite crystal.

Since the concentration effect of zeolites depends mainly on the ionic character and the related electrolytic strength of the crystal, it is of interest to compare the cracking of n-hexane over monacidic molecular sieves of strong and weak electrolytic character.

For contrast, the K-Y and Silicalite molecular sieves were chosen as catalyst candidates. The K-Y represents an aluminum rich, very strong electrolyte. On the other hand, Silicalite is substantially free of both aluminum and zeolite cations, and consequently it is a very weak electrolyte.

Figure 1 shows the reaction path of radical-type cracking for n-hexane and its cracked fragments. The measurement of the concentration effect rests upon a comparison between Reactions a_3 and a_2 because the relative rates of these reaction steps reveal the enhancement of the H-transfer reaction step over β-scission, respectively, on the same cracked fragment. Experiments showing the rates of Reactions a_2 and a_3 for K-Y and Silicalite are shown in Table 2. In addition, the table includes experimental data showing the effect of about tenfold helium diluent applied to the n-hexane feed during the hexane cracking experiments. The Silicalite used in the experiment was treated with potassium salt solution to remove traces of acidic hydrogen ions.

The data show that the application of helium gas diluent results in a sharp reduction of the rate of H-abstraction step (a_3) relative to the β-scission step (a_2) both with the reactor filled with quartz chips as well as with K-Y and Silicalite. This result is expected, because the addition of diluent to the reactant reduces the n-hexane concentration, resulting in less frequent interaction between n-hexane molecules required for the H-abstraction step.

The difference between the K-Y and Silicalite "catalysts" is very substantial. Silicalite, in sharp contrast to K-Y, performs like inert quartz chips. It has, within experimental error, no effect on the rate

299

FIGURE 1 - Decomposition network for n-hexane {7}

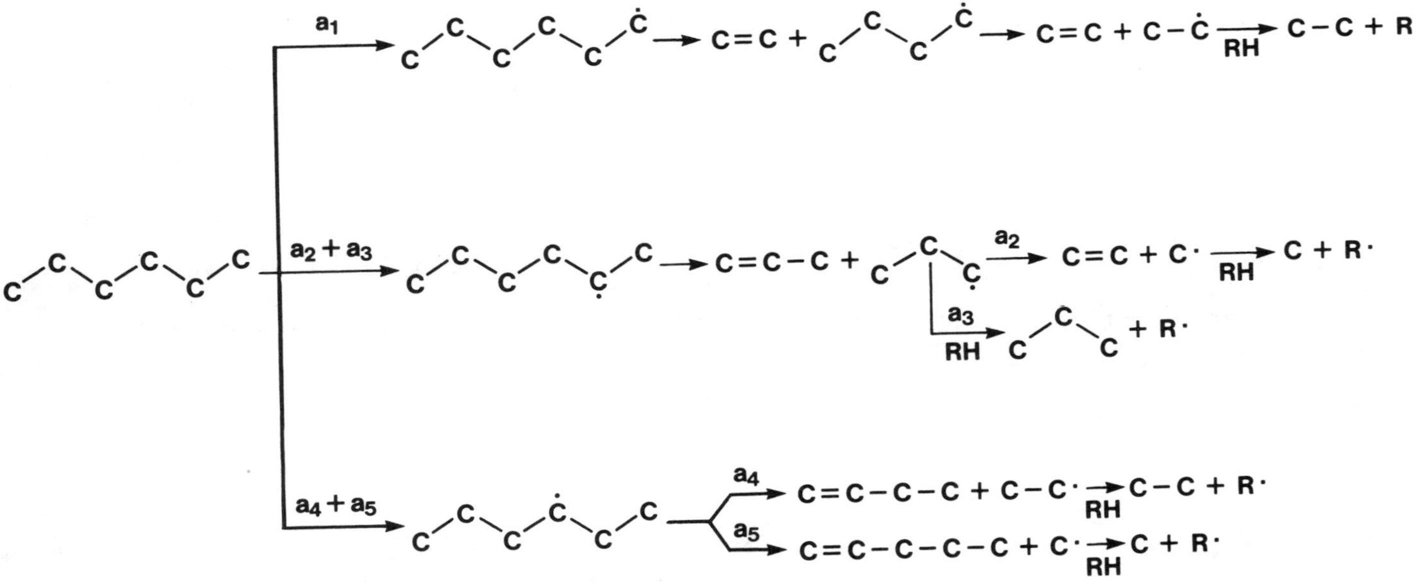

TABLE 2[a]

Rates of H-Abstraction (a$_3$) and β-Scission (a$_2$) Steps, See Figure 1[b]

Catalyst: Diluent:	None[c,d] -	None[c] He[e]	K-y[d] -	K-Y He[e]	Silicalite -	Silicalite He[e]
a$_2$	44	52	18	57	59	64
a$_3$	11	2	39	6	4	1

[a]Private communication from R. D. Bezman, I. R. Ladd and M. T. Staniulis.

[b]For experimental details, see Ref. 7.

[c]Reactor filled with quartz chips.

[d]From Ref. 7.

[e]About tenfold dilution.

of the H-abstraction step. This molecular sieve, in spite of its high surface area (~ 400 m^2/g), has no influence on the thermal cracking of n-hexane. Considering the strongly covalent character of the Si-O linkages forming the silicalite crystal, it is not surprising that the n-hexane cracking experiments show that Silicalite does not display the properties of zeolite electrolytes.

From the absence of skeletal isomerization in the cracking of n-hexane over K-Y, the absence of ionic (carbenium ion) reactions can be inferred. Hence the K-Y free of OH groups is not capable of ionizing either hexane or the radicals formed upon its fragmentation. This is somewhat surprising considering the facile ionization of occluded inorganic compounds (see former section). One has to recognize, however, that the ionization of hydrocarbons requires very high energy. For example, ionization of a tertiary—C—H bond to form a tertiary carbenium ion and hydride ion is endothermic by 240-250 kcal/mol while the ionization potential of a tertiary radical is ~ 170 kcal/mol.

Even if the absence of direct ionization of hydrocarbons is well demonstrated in these experiments, it is reasonable to expect that strong, zeolite-type electrolytes exert a strong influence upon carbenium ion reactions. First, by the concentration of the hydrocarbon reactants in the catalyst crystal, and second, by aiding the formation of hydrocarbon cations and by stabilizing these ionic species after formation.

Chemistry and Catalysis With H-Acid Zeolites

Formation of Acid Sites

The hydroxyl groups with very acidic hydrogen have been well recognized as the source of acidic catalytic activity in zeolites. Prior to the discovery of acidic zeolite catalysts, the most notable acid catalyst was the silica-alumina gel. Here the source of acidity is hydroxyl groups linked to aluminum or to silicon-aluminum sites. With silica-alumina gel type catalysts, all metal cations were recognized as poisons for acid sites. These cations replaced acidic hydroxyl hydrogens and fully removed acid activity. With this background in acid catalyst chemistry, it was surprising that zeolites NaX and NaY, when cation exchanged with bi- or multivalent metal cations, gave rise to acidic catalytic activity, well exceeding the acid strength of silica-alumina gel. It was also found that NaY, when NH_4^+ exchanged and activated at high temperatures to remove ammonia, fully retained the crystal framework of the original NaY, and it gave rise to very strong acidic catalytic activity, well exceeding in strength all oxide-type acid catalysts known before {30}. In contrast to this material, the NaX following NH_4^+ exchange and heat treatment showed complete loss of crystal structure, and it displayed no significant acidic catalytic activity. These phenomena, discovered in the late 1950s, were subsequently thoroughly investigated and further characterized.

The cause of the great differences in the application of bi- or multi-valent cations in X and Y-type zeolites versus silica-alumina gel was ascribed to the strong tendency of zeolites toward cation hydrolysis. It was found that the tendency to cation hydrolysis increases with the Si/Al ratio of the zeolite framework. The cause of this reaction is the increasing distance between the anion sites $(AlO_4)^-$ at higher Si/Al ratios. The increased distance between the singly charged anion sites renders the electrostatic shielding of highly charged cations ineffective. It was also suggested that Y zeolite is a very strong electrolyte, and consequently it favors reactions resulting in the formation of added ions in the zeolite crystal (hydrolysis) {9}. Upon cation hydrolysis in Y zeolite, two distinct hydroxyl groups are formed: one of them is attached to the zeolite cation while the other is linked to the framework cations {11}. It is also expected that cation hydrolysis in Y zeolites occurs first with cations at low coordination sites (Site 2). It was determined that, following hydrolysis, the hydroxylated cations shift from the low coordination sites to the sodalite cage {11, 12}. With two cations moving into the sodalite cage, a cation-oxygen/hydroxyl cluster of great stability is formed, enhancing the thermal stability of the Y zeolite crystal. The catalytically effective acid sites in these bi- or multivalent cation Y zeolites are the acidic hydroxyls linked to framework Si-Al sites.

Several important characteristics of bi- and multivalent cation Y zeolites have been identified. For example, the absence of strong acid activity with alkaline earth Y up to about 45% cation exchange was explained on the basis that bivalent cations corresponding to this cation exchange level occupy the octahydral sites in the hexagonal ring, and they do not undergo cation hydrolysis. Extensive infrared spectroscopic studies of the O-H bond in bivalent cation Y revealed an influence of the size of cations and the Si/Al ratio of the zeolite framework on the wave number of the O-H band ($\sim$3640 cm^{-1}) representing acidic hydroxyls {5}.

The chemical and structural changes occurring in Y zeolites upon NH$_4$ exchange followed by thermolysis of the NH$_4$ groups have been extensively investigated {13}. It is assumed that crystal retention of the activated NH$_4$-Y zeolite versus the similarly treated NH$_4$-X rests on the larger number of stable Si-O-Si linkages {30}. The acidity here, similar to bi/multivalent cation zeolites, is centered on hydroxyls attached to framework cations. Infrared spectroscopic investigations revealed several hydroxyl types. One of these categories is represented by species in the large cavity, giving rise to acidic behavior in catalysis while others are occluded in the sodalite cage or in the hexagonal ring.

Both thermal and steam treatments result in chemical changes in the NH$_4$Y zeolite. At the mildest thermal treatment the tetrahedral coordination of the Si and Al atoms associated with acid sites may be retained {14}. In the presence of nascent moisutre or added steam, an extensive hydrolysis of framework aluminum begins {15, 17-22}. Upon extensive hydrolysis the aluminum is linked with hydroxyl groups, leaving the zeolite framework to assume cation sites. Ultimately, hydroxylated aluminums form stable oxide/hydroxyl clusters occupying the sodalite cage. The framework sites vacated by aluminum are occupied by silicon which is rendered more mobile by the removal of adjacent aluminum. Ultimately, a rehealing of the crystal structure occurs with the silicon atoms filling up the tetrahedral framework sites vacated by aluminum. A reduction of the number of unstable defect sites (cation vacancy) also takes place in order to reestablish short-range crystal continuity. As a result of these processes, empty "pockets" are formed throughout the zeolite crystal. The recrystallized part of the Y zeolite framework is now enriched in silica, and the crystal is further stabilized by the aluminum oxide/hydroxyl clusters occupying sodalite cages.

The success of "steam stabilization" depends on a delicate balance between the rates of aluminum hydrolysis, silicon migration, and silicon resubstitution reactions. Inappropriate reaction conditions, causing rapid aluminum hydrolysis without adequate silicon substitution,readily result in a collapse of the crytal structure. Consequently, temperature, steam pressure, and time are all important variables.

Characterization of Acid Sites

Some of the chemical and structural aspects of steam stabilization remain unresolved in spite of extensive attempts at characterization using mainly infrared spectroscopy and crystallography. Furthermore, characteristic of samples presumably of the same type prepared by differe investigators, often show different behavior because of differences, no matter how small, in treatment and in crystal source.

The most important structural aspect of steam-stabilized Y zeolite is the full retention of crystallinit in spite of the extensive ionic migration involved in the process. Crystallographic studies, particularly the study of Si(Al)-O bond lengths, suggest that probably three out of the four crystallographically distinct oxygen atoms participate in the formation of acidic hydroxyl groups {12, 23, 24}. The silicon enrichment in the crystal framework of steam-stabilized Y is indicated by shrinking of the crystal unit cell, corresponding to the smaller length of Si-O bonds versus Al-O linkages. A similar relationship between unit cell size and increasing Si/Al ratio has been well recognized before, and it has been used to estimate the Si/Al ratio of the Y zeolite framework {25}. With steam-stabilized Y, the determination of the framework Si/Al ratio based on x-ray diffraction pattern alone is difficult. The unit cell size is affected both by silicon enrichment of the framework and by the stuffing of the sodalite cage with aluminum oxide/hydroxide clusters, and these two phenomena are expected to contribute to changes in unit cell size in opposing directions {1, 25}. Therefore, the determination of cation exchange capacity remains a more reliable technique to determine framework aluminum content.

The complexity of the chemistry and structure of stabilized Y is reflected in the infrared spectrum of O-H bonds. Here, depending on the degree of framework aluminum hydrolysis, several infrared O-H bands are detected, representing both "accessible" and "non-accessible" hydroxyls. Since the extinction coefficient for all these infrared bands is not known, it is not possible to determine the relative concentration of hydroxyls of different O-H infrared frequency with precision. Nevertheless, assignments of O-H bands to assumed structural hydroxyls have been made, and they are used frequently to monitor steam stabilization and acidity {13}.

It has been discussed above that the hydroxyls formed in activated NH_4Y represent several distinct species differing in linkage, coordination, and accessibility. Correspondingly, the determination of zeolite acid strength is difficult by any single measurement. The use of colored dyes often used with other solid acids is not possible because of the large size of these molecules. Titration with small bases such as NH_3 is useful. However, one must consider that NH_3 is strongly attached to both Lewis and Bronsted acid sites. Consequently, NH_3 sorption measures both acidic hydroxyls and accessible cations. One of the most reliable measurements of acid concentration rests on the inspection of the infrared spectrum of pyridine adsorbed on Y catalysts. Here, the infrared bands representing distinct protonated or polarized pyridine species give quantitative account of Bronsted and Lewis acidity {5, 13}. Steric hindrance introduced onto the pyridine by methyl- or ethyl-substitutions at appropriate positions gives further information on OH accessibility in the zeolite structure.

One of the shortcomings of N-base titration results from the high energy of interaction between strong bases and zeolite acid sites, rendering the interaction almost irreversible at low temperatures and preventing equilibration among sites of various acid strength. A good "compromise" technique uses thermometric titration {26-28} with amines applied in a solvent. Here an aromatic solvent serves to expedite the desorption and the equilibration of the amine applied. This results in aiding equilibration, first neutralizing the strongest and then gradually the weaker acid sites. Unfortunately, even this technique has steric or diffusion-related limitations, and reliable data are obtained only with crystals with three-dimensional pore systems (Y zeolite). These titrations are much less successful with zeolites of less open structure (mordenite) {29}.

A strong enhancement in crystal stability and acid activity with increasing Si/Al ratio was recognized in the early days of zeolite catalysis, using zeolites X and Y {30}. It was shown that based on electrostatic considerations, the charge density at a cation site increases with increasing Si/Al ratio. It was conceived that these phenomena are related to a reduction of electrostatic interaction between framework sites, and possibly to a difference in ordering of aluminum in the zeolite crystal {31}. Later, using titrations with various amines, it was found that the slope of the titration curve of NaH-Y zeolites is linear with cation exchange {32}. This slope

changes with the Si/Al ratio, and it reflects the degree
of "efficiency" of the average $(AlO_4)^-$ site {33}. Higher
aluminum contents result in "interference" between alumina
sites, and thus in less than a linear increase of titra-
table strong acidity {33}. Conversely, lowering the
alumina concentration in the Y zeolite framework first
results only in a depletion of the weaker acid sites
without affecting the strong acid sites (>90% H_2SO_4) {34}.
Only extensive framework aluminum hydrolysis, resulting
in less than 16 Al per unit cell, results in a.decline of
strong acid sites {13}.

The direct relationship between acid strength and
Si/Al ratio seems to apply in some cases even for zeolite of different structure. It was reported that the
acid sites present in H-ZSM-5 (Si/Al = 19.2) are stronger
than those present in H-zeolon (Si/Al = 5) {35}. In a
different approach it was found that, using an electronegativity model for calculating the residual charge for
hydrogen atoms in zeolites, the acid strength declines
with increasing aluminum content {36}. Catalytic data
consistent with these calculations confirm the chemical
tendency suggested by these calculations {37}.

Catalytic Properties of Acid Zeolites

The main catalytic characteristics of acid zeolites in hydrocarbon reactions are their selectivity
as molecular sieves, their high acid strength, and
their electrolytic properties which influence reaction
kinetics.

1. Molecular Sieve Effect. Shape Selectivity

In many acid-catalyzed reactions the reaction rate
depends on the rate of carbenium ion formation. Normall
tertiary carbons form carbenium ions easier than seconda
carbons. Primary carbons do not form carbenium ions
readily under mild conditions, and consequently tend to
be unreactive. The differences in the ease of ion formation are readily explained by similar differences in
the energy required for the removal of H^- ion from C-H
bonds formed with tertiary, secondary, primary carbon
atoms. For this reason, isoparaffins are normally much
more reactive than n-paraffins. The order of this well-
established reactivity for paraffins is, however, often
reversed in certain zeolites as a result of steric
hinderance or diffusion limitation imposed by the zeolite structure. Effects of this type are often referred
to as shape selectivity {38}.

With zeolites containing pores just large enough to
admit small, mono-substituted paraffins, the rate of
diffusion for the isoparaffins is very substantially less
than the rate of diffusion of n-paraffins. Additionally,
the diffusion rate of n-paraffins can differ by several
orders of magnitude between small pore zeolites and the
large pore zeolite Y. Changes in the diffusion rate by
several orders of magnitude can readily reverse the order
of reaction selectivity, resulting in higher reaction
rates for the molecule of smaller critical dimension {39}.
In the case of diffusion-controlled selectivity, the size
of the crystal also becomes an important selectivity
parameter.

Molecular sieve effects in catalysis can be distinguished with respect to whether the zeolite pore size
prevents the passage of the reactant or the product through
the zeolite pores, or whether the formation of a particular transition state required for product formation
is prevented. There are convincing examples for each of
these categories {39}. There are many examples of molecular sieve effects-shape selectivity-using mixtures of
small and larger olefins or alcohols, with the smaller
molecules reacting while the larger ones remain intact
{40}. Molecular sieve effects showing product-size
selectivity have been demonstrated in the isomerization
of n-paraffins with small-pore type acidic zeolites. Here,
in spite of the presence of acid sites, no isoparaffins
are released in the product, and the product contains
only a mixture of methane and ethane in addition to the
n-hexane reactant. Here the small zeolite pores prevent
the desorption of isoparaffins which are presumably formed
in the large cavities of the zeolite crystal.

An interesting example of product-size type selectivity is shown in the isomerization of n-hexane
with the medium pore silicalite {43}. For this
experiment the silicalite was synthesized with a small
aluminum impurity (calculated as 0.6 wt.% alumina).
Following treatment with an NH_4 salt solution and thermal
activation, the silicalite composition showed a modest
degree of acidity, as shown in Table 3.

TABLE 3

ISOMERIZATION OF n-HEXANE ON Pt-Y AND Pt-SILICALITE CATALYSTS
(pressure = 230 psig; H/HC ratio = 3, Pt loading ~0.5 wt.%)

Catalyst	Steam-Stabilized Y		Silicalite		
Temperature, $^{\circ}$F	522	504	620	642	665
Conversion of nC$_6$, %	81.44	77.39	74.74	75.58	76.88
Methyl pentanes in C$_6$ product, %	55.42	56.53	68.26	66.22	64.41
22-DMB in C$_6$ product, %	18.97	14.84	0.64	2.00	3.67

From the difference in reaction temperature required
to reach similar conversions for n-hexane, it is clear
that the acidity of the Silicalite sample is very inferior
relative to steam-stabilized Y. Nevertheless, at higher
temperatures the Silicalite readily produced a mixture
of methylpentanes, close to the equilibrium composition.
However, in sharp contrast to the steam-stabilized Y
catalyst, the Silicalite produced no significant amounts
of 2,2-dimethylbutane. The absence of the relatively
large size 2,2-dimethylbutane in the product obtained
with Silicalite is consistent with independent adsorption experiments, showing that silicalite does not
adsorb the similar size neopentane molecule.

Other examples of molecular sieve effects are related to transition state selectivity. In this case,
certain reactions are prevented because the transition-
state complex is too large for the pore or pore-cavity
system of a zeolite. In some of these cases, neither
the reactants nor the products are prevented from passing through the zeolite pores. Examples of such transition state selectivity are displayed in the transalkylation and certain isomerization reactions of alkyl-
benzenes {39, 41}.

2. The Cracking of Hydrocarbons Over Acidic Y Zeolites

The cracking of n-hexane over H-Y zeolite shows
markedly different behavior {42} from that shown by
alkali cation zeolites {7}. Here catalytic activity
is much higher, skeletal isomerization of fragments
becomes dominant, and the product distribution is
generally consistent with that occurring with other
strong Bronsted acid catalysts for which a carbenium
ion rather than free radical mechanism is generally
accepted. Significantly, the product contains a relatively low concentration of olefins, and it also contains significant amounts of aromatics.

The isomerization of the hexane isomers was studied
at low conversion levels using a palladium-loaded zeolite catalyst. Here, the primary products derived from
the individual hexane isomers cannot be explained in
terms of an intramolecular rearrangement of a carbenium
ion intermediate. It was observed that isomerization
is invariably accompanied by hydrocracking, but the
hydrocracked porducts are not consistent with a simple
cleavage of the hexane molecules. A bimolecular mechanism
was proposed which satisfactorily explains both the

observed products from both isomerization and hydrocracking reactions as well as the presence of heptanes in the
product {49}.

In the industrial catalytic cracking process, the
acidic Y zeolite-based catalysts show certain new features
not found with the earlier used acid catalysts {44, 45}.
First, the activity of Y zeolite-type catalysts is much
higher relative to silica-alumina gel. In addition,
zeolite catalysts produce substantially larger gasoline
yields. The composition of the gasoline produced over
Y zeolites contains much less olefins and substantially
more aromatics relative to the gasoline made with silica-
alumina gel (see Table 4). Mechanistic suggestions by
Weisz {46} regarding these interrelated differences in
gasoline yield and composition have been made to the
effect that there is a more efficient hydrogen redistribution between hydrocarbon molecules over the zeolite
catalyst. According to the gasoline selectivity model of
Weekman and Nace {47}, the primary cracking reaction
produces gasoline and gas, whereas the secondary cracking
reaction produces only gas, and thus lower gasoline
yield.

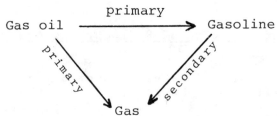

The difference in product distribution between zeolite
and silica-alumina can be explained quantitatively by
the greater occurrence of the following overall hydrogen
redistribution reaction in the zeolite:

Olefins + naphthenes ⟶ paraffins + aromatics

It was suggested {46} that this reaction effectively
reduces the secondary cracking in the cracking of gas
oil by converting the initially produced olefins and
naphthenes to more refractory paraffins and aromatics
before they crack further to gas. The high efficiency
of the conversion of olefins and naphthenes to paraffins
and aromatics may be explained by the superior hydrogen
redistribution via hydride ion shift between carbenium
ions and neutral hydrocarbons over the zeolite catalyst.

TABLE 4

GASOLINE COMPOSITIONS AS A FUNCTION OF CATALYST[a]

Catalyst	Composition (%)			
	Paraffin	Olefin	Naphthene	Aromatics
Silica-alumina	13	17	41	29
Zeolite	23	5	23	49
Differences	+10	-12	-18	+20

[a]Data from Ref. 46.

It is apparent from the information presented here that the cracking of hydrocarbons over K-Y and acidic Y zeolites show an important common feature. In both cases the efficiency of the hydrogen transfer step between reactant molecules is greatly enhanced by both types of zeolite catalysts. In the K-Y catalyzed cracking of n-hexane, this effect resulted in a substantial increase in the formation of propane and a similarly substantial decrease in the production of fragments smaller than C_3. In the case of gas oil cracking over acidic Y zeolite-based catalysts, an extensive redistribution of hydrogen is achieved to the extent that hydrogen is transferred from naphthenes to olefins forming paraffins and aromatics. This overall reaction, catalyzed by acidic Y zeolites, follows the thermodynamics prevailing at the catalytic cracking process conditions. One of the new chemical factors introduced by zeolite catalysts is that they are able to carry out both hydrogenation and dehydrogenation reactions efficiently via multiple hydride transfer steps. This phenomenon is quite similar to the enhancement of the hydrogen transfer steps found with the nonacidic K-Y.

Similar differences in product formation were reported recently in the direct synthesis of hydrocarbons from syn gas over two molecular sieve catalysts {48}. One of the catalysts consisted of a mixture of iron metal and H-ZSM-5 zeolite while the other consisted of a mixture of a similar fraction of iron metal and Silicalite. By comparison with a pure iron catalyst, both molecular-sieve-based catalysts substantially reduced the boiling range of the hydrocarbon product. The H-ZSM-5 catalyst displayed both higher activity and higher boiling-range selectivity than the silicalite-based catalyst. In spite of the similar influence of both ZSM-5 and Silicalite on product boiling range, the two catalysts produced hydrocarbons of entirely different compositions. The iron-H-ZSM-5 catalyst produced hydrocarbons rich in aromatics and very low in olefins while the iron-silicalite catalyst produced mainly olefins and only minor amounts of aromatics.

The determination as to what extent the acid strength, concentration, and electrolytic strength are responsible for the contrast in product compositions is not possible without establishing the intrinsic acid strength, acid concentration, and electrolyte strength in each catalyst. However, the similarity with the cracking data discussed above for the strong electrolyte Y zeolite versus silicalite suggests that in the hydrocarbon synthesis experiments the H-ZSM-5 plays the role of a strong acid as well as a relatively stronger electrolyte.

The experimental evidence strongly suggests that the cause of the hydrogen transfer enhancement with both K-Y and with acidic Y zeolites is common, both related to the high concentration of reactant hydrocarbons in the zeolite crystal relative to the surrounding gas phase. The high concentration of **reactant** molecules enhances the rate of bimolecular reaction steps-the hydrogen transfer step-with both K-Y as well as with acidic Y, over the unimolecular cracking step. The higher concentration of reactants, persisting even at high reaction temperatures, is the result of strong interaction between the polarizable hydrocarbons and the strongly polar intracrystalline zeolite surface. The degree of this interaction reflects the strength of the zeolite electrolyte.

CONCLUSIONS

A large body of experimental evidence in zeolite chemistry and zeolite catalysis suggests that the key distinguishing features of zeolite catalysts used in industrial applications are:

1. The catalytic selectivity based on molecular sieve effects or on diffusion limitations.

2. The high concentration of strongly ionic hydrogen (H^+) atoms attached to framework oxygen atoms.

3. The large enhancement of ionization reactions and the stabilization of carbenium ions.

4. The high concentration of hydrocarbon reactants within zeolite crystals, resulting in the enhancement of bimolecular reaction steps over unimolecular reaction steps.

314

REFERENCES

{1}   J. A. Rabo, in Zeolite Chemistry and Catalysis
       (ACS Monograph 171), Am. Chem. Soc., 1976, Chap. 5.
{2}   P. H. Kasai and R. J. Bishop, Jr., in Zeolite
       Chemistry and Catalysis (ACS Monograph 171),
       Am. Chem. Soc., 1976, Chap. 6
{3}   R. M. Barrer and W. M. Meier, J. Chem. Soc., 58,
       299 (1958).
{4}   C. J. Planck, Proc. Int. Congr. Catal., 3rd,
       Amsterdam, 1964, p. 727.
{5}   J. W. Ward, in Zeolite Chemistry and Catalysis
       (ACS Monograph 171), Am. Chem. Soc., 1976, Chap. 3.
{6}   J. A. Rabo, C. L. Angell, P. H. Kasai, and
       V. Schomaker, Discuss. Faraday Soc., 41, 328-349
       (1966).
{7}   M. L. Poutsma and S. R. Schaffer, J. Phys. Chem.,
       77, 158 (1973).
{8}   J. A. Rabo and M. L. Poutsma, Adv. Chem., 102,
       284 (1971).
{9}   J. A. Rabo, R. D. Bezman, and M. L. Poutsma,
       Acta Phys. Chem., pp. 39-52 (1978).
{10}   M. L. Poutsma, in Zeolite Chemistry and
       Catalysis (ACS Monograph 171), Am. Chem. Soc.,
       1976, Chap. 8
{11}   J. A. Rabo, C. L. Angell, and V. Schomaker,
       IV Int. Congr. Catal., Moscow, 1968, p. 966.
{12}   J. V. Smith in Zeolite Chemistry and Catalysis
       (ACS Monograph 171), Am. Chem. Soc., 1976,
       Chap. 1.
{13}   P. A. Jocobs, Carboniogenic Activity of
       Zeolites, Elsevier, 1977.
{14}   R. L. Stevenson, J. Catal., 21, 113 (1971).
{15}   C. V. McDaniel and P. K. Maher, in Zeolite
       Chemistry and Catalysis (ACS Monograph 171),
       Am. Chem. Soc., 1976, Chap. 4.

{16}   G. C. Bond, P. B. Wells, and F. C. Tompkins, *Proc. Sixth Int. Congr. Catal.*, Chemical Society, London, 1976.

{17}   D. W. Breck and G. W. Skeels, *ACS Symp. Ser.*, 40, 271 (1971).

{18}   D. W. Breck and G. W. Skeels, *Proc. Fifth Int. Conf. Zeolites, Italy*, 1980, p. 335.

{19}   C. V. McDaniel and P. K. Maher, *Molecular Sieves*, Society of Chemical Industry, London, 1968, p. 186.

{20}   G. T. Kerr, *J. Catal.*, 15, 200 (1969).

{21}   G. T. Kerr, *J. PHys. Chem.*, 71, 4155 (1967).

{22}   G. T. Kerr, *Mol. Sieves, Internat. Conf. 3rd*, (Adv. Chem. Ser., 121), Am. Chem. Soc., 1973, p. 219.

{23}   D. H. Olson and E. Dempsey, *J. Catal.*, 13, 221 (1969).

{24}   W. J. Mortier, J. J. Pluth, and J. V. Smith, *Ibid.*, To be published.

{25}   D. W. Breck, *Zeolite Molecular Sieves*, Wiley-Interscience, New York, 1974.

{26}   Y. Okamoto, T. Imanaka, and S. Teranishi, *Bull. Chem. Soc. Jpn.*, 43, 3353 (1970).

{27}   T. R. Brueva, A. L. Klyachko-Gurvich, and A. M. Rubinhstein, *Izv. Akad. Nauk USSR, Ser. Khim.*, p. 2807 (1972).

{28}   B. V. Romanovskii, K. V. Topchieva, L. V. Stolyarova, and A. M. Alekseev, *Kinet. Katal.*, 12, 890 (1971).

{29}   R. D. Bezman, *Characterization of Acidic Zeolites by Thermometric Titration*, To be published.

{30}   J. A. Rabo, P. E. Pickert, D. N. Stamires, and J. E. Boyle, *Actes du Duxieme Congress International de Catalyse, Paris*, 1960, pp. 2055-2074.

{31}   P. K. Pickert, J. A. Rabo, E. Dempsey, and V. Schomaker, *Proceedings of the Third International Congress on Catalysis, Amsterdam*, 1964, pp. 714-728.

{32}   R. Beaumont, D. Barthomeuf, and Y. Trambouze, *Adv. Chem. Ser.*, 102, 813 (1971).

{33}   R. Beaumont and D. Barthomeuf, *J. Catal.*, 26, 218 (1972).

{34}   R. Beaumont and D. Barthomeuf, *Ibid.*, 27, 45 (1972).

{35}   J. C. Vedrine, A. Auroux, V. Bolis, P. Dejaifve, C. Naccache, P. Wierzchowski, E. G. Derouane, J. B. Nagy, J.-P. Gilson, J. H. C. vonHooff, J. P. van den Berg, and J. Wolthuizen, *Ibid.*, 59, 248-262 (1979).

316

{36}   W. J. Mortier, Ibid., 55, 33 (1978).

{37}   P. A. Jacobs, W. F. Mortier, and J. B. Uytterhoeven, J. Inorg. Nucl. Chem., 40, 1919 (1978).

{38}   P. B. Weisz and V. J. Frilette, J. Phys. Chem., 64, 382 (1960).

{39}   S. M. Csicsery, Utah Conference on Catalysis, 1979.

{40}   P. B. Weisz, V. J. Frilette, R. W. Maatman, and E. B. Mower, J. Catal., 1, 307 (1962).

{41}   S. M. Scicsery, J. Org. Chem., 34, 3338 (1968).

{42}   S. E. Tung and E. J. McIninch, J. Catal., 10, 166 (1968).

{43}   C.-L. Yang, A. P. Risch, and G. N. Long, Union Carbide Corp., Private Communication.

{44}   P. E. Venuto and E. T. Habib, Jr., Fluid Catalytic Cracking with Zeolite Catalysts, Dekker, New York, 1979.

{45}   J. S. Magee and J. J. Blazek, in Zeolite Chemistry and Catalysis (ACS Monograph 171), Am. Chem. Soc., 1976, Chap. 11.

{46}   P. B. Weisz, Chem. Technol., p. 498 (1973).

{47}   V. W. Weekman, Jr., and D. M. Nace, AIChE J., 16 (1970), V. W. Weekman, Jr., Ind. Eng. Chem., Process Des. Dev., 8 (1960).

{48}   V. U. S. Rao, R. J. Gormley, H. W. Pennline, L. C. Schneider, and R. Obermyer, ACS Meeting, Div. Fuel Chem., 25, 119-126 (1980).

{49}   A. P. Bolton and M. A. Lanewala, J. Catal., 18, 1 (1970).

ACIDIC CATALYSIS WITH ZEOLITES

Denise Barthomeuf

Exxon Research and Engineering Company
Linden, New Jersey 07036

1. Uses of Zeolites as Acidic Catalysis

After the work of Barrer, zeolites have been used as selective adsorbents. The discovery of their catalytic properties in 1960 started a new era in the field of catalysis.(1,2) Less than ten years later, 90% of catalytic cracking units in the United States used zeolites. A continuous improvement of zeolites properties led to a simultaneous innovation and competiton in units design. At the present time the four main industrial applications of zeolites involve their acidic properties. The most important is catalytic cracking which uses zeolites thermally stabilized by various treatments. The hydrocracking process superimposes the cracking due to the acid properties of the zeolite with the hydrogenation due to noble metal. The selectoforming adds a shape selective effect of the zeolite cages to the hydrocracking by the choice of zeolites (erionite, offretite) which selectively convert only C_5 to C_9 n-paraffins from naphthas and reformates.(3) Oil dewaxing also uses a shape selective catalyst for the hydrocracking of n-paraffins. Besides those industrial uses, zeolites are also very good catalysts for other reactions involving acidic catalysis such as isomerization, hydration-dehydration, alkylation, isomerization and disproportionation of alkylbenzene, etc.

Zeolites have been able to replace old catalysts as in catalytic cracking or generate novel applications as in selectoforming or oil dewaxing mainly because of their very high selectivity. This is strongly related to their open structure which may induce several specific and different types of selectivity. The shape selective catalysis known in ZSM-5 or mordenite is discussed in details elsewhere in this course. A selectivity related to chemical effects of the cages is involved in zeolites with larger cavities. Zeolites give a greater amount of paraffins and aromatics than amorphous silica alumina catalysts in industrial cracking. This is explained by an overall hydrogen redistribution reaction in the zeolite.(4)

Scheme 1 (from Ref. 4)

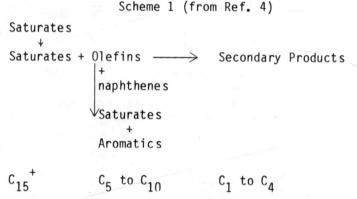

This product distribution leads to high gasoline yields and low coke formation. Usually, to take advantage of those properties, zeolites are mixed with a matrix in a proportion close to 15%. This optimizes the percent conversion and the gasoline and coke formations. It also gives to the catalyst other important properties such as resistance to attrition, lower cost, improved heat transfer performances...

2. Main Parameters Which Determine Catalytic Properties in Acidic Zeolites

The catalytic properties (% conversion, selectivity) depend on zeolite structure, chemical composition and various pretreatments which may change locally the composition or atom location in the zeolites.

2.1 Al Content (Si/Al Ratio)

The Al content of a zeolite may be expressed as the ratio Al/Al+Si (molar fraction of AlO_4 tetrahedra in the total number of AlO_4 and SiO_4 tetrahedra) or as the ratio Si/Al. Since the acidity arises from the replacement of a tetravalent Si^{4+} atom by a trivalent Al^{3+} atom in the alumino-silicate framework, the number of acid sites generated should parallel very strongly the Al content. Any change in Si/Al ratio should then be reflected in catalytic properties. In fact an increase in the Si/Al ratio improves several zeolite properties. It increases the conversion of hydrocarbons, the thermal stability and the acidity strength.

With regards to catalytic activity, in the faujasite series the X zeolites (Si/Al = 1.25, Al/Al+Si = 0.44) are less active than Y types (Si/Al = 2.4, Al/Al+Si = 0.29) despite their higher theoretical number of acid sites.(1,5-7) Similar kinds of increase with the Si/Al ratio are obtained in cumene cracking,(5,7-12) o-xylene isomerization,(6) gas oil cracking(13) or cyclopentane isomerization.(14) They can be explained by a higher electrostatic field in Y than in X zeolites(15) or by an increase in the so-called efficiency of acid sites from X to Y.(16) By contrast reverse results are obtained in the low Al range: the catalytic activity increases, as one could expect, with the Al content. Recent results have been published for ZSM-5 zeolite in a very large range of composition at a low Al level ranging from 10 to 10,000 ppm.(17,18) A linear relationship is observed between the increase in catalytic activity for n-hexane cracking and the Al content. For all the types of reaction mentioned above which all need strong acid sites, one can then expect a maximum in the catalytic properties for intermediate Al contents. Values of Si/Al between 4 and 8 have been found for faujasite structure (Figure 1)(7-12,14) or between 8 and 25 for mordenites.(19,20)

For reactions which involve only weak acid sites such as isopropanol dehydration a linear relationship is found in a wide range of composition(21,22) for very different zeolites structures when considering the turn-over number versus the Sanderson electronegativity which reflects acidity strength. A similar relationship is also observed for the hydroconversion of n-decane.(22) The Sanderson electronegativity of zeolites discussed later in details is very helpful to predict catalytic properties in such reactions. It is nevertheless not able to depict catalytic properties at very low Al content in the range where location of sites and small changes in their number are of prime importance.

Of very great industrial importance are the ultrastable zeolites (U.S.) in which the Si/Al ratio is increased by special treatment in order to remove Al atoms from the framework. This can be done either by chemical extraction with chelating agents such as EDTA or acetylacetone(8,23-25) or by procedures involving steaming of ammonium zeolites forms.(26) These zeolite modifications lead to a unit cell shrinkage and a considerable increase in stability(26,27) which is of major improvement for practical use. No significant improvements in catalytic activity or selectivity have been described. A

320

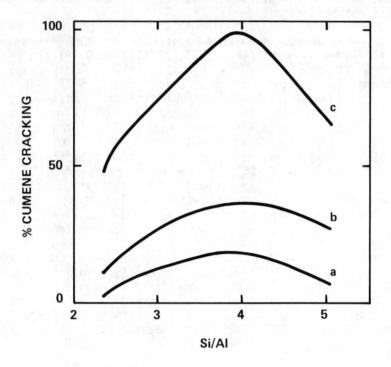

FIGURE 1: Change in cumene cracking as a function of Si/Al
in faujasites

(a) La^{3+} form (8-12)
(b) Ca^{2+} form (8-12)
(c) H^+ form (7-12)

relatively low amount of fundamental results has been reported
on those U.S. zeolites Y as opposed to the usual not stabilized
Y zeolites of less practical value.

A good thermal stability of zeolites is needed for their
use in reactors and for their regeneration usually by coke
residue burning. The acidic forms obtained by replacement of
exchangeable cations by protons are usually less stable than
the parent forms. An increase in the Si/Al ratio improves
greatly the thermal or hydrothermal stability.(27) This has
been ascribed to the lower density of hydroxyl groups which
parallel that of Al content. A longer distance between
hydroxyls decreases the probability of dehydroxylation then
also that of defect generation.(28)

2.2 Cation Content and Identity

The acid catalysis in zeolites is strongly dependent on the proton content and the acidity strength. Both parameters vary with remaining cations. The catalytic activity in cracking, isomerization etc. reactions increases as the sodium content is decreased. Figure 2 is typical of results obtained with Y zeolites in reactions such as cumene cracking,(29) isooctane cracking(30,31) or o-xylene isomerization(32) which require medium and strong acid sites. Similar behavior is observed for other zeolite structures such as offretite(33) but it shows some exceptions. For instance Mg HY zeolites show a maximum in catalytic properties related to the formation of very strong acid sites.(34) The L type zeolite also gives a maximum at 50% exchange of the cations which cannot be explained by any crystallinity loss or ultrastabilization effect.(35) Nevertheless the usual explanation for Figure 2's results correlates the increase in activity with the higher acid strength of sites generated in the highly protonated zeolites.

The exchange of monovalent ions by polyvalent cations improves the catalytic properties.(1,2) Those highly charged cations create very acidic centers by a hydrolysis phenomenon(36)

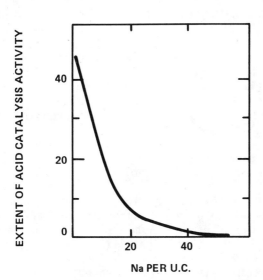

FIGURE 2:   Change in acid catalysis activity
as a function of Na[+] content
in Y zeolite

$$Me^{n+} + xH_2O \longrightarrow Me(OH)_x^{(n-x)+} + xH^+ \qquad (2)$$

The new acid centers formed increase catalytic activity in many reactions such as cumene cracking(7) or o-xylene isomerization.(37) The rare earth zeolites appear to be very attractive, more than calcium forms for instance. They generate a higher activity, a higher thermal stability and a lower site deactivation.

2.3   Influence of Pretreatment Temperatures

For most acid catalyzed reactions the activity increases up to a maximum for pretreatment temperatures comprised between 700 and 900K.(38-41) A selectivity change is observed for 2-propanol dehydration over various Y zeolites after degassing at 673K.(42) Ether is formed preferentially at the maximum hydroxyl concentration confirming the hypothesis that its formation requires a pair of hydroxyl groups.(43) For paraffin cracking the maximum in activity occurs after pretreatment at temperatures 100 to 300 degrees higher than the maximum hydroxyl concentration. Figure 3 gives as an example the change in C_3 formation from n-hexane at 623K(40) and the total zeolite hydroxyl concentration (3660 and 3550 cm^{-1} bands) as determined by infrared spectrometry(44) as a function of pretreatment temperatures of NH_4Y. The explanation usually accepted relies on the fact that only the strongest acid sites are required for catalysis and they are generated at high temperatures.(40) A similar behavior has been observed for different zeolite structures.(38) The temperature for the maximum of catalytic properties also depends on which acid strength is needed, i.e. which catalytic reaction is under study. Jacobs(41) gives a classification of reactions in increasing order of activation temperature which correlates with a decreasing number of sites and an increasing strength of their acidity (Table 1).

3.   Correlations Between Acidic and Catalytic Properties in Zeolites

The importance of acidity for catalytic properties has been pointed out in the previous paragraphs. Many correlations have been published between both of those properties (for example 45,46). Most of acid catalytic reactions are related to the presence of Bronsted acidity. For instance it has been shown that the maximum in catalytic properties as a function of pretreatment temperature correlates rather well with Bronsted acidity which shows a maximum in the same temperature range. There is no obvious correlation with Lewis acidity.(44) Nevertheless it has not been possible up to now to find a unique

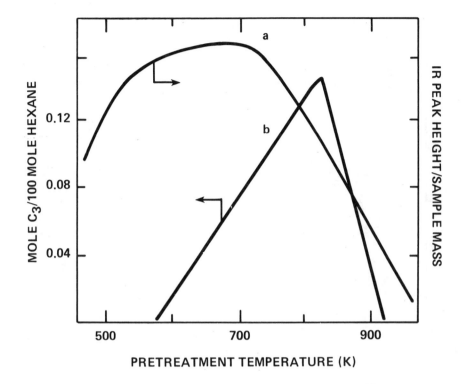

**FIGURE 3: a: Total OH content (44) and b: n-Hexane cracking
(40) as a function of pretreatment temperature**

correlation between all the different types of acid catalyzed
reactions and the protonic zeolite acidity. Moreover the
protonic acidity itself depends on several parameters. This
explains why so much work has been devoted to the study of
zeolite acidity. A better understanding of this property
should be a basis for a more valuable prediction of catalytic
properties.

4. Zeolite Acidity

The main difference between zeolites and other oxides is
their very open structure which makes the larger part of the
framework AlO_4 and SiO_4 tetrahedra accessible to adsorbed
molecules. The tridimensional network creates cavities and
channels in which molecules may undergo catalytic reactions.
Then the surface of a zeolite is in fact inside the bulk of the
crystal and one may expect that the surface properties, i.e.
the cavity wall properties, are strongly dependent upon the
framework constituting atoms. This dependence should be

Table 1

Number and Strength of the Active Sites on HY Zeolite (from Ref. 41)

Reaction	Temp. (K) for $r^{(1)}$ $> 10^6$ mole·$g^{-1}s^{-1}$	Number Sites g^{-1}	Strength (% H_2SO_4)	Outgassing Temp. (K) for Maximum Activity
1. Alcohol dehydration	373	$\sim 10^{21}$	w. $(3\ 10^{-4})$	673
2. Olefin isomerization	400	$\sim 10^{21}$	w. $(3\ 10^{-4})$	673
3. Alkylation aromatics	500			723
4. Isomerization alkylaromatics	550			823
5. Transalkylation alkylaromatics				823
6. Cracking alkylaromatics	573	5-10 10^{19}	s. (70)	873
7. Cracking paraffins	673	10^{17}	v.s. (88)	770-873

(1) r = reaction rates

(2) w. = weak; s. = strong; v.s. = very strong

effective not only on a short range but also on a rather long range all over the framework. In the last few years research on zeolite acidity has been moved from the characterization of the concentration and strength of isolated sites towards attempts to present an overall view of the acid properties. Before considering those approaches it is useful to point out the main parameters which characterize acidity in zeolites and the methods which are used to study these properties.

4.1 Acidity Study Methods

The ideal method of acidity measurement should give information on several parameters: the number, nature, strength, location, environment and mean lifetime of acid sites. Hence it should be able to characterize acid centers precisely enough to assign one type to the selective transformation of a reactant molecule. In fact each method gives information but none fully describes the acid sites.

Spectrometric methods have been widely used. Certainly IR spectrometry is the more usual. It has given a large number of results related to hydroxyl characterization and to Bronsted and Lewis acidity.(47,48) Pyridine and to a lesser extent NH_3 are the bases commonly used for semi-quantitative or quantitative evaluations of strength or concentration.(49,50) Table 2 gives the assignments of pyridine absorption bands. 2-6 dimethylpyridne has been suggested as a base specific for protonic centers characterization since steric hindrance of the nitrogen atom prevents its coordination to Al atoms.(51,52) Recently the Bronsted acid strength has been evaluated in zeolites from the frequency shift of hydroxyl groups upon interaction with hydrogen bond acceptor molecules such as benzene.(22,53,54) UV spectroscopy was used to study the adsorption of various molecules on zeolites.(55,56) In the case of pyridine on X zeolite with sodium cations(56) three kinds of adsorbed species characterized respectively the interactions with cations, protons and Lewis sites. It was noted that UV methods have the advantage to be very sensitive but it is difficult to distinguish the positions of the respective peak maxima. Optical electronic spectroscopy was also used to differentiate protonic and non-protonic sites.(57) Besides the study of redox properties of zeolites,(58) ESR has been applied to study atomic hydrogen formed on γ irradiation and which is related to protonic acidity.(59) Relaxation and linewidth studies in NMR have been employed to characterize the proton mobility and interactions with cations.(60,61) These studies afford a further insight into the nature of protonic acidity. It is proposed that the higher is the jump frequency (inverse mean life time at lattice

Table 2

Assignment. IR Bands of Pyridine (Py) Adsorbed on
Bronsted (H^+) or Lewis Sites (L)

Vibrational Mode	Py H^+	Py L
8a $\nu_{CC(N)}$	1655	1595
8b $\nu_{CC(N)}$	1627	1575
19a $\nu_{CC(N)}$	1490	1490
19b $\nu_{CC(N)}$	1550	1455-1442

oxygen atoms), the higher is the strength of the proton.(60,61) It is shown that an adsorbed base such as pyridine increases (up to 60 times at 200°) the proton mobility probably because during pyridinium ion formation and decomposition a hydroxyl proton must be first attached to the pyridine molecule and then given back to another oxygen atom. It is hence suggested that investigations of proton mobility can lead to conclusions on acidity only if the study is made in the presence of basic molecules.(61) NMR measurements also permits the computation of an elementary "proton capture probability" by the acceptor molecule diffusing upon the surface. This probability decreases from 1 to 0.06 after a long outgassing performed between 0° and 300°C(60) (for NH_3).

Besides the spectroscopic methods several approaches have been used to characterize acidic properties. Minachev, Bremer et al. performed several works using H_2-O_2 exchange to study the proton mobility.(62-64) Several thermal methods have been used to study the interactions of bases with acid sites (DTA,(65) calorimetry and chromatography).(66-68) It was shown that NH_3 or butylamine give a small heat of adsorption on cations. The distribution of the strengths of acid sites could be obtained from the heat of adsorption of benzene on progressively pyridine poisoned samples.(66) A method based on the determination of the amount of oxygen used for the oxidation of NH_3 in the ammonium forms of zeolites has been described to measure the number of acid centers. It distinguishes between Bronsted and Lewis acidities.(69) After the early work of Hirschler,(70) titration with butylamine and colored indicators has been used.(25,71-75) The method allows an easy determination of acid site concentration and strength to be done but it does not give information on the precise nature of acid centers. The question arises also as to the size of the

reactants may modify the results. In fact in Y zeolites, the number of acid sites determined in this way is close to that deduced from IR experiments using pyridine.(49) The results performed with various bases, indicators and zeolites suggest that with regard to the aperture of the zeolite channels, the size of the base molecule is more critical than that of the indicator. Assuming that at the equilibrium the base, small enough to move in the channels, neutralizes the same fraction of acid sites wherever they are located (inside or outside the particles), the large indicators molecules may detect the end of neutralization from the reaction with the only accessible sites.(74,75)

Among all these methods, infra-red spectroscopy is the most powerful since it gives a very large number of informations on the acid sites. However the difficulty to correlate precisely infra-red results with other properties such as catalytic behavior, points out the fact that the characterization of static and definite hydroxyl groups is not precise enough to explain which protons act in the dynamic catalytic processes.

4.2 Nature and Generation of Acid Sites

The negative charges in excess due to the replacement of SiO_4 tetrahedra by AlO_4^- tetrahedra in the framework are neutralized by protons or other cations. The protonic acid centers are generated in various ways.

i. The thermal decomposition of ammonium exchanged zeolites yields the hydrogen form. Deammination in anhydrous conditions of alkylammonium, piperidinium or pyridinium Y zeolites produces a stoichiometric hydrogen zeolite only in the case of primary alkylammonium ions. With other cations a considerable dehydroxylation is observed producing a so-called dehydroxylated zeolite with Lewis acid sites.(76) This dehydroxylation effect is also observed during the adsorption of amines, particularly with pyridine.(77,78) One can wonder whether the high proton mobility in the presence of pyridine(61) does not facilitate the dehydroxylation phenomena.

ii. The Brönsted acidity due to the water ionization on polyvalent cations (36) already described (Scheme 2) has been studied by various methods. NMR was applied to the calculation of the interproton distance (d_{H_2O}) in water coordinately bonded to the cations.(79)

iii. The reduction by hydrogen of transition metal cations in zeolites was supposed to form a hydrogen zeolite.(80,81) Such Bronsted acidity has been observed in an IR study of a hydrogen reduced $Cu^{2+}Y$ zeolite.(82)

$$2\ Cu^{2+}\ +\ H_2\ \rightarrow\ 2\ Cu^+\ +\ 2\ H^+ \tag{3}$$

$$2\ Cu^+\ +\ H_2\ \rightarrow\ 2\ Cu^0\ +\ 2\ H^+$$

The reduction of Cu^{2+} cation with $CO^{(82)}$ or its self-reduction(83) gives cuprous ions and Lewis acidity. The concentration of OH groups of Y zeolites containing Ni, Co or Cu was noted to increase by reduction with hydrogen at 250-450° and to increase with the rise of the reduction temperature.(84) A reduction by hydrocarbons of cations to metals with formation of protonic acidity has been shown in the case of Ni, Fe and Co zeolites during the cumene cracking. A similar reduction is postulated with Cr and Cd zeolites:(85)

$$Ni^{2+}\ \xrightarrow{2RH}\ Ni^0\ +\ 2\ H^+ \tag{4}$$

iv. Bronsted acid sites are also generated in bivalent cation-containing Y zeolites on exposure at room temperature to halide compounds(86) or at 150-400°C to CO_2.(87)

The various and independent ways of acid site generation show that the experimental conditions of zeolite pre-treatments or acidity measurements could modify greatly the intrinsic acidity. Further this suggests that zeolite acidity may be changed by the presence of catalytic reagents (dehydroxylation by reactants, reduction of reducible cations by hydrocarbons, reaction with acidic compounds). Hence independently of aging effects, there may be large differences in acidity between the fresh and the actual catalyst.

Scheme (5) depicts the formation of Lewis acidity from Bronsted sites(88):

$$2\ \underset{\displaystyle Si\ \ Al\ \ Si}{\overset{\displaystyle O\ \ \overset{H}{O}}{}}\ \xrightarrow{\Delta t}\ \underset{\displaystyle Si\ \ Al\ \ ^+Si}{\overset{\displaystyle O}{}}\ +\ \underset{\displaystyle Si\ \ Al\ \ Si}{\overset{\displaystyle O\ \ \ O}{}}\ +\ H_2O \tag{5}$$

The Brönsted (OH) and Lewis (-Al-) sites can be present simultaneously in the structure at high temperature. In faujasites, the dehydration reaction occurs above 873K which decreases the number of protons and increases that of Lewis sites.(44) Scheme 5 has been confirmed by IR spectrometry. At temperatures higher than 673K the sum of the number of Bronsted sites plus twice that of Lewis sites is almost constant.(44)

4.3 Hydroxyl Groups in Zeolites

In the early works on zeolite acidity the formation of two different hydroxyl groups has been reported in faujasites upon ammonium ion decomposition.(47,88,89) They are asigned to OH vibrating in two different cages, the supercage and the sodalite. For most zeolites several hydroxyl bands are reported corresponding to OH groups vibrating in different cavities (Table 3).(90) The hydroxyls in very small cavities are not accessible to hydrocarbons or base molecules. If they are acidic, they can nevertheless interact with those molecules, the proton being attracted in the large cages. This occurs for the 3550 cm^{-1} hydroxyls in faujasites which move from the sodalite cage to the supercage upon base adsorption. The question then arises to know which hydroxyl is involved in the catalytic process. It was shown that in the case of faujasite the active sites for cumene cracking are the hydroxyls vibrating at 3650 cm^{-1}. The low frequency hydroxyls (3550 cm^{-1}) start to interact with the hydrocarbon only at higher temperatures (>365°C) when they become sufficiently activated.(91)

Each of the two hydroxyl groups in faujasites behaves separately. Exchange of protons with increasing amounts of cations such as sodium decreases more rapidly the intensity of the 3540 cm^{-1} hydroxyl band because of the preferential location of the first cations in the hexagonal prisms and the sodalite cages.(32) Each of the hydroxyls also includes a large range of protonic acid strengths. The weaker sites are neutralized first upon the exchange of protons by other cations. It has been reported that approximately 30% of the sites are weak.(71)

The origin of the difference in wavenumber according to the hydroxyl group location has been for a long time a matter of speculation. The existence of hydrogen bonding with close oxygen atoms has been postulated.(28,47) Recently it has been shown that the high wavenumber band relates to an unperturbed hydroxyl.(90) The shift to a lower wavenumber for the hydroxyls in small cavities arises from their disturbance by electrostatic field created by the nearest oxygens. The

Table 3

Hydroxyl Stretching Frequencies in Hydrogen-Zeolites
and Their Proposed Assignment (from Ref. 90)

Zeolite	OH Frequency (cm^{-1})	Assignment
FAU	3659^a	O_1-H in supercages
	3584	O_3-H in sodalite cages (6-memb. ring, site 1*)
	3578	$O_2(O_4)$-H in 6-memb. rings (site II)
FAU*	3562^a	O_1-H
	3570	O_3-H
	3547	$O_2(O_4)$-H
FAU**	3647^a	O_1-H
	3565	O_3-H
	3551	$O_2(O_4)$-H
RE-FAU	3628^a	O_1-H
	3530	La-OH (?)
HEU	3620^a	Crystal terminating, influenced by sample composition
	3560	In 8-memb. rings (pores)
CHA	3630^a	8-memb. ring
	3540	6-memb. ring
STI	(3650)	Amorphous phase with STI composition
	3620^a	Crystal terminating, influenced by sample composition
	3575	8-memb. rings
ERI	3612^a	8-memb. rings
	3563	6-memb. rings
MOR	3720	SiOH of unidentified nature
	3650	Occluded impurities

Zeolite	OH Frequency (cm^{-1})	Assignment
	3610[a]	In pores
MFI	3720 3601[a]	Extra lattice Si-OH In pore intersections
MEL	3720 3605(a)	Extra-lattice Si-OH In pore intersections
MAZ	3624[a]	In pores
LTL	3630[a]	In pores
FER	3612[a]	In pores (10-memb. rings)
OFF	3618[a]	In pores (?)
RHO	3612[a]	In pores (?) (8-memb. rings)

[a] These frequencies represent OH groups, vibrating in the largest pores and/or cages of the respective zeolites

For references see (90)

frequency shift from the unperturbed hydroxyls follows a linear relationship with the inverse of the squared distance of the proton to the nearest oxygen for a series of zeolites containing the perturbed hydroxyls in 6-. or 8-membered rings.

4.4 Number of Sites

The potential number of acid sites in zeolites equals that of Al atoms per any reference unit (g, cm^2...). The number of acid sites present in a sample is usually lower since it depends on many parameters: degree of crystallinity, dehydroxylation, partial neutralization with cations or bases etc. The number of acid sites active in a given catalytic reaction can be even smaller due for instance to the inaccessibility of sites (for instance OH groups in small cages) or to the requirement for the right acid strength (Table 1). A quantitative evaluation of both types of hydroxyls in faujasites gives a maximum value of 16 hydroxyls per unit cell

vibrating in the supercage at 3650 cm^{-1},(50,92) i.e. one
hydroxyl group on average per hexagonal prism. Titration with
pyridine gives close to 35 OH per unit cell in the faujasite
supercage(92) which is close to the number of pyridinium
formed. This might arise from attraction by pyridine of some
of the 3550 cm^{-1} hydroxyls in the supercage.

A large number of studies have been performed using base
titration with colored indicators. They give the total number
of sites Bronsted and Lewis. A detailed study of a faujasite
type series as a function of cation content (Na, K, Ca, La) or
Si/Al ratio showed that for each cation exchanged by one proton
only one fraction of an acid site could be titrated. This
fraction α_0 is small in highly alumineous zeolites such as X
and it increases at lower aluminum content (Figure 4).(25)
Such an efficiency or self-inhibition coefficient should
reflect the high density of acid sites at high Al content. It
could work as an activity coefficient in concentrated solutions.(16) This would explain the lower catalytic activity of
X zeolites compared to Y mentioned earlier.

4.5 Acidity Strength

The acid strength of sites in zeolites depends on the
Si/Al ratio. From all the results published two ideas emerge
emphasizing the superimposition of short range and long range
effects in determining the acid strength.

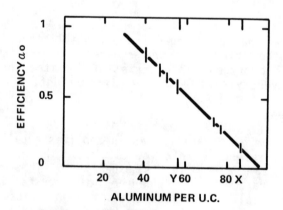

FIGURE 4: Dependence of efficiency of acid
site α_0 on the aluminum content
in faujasites (25)

4.5.1 Short Range Interactions

The question of whether the Al ordering is the same for X and Y zeolites was raised a long time ago.(93) An acidity study of these zeolites in which Al atoms were progressively removed from the framework by dealumination suggested that the acid strength of sites was dependent upon the Si or Al atom environment, despite the fact that all the T (Al or Si) positions are structurally equivalent.(25,30,71) Several attempts have been made to account for the heterogeneity of acid strengths. The general idea of these approaches is that the protonic acidity strength associated with an AlO_4^- tetrahedron is highest for the smallest number of close Al neighbors. Since, except for zeolites with a Si/Al ratio of 1 (A or X type), the Al distribution is not perfectly homogeneous a range of acid strengths is expected to occur. Dempsey was the first to relate quantitatively the acid strength of the proton to the geometry of the structure and the environment of the Al atoms.(94) In a more general model the strength of the protons was derived from the statistical distribution of Al atoms in the faujasite structure.(95) A further extension considers the next nearest neighbors of each Al atom and the buffering action of the cations to explain the known changes in proton acid strength and the thermal stability of hydroxyl groups.(96) In this model the parameters of importance are the distance and number of close Al atoms, of cations and of hydroxyls.

Until recently no information was available on the Al distribution in the framework. The use of ^{29}Si NMR(97) and ^{27}Al NMR(98) has made possible the determination of the arrangement of Si and Al atoms. This is a very valuable source of information for calculations on the acid strength distribution of protons existing in a given structure. The existence of such data for variously treated zeolites, for instance dealuminated Y,(99) and ultrastable(98) may help in the understanding of their changes in acidic properties. Also the presence of different acid strengths observed by Jacobs et al.(54) in highly siliceous zeolites of similar chemical composition (ZSM-5, ZSM-11 and dealuminated faujasite) could be related to environmental effects detectable by ^{29}Si NMR or ^{27}Al NMR.

Besides the effects due to the geometry of the Al distribution, other effects are well known to modify the acid strength of protons.(48) The most important is the exchange of protons by cations. The strongest sites are neutralized first and the acid site distribution moves to weaker acidity. Polyvalent cations generate strong acid sites by water hydrolysis. These acidity changes are due to localized effects (chemical reactions) and are then also related to short range interactions.

4.5.2 Overall Interactions

The idea of an influence of all the Al atoms on the acidic and catalytic properties of zeolites has grown progressively. Attempts have been made to quantify it. (14,16,21,22,25,71,90,100)

In contrast to the previous models which deduce the distribution of acid strengths from the nature of the close neighbors atoms (Al or Si), models taking into account a very large number of atoms only calculate the effects of the average environment.(21,22,100-102) They give a mean value of acid strength for a given Al content, i.e. a given density of charges in the structure.

A very successful approach uses the Sanderson electro-negativity equalization concept.(103) It has been applied to the calculation of the zeolite electronegativity and the charges on various framework atoms and on cations.(21,22,100) This intermediate electronegativity is postulated to be the geometric mean of the compound atoms of the molecule, i.e. of a given portion of a zeolite network. Very interestingly it has been shown that the charge on the proton increases as the Al content (Al/Al+Si) decreases for faujasites with various Al contents or for various zeolite structures (L, mordenite, clinoptilolite.(21) This follows the experimental order observed for the increase in strong protonic acidity for the same zeolite series.(16,41) Correlations with catalytic properties give a good relationship only with reactions which involve all the hydroxyl groups, i.e. reactions catalyzed by the protonic sites regardless of their strength (isopropanol dehydration) or reactions which would involve a constant fraction of hydroxyls (n-decane hydroconversion).(22) For instance the isopropanol dehydration turn-over numbers(21,22) are proportional to the calculated charge on the proton. All these studies show that the mean charge on the proton is shifted regularly towards higher values as the Al content decreases. Simultaneously the total number of acidic hydroxyls, governed by the Al content, has to decrease. This strongly suggests that the entire acid strength distribution (weak, medium, strong sites) is shifted towards stronger values. The weaker acid sites should become stronger with the decrease in the Al content. This is in fact observed. The infrared wavenumber of acidic hydroxyls (high frequency band), for a large number of zeolite structures, has been shown to decrease with the Al content.(41,104) The corresponding decrease in the force constant characterizes an increase in acid strength. More precisely, information may be obtained on the weakest acid sites. Starting from a zeolite in a fully cationated form, Na for instance, the exchange of the first cations creates weak acidic OH groups (high frequency band) detectable by IR spectroscopy. The $\bar{\nu}_{OH}$ values of these first hydroxyls formed decrease with the Al content indicating an increase in their acid strength.

Both CNDO(101-102) and Sanderson calculations give only an average protonic acid strength and little information on the strongest sites. It has been known for a long time that the strength of strong protonic sites increases as the Al content decreases. For instance, this has been shown experimentally from the changes in the strength of base adsorption(16,41 and references therein), the wavenumber of acidic OH(41,104) the integrated extinction coefficient of OH groups(22) or the $\bar{\nu}_{OH}$ shift upon the interaction of the hydroxyls with a hydrogen-bond-acceptor molecule.(22,53)

In trying to understand how a change in composition may increase the acid strength two features have to be considered. Firstly zeolites are inorganic acids, secondly they are polymeric acids. These two properties modify the acid strength separately but always simultaneously. In an attempt to clarify these points, it is possible to distinguish the two contributions in an analytical approach. As oxyacids, zeolite formulas may be written as $TO_n(OH)_m$, (T = Si or Al) with m = Al/Al+Si and n = 2-m.(19) This is comparable to the way of writing, for instance, the series of oxychloroacids $ClO_n(OH)_m$. It is well known that their acid strength increases with n in the series

$$Cl(OH) < ClO(OH) < ClO_2(OH) < ClO_3(OH)$$

Explanations based either on electrostatic or electron delocalization considerations have been proposed. Calculations based on the Sanderson electronegativity equalization principle give the partial charge on the proton which parallels the increase in n and the pK_A of these acids. In zeolites n varies from 1.5 (Al/Al+Si = 0.5) to 2 (Al/Al+Si = 0). The increase in n, i.e. increase in acid strength, parallels the decrease in Al content which is in line with the strengths of the oxychloroacids. The value of n, limited to between 1.5 and 2, close to that of sulfuric acid $SO_2(OH)_2$, is in agreement with what is known since zeolites are considered as strong as sulfuric acid. This analogy with oxychloroacids points out the importance of the Al content which determines n and then the acid strength.(16)

This contribution to the acid strength may be called a composition contribution.

As to the second point, in polymeric acids, the charge on the proton depends on the interactions between the OH groups themselves and/or the protons and the surrounding molecules. This effect is comparable in some ways with what happens in concentrated solutions where empirical activity coefficients have to be considered. In zeolites, the higher the number of anions in the polymer, i.e. the higher the Al content, the greater the OH-OH and OH/framework interactions and the weaker is the acid. The importance of these interactions suggests the necessity to consider activity coefficients in zeolites; these would reduce the efficiency of protons in catalysis.(16) This contribution to acid strength may be called the concentration contribution.

Increasing the Al content in a zeolite simultaneously modifies the nature of the moiety (TO_n) and its concentration. Both effects decrease the acid strength of the m protons and are concomittent. The question then arises as to whether or not the charge on the proton, calculated from the Sanderson electronegativity model, represents the true acid strength of the zeolite. It obviously takes into account the so-called composition contribution but it is not clear if it includes the concentration contribution. The calculation involves the composition of only one isolated molecule. For instance in zeolites the calculated proton charge is exactly the same for one fragment of structure containing one proton or for any larger domain considered, having a large number of interacting OH groups.

In spite of these uncertainties, the theroetical calculations allowed much progress to be made in our knowledge of zeolite overall properties. The CNDO/2 calculation type is general in its concept since it considers the Si, Al distribution but it has only been applied to clusters simulating the faujasite structure in its X or Y forms.(101,102) No results are available for zeolites in which the Al/Al+Si ratio lies between 0 and 1/6 since the total number of (Al+Si) atoms in the cluster is 6. Very large clusters should be used to represent highly siliceous zeolites in which the Al/Al+Si ratio varies from 0.1 to 0.01 or even 0.005 (for instance ZSM-5 with Si/Al ratio from 10 to 200). These limits restrict the generalization to various structures and to low Al contents which are both of great interest for catalytic purposes. The Sanderson electronegativity type calculations are also general in the sense that they are based only on the chemical composition, and are independent of the structure. At low Al content the results obtained are less significant since the calculation does not distinguish between two different Al distributions which may give very different acid strengths. The Sanderson electronegativity, which appears to be extremely helpful(21,22,90,100) in rationalizing overall acid properties and some catalytic properties only allows a comparison of zeolites with similar acid strength distribution. No model is quite satisfactory to describe the properties of the highly siliceous zeolites. Another point concerns the wavenumber of acid hydroxyls (high frequency band). The observed, almost constant, value of $\bar{\nu}_{OH}$ for Al/Al+Si ratios lower than 0.17-0.15 was proposed to show the absence of any significant interactions between the acid sites. The activity coefficients would be 1 for the protons in these zeolites.(105) In fact, the Sanderson electronegativity of these zeolites correlates well with the $\bar{\nu}_{OH}$ values but not with the catalytic properties or with the acid strength measured in a different way.(22) An attempt was made to correlate the $\bar{\nu}_{OH}$ shift to a field parameter(106) but the relationship is not linear at low Al contents. In view of these recent results it becomes more and more doubtful that in the range of Al/Al+Si ratios lower than 0.16 overall properties can be deduced from simple models. In the faujasite structure this value corresponds to one Al per six-membered ring. For such "dilute" composition, local environment effects (gradient of composition, pairs of sites...) would prevail over the whole chemical composition. In this range, the overall properties would still be of significance only for zeolites which would show comparable Al distribution in the framework.

4.5.3 Connection Between Short Range and Long Range Protonic Acid Strength

From the evidence provided by the short range and the long range approach it is suggested that two kinds of protonic strengths have to be considered. An overall acid strength is a characteristic of the zeolite and is determined by an intrinsic parameter such as the Al/Al+Si ratio or the Sanderson electronegativity. An environmental acid strength modifies the first one by taking into account the local neighboring effects and gives the distribution of strengths around the average overall value. The first originates from overall properties and the second from short range interactions.

At low Al contents (Al/Al+Si < 0.15-0.17) small changes in Al distribution would greatly modify the environmental acid strength. In the intermediate and high Al content range a decrease in Al/Al+Si ratio shifts the whole acid strength

towards strongest acidity. The absolute overall acid strength
of each proton is increased.

4.5.4 Lewis Acidity Strength - Interaction Between Close Sites

Since the importance of Lewis acidity in catalysis is much
less than that of Bronsted acidity, no detailed studies have
been performed to characterize it. It is usually considered as
a "by-product" of protonic acidity. The connections between
the two types of sites make it interesting to look at their
changes in strength upon various treatments.

It has been known for a long time that for a given acidic
zeolite the Lewis acidity is always stronger than the Bronsted
acidity (100-150° difference in the temperature of complete
pyridine evacuation).(48) Such a parallelism still exists when
the protonic acidity strength is changed upon various treatments. The strength of Lewis sites increase when the Al content
decreases. It varies as the Bronsted sites strength.(107) A
similar correlation between the simultaneous decrease in
Bronsted and Lewis acidity strengths upon dehydroxylation in
NaHY zeolites has been pointed out very recently.(108)

All the results point out a very strong interdependence
between the strength of the two types of site at least for a
constant Al/Al+Si ratio. Lunsford proposed that the strength
of Bronsted sites may be increased by electron attraction by
Lewis sites.(109) From his new data, Datka(108) suggests that
the inverse effect also exists; the strength of both types of
site depends on the lattice polarization. In contrast to
previous models, this one takes into account the interactions
between sites of a very different nature for a given Al/Al+Si
ratio. It implies that sites are close enough so that the
polarization effect can occur. It explains the presence of
very strong acid sites in steamed mordenites(110) which are
supposed to be formed as superacids in solution, (AlO)p
deposits acting as Lewis sites. In this hypothesis of a reciprocal influence of sites, the Lewis acid strength should also
be very high. This is in good agreement with a calculation of
Lewis site strength of extra framework cationic aluminum.(111)

As a result of the idea of related strengths, it may be
expected that the adsorption of a molecule on one site will
modify the equilibrium between the strength of both types of
site. The acid strength of the site which is adsorbing the
molecule will itself be disturbed. This may happen in acidity
measurements and also during a catalytic reaction.

In a similar situation it has been suggested that interactions between close oxidizing and reducing sites exist in
zeolites. It has been shown that zeolites possess electron-
acceptor and electron-donor centers which interact with elec-
tron-donors (perylene for instance) or electron-acceptors
(tetracyanoethylene) to form paramagnetic positive (Pe^+) or
negative ($TCNE^-$) ions, respectively.(58) Strong interaction
between the two types of sites is shown by enhancement (up to
tenfold) of the reducing power of zeolite samples when certain
electron-donor molecules are adsorbed on the surface. Studies
of the effect of the site density or strength (ionization
potential) of the electron-donor molecule show that the
enhancement requires neighboring sites of not too high
energy. The same explanation has been used to describe the
enhancement of the zeolite electron-donor properties upon
interaction of small Pt particles with the oxidizing
sites.(112) The metal particles become electron-deficient.
Various proofs have been given of the electrophilic character
of these small Pt particles encaged in the zeolite channels.

In summary, both acidity results and the behavior of
reducing and oxidizing sites provide good evidence of a strong
interdependence between sites in close proximity.

5. Overall Concepts of Zeolite Properties

Several unifying principles emerged from underlying
physicochemical features of zeolites.

Detailed studies on ionizing properties of zeolite(113-
115) led Rabo to present an overall view of zeolites considered
as electrolytes.(116) Strong interactions between the polarizable hydrocarbons and the strongly polar intracrystalline
surface give rise to a high concentration of reactants persisting even at high temperatures. As a consequence, the rate of
bimolecular reaction steps is enhanced over that of unimolecular reaction steps. An example is the hydrogen transfer step
(Scheme 1) from cycloalkanes to olefins giving the high
aromatics + paraffins yield characteristic of cracking in
zeolites. Hydrogen transfer reactions are observed with both
KY or acidic HY. They are then not related to the Bronsted
acidity only but to a cage effect. The behavior of zeolites as
electrolytes is also reflected in a large enhancement of
ionization reactions and in the stabilization of carbonium
ions.

Studies on transition metal complexes in zeolites(117,118)
showed that zeolites exhibit some properties of conventional
solvents and behave as a solid matrix. Nevertheless the cage

geometry is still very important for the complex formation and stabilization and it distinguishes zeolites from usual solvents.

Calculations, using the Sanderson equalization principle, on charges on atoms were made in order to rationalize the properties of zeolites.(21,100) Further calculations on a large variety of zeolite structures with different chemical compositions gave a strong basis for a unifying concept.(22) Some major zeolite properties (wavenumber of OH groups, acidity strength, turn over number in isopropanol decomposition or n-decane hydroconversion) can be related to the zeolite Sanderson electronegativity(22) which can be identified to the negative chemical potential(119) (Figure 5). This gives an important tool for the prediction and the understanding of acidic and catalytic properties. Further improvements of the model are needed in the range of low Al content since the Sanderson electronegativity does not take into account the great importance of Al distributions, i.e. of local geometry.

The finding that self-inhibition coefficients in acidity changed uniformly with the zeolite Al content(25) generated the idea that zeolites behave as solutions.(16,120) Activity coefficients should exist and would greatly reduce any catalytic rate at high acidity concentration i.e. high Al content. By contrast at low Al level, as in dilute solutions, no interaction should decrease the reaction rate and the acid sites should behave as if they were fully isolated. The existence of such activity coefficients explains the maxima observed in various reactions (Figure 1) as a function of the Si/Al ratio. The linear increase in n-hexane conversion with the ZSM-5 zeolite Al content(17,18) is also in line with a constant value of 1 for the activity coefficient in the low Al level range. The calculated probability for having no close neighbor in 4-rings faujasite structure, i.e. no close site interactions, follows the self-inhibition coefficient curves as a function of Al content as given in Figure 4.(14) The analogy of zeolites with solutions may also be extended to electrochemistry for metal-loaded zeolites.

6. Conclusions

Many reactions involving carbonium ions intermediates are catalyzed by acidic zeolites.(121) With respect to a purely chemical standpoint the reaction mechanisms are not fundamentally different with zeolites or with any other acidic oxides. What zeolites add are cage effects, even in the absence of geometrical shape selectivity, and overall properties. The cage effects, arising from the highly ionizing power

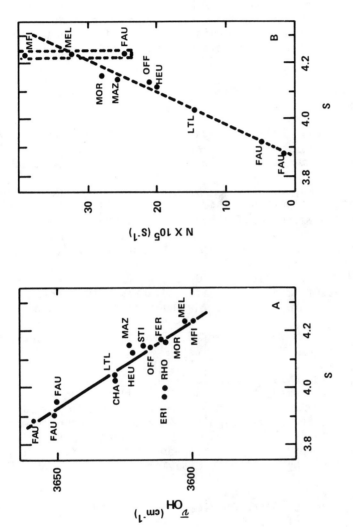

FIGURE 5: OH stretching frequency (A) and Turnover frequency of isopropanol decomposition at 423K (B) on H-zeolites as a function of zeolite Sanderson electronegativity (22)

342

of zeolites,(116) are probably responsible for the large variety of selectivities observed. It seems rather difficult to rationalize and predict those effects in the very near future. The overall properties have been more deeply understood in the last years.(16,22) They allow a better understanding and improved predictions of the extent of transformation of any reactant. They also afford a further insight into the nature of acidity and its correlation with catalytic properties. Despite all the progress made, these are still opportunities for new discoveries in the fundamental and applied fields of the acid catalysis with zeolites.

References

1. J. A. Rabo, P. E. Pickert, D. N. Stamires and J. Boyle, Int. Congr. Catal. 2nd, Paris, 1960, p. 2055
2. P. B. Weisz and V. J. Frilette, J. Phys. Chem. 1960, 64, 382.
3. N. Y. Chen, J. Maziuk, A. B. Schwartz and P. B. Weisz, Oil Gas J. 1968, 66, 154.
4. P. B. Weisz, Chem. Technol. 1973, p. 498.
5. K. V. Topchieva, B. V. Romanovsky, L. I. Pigusova, H. S. Thuoang and Y. W. Bizreh, Fourth Inter. Congr. Catal. Moscow, 1968, Paper 57.
6. J. W. Ward, J. Catal. 1970, 17, 355.
7. K. Tsutsumi and H. Takahashi, J. Catal. 1972, 24, 1.
8. K. V. Topchieva and H. S. Thuoang, Kin. i. Katal. 1970, 11, 490.
9. K. V. Topchieva and H. S. Thuoang, Dokl. Akad. Nauk. SSSR, 1970, 193, 641.
10. id., Kin. i. Katal. 1970, 11, 406.
11. id., ibid. 1971, 12, 1203.
12. id., Dokl. Akad. Nauk. SSSR 1973, 211, 870.
13. P. E. Pickert, American Petroleum Institute, 33rd Mid Year Meeting, Philadelphia, PA, May 1968, Paper 26-28.
14. S. H. Abbas, T. K. Ai-Dawood, J. Dwyer, F. R. Fitch, A. Georgopoulos, F. J. Machado and S. M. Smyth, in "Catalysis by Zeolites", Elsevier, Amsterdam, 1980, 127.
15. E. Dempsey, in "Molecular Sieves", Society of Chemical Industry, London, 1968, 293.
16. D. Barthomeuf, Acta Physica et Chemica 1978, 24, 71.
D. Barthomeuf, J. Phys. Chem. 1979, 83, 249.
17. D. H. Olson, W. O. Haag and R. M. Lago, J. Catal. 1980, 61, 390.
18. S. L. Meisel and P. B. Weisz, Adv. in Catalytic Chemistry II Symposium, Salt Lake City, May 1982.

19. P. B. Koradia, J. R. Kiovsky and M. Y. Asim, J. Catal. 1980, 66, 290.
20. A. M. Tsybulevskii, A. L. Klyachko, V. I. Berezhanaya, M. F. Pluzhnikova, T. R. Brueva and G. I. Kapustin, Izv. Akad. Nauk. SSSR, Ser. Khim. 1980, 12, 2690.
21. P. A. Jacobs, W. J. Mortier and J. B. Uytterhoeven, J. Inorg. Nucl. Chem. 1978, 40, 1919.
22. P. A. Jacobs, Catal. Rev. Sci. Eng. 1982, 24, 415.
23. G. T. Kerr, J. Phys. Chem. 1968, 72, 2594.
24. G. T. Kerr, J. Phys. Chem. 1969, 73, 2380.
25. R. Beaumont Thesis, Lyon, 1971.
    R. Beaumont and D. Barthomeuf, J. Catal. 1972, 26, 218.
26. C. V. McDaniel and P. K. Maher, "Molecular Sieves", Society of Chemical Industry, London, 1968, p. 186.
27. C. V. McDaniel and P. K. Maher, in "Zeolite Chemistry and Catalysis", (J. A. Rabo ed.), ACS Monograph 1976, 171, p. 285.
28. J. B. Uytterhoeven, P. A. Jacobs, K. Makay and R. Schoonheydt, J. Phys. Chem. 1968, 72, 1768.
29. J. Turkevich, Y. Murakami, F. Nozaki and S. Ciborowski, Kin. and Catal. Chem. Eng. Progress, Symposium Series, 1967, 73, Vol. 63, 75.
30. D. Barthomeuf and R. Beaumont, J. Catal. 1973, 30, 288.
31. R. Beaumont, D. Barthomeuf and Y. Trambouze, Adv. Chem. Ser. 1971, 102, 813.
32. J. W. Ward and R. C. Hansford, J. Catal. 1969, 13, 364.
33. C. Mirodatos and D. Barthomeuf, J. Catal. 1979, 57, 136.
34. R. C. Hansford and J. W. Ward, Adv. Chem. Ser. 1971, 102, 354.
35. C. Franco Parra, D. Ballivet and D. Barthomeuf, J. Catal. 1975, 40, 52.
36. C. J. Planck, Proc. Intern. Congr. Catal., 3rd Amsterdam, 1964, 1, 727.
37. R. C. Hansford and J. C. Ward, J. Catal. 1969, 13, 316.
38. H. A. Benesi, J. Catal. 1967, 8, 368.
39. D. A. Hickson and S. M. Csicsery, J. Catal. 1968, 10, 27.
40. P. D. Hopkins, J. Catal. 1968, 12, 325.
41. P. A. Jacobs, "Carboniogenic Activity of Zeolites", Elsevier Amsterdam, 1977.
42. P. A. Jacobs, M. Tielen and J. B. Uytterhoeven, unpublished results (see ref. 41).
43. S. J. Gentry and R. Rudham, J. C. S. Faraday I, 1974, 70, 1685.
44. J. W. Ward, J. Catal. 1968, 11, 251.
45. J. W. Ward, J. Catal. 1969, 13 321.
46. R. Beaumont and D. Barthomeuf, C.R. Acad. Sci. Paris 1971, 272, 363.
47. T. R. Hughes and H. M. White, J. Phys. Chem. 1967, 71, 2192.

344

48. J. W. Ward in "Zeolite Chemistry and Catalysis", (J. A. Rabo ed.) ACS Monograph 1976, 171, 118.
49. P. A. Jacobs, H. E. Leeman and J. B. Uytterhoeven, J. Catal. 1974, 33, 17 and 31.
50. A. Bielanski, J. M. Berak, E. Czerwinska, J. Datka and A. Drelinkiewicz, BUll. Acad. Pol. Sci., Ser. Sci. Chim. 1975, 23, 445.
51. H. A. Benesi, J. Catal. 1973, 28, 176.
52. P. A. Jacobs and C. F. Heylen, J. Catal. 1974, 343, 267.
53. J. Datka, J.C.S. Faraday Trans. I, 1981, 77, 511.
54. P. A. Jacobs, J. A. Martens, J. Weitkamp and H. K. Beyer, Faraday Discussion 1981, 72, Paper 21.
55. A. V. Kiselev, D. G. Kitiashvili and V. I. Lygin, Kim. i. Katal. 1971, 12, 1075 and 1973, 14, 262.
56. Y. Kageyama, T. Yotsuyanagi and K. Aomura, J. Catal. 1975, 36, 1.
57. S. P. Zhdanov and E. I. Kotov, Adv. Chem. Ser. 1973, 121, 240.
58. B. D. Flockhart, M. C. Megarry and R. C. Pink, Adv. Chem. Ser. 1973, 121, 509.
59. A. Abou Kaïs, J. Vedrine, J. Massardier and G. Dalmai Imelik, J. Catal. 1974, 34, 317.
    D. Ballivet, J. C. Vedrine and D. Barthomeuf, C.R. Acad. Sci. 1976, 283, C, 429.
60. M. M. Mestdagh, W. E. E. Stone and J. J. Fripiat, J. Phys. Chem. 1972, 76, 1220, J. Catal. 1975, 38, 358 and J.C.S. Faraday Trans. I, 1976, 1, 154.
61. D. Freude, W. Oehme, W. Schmiedel and B. Staudte, J. Catal. 1974, 32, 137.
62. C. F. Heylen and P. A. Jacobs, Adv. Chem. Ser. 1973, 121, 490.
63. Kh.M. Minachev, H. Bremer, R. V. Dmitriev, K. H. Steinberg, Ya. I. Isakov and A. N. Detyuk, Izv. Akad. Nauk. SSSR, Ser. Khim. 1974, 2, 289.
64. K. H. Steinberg, H. Bremer, F. Hofmann, Kh.M. Minachev, R. V. Dmitriev and A. N. Detyuk, Z. Anorg. Allg. Chem. 1974, 404, 129 and 142.
65. K. H. Steinberg, H. Bremer and P. Falke, Z. Chem. 1974, 14, 110.
66. M. D. Navalikhina, B. V. Romanovskii and K. V. Topchieva, Kin. i. Katal. 1971, 12, 1062.
67. H. S. Thuoang, K. V. Topchieva and B. V. Romanovskii, Kim. i. Katal. 1974, 15, 1053.
68. T. R. Brueva, A. L. Klachko-Gurvich, I. V. Mishin, A. M. Rubinshtein, Izv. Akad. Nauk. SSSR, Ser. Khim. 1974, 6, 1254 and 1975, 4, 939.
69. R. A. Hildebrandt and H. Skala, J. Catal. 1968, 12, 61.
70. A. E. Hirschler, J. Catal. 1963, 2, 428 and 1968, 11, 274.

71. R. Beaumont and D. Barthomeuf, J. Catal. 1972, 27, 45.
72. K. V. Topchieva and H. S. Thuoang, Zh. Fiz. Khim. 1973, 47, 2103.
73. L. Moscou and R. Mone, J. Catal. 1973, 30, 417.
74. W. F. Kladnig, J. Phys. Chem. 1979, 83, 765 and 1976, 80, 262.
75. D. Barthomeuf, J. Phys. Chem. 1979, 83, 766.
76. P. A. Jacobs and J. B. Uytterhoeven, J. Catal. 1972, 26, 175.
77. P. A. Jacobs, B. K. G. Theng and J. B. Uytterhoeven, J. Catal. 1972, 26, 191.
78. M. F. Guilleux, J. F. Tempere and D. Delafosse, J. Chem. Phys. 1974, 6, 963.
79. V. G. Gvakariya, V. I. Kulividze and G. V. Tsitsishvili, Dokl. Akad. Naurk. SSSR 1975, 223, 273.
80. J. A. Rabo, V. Schomaker and P. E. Pickert, Proc. Interm. Cong. Catal., 3rd Amsterdam, 1964, 2, 1264.
81. D. M. Breck, C. R. Castor and R. M. Milton, U.S. Patent 3,013,990 (1961).
82. C. Naccache and Y. Ben Taarit, J. Catal. 1971, 22, 171.
83. P. A. Jacobs and H. K. Beyer, J. Phys. Chem. 1979, 83, 1174.
84. K. H. Steinberg, Kh.M. Minachev, H. Bremer, R. V. Dmitriev and A. N. Detyuk, Z. Chem. 1975, 15, 372.
85. K. Tsutsumi, S. Fuji and H. Takahashi, J. Catal. 1972, 24, 8.
86. C. L. Angell and M. V. Howell, J. Phys. Chem. 1970, 74, 2737.
87. C. Mirodatos, P. Pichat and D. Barthomeuf, J. Phys. Chem. 1976, 80, 1335.
88. J. B. Uytterhoeven, L. G. Christner and W. K. Hall, J. Phys. Chem. 1965, 69, 2117.
89. J. W. Ward, J. Catal. 1967, 9, 225.
90. P. A. Jacobs and W. J. Mortier, Zeolites 1982, 2, 266.
91. J. W. Ward, J. Catal. 1968, 11, 259.
92. P. A. Jacobs, L. J. Declerk, L. J. Vandamme and J. B. Uytterhoeven, J.C.S. Faraday Trans. I,, 1975, 71, 1545.
93. P. E. Pickert, J. A. Rabo, E. Dempsey and V. Schomaker, Proceed. 3rd Int. Cong. Catal. Amsterdam, 1964, 714.
94. E. J. Dempsey, J. Catal. 1974, 33, 497 and 1975, 39, 155.
95. R. J. Mikowsky and J. F. Marshall, J. Catal. 1976, 44, 170.
96. W. Wachter, British Zeolite Ass. Meeting, Chislehurst, 1981.
97. G. Engelhardt, D. Zeigan, E. Lipmaa and M. Magi, Z. Anorg. Allg. Chem. 1980, 468, 35.
98. J. Klinowski, J. M. Thomas, C. A. Fyfe and G. C. Gobbi, Nature 1982, 296, 533.

99. G. Engelhardt, U. Lohse, A. Samoson, M. Magi, M. Tarmak and E. Lippmaa, Zeolites 1982, 2, 59.
100. W. J. Mortier, J. Catal. 1978, 55, 138.
101. S. Beran and J. Dubsky, J. Phys. Chem. 1979, 83, 2538.
102. W. J. Mortier and P. Geerlings, J. Phys. Chem. 1980, 84, 1982.
103. R. T. Sanderson, "Chemical Bonds and Bond Energy", Academic Press, New York, 1976.
104. D. Barthomeuf, J.C.S. Chem. Comm. 1977, 743.
105. D. Barthomeuf in "Catalysis by Zeolites", Elsevier, Amsterdam, 1980, 55.
106. N. Y. Topsoe, K. Pedersen and E. G. Derouane, J. Catal. 1981, 70, 41.
107. R. Beaumont, P. Pichat, D. Barthomeuf and Y. Trambouze, "Catalysis", J. W. Hightower, ed., North Holland, Amsterdam, 1973, 1, 343.
108. J. Datka, J.C.S. Farad. Trans. I, 1981, 77, 2877.
109. J. H. Lunsford, J. Phys. Chem.. 1968, 72, 4163.
110. C. Mirodatos and D. Barthomeuf, Chem. Comm. 1981, 39.
111. S. Beran, P. Jiru and B. Wichterlova, J. Phys. Chem. 1981, 85, 1581.
112. J. C. Vedrine, M. Dufaux, C. Naccache and B. Imelik, J.C.S. Farad. Trans. I, 1978, 74, 440.
113. P. H. Kasai and R. J. Bishop, J. Phys. Chem. 1973, 77, 2308.
114. J. A. Rabo and P. H. Kasai, Prog. Solid State Chem. 1975, 9, 1.
115. J. A. Rabo in "Zeolite Chemistry and Catalysis", J. A. Rabo ed., ACS, Washington, DC, 1976, 332.
116. J. A. Rabo, Catal. Rev. Sci. Eng., 1981, 23, 293.
117. J. H. Lunsford, Catal. Rev. Sci. Eng. 1976, 12, 137.
118. Y. Ben Taarit and M. Che in "Catalysis by Zeolites", (B. Imelik et al., ed.) Elsevier, Amsterdam, 1980, 5, 167.
119. R. G. Parr, R. A. Donnelly, M. Levey and W. E. Palke, J. Chem. Phys. 1978, 68, 3801.
120. D. Barthomeuf, C.R. Acad. Sci, Paris, Ser. C, 1978, 286, 181.
121. M. L. Poutsma in "Zeolite Chemistry and Catalysis", (J. A. Rabo ed.) ACS Monograph 1976, 171, 437.

MOLECULAR SHAPE-SELECTIVE CATALYSIS BY ZEOLITES

Eric G. Derouane

Mobil Research and Development Corporation
Central Research Division
P. O. Box 1025
Princeton, New Jersey 08540   U.S.A.

1.   INTRODUCTION

Zeolites are three-dimensional framework aluminosilicates presenting a large intracrystalline free volume which consists of cavities and/or pores. Building-in catalytically active sites within such structures is the essence of molecular shape-selective catalysis. These active sites can pertain to the zeolite framework itself: Brønsted acidic sites are associated to the presence of framework aluminum atoms; they can be converted into Lewis acidic sites by dehydroxylation. Catalytic centers can also be metals or their ions which can be introduced into the zeolite framework by a variety of techniques, ion-exchange being most commonly used.

Weisz and Frilette (1) were first to report 23 years ago on molecular shape-selective catalysis. It is now recognized that molecular shape-selectivity can be achieved by virtue of diffusional effects, steric constraints, and even coulombic field interactions. Since the original article by these authors, more than 300 essential contributions have appeared in the journal and patent literatures. Critical discussions and reviews of molecular shape-selective effects in catalysis have been proposed recently by Csicsery (2), Weisz (3), and Derouane (4,5).

The concept of molecular shape-selective catalysis (3) is based on the action of catalytically active sites, internal to the zeolitic framework, on molecular structure(s) which can exist and/ or diffuse in the structural environment where such sites have been generated. Clearly, it implies an intimate interaction between the shape, size, and configuration of the molecules taking part in the reaction, and the dimension, geometry, and tortuosity of the channels and cages of the zeolite used as catalyst or catalytic support. Molecular shape-selective catalysis must therefore be considered as one of the first, and an outstanding, illustration of "molecular engineering" (6).

Several types of molecular shape-selective effects exist:

(1) REACTANT OR CHARGE SELECTIVITY will take place if the reactants can be divided into two or more classes of molecules, of which one, at least, will not be able to enter or diffuse freely within the intracrystalline volume of the zeolite because of diffusion constraints, selective sorption, or molecular sieving effects. Molecular shape-selective cracking, hydrocracking, and selectoforming are typical processes which take advantage of this property.

(2) PRODUCT SELECTIVITY occurs when similar restrictions apply to the product molecules. It plays an important role in the selective production of para-aromatic compounds over ZSM-5 zeolite based catalyst and also affects the deactivation by coking of zeolite catalysts in general.

(3) RESTRICTED TRANSITION-STATE MOLECULAR SHAPE SELECTIVITY is observed when local configuration constraints, acting in the direct environment of the catalytically active sites, will prevent or decrease the occurrence probability of a given (bimolecular, for example) transition state, characteristic of an elementary catalytic step. It can act directly on the reactants as proposed in the cracking of paraffins (7) or by impeding the formation of bimolecular complexes, as claimed to justify the near-absence of transalkylation in the isomerization of the xylenes (8) and the typical hydrocarbon distribution from the methanol conversion (9) over ZSM-5 based catalysts.

The former molecular shape-selective effects are schematized in Figure 1.

Other manifestations of molecular shape-selectivity are the concentration effect proposed by Rabo (10), the general concept of molecular traffic control (11-13) which still needs definitive support, and the importance of molecular circulation in the internal free volume of zeolites such as erionite and offretite (14,15).

Clearly, the discussion of molecular shape-selective catalysis hence requires (a) a brief description of the relevant zeolite frameworks and pore systems (5,16), (b) the delineation of the essential factors which govern intracrystalline diffusion (3,5,17), and (c) a classification and an illustration of typical molecular shape-selective effects and their uses. Table 1 lists some of the major industrial processes based on molecular shape-selective zeolites (3).

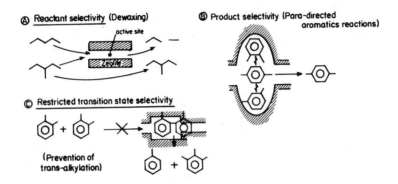

Figure 1. Idealized representation of typical molecular shape-selective effects in zeolite catalysis (from reference 4).

TABLE 1

Industrial Molecular Shape-Selective Processes
(adapted from reference 3)

Process	Objective	Major Chemical/ Process Characteristics
Selectoforming	Octane number increase in gasoline; LPG production	Selective n-paraffin cracking
M-Forming	Octane number increase in gasoline	Cracking depending on degree of branching; aromatics alkylation by cracked fragments
Dewaxing	Light fuel from heavy fuel oil; reduction of lubes pour point	Cracking of high molecular weight n- and mono-methyl paraffins
Xylene Isomerization	High yield para-xylene production	High yield, long cycle life; suppression of side reactions
Ethyl Benzene	High yield ethyl benzene production	
Toluene Disproportionation	Benzene and xylenes from toluene	
Methanol-to-Gasoline	Methanol conversion to high grade gasoline	Synthesis of hydrocarbons restricted to gasoline range, including aromatics

2. PORE SYSTEM CHARACTERISTICS OF INDUSTRIALLY IMPORTANT ZEOLITES

Zeolites of which the industrial importance has been widely recognized are the A, X, and Y zeolites, ZSM-5, erionite, offretite, and mordenite. Figure 2 summarizes the major features of their pore structure (5) and compare their critical dimensions to those of typical hydrocarbon molecules.

Zeolites A, X, Y, erionite and offretite have both channels and cages while ZSM-5 and mordenite only have channels. The intersecting channels of ZSM-5 may allow a three-dimensional motion for molecules of the proper size whilst the differentiated pores of mordenite renders the latter structure essentially unidimensional with respect to the diffusion of hydrocarbon molecules. As evidenced from Figure 2, zeolite ZSM-5 (as well as ZSM-11) has unique channel dimensions (ca. 0.55 nm) and bridges the two classical zeolite categories, i.e., the large pore mordenite and faujasite-structure (X,Y) materials which accept in their free intracrystalline space most simple organic molecules (linear and branched aliphatics and single ring aromatics) and the small pore structures which only adsorb linear aliphatics. Zeolite ZSM-5 can adsorb, by decreasing order of preference, normal paraffins, isoparaffins, other monomethyl-substituted paraffins, and single ring aromatic hydrocarbons containing up to ca. 10 C-atoms (18). Detailed descriptions of the ZSM-5 zeolite adsorptive properties have been given by Dessau (19), Olson (20), Gabelica (21), and Jacobs (22).

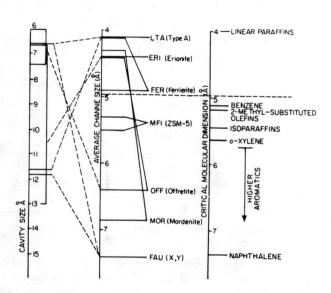

Figure 2. Pore structure of industrially important zeolites (from reference 5).

351

Table 2 compares hydrocarbon sorptions by zeolites ZSM-5 and
ZSM-11 (22). As expected from its structural characteristics,
ZSM-5 has a larger sorption capacity. The comparison of methylnonanes adsorptions over both zeolites also indicates that the
substituted carbon atom and its methyl side-chain prefers to sit at
the channel intersections. Such preferential molecular configurations affect the shape-selective behavior of these zeolites.

TABLE 2

Hydrocarbon Adsorptions over Zeolites HZSM-5 and HZSM-11
(adapted from reference 22)

Sorbate	Sorption Temperature (K)	Maximum number of molecules per unit cell	
		HZSM-5	HZSM-11
C3*	195	13.1	7.4
C4*	273	10.9	5.8
i-C4*	259	11.2	6.5
C5*	293	9.9	5.0
i-C5*	293	9.6	5.3
C6*	293	7.9	4.6
C7**	343	6.9	3.3
C8**	373	5.9	2.8
C9**	373	3.8	2.9
C10**	423	3.1	2.8
2MC9**	423	1.8	1.5
3MC9**	423	2.1	1.6
4MC9**	423	3.1	2.8
5MC9**	423	3.7	2.8
Neopentane*	273	0.05	0.04
p-Xylene**	333	5.6	4.2
m-Xylene**	333	3.8	3.9
o-Xylene**	333	0.8	0.7

*Derived from adsorption isotherms.

**Derived from thermogravimetric data at p/p_o = 0.5.

3. DIFFUSION IN ZEOLITES

The classical theory of diffusion considers two regimes: the normal diffusion regime in which the pore size of the host material is greater than the mean free path of the diffusing molecules and the Knudsen regime for which the diffusivity decreases with the pore dimension. As emphasized by Weisz, a new diffusion regime exists in zeolites, i.e., configurational diffusion (23). It implies that molecular migration within the zeolite framework necessitates the matching of size, shape, and configuration of the diffusing species to the corresponding parameters of the zeolite. Figure 3 illustrates the various diffusion processes encountered in porous solids.

Barrer (24) has reviewed the major features of diffusion in zeolites. Some of those were also discussed by Derouane (5) who insisted on the distinction to be made between classical non-equilibrium measurements and self-diffusion (at equilibrium) experiments (using NMR pulsed field-gradient techniques for example (25)). Self-diffusion becomes equivalent to counterdiffusion when all the diffusing molecules are identical.

Weisz and Prater (26) demonstrated that the observed rate of catalytic reactions in zeolites are moderated by an effectiveness parameter η which is itself a function of a dimensionless variable Φ defined as:

$$\Phi = \frac{dn}{dt} \cdot \frac{1}{C} \cdot \frac{R^2}{D}$$

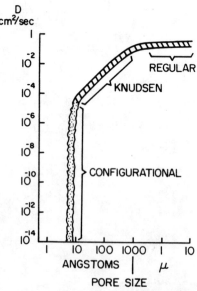

Figure 3. Diffusion mechanisms, diffusivity vs. pore size, in porous materials (from reference 23).

D being the diffusivity coefficient in the particle of equivalent radius R, C the concentration of the reactant(s), and dn/dt the observed reaction rate (27). Figure 4 illustrates the former dependence.

The observed rate constant, k_{obs}, is then related to the intrinsic rate constant, k, by $k_{obs} = \eta \cdot k$. The interaction between diffusion and kinetics is then obvious: reactions characterized by a small η value (formation or transformation of antiselective species) will be selectively retarded with respect to those having a higher effectiveness factor (formation or transformation of proselective species).

4. GEOMETRIC VS. ELECTROSTATIC EFFECTS

Geometric effects (pore size, shape, and tortuosity) are generally claimed as the major factors affecting the molecular shape-selective behavior of zeolites. As the SiO_2/Al_2O_3 ratio decreases, however, the influence of electrostatic fields will become more noticeable as a result of the higher framework charge and the larger concentration in counterions, offering thereby a better discrimination between polar molecules (6,28) and leading to selective sorption.

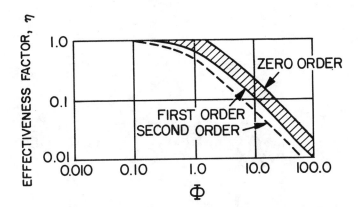

Figure 4. Effectiveness factor, η, as a function of the dimensionless variable Φ (from references 26 and 27).

Electrostatic interactions may take place between the net dipolar moment of the reactant molecules and the electrostatic field at the pore mouths (windows), orienting the diffusing molecule in a favorable or critical position with respect to the channel (window). The role of electrostatic effects has been demonstrated most convincingly for type A molecular sieves, in particular for the isomerization of 1-butene to cis- and trans-2-butenes (28) (see Figure 5).

5. MOLECULAR SHAPE-SELECTIVITY MODIFICATIONS

Before discussing specifically molecular shape-selective effects, it is necessary to make a formal distinction between the molecular shape-selective active sites present in the intracrystal-line volume of the zeolite and those present on the external sur-face of the crystallites. Those will also show activity but no shape-selectivity. Typically, the "external" surface area will represent about one percent of the total zeolite surface area for a crystallite size of one micron (29).

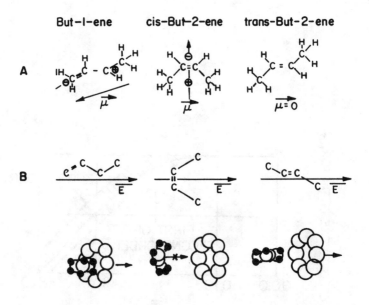

Figure 5. Electrostatic effects in molecular shape-selective catalysis. (A) Dipolar moment orientation in the butene isomers; (B) molecular orientation with respect to the pore openings (from reference 28).

That reactions occur in the intracrystalline volume of zeolites was demonstrated in the very early stages of molecular shape-selective catalysis (6,30). Zeolite Linde type A was found to crack selectively linear paraffins (30). A (Pt,Na)-mordenite hydrogenation catalyst was able to remove selectively ethylene from mixed propylene-ethylene feeds by converting the latter to ethane (6).

Small crystallites, with a larger external surface, will have decreased molecular shape-selectivity as it can be illustrated, for example, by the increased production of unwanted durene in the methanol-to-gasoline conversion (MTG) over HZSM-5 type catalysts (see Table 3).

It is then easily conceived that the molecular shape-selective properties of a zeolite are maximized when it is feasible to deactivate its external surface. Several such improvements have been reported recently for ZSM-5 catalysts. The active ZSM-5 phase can be bound by a preferably alumina-free material (32) or its surface coated by a metacarborane-siloxane polymer (33). The most elegant way to deactivate the ZSM-5 zeolite external surface is, however, to take advantage of the fact that it can be prepared with nearly infinite SiO_2/Al_2O_3 ratio. ZSM-5 crystallites terminated by a virtually aluminum-free outer shell are more selective for the production of para-aromatic compounds: for example, the para/meta-xylene ratio in the products is nearly doubled in the conversion of a mixed C_6-C_9 aliphatic-aromatic feed (315°C, 14 atm., hydrogen/hydrocarbons = 3.6) (34).

As also demonstrated for zeolite ZSM-5, a fine tuning of the molecular shape-selective properties can also be achieved by selective coke deposition which may restrict the pore mouths in addition to deactivating the external surface or by chemical modifications with P (35), Sb (36,37), B (38), Mg (39) containing compounds. Large cations such as Cs^+ and Ba^{2+} were claimed to increase the ethylene production selectivity in the methanol conversion (40) while the addition of Na^+ cations or of a group Va element maximizes the C_5^+-hydrocarbon yield in certain operating conditions (41). Most of these observations are understandable in terms of a modification of the catalyst diffusion characteristics, notwithstanding, however, secondary effects on its acidity.

TABLE 3

Durene Production in the MTG Conversion Over HZSM-5
Catalysts of Varying Crystallite Size (31)
(Temperature = 371°C; SiO_2/Al_2O_3 = 185; WHSV (h^{-1}) = 2)

Crystallite Size (microns)	Durene in HC Product (wt.%)
0.02	5.9
2-5	2.6

356

6. REACTANT MOLECULAR SHAPE-SELECTIVITY

A major application of molecular shape-selective catalysis
is the removal of linear paraffins from liquid reformates to improve
their octane number, or from distillates to lower their viscosity,
pour point, and freezing point.

In Mobil's selectoforming, linear paraffins are hydrocracked
selectively from a mixture of paraffinic and aromatic hydrocarbons
using a low potassium (Ni,H)-erionite zeolite as catalyst (42).
Branched and cycloparaffins and aromatic hydrocarbons are not
affected. As described by Chen and Garwood (43-45), linear chain
hydrocarbons only can enter the erionite framework. Maxima are
observed in the catalytic activity pattern for the cracking of C_6
and C_{10-11} hydrocarbons. They are attributed to a "cage" effect
which is analogous in essence to the "window effect" described by
Gorring (46) to justify the product distribution from the cracking
of n-tricosane over H-erionite and the variation of the n-paraffins
diffusion coefficients in zeolite T.

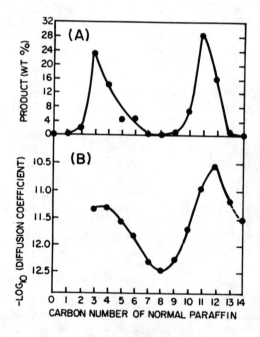

Figure 6. Product distribution from the cracking of n-tricosane
over H-erionite (A) and diffusion coefficients of linear paraffins
in zeolite T (B) (both at 340°C) (from reference 46).

In the latter case, there is obviously a close relationship between the parameters which govern intracrystalline diffusion and the catalytic selectivity (see Figure 6). Zeolite T is mainly offretite with a minor intergrowth of erionite; the erionite structure being less open than the offretite structure, erionite will limit the diffusion behavior of hydrocarbons in the intracrystalline free space of zeolite T. The erionite structure has a "window of high transmittance" for molecules (C_{3-5} and C_{9-13}) which can either orient themselves quickly with respect to the 8-membered ring window of the erionite cage or retain some orientation because they extend through the window limiting the cage. As a maximum is observed in the product distribution (C_{6-9}) when large pore zeolites or silica-alumina are used as catalysts (47), this observation demonstrates the superposition of a shape-selective pattern of diffusivities onto an intrinsically continuous reaction product distribution.

As demonstrated recently by Haag et al. (48), diffusion inhibition effects can be dissociated from the action of steric constraints on the transition state complex by considering zeolite crystallites with different sizes and activities. Table 4 lists pertinent data for the cracking of paraffinic and olefinic hydrocarbons over two HZSM-5 catalysts with different particle size.

TABLE 4

Observed and Intrinsic Rate Constants for the Cracking of Hydrocarbons over HZSM-5 Catalysts (at 538°C)*

CRYSTAL SIZE, R(μm)	k_{obs}		k	η		R^2k
	0.025	1.35		0.025	1.35	1.35
COMPOUND						
Hexane	29	28	29	1	1	5.2×10^{-7}
3-Methylpentane	19	20	19	1	1	3.5×10^{-7}
2,2-Dimethylbutane	12	3.6	12	1	0.30	2.2×10^{-7}
Octane	54	–	–	–	–	–
2-Methylheptane	37	–	–	–	–	–
Nonane	93	93	93	1	1	1.7×10^{-6}
2,2-Dimethylheptane	63	8.4	63	1	0.13	1.1×10^{-6}
Dodecane	663	662	663	1	1	1.2×10^{-5}
1-Hexene	7530	6480	7530	1	0.86	1.4×10^{-4}
3-Methyl-2-Pentene	7420	3610	7420	1	0.50	1.4×10^{-4}
3,3-Dimethyl-1-Butene	4350	141	4950	0.86	0.028	9.0×10^{-5}

*See text for definition of symbols; from reference 48.

They indicate that mass-transport limitations occur in the cracking of hexenes and of gem-dimethyl-paraffin isomers. Branching of the aliphatic chain is the essential factor which affects the relative effective diffusivities of the reactants at steady-state reaction conditions.

The effects of chain length and branching on the relative cracking rates of C_{5-7} paraffins have been described in detail by Chen and Garwood (49). As seen from Table 5, the following trends hold in the cracking of these paraffins:

(a) $n-C_7 > n-C_6 > n-C_5$

(b) straight chain > 2-methyl > 3-methyl > dimethyl-
or ethyl substituted.

In contrast to the "window" or "cage" effect which is observed in erionite (46,47), pore size has more importance for ZSM-5 cracking catalysts than the actual channel tortuosity.

These unique molecular shape-selective properties of ZSM-5 catalysts constitute the essence of the Mobil distillate dewaxing (MDDW) process (50,51) in which a mixed feed of linear paraffins, isoparaffins, highly branched paraffins, and aromatics (gas-oil distillate) is selectively hydrocracked. Linear and isoparaffins react preferentially, as illustrated in Figure 7, and the freeze and pour points of the distillate are lowered, thereby enabling one to adapt the properties of the product to climatic or utilization conditions requirements.

TABLE 5

Reactant Molecular Shape-Selective Effects in the
Cracking of Paraffins over HZSM-5 (from reference 49)
(Temperature = 340°C, Pressure = 35 atm., LHSV (h^{-1}) = 1.4)

Paraffin	Relative Cracking Rate
n-Heptane	1
2-Methylhexane	0.52
3-Methylhexane	0.38
2,3-Dimethylpentane	0.09
n-Hexane	0.71
2-Methylpentane	0.38
3-Methylpentane	0.22
2,3-Dimethylbutane	0.09

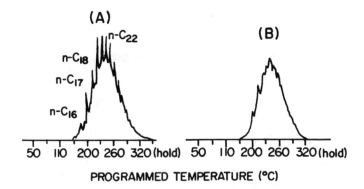

Figure 7. Dewaxing of a midcontinent distillate (340-390°C) on a ZSM-5 based catalyst: (A) before hydrocracking; (B) after processing, the n-C_{16-28} paraffins being selectively removed (from reference 50).

Reactant molecular shape-selectivity can also play a role in metal-catalyzed hydrogenation and oxidation (52). It has been observed, for example, that a Pt-ZSM-5 catalyst hydrogenates preferentially linear olefins while a Cu-ZSM-5 catalyst oxidizes mostly para-xylene when it is admixed with its ortho-isomer, the latter of course because of the higher diffusivity of para-xylene in ZSM-5.

7. PRODUCT MOLECULAR SHAPE-SELECTIVITY

The striking analogy of the hydrocarbon product distributions stemming from the methanol and several triglycerides conversions on ZSM-5 illustrates product molecular shape-selectivity (see Figure 8) (3,53).

The methanol-to-gasoline (MTG) conversion is a large scale application of the concept of product molecular shape-selectivity. It is discussed at length elsewhere in this volume (54). The effect of pore size on the selectivity of the methanol or dimethylether conversion to olefins has been identified by Cormerais et al. (55) who compared the activities and selectivities of H-erionite, HZSM-5, and zeolite H-Y. The smaller the pore opening, the higher is the yield in C_{2-3} olefins.

The best demonstration of product molecular shape-selectivity is found in a variety of reactions which use ZSM-5 based catalysts and aim at the selective preparation of para-aromatic compounds. Such processes are the disproportionation and alkylation (by methanol) of toluene (56-60) and the xylenes isomerization (56).

360

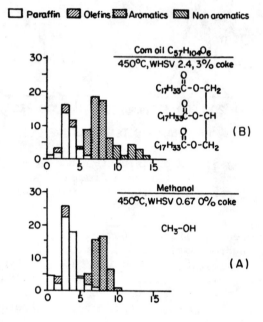

Figure 8. Hydrocarbon product distributions from the conversions of methanol (A) and corn oil triglyceride, $C_{57}H_{104}O_6$ (B) on HZSM-5 (from reference 3).

Yields in para-xylene, exceeding the expected equilibrium values, can be observed because of diffusion/reaction interaction. Factors which increase the diffusion path length (larger crystals) or decrease the effective pore size of the ZSM-5 catalyst (bulky atoms such as P, bulky counterions, presence of inorganic fillers) and lower the activity of its non-selective external surface, favor the formation of the para-isomer (56-60). Considering the dispro-portionation of toluene, Haag and Olson (61) have correlated the para-xylene selectivity of ZSM-5 catalysts of different crystal sizes and pore tortuosities (because of the presence of inorganic salts or coke plugging) with a diffusion residence time for ortho-xylene obtained from separate sorption measurements (see Figure 9).

Para-xylene selectivity is noticeably enhanced when diffusion/ reaction interactions increase. Young et al. (60) have recently analyzed in detail the reaction paths which lead to the formation of para-xylene in the toluene disproportionation or its alkylation by methanol and the xylenes isomerization. Figure 10 shows the

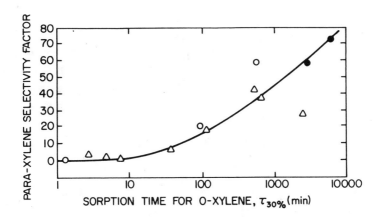

Figure 9. Selectivity to para-xylene in the toluene disproportionation reaction vs. ortho-xylene sorption time for various ZSM-5 based
catalysts. O = different crystal sizes; Δ = tortuosity increased
by inorganic salts; ● = coked catalysts (from references 3 and 61).

reaction paths for the isomerization of the pure xylene isomers over
non-modified and modified ZSM-5 catalysts as well as over non-
zeolitic catalysts. Paths A, B, and C correspond to non-shape
selective catalysts such as silica-alumina or phosphoric acid on
Kieselguhr while paths A', B', and C' are those followed with shape-
selective catalysts. These are Mg and P-modified ZSM-5; as readily
seen they yield a para-xylene concentration in excess of the expected
thermodynamic equilibrium value.

The same authors have compared the relative activities for
toluene alkylation and xylene isomerization of the same catalysts.
Modified shape-selective catalysts are characterized by a toluene-
methanol alkylation rate which is about 2.5 to 15 times that of
xylene isomerization. In contrast, non-modified HZSM-5 is ten times
more active for xylene isomerization than for toluene alkylation.
These observations are explained by an increased diffusion resistance
in the shape-selective catalysts and by the relative diffusivities
of toluene and methanol (high), para-xylene (medium), and ortho-
and meta-xylenes (low).

The selectivity of these reactions to yield para-xylene is
favored additionally by restricted transition-state molecular shape
selective constraints as discussed in the next section. These
constraints apparently prevent the formation of the bimolecular
transition state which is necessary to transalkylate toluene within
the zeolite and explain the high value of the rate ratio xylene
isomerization/toluene disportionation, i.e., 100-1000 over non-
modified HZSM-5 (60,62).

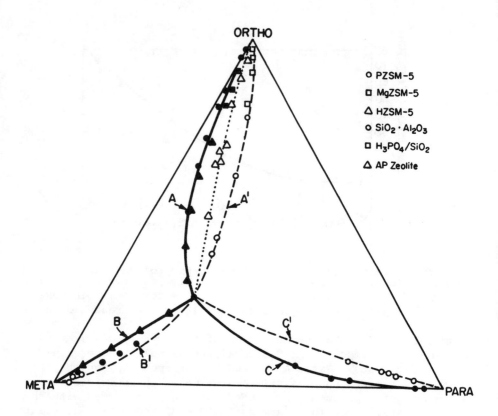

Figure 10.   Reaction paths for the isomerization of the pure xylenes over shape-selective (A',B',C') and non-shape-selective (A,B,C) catalysts (from reference 60).

8.   RESTRICTED TRANSITION STATE MOLECULAR SHAPE SELECTIVITY

      Restricted transition state molecular shape-selectivity is observed when steric constraints in the environment of the catalytic site affect or prevent the formation of intermediate complex structures.   The inability to reach a given transition state will affect both monomolecular and bimolecular reactions; it can, of course, also discriminate between mono- and bimolecular complexes which occur along the various possible reaction paths for a given reaction (3,4).   These constraints will act on intrinsic kinetics rather than by diffusion/reaction interaction.

      As mentioned in Section 7, this type of molecular shape-selectivity explains the low xylene disproportionation activity of HZSM-5.   Haag and Dwyer (62) and Gnep et al. (63) have correlated the activity of various zeolites (ZSM-5, ZSM-4, mordenite, and

Type Y) for the former reaction with their effective pore size.
As expected, the transalkylation is dramatically inhibited as the
pores become more restrained.

Restricted transition state molecular shape-selectivity is
essential to account for the high ethylbenzene selectivity which
characterizes the formation of ethylbenzene by the Mobil-Badger
process (64,65). In contrast to other zeolites such as mordenite
or faujasite which rapidly deactivate as a consequence of coking,
ZSM-5 based catalysts have a stable activity for cycle lengths of
several weeks and yield ethylbenzene almost stoichiometrically.
Further alkylation of ethylbenzene is prevented by the complementary
actions of restricted transition state and diffusion constraints.
The same argumentation and advantages hold to describe and justify
the high para-methylstyrene yields obtained in the direct alkylation
of toluene by ethylene over ZSM-5 class catalysts (66).

This particular type of selectivity, as discussed elsewhere
(4,54,67,68), explains probably partially the selectivity of the
formation of aromatic compounds (cut-off at C_{10}) in the methanol
conversion using HZSM-5 catalysts. The bimolecular cyclo-addition
of an olefin and a carbenium ion is only possible at the channel
intersections of ZSM-5, the smaller size products (following further
dehydrogenation, alkylation, and isomerization) being able to
diffuse out through its channels.

Constraint index measurements which consist in the evaluation
of the ratio of the cracking rates of n-hexane and 3-methylpentane,
are recognized as means to characterize zeolites (69). It has been
demonstrated recently (48,70) that shape-selective constraints on
the local kinetics govern these reactions. Indeed, the relative
cracking rates of these two hydrocarbons are independent of crystal
size (as shown for HZSM-5 catalysts) and then, apparently, free of
diffusional effects (70). The cracking mechanism implies hydrogen
transfer between the reacting molecule and a carbenium ion as
schematized in Figure 11. It is obvious that the larger transition
state required for 3-methylpentane will lead to more severe steric
inhibitions and lower conversions, in particular over ZSM-5 catalysts
with critical pore diameter of ca. 0.6 nm. Constraint index measurements can then be considered as indirect evaluations of the free
space available in the direct environment of the catalytic sites,
which give support to their use as zeolite characterization means.

9.   MOLECULAR SHAPE-SELECTIVE EFFECTS IN THE COKING AND
     AGING OF ZEOLITES

Coke deposition in zeolites originates mainly from olefinic
(71) and aromatic (72) compounds condensation and dehydrogenation
reactions. The formation of coke can be viewed as follows (5,73):

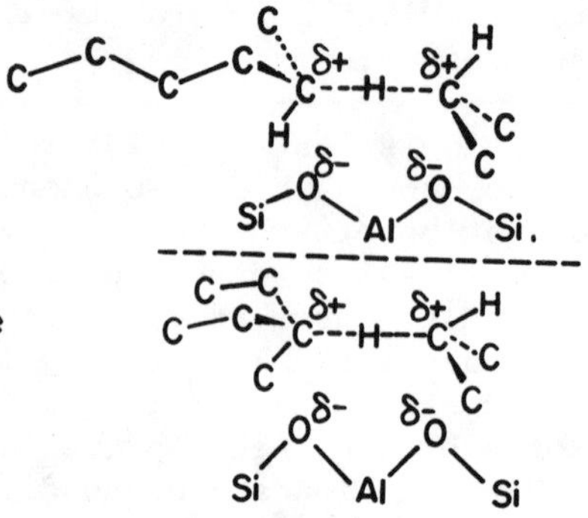

n-hexane

cross-section

4.9 x 6Å

3-methylpentane

6 x 7Å

Figure 11. Transition states in the cracking of the hexane isomers (from reference 48).

olefins (possibly resulting from paraffins dehydrogenation) are first cyclo-oligomerized to naphthenes which, in turn, can be converted to aromatics by successive dehydrogenation and hydrogen transfer steps; these aromatic compounds can be further alkylated and dehydrogenated to yield fused-ring aromatic compounds. Ultimately, those are progressively dehydrogenated into coke.

Rollmann and Walsh (74-77) have investigated in detail the deposition of coke over a variety of zeolite catalysts and proposed an impressive correlation between coking activity and molecular shape-selectivity (77) (measured by the ratio of the cracking rates for n-hexane and 3-methylpentane). The latter plot, which is shown in Figure 12, was obtained by reacting a mixed feed of C_6 hydrocarbons over various zeolites (425°C, 15 atm, $H_2/HC = 3$). Intracrystalline coking clearly depends on the pore structure of the zeolite. The alkylation of aromatic compounds which can be converted to fused-ring products at a later stage, was found to be the initial and decisive step in the coking of mordenite and zeolite H-Y. For the small pore structures such as ferrierite and erionite, the low coking activity seems to be related to constraints acting on the formation of cyclic coke precursors (naphthenes and cycloparaffins) from aliphatic reactants. The unusually low coking activity of ZSM-5 is attributed to restricted transition-state shape-selectivity which prohibits secondary reactions of alkylaromatics in its intermediate pore-size channels (74).

As discussed by Derouane (5) and Dejaifve et al. (78), zeolite aging because of the deposition of carbonaceous residues is a function of two factors, namely the probability P(t) that an active site is accessible at time t and the corresponding conditional probability S(t) that it is not poisoned at the same time (79,80).

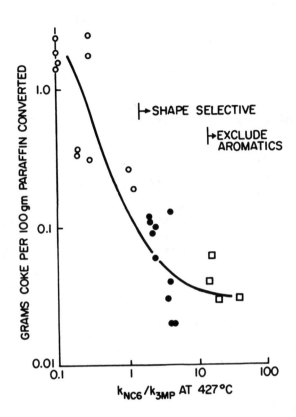

Figure 12. Coke yield vs. molecular shape-selectivity (relative cracking rate of n-hexane vs. 3-methylpentane) in the conversion of hydrocarbons over zeolite catalysts (from reference 77).

P(t) is a function of the channel network geometry while S(t) is related to the characteristics (activity and molecular shape-selectivity) of the active site and its environment. A comparison of the aging and rejuvenation behaviors of HZSM-5, offretite, and mordenite indicated that they were intimately depending on the actual pore system structure. It also confirmed the proposal (5) that aging was less rapid for zeolites possessing interconnected channels. The initial coking activity was found to be directly related to the availability of the acid catalytic sites (78).

When cavities are present, products can be formed which have a size too large to be desorbed through the windows leading to other cages or to the channels (a reversed product molecular shape-selective effect). Such bulky molecules will act as coke precursors and deactivate rapidly the zeolite. This situation is illustrated by the "Faujasite Trap", described by Venuto et al., which occurs in the isomerization-oligomerization of 1-hexene over rare-earth exchanged zeolite-X (81).

10. MISCELLANEOUS MOLECULAR SHAPE-SELECTIVITY EFFECTS

Acidic Y-zeolite based catalysts show an increased H-abstraction rate compared to that of β-scission when compared to silica-alumina gels. While the latter reaction is a monomolecular process, the former one is a bimolecular event which involves a hydride shift from a neutral hydrocarbon to a carbenium ion. Rabo et al. (82) rationalized this observation by proposing that "zeolites concentrate hydrocarbon reactants to a large extent...this *concentration effect* enhancing the rate of bimolecular reaction steps...over unimolecular (fragmentation) steps". Consequently, secondary cracking reactions become less important as olefins and naphthenes are more readily converted into more refractory (paraffinic) products, and larger net gasoline yields are observed.

Jacobs et al. (83) have used the *bifunctional conversion* of n-decane over Pt-loaded zeolites to characterize the catalyst molecular shape-selective properties. They demonstrated that the distribution of the feed isomers, in particular the yield of 2-methyl-nonane, could be used to discriminate between the ZSM-5 and the ZSM-11 structures. The conversion of n-decane over Pt-ZSM-5 shows a high (ca. 60%) and unusual production of the above-cited decane isomer, far above equilibrium and at the expense of the other methylnonanes. Protonated cyclopropane structures were postulated as reaction intermediates and possibly explain the observed differences between ZSM-5 and ZSM-11 based catalysts. A preliminary evaluation of these results indicate that this type of test could be used to evidence intergrowths in pentasil zeolites and to detect inhomogeneities in the distribution of aluminum through the zeolite crystallites.

Recently, the concept of *molecular traffic control* was proposed (11) on the basis of sorption measurements on zeolite ZSM-5 and generalized (12) to include transformations in which (some of) the reactants reach the active sites through diffusion pathways less readily accessible to (some of the) products, or vice versa. Principally, molecular traffic control can occur in zeolite structures presenting non-equivalent but intersecting channels and should be effective when diffusion-limited kinetics take place. Although derived from near-equilibrium sorption measurements, the concept of molecular traffic control obviously applies only to dynamic systems (13). The alkylation of para-xylene by methanol over HZSM-5 (linear and zig-zag channels) and HZSM-11 (straight channels only) has been used to test this concept (84). The higher aromatic-ring alkylation activity of HZSM-11, compared to HZSM-5, can be explained by the absence of molecular traffic control although differences in these zeolite acid strengths (83) could also play a non-negligible role. The molecular traffic control concept clearly deserves more attention. Support for its existence should eventually be gained in dynamic reaction conditions, at or near stationary state.

Mirodatos and Barthomeuf (85,86) have proposed from comparative studies of the cracking of n-heptane and n-octane on offretite and mordenite that *molecular circulation* was an important factor affecting the catalytic activity of zeolites. If molecular circulation is impeded, for example, at high cationic content levels or by the presence of coke, molecules that enter cages or channels can undergo more severe transformations. The molecular circulation effect seems apparented to the "window effect" put forward by Gorring (46).

11. CONCLUSIONS

The matching of size and configuration of diffusing (reacting) species to those of the zeolite channels affect the kinetics of their catalytic conversion. Conversely, the diffusion of product molecules is influenced by the same factors.

A more subtle molecular shape-selectivity effect is that of restricted transition-state shape-selectivity which implies local constraints on the active site kinetics, possibly leading to discrimination between unimolecular and bimolecular transition states.

The above factors give zeolites unique catalytic properties: their high activity which results from their ability to operate at high temperature with minimal deactivation is often complemented by an unusual selectivity because of their particular structural peculiarities.

ACKNOWLEDGMENTS

The author thanks the following publishers for having released their copyrights on the following figures and tables:

Elsevier Publishing Company (Figure 1); Academic Press (Figures 2, 6, 10, and 12); American Chemical Society (Figure 3); American Association for the Advancement of Science (Figure 4); Petroleum Publishing Company (Figure 7); Pergamon Press (Figures 8 and 9, and Table 1); The Royal Society of Chemistry (Figure 11 and Table 4); Butterworth & Co., Publishers (Table 2).

REFERENCES

1. P.B. Weisz and V.J. Frilette, J. Phys. Chem., 64 (1960) 382
2. S.M. Csicsery, in ACS Monograph Nr. 171, "Zeolite Chemistry and Catalysis", J.A. Rabo, ed., (American Chemical Society, Washington, 1976); p. 680.
3. P.B.Weisz, Pure Appl. Chem., 52 (1980) 2091.
4. E.G. Derouane, in "Catalysis by Zeolites", Stud. Surf. Sci. Catal., Vol. 4, B Imelik et al., eds., (Elsevier Sci. Pub. Co., Amsterdam, 1980); p.5.
5. E.G.Derouane, in "Intercalation Chemistry", M.S. Whittingham and A.J. Jacobson, eds., (Academic Press, New York, 1982); p. 101.
6. N.Y. Chen and P.B. Weisz, Chem. Eng. Progr. Symp. Ser., 63 (1967) 86.
7. V.J. Frilette, W.O. Haag, and R.M. Lago, J. Catal., 67 (1981) 218.
8. S.L. Meisel, J.P. McCullough, C.H. Lechthaler, and P.B. Weisz, Leo Friend Symposium, American Chemical Society, Chicago, August 30, 1977
9. E.G. Derouane and J.C. Vedrine, J. Molec. Catal., 8 (1980) 479.
10. J.A. Rabo, R.D. Bezman, and M.L. Poutsma, Acta Phys. Chem., 24 (1978) 39.
11. E.G. Derouane and Z. Gabelica, J. Catal., 65 (1980) 486.
12. E.G. Derouane, Z. Gabelica, and P.A. Jacobs, J. Catal., 70 (1981) 238; reply to B.M. Lowe, D.A. Whan, and M.S. Spencer, J. Catal., 70 (1981) 236.
13. E.G. Derouane, J. Catal., 72 (1981) 177; reply to C.G. Pope, J. Catal., 72 (1981) 174.
14. C. Mirodatos and D. Barthomeuf, J. Catal., 57 (1979) 136.
15. C. Mirodatos and D. Barthomeuf, Abstracts 7th North American Mtg., Catalysis Society, Boston, 1981, paper E-5
16. See lecture by R.M. Barrer on "Zeolite Structure" in this volume.
17. See lecture by R.M.Barrer on "Diffusion and Sorption Mechanisms in Zeolites" in this volume.
18. N.Y. Chen and W.E. Garwood, J. Catal., 52 (1978) 453.
19. R.M. Dessau, in "Adsorption and Ion-Exchange with Synthetic Zeolites", W.H. Flank, ed., A.C.S. Symposium Series, Vol. 135 (American Chemical Society, Washington, D.C., 1980); p. 123.
20. D.H. Olson, G.T. Kokotailo, S.L. Lawton, and W.M. Meier, J. Phys. Chem., 85 (1981) 2238.
21. Z. Gabelica, J.P. Gilson, and E.G. Derouane, in "Proc. 2nd European Symp. Thermal Anal.", Aberdeen, 1981, D. Dollimore, ed. (Heyden and Sons, Amsterdam, 1981); p. 434.
22. P.A. Jacobs, J. Valyon, and H.K.Beyer, Zeolites, 1 (1981) 161.

23. P.B. Weisz, Chem. Technol., 3 (1973) 498.
24. R.M. Barrer, Adv. Chem. Ser., 102 (1971) 1.
25. For example, and reference cited therein; J. Karger, M.B. Low, and P. Lorenz, J. Colloid Interf. Sci., 65 (1978) 181.
26. P.B. Weisz and C.D. Prater, Adv. Catal. Relat. Subj., 6 (1954) 143.
27. P.B. Weisz, Science, 179 (1973) 433.
28. J.F. Tempere and B. Imelik, Buul. Soc. Cmi. Fr., (1970) 4227.
29. P.B. Venuto and P.S. Landis, Adv. Catal. Relat. Subj., 18 (1968) 259.
30. P.B. Weisz, V.J. Frilette, R. W. Maatman, and E.B. Mower, J. Catal., 1 (1962) 307.
31. B.P. Pelrine, U.S. Patent 4,100,262 (1978).
32. C.D. Chang and W.H. Lang, U.S. Patent 4,013,732 (1977).
33. C.C. Chu, U.S. Patent 3,965,210 (1976).
34. L.D. Rollmann, U.S. Patent 4,148,713 (1979).
35. S.A. Butter and W.W. Kaeding, U.S. Patent 3,965,208 (1976)
36. S.A. Butter, U.S. Patent 4,007,231 (1977).
37. S.A. Butter, U.S. Patent 3,979,472 (1976) and Brit. Pat. 1,528,674 (1978).
38. W.W. Kaeding, U.S. Patent 4,029,716 (1977).
39. W.W. Kaeding and L.B. Young, U.S. Patent 4,034,053 (1977).
40. P.G. Rodewald, U.S. Patent 4,066,714 (1978).
41. C.D. Chang and W.H. Lang, U.S. Patent 3,899,544 (1975).
42. N.Y. Chen, J. Maziuk, A.B. Schwartz, and P.B. Weisz, Oil and Gas J., 66 (1968) 154.
43. N.Y. Chen and W.E. Garwood, Ind. Eng. Chem. Process Res. Devel., 17 (1978) 513.
44. N.Y. Chen and W.E. Garwood, Advanc. Chem. Ser., 121 (1973) 575.
45. N.Y. Chen and W.E. Garwood, J. Catal., 53 (1978) 284.
46. R.L. Gorring, J. Catal., 31 (1973) 13.
47. N.Y. Chen, S.J. Lucki, and E.B. Mower, J. Catal., 13 (1969) 329.
48. W.O. Haag, R.M. Lago, and P.B. Weisz, Faraday Disc. Chem. Soc., 72 (1981) 317.
49. N.Y. Chen and W.E. Garwood, J. Catal., 52 (1978) 453.
50. N.Y. Chen, R.L. Gorring, H.R. Ireland, and T.R. Stein, Oil Gas J., 75 (23) (1977) 165.
51. S.P. Donnelly and J.R. Green, Oil Gas J., 77 (1980).
52. R.M. Dessau, J. Catal., 77 (1982) 304.
53. P.B. Weisz, W.O. Haag, and P.G. Rodewald, Science, 206 (1979) 57.
54. See lecture by E.G. Derouane on "Conversion of Methanol to Gasoline Over Zeolite Catalysts" in this volume.
55. F.X. Cormerais, G. Perot, and M. Guisnet, Zeolites, 1 (1981) 141.

56. N.Y. Chen, W.W. Kaeding, and F.G. Dwyer, J. Amer. Chem. Soc., 101 (1979) 6783.
57. S.A. Butter, U.S. Patent 4,007,231 (1977).
58. W.W.Kaeding, C. Chu, L.B. Young, B. Weinstein, and S.A. Butter, J. Catal., 67 (1981) 159.
59. W.W. Kaeding, C. Chu, L.B. Young, and S.A. Butter, J.Catal., 69 (1981) 392.
60. L.B.Young, S.A. Butter, and W.W. Kaeding, J. Catal., 76 (1982) 418.
61. W.O. Haag and D.H. Olson, U.S. Patent 4,117,026 (1978).
62. W.O. Haag and F.G. Dwyer, Aromatics Processing with Intermediate Pore Size Zeolite Catalysts, Am. Inst. Chem. Eng. An. Mtng., Boston, August 1979.
63. N.S. Gnep, J. Tejada, and M. Guisnet, Bull. Soc. Chim. France, I-5(1982).
64. L.B. Young, U.S. Patent 3,962,364 (1976).
65. F.G. Dwyer, J.P. Lewis, and F.H. Schneider, Chem. Eng., January 5, 1976.
66. W.W. Kaeding, L.B. Young, and A.G. Prapas, Chem. Technol., 12 (1982) 556.
67. E.G. Derouane and J.C. Vedrine, J. Molec. Catal., 8 (1980) 479.
68. P. Dejaifve, J.C. Vedrine, V. Bolis, and E.G. Derouane, J. Catal., 63 (1980) 331.
69. For example, C.D. Chang, A.J. Silvestri, and R.L. Smith, U.S.Patent 3,928,482 (1975).
70. V.J. Frilette, W.O. Haag, and R.M. Lago, J. Catal., 67 (1981) 218.
71. P.B. Venuto, in "Catalysis in Organic Synthesis", G.V. Smith, ed., (Academic Press, New York, N.Y.,1977) p. 67.
72. W.G. Appleby, J.W. Gorbson, and G.M. Good, Ind. Eng. Chem. Process Des. Devel., 1(1962) 102.
73. P.B. Venuto and E.T. Habib, J. Catal. Rev.-Sci. Eng., 18 (1978) 1.
74. D.E. Walsh and L.D. Rollmann, J. Catal., 56 (1979) 195.
75. L.D. Rollman, J. Catal., 47 (1977) 113.
76. D.E.Walsh and L.D. Rollmann, J. Catal., 49 (1977) 369.
77. L.D. Rolmann and D.E. Walsh, J. Catal., 56 (1979) 139.
78. P. Dejaifve, J.C. Vedrine, A.Auroux, P.C. Gravelle, Z. Gabelica, and E.G. Derouane, J. Catal., 70 (1981) 123.
79. J.W. Beeckman and G.F. Froment, Ind. Eng. Chem. Fundam., 18 (1979) 245.
80. J.W. Beeckman and G.F. Froment, Chem. Eng. Sci., 35 (1980) 805.
81. P.B. Venuto, L.A. Hamilton, and P.S. Landis, J. Catal., 5 (1966) 484.
82. J.A. Rabo, R.D. Bezman, and M.L. Poutsma, Acta Phys. Chem., 24 (1978) 39.

83. P.A. Jacobs, J.A. Martens, J.Weitkamp, and H.K. Beyer, Faraday Disc. Chem. Soc., 72 (1981) 353.
84. E.G. Derouane, P. Dejaifve, Z. Gabelica, and J.C. Vedrine, Faraday Disc. Chem. Soc., 72 (1981) 331.
85. C. Mirodatos and D. Barthomeuf, J. Catal., 57 (1979) 136.
86. C. Mirodatos and D. Barthomeuf, Abstracts 7th North American Catalysis Society Mtg., Boston, 1981; paper E -5.

TRANSITION METAL EXCHANGED ZEOLITES : PHYSICAL AND CATALYTIC
PROPERTIES

Claude Naccache and Younès Ben Taarit

Institut de Recherches sur la Catalyse, CNRS
2, avenue A. Einstein, 69626 Villeurbanne, Cédex, France

INTRODUCTION

Catalysis by transition metal compounds in solution or supported
on solids is an active field of research of considerable interest.
It is worthwile to recall that transition metal ions were found active and selective for a great number of reactions among them, oxidation of ethylene to acetaldehyde, hydroformylation of olefins to
aldehydes, carbonylation and homologation of alcohols, hydrogenation
and isomerisation of olefins, oligomerisation and cyclodimerisation
of olefins, water gas shift reaction ... etc. In recent years a
new area of research developed which consisted to anchor or immo-
bilize to a solid support a soluble transition metal complex to produce a potential active and selective new type of heterogeneous
catalyst. In addition it is thought that relationship between homogeneous and heterogeneous catalysis may be found through the studies of such heterogeneized catalysts. Several distinct ways for
"heterogeneizing homogeneous catalysts" have been proposed. One
approach was to anchor the soluble metal complex to an oxide surface either through surface oxygen bond resulting from the reaction
of the metal ligands with hydroxyl groups, or by ligand exchange
with functionalized oxide surface. An alternative means for converting homogeneous metal complexes into heterogeneous catalysts is to
introduce the active complex into the intercrystal space of a layer
lattice silicate by exchanging the Na^+ cations of the layer silicate
by cationic transition metal ions. Zeolites contain also exchangeable cations, thus they can be used for anchoring soluble transition
metal complexes. The zeolite structure thus behaves as a "solid
solvent" and leads to a new class of catalysts when the material is
exchanged with transition metal ions. Extensive studies on transition metal exchanged zeolites have been performed during the last

two decades. Several excellent reviews have been published in recent years (1-8). In this review we have not attempted to include every materials of interest concerning transition metal ions in zeolites, but rather we have put our efforts to provide a comprehensive survey on the chemistry of transition metal ions in zeolites and to show how such chemistry has opened a very promising new area in heterogeneous catalysis. It is our hope that the few examples assembled in these lectures will stimulate the interest of catalyst scientists in these materials. It is worthwhile to note that most of the materials that will be discussed are of the faujasite-like structure, which among other type zeolites offer the advantages of having a three dimensional pore arrangement, and relatively large cavities.

1 - Procedures for the preparation of zeolite-supported transition
    metal ions.

    Several methods have been used to prepare zeolites containing transition metal ions. The most common method is the well known conventional ion exchange technique. Other methods such as reaction with chloride salts, adsorption from the vapor phase or in solution of metal carbonyl compounds, reaction of cations present in the zeolite cavities with anionic metal complexes, impregnation with a solution of the metal salts. The dispersion of the cations and their localization within the zeolite framework or/and the external surface depend strongly on the method employed. This paragraph will refer to these various methods.

    1-1- Preparation by ion exchange technique. The procedure which is certainly the most suitable to introduce cations into the zeolite framework consists of exchanging the Na^+ cations which equilibrate the negative charge beared by AlO_4 tetrahedra, with a solution of the metal salt, through conventional ion exchange technique. This procedure has been successfully applied in most cases, especially with alkali, alkali-earth and rare-earth cations. In recent years several studies on the kinetics and the equilibrium of the exchange reaction have been published. The general conclusions which may be given are that the exchange reaction equilibrium depends on the concentration of the exchanging metal salt solution, on the temperature. The general features of the equilibrium and kinetic aspects have been reviewed recently (9). It was shown that many exchange reactions in X and Y zeolites fail to proceed to completion and it was suggested that the lack of total sodium exchange is due to the difficulty of displacing the residual Na^+ cations present in the sodalite cages, in NaY zeolite about 16 Na^+ are localized in the sodalite cages. Since the exchangeable cations are generally hydrated, the free diameter of such solvated cations is generally larger than the diameter of the 6-membered ring window of the sodalite cage, about 0.22 nm, thus the Na^+ in the sodalite cages are not accessible by the exchangeable hydrated cations. It results that a complete exchange would occur if there is a redistribution of the cations between the sodalite and the supercages. This will necessitate a partial dehydration of the solvated exchangeable cation. The results on lanthanum ion exchange are significative of this respect. Hydrated La^{3+} cations were found to replace Na^+ present in the supercage on NaY zeolite only. However at 180°C the exchange proceeds to completion, Na^+ in the sodalite cages being removed by La^{3+}. It was concluded that at 180°C $La-H_2O$ bond was weakened thus allowing the splitting of H_2O ligands from the $La^{3+}(H_2O)_9$ complex. This results in a facile diffusion of the dehydrated La^{3+} cations in the sodalite cages at 180°C where exchange with Na^+ takes place (10-11). In conclusion of these data it is now well established that hydrated cations, hydrated La^{3+} has a diameter of 3.96 Å, cannot penetrate the sodalite cages of X and Y zeolites. Only at high temperature the hydrated cations lose their coordinated H_2O molecules which allow them to penetrate the sodalite cages. However during this process hydroxylation of the cation through water ionization may occur resulting in the formation of $Me^{n+} - OH_x$ entities. At high temperature OH condensation between $Me^{n+} - OH_x$ species would result in the formation of Me-O-Me bonds. According to this OH condensation oxide inside the zeolite framework may be formed thus leading to cation collapse during the subsequent calcination of the zeolite. It is clear that this undesired cation clustering requires the existence of hydroxometal cations, which are produced by hydrolysis of the exchangeable cations, in sufficient number.

To summarize the above discussion one may conclude that although it is relatively easy to change the nature of the cations in zeolites by ion exchange, the diffusion of the bulky hydrated cations will be limited by the zeolite pore size. Reduction in solvated cation size would be necessary in order for the ions to pass freely through the oxygen membered rings. It is also known that cation exchange into hydrogen form zeolite, is often a difficult process, due to the strenght of the bonding of the protons with the lattice oxygen. To overcome the proton exchange limitation it is recommended to transform the hydrogen form into ammonium form, NH_4^+ cations will behave similarly to Na^+ cations.

The study of transition metal ions exchanged zeolite has revealed that the state of the transition metal cations is strongly dependent on the method and on the experimental conditions employed for the ion exchange (11). It was shown that sodium form zeolites in solution give a basic reaction with the subsequent increase of the pH of the exchanging solution. Since transition metal ions in basic solution are easily hydrolyzed, the ion exchange can be accompanied with hydrolysis of the transition metal ions and the subsequent precipitation of the hydroxy anion transition metal. Thus the transition metal ions would be adsorbed in the form of metal hydroxide either inside the zeolite cavities or on the external surface. Precipitation of hydroxide species of transition metal ions in, in/on the zeolite would be avoided is the pH of the solution is maintained low enough such that no cation hydrolysis occurs. However at low pH, several type zeolites such as A, X, Y are relatively unstable, their structures collapse in acidic solution, pH < 4. The extent to which the material loses its crystallinity in acidic medium depends on the nature of the zeolite and the extent to which cation hydrolysis occurs depends on the pH of the solution and on the nature of the transition metal ion, the second and third series of transition metal ions being more easily hydrolyzed than the first series. The general equation for ion exchange is

$$(Na)_n^+ \text{-Zeol} + Me^{n+}(H_2O)_x \rightarrow Me^{n+}(H_2O)_x\text{-Zeol} + nNa^+$$

at high pH hydrolysis occurs following

$$Me^{n+}(OH_2)_x \rightarrow \left[Me(OH)_m\right]^{(n-m^+)} + m\ H^+$$

Copper ion exchange process on X and Y zeolites has been studied extensively. The intensity of the esr signal of Cu^{2+} exchanged Y-zeolite has been followed as a function of the pH of the exchanging solution (12). The pH was varied from 3 to 10. Ammonia solution was used for samples prepared at high pH. As the pH increased the esr signal intensity decreased until a minimum was reached at a pH of 8-9. The decrease of the esr signal intensity was attributed to a decrease of copper ion dispersion which produced a esr line broadening through dipole-dipole interaction. The lower Cu^{2+} dispersion as the pH of the solution increased is due to cation hydrolysis producing $Cu^{2+}(OH)$ species with the subsequent formation of Cu-O-Cu bridges upon dehydration. Thus when the exchange was carried out at high pH, upon dehydration of the material, Cu^{2+} exists in the zeolite in the form of a polymeric hydroxy anion copper species which exhibit a relatively broad esr signal. However at very high pH, higher than 10, in NH_4OH medium the esr signal increased abruptly, which indicated that the formation of Cu-O-Cu bridges was hindered. The inhibition was the consequence of the formation of ammonia-copper complexes $Cu^{2+}(NH_3)_4$ which allowed cupric ions to remain dispersed in the zeolite. Cobalt and nickel have the same characteristic features as those exhibited by copper in zeolite. It is possible for both Ni^{2+} and Co^{2+} to be stabilized in the zeolite framework as isolated cations by mixing the sodium form of the zeolite with a solution of the metal salt, provided the pH of the solution be kept low enough. By contrast the exchange conditions must be more carefully controlled when preparing chromium, iron or aluminium exchanged zeolites, particularly with X or Y-type zeolites. Indeed these cations form very easily hydroxy anions which polymerize to form clusters. Furthermore extensive exchange with these cations produces often an appreciable lattice destruction. An important loss of lattice crystallinity was observed for X, Y zeolites extensively exchanged (13, 14) while when Cr^{3+} exchange level was low, less than 25 % of Na^+ being exchanged, no framework destruction was observed by X-ray analysis (15). Similarly it has been shown that iron-exchanged zeolites were unstable toward high temperature treatment (16).

Group VIII metal ion-exchanged zeolites have been widely used as starting materials to prepare highly dispersed supported noble metal catalysts. While in the case of the first series of transition metal ions one can use various metal salts, chloride, nitrate, sulfate, oxalate, in aqueous solution for exchange, group VIII metal ions were generally introduced in the zeolite framework by ion exchange using metal ammine complexes. Platinum exchanged faujasite-type zeolite was obtained when NaY sample was treated with a $Pt^{2+}(NH_3)_4$ solution (17). Since $Pt^{2+}(NH_3)_4$ cation is stable and not subjected to hydrolysis over a wide pH range the ion-exchange may be performed over a wide range pH values, solutions with pH up to 9 have been used. The exchange process is represented by the following expression $Pt^{2+}(NH_3)_4 + 2Na^+Z \rightarrow Pt^{2+}(NH_3)_4 - 2Z + 2Na^+$ when $Pt^{2+}(NH_3)_4$-Z is heated in oxygen the complex decomposed with the subsequent removal of NH_3 ligand. However in the temperature range 200-600°C no cation hydrolysis occurs. It has been shown by X-ray diffraction analysis that Pt^{2+} ions remain dispersed in the zeolite framework (18).

Palladium exchanged zeolites were prepared using the same procedure as for Pt^{2+}-NaY, using a solution of $Pd^{2+}(NH_3)_4$ for the exchange (19). It appeared that $Pd^{2+}(NH_3)_4$ cations did not experience hydrolysis in the pH range 6-9. Furthermore, as it has been observed for platinum exchanged zeolite, Pd^{2+} cations in NaY do not show a tendency to hydrolysis. The palladium ammine complex $Pd^{2+}(NH_3)_4$, following thermal treatment loses it NH_3 ligands and thus migrates from the supercage towards the sodalite cage. The localization of Pt^{2+} and Pd^{2+} has been determined by X-ray diffraction analysis (18-20).

Rhodium, iridium and ruthenium cations have been exchanged in zeolites using their ammine complexes. However it was also claimed that, at least for rhodium and ruthenium, aquo complexes could be used. However since these Rh, Ir and Ru cationic forms are easily hydrolyzed it is necessary to carry out the exchange reaction in well controlled experimental conditions. Rh^{3+} cations can be either uniformly distributed through the zeolite framework when proper rhodium salts and experimental conditions are used or predominantly supported on the external surface. $RhCl_3, 3H_2O$ in aqueous solution forms several rhodium species such as $[RhCl_6]^{3-}$, $[Rh(H_2O)_4Cl_2]^+$, $[Rh(H_2O)_5Cl]^+$ ... etc, their relative concentrations depend on the pH and the temperature of the solution. The cationic chloro aquo rhodium complexes are favoured at 80-90°C, thus rhodium exchange will be favoured when the reaction is carried out at 80-90°C. Rh-NaY samples have been prepared by mixing NaY zeolite with a $RhCl_3, 3H_2O$ solution at 80-90°C (21-22). From the diffuse reflectance spectrum of Rh^{3+} and the Xps measurement of atomic ratio Cl/Rh it was concluded that indeed Rh ions are exchanged as $\left[Rh(H_2O)_xCl\right]^{2+}$ and $\left[Rh(H_2O)_x\right]^{3+}$. $\left[Rh(NH_3)_5Cl\right]^{2+}$ complex has been used also to prepare Rh-NaY by ion exchange. Ion exchange was carried out either at room temperature or at 80°C (23). Pentammine chloro iridium chloride salt $\left[Ir(NH_3)_5Cl\right]$ Cl_2 has also been used to prepare iridium exchanged zeolites (24). Although $RuCl_3$, H_2O salt has been used in aqueous solution to prepare Ru exchanged zeolite it appeared that this salt was not suitable for exchange since rapid hydrolysis occured. In aqueous solution the ruthenium salt most often used for exchange is $Ru(NH_3)_6Cl_3$ (25).

1-2- Ion exchange by interaction of anhydrous metal salts with the zeolite framework. Ion exchange in liquid phase occurs readily when the exchanging metal is in its cationic form in the solution. It was found that ion exchange occured by interaction between anhydrous salts and zeolites. Heating at 300°C NH_4Cl with sodium form zeolite resulted in the replacement of Na^+ by NH_4^+ (26). Similarly it was shown that zirconium phosphate $Zr(HPO_4)_2$ in the dry state exchanged H^+ with cations when heated in the presence of metal chloride, HCl gas evolved during the reaction (27). Transition metal ions such as titanium, chromium, molybdenum are difficult to exchange into zeolites because they are stable in their cationic form at low pH where in general zeolites decompose. Effective exchange is possible through reaction of the H-form zeolite with metal chloride. The exchange procedure consists in :i) exchange of the sodium form with NH_4^+ following by heat treatment to produce H-form zeolite

$$Na-Zeol + NH_4^+ \rightarrow NH_4 - Zeol$$

$$NH_4-Zeol \xrightarrow{heat} H-Zeol + NH_3$$

ii) reaction of H-Zeol with metal chloride :

$$n \ H-Zeol + Me \ Cl_n \rightarrow Me^{n+} Zeol + n \ HCl$$

Titanium exchanged Y zeolites were prepared by this method. Recently molybdenum containing NaY zeolites were prepared by solid-solid reaction (28). The NH_4-Y form was deaminated by heating the solid at 350°C. $MoCl_5$ was ground with H-Y which formed a H-Y zeolite supported $MoOCl_4$. When the mixture was heated at 400°C exchange occured following the reaction $MoOCl_4 + 4H-O-Zeol \rightarrow O=Mo (O-Zeol)_4 + 4HCl$.

1-3- Zeolite-supported transition metal ions by reaction with metal complexes.

A - Uniform distribution of the metal ions on the zeolite could be obtained by the reaction between the metal exchanged zeolite and a metal-containing coordination compound such as a metal cyanide complex (29). The method consists to react the metal exchanged zeolite, iron zeolite for example, with a soluble iron cyanide complex in solution :

$$Fe^{2+}Z + (NH_4)_3 \left[Fe(CN)_6\right] \rightarrow NH_4Z + Fe_3 \left[Fe(CN)_6\right]_2$$

The iron ferro cyanide resulting is insoluble and thus precipitated within the zeolite cavities. Subsequently it resulted in a uniform distribution of the complex within the zeolite framework. Iron potassium Y zeolite was obtained by reacting $K_4 \left[Fe(CN)_6\right]$ with Fe, NH_4-Y zeolite. Iron cobalt-Y was obtained through the reaction of Co, NH_4-Y with $(NH_4)_4 \left[Fe(CN)_6\right]$

B - Reaction of the zeolite with metal carbonyl compounds. The adsorption of metal carbonyl compounds on zeolite has been used as an alternative route to prepare zerovalent transition metal supported on zeolite. The method consisted in adsorbing zerovalent metal carbonyl compounds within the zeolite cavities followed by a thermal decomposition to remove the carbonyl ligands (30). This procedure was first applied with $Mo(CO)_6$ and $Ru_3(CO)_{12}$. $Fe(CO)_5$, $Fe_2(CO)_8$ $Fe_3(CO)_{12}$ supported on Y zeolite samples have been prepared (3). It was shown that upon thermal decomposition of the carbonyl, the final oxidation state of the metal depended on the acidity of the starting zeolite on H-form Y zeolite the metal was oxidized by the protons with the subsequent evolution of hydrogen

$$2\ M(O) + 2n\ H^+ \underset{\leftarrow}{\overset{\rightarrow}{}} 2\ M^{n+} + n\ H_2$$

The interaction of iron carbonyl with HY zeolite has been studied by IR and gravimetric methods (31, 32). It was shown than in HY, $Fe(CO)_5$ interacted strongly with the OH groups present in the supercages (32). During the thermal decomposition the iron is partially oxidized following the reaction

$$Fe(CO)_5 + 2\ H^+ \rightarrow H_2 + Fe^{2+} + 5\ CO$$

1-4- Location and oxidation state of transition metal ions in zeolites.

Since the catalytic behaviour of zeolite-contained transition metal ions is dependent of the location and the oxidation state of the cations, it is important to establish the degree of dispersion of the transition metal ions, their location within the zeolite cavities or on the external surface of the zeolite crystal, their oxydation state. Several methods have been used for these studies such as X-ray diffraction, UV spectroscopy infrared, electron spin resonance, Mössbauer spectroscopy, xps. Among these techniques X-ray photoelectron spectroscopy was successfully employed to determine both the oxidation state and the location of cations. Thus Minachev et al (33,34) have shown that upon reduction of Ni^{2+}, Cu^{2+}, Ag^{2+} exchanged zeolite the cations are stabilized to lower oxidation state and also that they migrate to the external surface.
The xps spectra are characterized by xps peaks which positions are characteristic of the transition metal ion and its oxidation state,

and intensities characteristic of the concentration of the cations
on the surface ; depending of the element examined the xps will reveal a surface depth of about 2-5 nm. The peak intensity is given
by the relation :

$$I = n \; \sigma \; \lambda \; F \; K_{sp}$$

where n is the concentration of the element, σ the cross section for
photoelectron emission from the level, λ is the escape depth, F et K
parameters depending of the spectrometer. Thus the relative surface
concentration of two elements will be given by the relation :

$$n_1/n_2 \; = \; I_1 \; \sigma_2 \; / \; I_2 \; \sigma_1$$

where I_1 et I_2 are the peak areas $\left[Rh(NH_3)_5 Cl \right]^{2+}$ - NaY has been
studied by xps (35). The binding energies of Rh 3d3/2 and Rh 3d5/2
were found equal to 310.8 and 315.7 eV as expected for Rh^{3+}. Furthermore atomic ratios determined by xps and chemical analysis are very
close as shown in table I :

Table I : xps results for $Rh(NH_3)_5 Cl$ NaY

9.7 Rh u.c.	Cl/Rh	N/Rh	Si/Rh
xps	1	4.8	14.6
Chemical analysis	1	5	14

These results indicate that the rhodium complex is introduced without
decomposition and homogeneously distributed over the zeolite framework. Further investigations of rhodium exchanged zeolite by xps
have been made (36). The data given in (36) confirmed that the atomic ratio Rh : N : Cl found in rhodium pentammine exchange X zeolite
is as expected for the cation $\left[Rh(NH_3)_5 Cl \right]$. The Si/Rh ratio determined by xps is close to that predicted on the basis of the bulk
composition for homogeneous distribution of the cation into the zeolite. Additional xps evidence of homogeneous distribution of Rh cations in NaY zeolite, when $\left[Rh(NH_3)_5 Cl \right]^{2+}$ was used for ion exchange was given in (37). The atomic Rh/Si ratios measured from xps experiments are identical to the theoretical values assuming true ion
exchange. In contrast it was found that when the ion exchange was
carried out in a solution of $Rh(NO_3)_3$, $2H_2O$ the metal ions were predominantly localized on the external surface of the zeolite. Indeed
for these samples the xps atomic ratios Rh/Si were one order of magnitude higher than the theoretical values. Molybdenum-containing
zeolites prepared by solid-solid reaction between $MoOCl_4$ and HY
zeolite were examined by xps (28). The binding energies for the
Mo(3d3/2) and Mo(3d5/2) were respectively 235.8 and 232.7 eV

indicating that molybdenum ions were present as Mo(VI) in the zeolites. Mo-exchanged zeolite obtained by solid-solid reaction between $MoOCl_4$ and HY exhibited a xps Mo/Si ratio close to the theoretical value, which suggests that Mo ions were homogeneously distributed in the zeolite framework, while the Mo/Si ratio in samples prepared by impregnation of NaY with $MoCl_5$ solution is about sevenfold greater than the theoretical value, molybdenum being mainly on the zeolite external surface (28).

2 - Reactivity of transition metal ions in zeolites

The intracrystalline pore and cavity system of zeolites and the important lattice induce unusual reactivity to the entrapped transition metal ions. In addition the reactivity of these ions in zeolites appeared to be very similar to their homologous in solution. This paragraph will provide some interesting reactivities of transition metal ions exchanged zeolites.

2-1- Ionisation of molecules in zeolites.

The very strong ionisation properties of zeolites are responsible for the ionisation of water by multivalent exchanged cations. There are now large number of experimental evidences showing the ionisation of H_2O with the subsequent formation of $Me^{n+}(OH)_x$ and acidic OH groups (38). When Me^{n+} is a transition metal ion, thermal decomposition of the hydrated sample often led to the reduction of the cation. It has been shown that dehydration of Fe^{3+}-NaY zeolite produced Fe^{2+} ions (39). From Mössbauer spectroscopy studies and quantitative measurements of O_2 adsorbed it has been assumed that the following reactions occured on ferrous ion-exchanged zeolite (40)

$$2 \ Fe^{2+} + 1/2 \ O_2 \rightarrow Fe^{3+} - O^{2-} - Fe^{3+}$$

$$Fe^{3+} - O^{2-} - Fe^{3+} + H_2O \rightarrow 2\left[Fe^{3+}(OH)^-\right]$$

These observations were further extended to several transition metal ions and it was concluded that the zeolites have the remarkable properties to decompose water into oxygen and hydrogen following a thermochemical cycle (41). The reactions producing water splitting are the following :

H_2O ionisation : $2 \ Cu^{2+} + 2 \ H_2O \rightarrow 2 \ Cu-OH^+ + 2H^+$

O_2 and H_2O desorption

$$2 \ OH^- \rightarrow H_2O + 1/2 \ O_2 + 2e^-$$

Reduction $\quad\quad 2 \ Cu^{2+} + 2 \ e^- \rightarrow 2 \ Cu^+$

Rehydration $\quad 2 \ Cu^+ + 2 \ H^+ + H_2O \rightarrow 2 \ (Cu-OH)^+ + H_2$

The ionisation property of zeolites appeared to be responsible for the hydrolysis of group VIII transition metal ions in zeolite.

$Ru^{3+}(NH_3)_6$-exchanged zeolites have been investigated by esr and IR (42). The white sample $Ru^{3+}(NH_3)_6$-NaY showed an esr spectrum with g = 2.20 and was attributed to $Ru^{3+}(NH_3)_6$. Reflectance spectroscopy showed a band at 38.000 cm^{-1} (43) and IR showed a band at 1360 cm^{-1} (42,43) due to NH_3 ligands in $Ru^{3+}(NH_3)_6$ complex. Thus the results presented confirmed that on a freshly prepared ruthenium exchanged zeolite from hexammine ruthenium solution, the complex is in the form of $Ru^{3+}(NH_3)_6$ localized in the supercage. When the sample was outgassed the sample turned progressively red-wine along with a drastic change both of the esr spectrum of Ru^{3+} ions and the appearance of IR band at 1460 cm^{-1} due to the formation of NH_4^+ (42). It was concluded that $(Ru(NH_3)_5OH)^{2+}$, and ruthenium red were formed following thermal treatment a low temperature (42,43,25,44)

$$Ru^{3+}(NH_3)_6 + H_2O \rightarrow Ru(NH_3)_5OH^{2+} + NH_4^+ \text{ polymerisation of}$$

$[Ru(NH_3)_5OH]^{2+}$ would lead to ruthenium red :

$$(NH_3)_5 Ru^{3+} - O - Ru^{4+}(NH_3)_4 - O - Ru^{3+}(NH_3)_5$$

Thus it appeared that upon outgassing hydrolysis of the ruthenium ammine complex occured. Similarly when NH_3 is adsorbed on a freshly prepared $Ru^{3+}(NH_3)_6$-NaY the sample immediatly showed an esr spectrum attributed to $[Ru(NH_3)_5OH]^{2+}$ and IR spectra indicated the formation of NH_4^+ (45). The fact that outgassing $Ru(NH_3)_6$-Y and adsorbing NH_3 on this sample produced the same effect that is partial hydrolysis of the ruthenium hexammine complex was interpreted in the following manner :

$Ru^{3+}(NH_3)_6$ is relatively unstable in basic media. However when introduced in the supercage of the zeolite and when the cavity is fully hydrated, $Ru^{3+}(NH_3)_6$ is protected from the ionizing power of zeolite by the water molecules filling the cavity. Upon dehydration following outgassing, the complex is subjected to ionisation with the subsequent hydrolysis. Similarly by adsorbing NH_3, NH_3 forms with the water molecules present in the cavity NH_4OH, which is immediately ionized forming high concentration of OH^- groups. In the presence of these OH^- groups hydrolysis of $Ru^{3+}(NH_3)_6$ occurs.

This strong ionizing property of zeolite is thus responsible for the facile hydrolysis of group VIII transition metal ions in zeolite. Upon dehydration at high temperature the hydrolyzed group VIII transition metal ions form rapidly metal oxides which migrate on the external surface of the zeolite.

2-2- Oxygen-transition metal ions

It is well known that tetraphenyl-porphyrin cobalt (II) adsorbs molecular oxygen, the electron configuration in this oxygen-cobalt adduct approaches Co(III) - O_2^-, one electron being transfered from the cobalt orbital to the oxygen molecule. Pentaammine Co(II) forms also with oxygen a μ-peroxodicobalt ammine complex $(NH_3)_5$ Co-O_2-Co$(NH_3)_5$ the oxygen molecule bridging two cobalt complexes. Identical reactions occured with Co(II) exchanged NaY zeolite (46). The μ-peroxodicobalt complex was formed when O_2 reacted with Co(II)$(NH_3)_5$ entrapped complex. In contrast only the monomeric oxygen cobalt species was formed when NH_3 ligands were replaced by propylene diammine ligand ; the structure of the oxygen adduct was :

$$\left[(CH_3\ CH_2\ CH_2-NH_2)\qquad Co(III)-O_2^- \right]$$

Rhodium II porphyrin in dimethyl-formamide adsorbs H_2 dissociatively (47) following the reaction :

$$2\ Rh(II)\ +\ H_2\quad \rightarrow\quad 2\ Rh(I)\ +\ 2\ H^+$$

Rh(I) was very sensitive to oxygen and was oxidized with the subsequent formation of Rh(II) and H_2O :

$$2\ Rh(I)\ +\ 2\ H^+\ +\ 1/2\ O_2\ \rightarrow\quad 2\ Rh(II)\ +\ H_2O$$

Rhodium exchanged zeolites behave sililarly. Esr study of oxygen adsorption in activated Rh-NaY revealed the formation of a μ-peroxodirhodium adduct (23). It was suggested that following thermal activation of Rh(III)-NaY self reduction of Rh(III) to Rh(I) occured Rh(I) ions were bound to lattice oxygen ions. Upon addition of O_2 molecules the μ-peroxodirhodium complex Rh(II) - O_2^{2-} - Rh(II) was formed one electron from each Rh(I) ion being transferred to the O_2 molecule.

Zeolite was found to stabilize new type of oxygen adducts which were not revealed in solution. Example is given by palladium exchanged zeolite (47). When Pd(II)-mordenite was activated in vacuum esr studies showed the formation of Pd(I) ions due to self reduction of the Pd(II) ions. The adsorption of O_2 at room temperature was studied by esr, using ^{17}O enriched oxygen gas. It was suggested that O_2 molecule was trapped between three Pd(I) ions each Pd(I) transferring to the oxygen molecule one electron. The resulting oxygen species was described as a $\left[Pd(II) \right]_3 O_2^{3-}$ complex, the electron configuration of the charged O_2 molecule being similar to Cl_2^- ion.

2-3- Nitric oxide adducts

Transition metal ion exchanged zeolites showed high affinity for nitric oxide and it was thought that these materials could be advantageously used as catalysts in the dissociation and reduction of nitric oxide. This has prompted several studies of the formation of nitrosyl complexes within the zeolite cavities. Esr and infrared techniques were used to investigate the electronic structure of the metal nitrosyl complexes. Although the ionization potential of NO is

relatively high (9.3 eV) in several cases, due to the intracrystalline ionizing power of the zeolite, the formation of the nitrosyl
complex was accompanied by a transfer of electron from NO to the
transition metal ion with its subsequent reduction.

Nickel exchanged Y zeolite (Ni^{2+}-Y) does not exhibit an esr signal.
When $Ni^{2+}Y$ is contacted with NO a strong esr signal is observed.
This signal was attributed to Ni^+ ion ($3d^9$ electron configuration).
Furthermore IR data indicated the presence of a nitrosyl adduct
(νNO = 1892 cm^{-1}) (48-49). The interesting aspect of these studies
is that NO coordinates to Ni^{++} in zeolite to form Ni^+NO^+ complex.
This complex Ni^+NO^+ was chemically inert toward oxygen (48-49).
Nitric oxide forms also with Cr^{2+} in NaY zeolite (Cr^+NO^+) complex
which was also inert toward oxygen (49). Cr^{2+}-NaY was obtained by
H_2-reduction of Cr^{3+}-NaY. Nitric oxide reacts with ferrous ions in
iron exchanged zeolite to form an iron nitrosyl complex (50). Both
low spin (s = 1/2) and high spin (s = 3/2) $Fe(I)NO^+$ were present
in the zeolite cavities and were identified by esr and infrared. The
high spin iron nitrosyl complex showed an esr spectrum with $g_\perp$ = 4.07
and $g_{//}$ = 2.003 and an IR band at 1890 cm^{-1}. The study of the interaction of NO with transition metal ion exchanged zeolites has rein-
forced the idea concerning the high ionizing property of zeolite
which facilitates electron transfer reactions. Furthermore metal nitrosyl complexes exhibited an unusually high stability toward oxidation by O_2.

2-4- Reactivity with CO

The reactivity of 1st series transition metal ions exchanged
zeolites has been carried out partly to investigate the location
of the transition metal ions in the zeolite structure (for example
cations within the sodalite or the hexagonal prism of NaY will be
hidden from CO interaction) and partly to probe the oxidation state
of the transition metal ion from the CO IR frequencies. In general
CO forms weak bonds with the 1st series transition metal ions and
the Me^{n+}-CO complex is easily destroyed by outgassing the samples.
The IR studies of carbon monoxide adsorbed on transition metal ion
exchanged zeolites have been reviewed in (51). More recent spectroscopic studies of the interaction of carbon monoxide with group VIII
transition metal ion-exchanged zeolites have shown that well defined
metal carbonyl compounds were formed within the zeolite cavities (8).
It appears interesting to describe the formation and identification
of the mononuclear and polynuclear metal carbonyl complexes which
were synthesized in the supercages of the faujasite like structure
NaY zeolite i) zeolite entrapped mononuclear carbonyl compounds
Rh(III)-NaY reacts with CO at room temperature forming Rh(I) dicar-
bonyl compound which structure was identified by IR spectroscopy.
In addition xps measurements were obtained to further confirm the
reduction of Rh(III) to Rh(I) during the reaction with CO. IR spectra of the carbonyl complex formed within the zeolite cavities
showed bands at 2100-2020 cm^{-1} as expected for $Rh(I)(CO)_2$ species.

The general equation for the reaction is (35)

$$Rh(III) + CO \xrightarrow{H_2O} Rh(I) + 2H^+ + CO_2$$

$$Rh(I) + 2CO \rightarrow Rh(I)(CO)_2$$

Similarly Ir(III)-NaY reacts with CO to form monovalent iridium carbonyl species (52). Quantitative measurement of CO uptake at 170°C
indicated a CO/Ir ratio of about 4. Furthermore IR bands at 2086 and
2001 cm^{-1} were observed. The data were interpreted in the same manner as for Rh-NaY, Ir(III) being reduced by CO into Ir(I) with the
subsequent formation of $Ir(I)(CO)_3$ carbonyl species.

ii) Zeolite-entrapped metal carbonyl clusters. Recently there has
been interest in supported metal carbonyl clusters. The investigations have been directed towards : heterogenizing soluble catalysts,
preparing highly dispersed metal catalysts, preparing supported sub-
carbonyl metal clusters showing unique catalytic properties.
In general, it was found that the carbonyl metal cluster immobilized
on solid surfaces such as silica, alumina, maintain their molecular
structure only when the surface was highly dehydroxylated or when
the surface was functionalized by phosphine ligands. The NaY faujasite type zeolite appeared an interesting material for stabilizing
within the zeolite cavities metal carbonyl clusters such as $Rh_6(CO)_{16}$,
$Ir_4(CO)_{12}$. Since the critical dimension of the carbonyl clusters are
generally larger than the zeolite cavity-windows, it appeared necessary to synthesize the carbonyl metal cluster directly within the
cavities (6).
Rhodium and iridium carbonyl clusters are potentially active catalysts
for reactions such as oxidation, hydrogenation and hydroformylation
of olefins. It is expected that immobilization of the carbonyl metal cluster increases its stability toward aggregation. This was accomplished in two different ways :
a) a direct synthesis of the metal cluster within the zeolite cavities,
b) an insertion of the metal cluster within the cavities by sublimation.
Our results have shown that the first procedure is more suitable to
obtain highly dispersed metal carbonyl clusters

$Rh_6(CO)_{16}$ entrapped within the NaY zeolite cavities was obtained
in the following ways (53, 8). NaY was exchanged with Rh^{3+}. After
dehydration of the RhNaY sample at 573 K, the zeolite was treated
with a mixture of CO : H_2 at about 15 atmospheres and at 300 K for
few hours. The samples turned colored and exhibited a well defined
infrared spectrum with two strong carbonyl bands at 2095 and
1765 cm^{-1}. By comparison with other rhodium carbonyl compounds this
infrared spectrum was ascribed to $Rh_6(CO)_{16}$ cluster formed and stabilized within the zeolite cavities. Identical species were also
stabilized within the zeolite when $Rh_6(CO)_{16}$ was first supported
on the zeolite surface by sublimation a 373 K followed by a decar-
bonylation and recarbonylation at 373 K (54). The infrared spectrum

of this sample also showed two strong bands at 2095 and 1765 cm^{-1} attributed to $Rh_6(CO)_{16}$ entrapped in the zeolite. The stability of the rhodium carbonyl cluster was further investigated. The infrared results indicated that zeolite-entrapped $Rh_6(CO)_{16}$ may be decarbonylated, without significant aggregation, by reacting the sample at 373 K with H_2 or O_2. Recarbonylation by CO at 373 K regenerated the infrared bands at 2095 and 1765 cm^{-1}. These studies show that the retention of the cluster integrity is possible by using a zeolite support.

$Ir_4(CO)_{12}$, $Ir_6(CO)_{16}$ have also been synthesized and stabilized within NaY zeolites (8). The exchanged Ir^{3+} NaY zeolite was reduced by a mixture of $CO : H_2$ at atmospheric pressure and at 443 K. The sample turned colored and showed in the carbonyl stretching region infrared bands at 2086, 2040 and 1813 cm^{-1}. An identical infrared spectrum was observed when $Ir_4(CO)_{12}$ was deposited on the zeolite from solution. These results show clearly that the zeolite matrix is particularly suitable not only for the preparation of highly dispersed metal catalysts but also for stabilizing well defined metal clusters.

2-5- Reactivity with H_2 : formation of small metal particles

Since the importance of metal dispersion in the efficient use of metal catalysts has been well established extensive work has been carried out to develop methods of preparation that should produce metal catalysts finely dispersed. The highest degree of metal dispersion was obtained by fixing metal cations on a carrier, generally silica or alumina, by ion exchange technique, following by reduction in hydrogen (55-56). Because of their particular structures and their ion-exchange properties, zeolites appeared particularly appropriate for stabilizing finely dispersed metal particles. Rabo et al (57) were among the first to report on zeolite-supported platinum. From their results they concluded that platinum was almost atomically dispersed within the zeolite-framework. Further investigations on zeolite-supported platinum were performed by several authors (58-60). Generally it was concluded that the platinum particle-size depended strongly on the pretreatment conditions of the materials before H_2-reduction. Gallezot et al (59) from their electron microscopy and X-ray diffraction studies concluded that the metal dispersion would depend on the position of Pt^{2+} cations in the zeolite cages. Since much of the work concerned platinum catalysts, we have investigated the behaviour of other zeolite-supported group VIII metals in order to provide a general trend on the preparation and the properties of zeolite-supported group VIII metals.

The preparation of zeolite-supported noble metals involved the substitution of Na$^+$ cations in NaY, by ion-exchange with group VIII metals in their cationic form. For ion exchange we have used aqueous solutions of $Pt^{2+}(NH_3)_4$ $Pd^{2+}(NH_3)_4$, $Ru^{3+}(NH_3)_6$, $(Rh(NH_3)_5Cl)^{2+}$, $(Ir(NH_3)_5Cl)^{2+}$. After being carefully washed and dried the exchanged faujasite-type zeolites were first calcined in oxygen in the

temperature range 473-773 K, followed by H_2-reduced in the temperature range 383-773 K. Metal dispersion and particule size measurements were obtained by H_2 adsorption and transmission electron microscopy. X-ray diffraction, electron spin resonance, Xps and infrared
were used to investigate the state of the metal cations within the
zeolite framework before H_2- reduction, since it has been already
stated for Pt that this parameter may strongly influence the metal
dispersion.

As already observed by others (58-60) the average particle diameter
obtained from electron micrographs and that calculated from hydrogen
adsorption data are in good agreement in the case of Pt-NaY samples
calcined in oxygen at 573 K before H_2-reduced. In contrast Pt-NaY
calcined in oxygen at 773 K and then H_2-reduced at 773 K showed by
H_2-adsorption an apparent particle size of about 5nm while the electron micrographs indicated that the particle size were around 2 nm.
This discrepancy was interpreted in terms of the existence of platinum atomically dispersed in the sodalite cages, which did not adsorb
hydrogen. X-ray diffraction analysis (61) have indicated that the
increase of the activation temperature prior to reduction produced
a migration of Pt^{2+} cations from the supercages to the sodalite cages. Thus reduction of Pt^{2+} cations present in the supercages produced platinum aggregates of about 1 nm in diameter, while the reduction of Pt^{2+} cations in the sodalite cages would lead to atomically
dispersed platinum.

Ruthenium, rhodium or iridium exchanged zeolites behaved differently.
Samples precalcined in oxygen below 510 K prior to H_2-reduction did
form highly dispersed metal catalysts. The particle sizes as determined by electron microscopy and by H_2 adsorption were in good
agreement and in the range 1-1.5 nm. Furthermore the electron micrographs indicated that the metal particles were located within the
zeolite cavities, probably in the supercages. In contrast very large
metal particle (10-20 nm) were formed for those samples precalcined
in oxygen at 773 K. Electron micrographs showed that these particles
were on the external surface of the zeolite. These results clearly
indicate a different behaviour of Ru, Rh, Ir exchanged NaY zeolites
compared with Pt-NaY samples. To better understand the effect of
oxygen pretreatment on the Ru, Rh, Ir dispersion, the state of the
metal precursor before H_2 - reduction, was investigated.

It was shown (3) that upon activation of Ru-exchanged Y zeolite the
$\left[Ru(NH_3)_6 \right]^{3+}$ complex was progressively transformed into
$\left[Ru(NH_3)_x (OH)_y \right]$ species, following hydrolysis of the hexammine
ruthenium complex. Hence ruthenium in its cationic form is progressively transformed in an anionic species which will be no more bound to
the zeolite lattice by electrostatic field. Calcination at a temperature of 773 K produced a dehydration of the ruthenium hydroxyanions and the subsequent formation of Ru_2O_3, RuO_2 on the external
surface of the zeolite. The effect of the oxygen treatment of
$Ru(NH_3)_6$ -NaY zeolites prior H_2-reduction on the final metal particles is now well understandable : up to a temperature of about 523 K,

Ru^{3+} complexes remained well dispersed in the supercages of the zeolite. These well dispersed species formed upon H_2-reduction very small metal particles. At higher temperature the hydroxyanions of Ru^{3+} were dehydrated into large Ru_2O_3 crystallites, located on the external surface of the zeolite. Obviously the H_2-reduction of these large oxide crystallites generated large metal particles.

Similar studies were carried with $(Ir(NH_3)_5Cl)^{2+}$ NaY. X-ray diffraction and infrared studies incidated again that upon calcination in oxygen $(Ir(NH_3)_5Cl)^{2+}$ cations were progressively transformed into hydroxyanions $(Ir(NH_3)_{5-x}(OH)_xCl)$ which upon complete oxidation at 773 K formed large IrO_2 crystallites.

The general conclusion which may be derived from this study is the following : zeolites are suitable materials to prepare and to stabilize highly dispersed metal catalysts, provided that the metal precursors remain highly dispersed in the zeolite cavities before H_2-reduction.

Group VIII transition metals may be classified into two groups :

i) those for which the cationic form is relatively stable, such as platinum, palladium ; these cations remain always highly dispersed in the zeolite cavities, thus form highly dispersed metal catalysts.

ii) those for which the cationic form is unstable, rhodium, ruthenium, iridium, and easily transformed into an hydroxyanion. As long as the hydroxyanions remain dispersed in the zeolite framework, one obtains upon H_2-reduction highly dispersed metal catalysts. However since at high temperature the hydroxyanions are dehydrated into the oxide form, one should avoid such formation of large oxide crystallite in order to form highly dispersed metal catalysts.

3 - Catalytic properties of transition metal ion-exchanged zeolites

It is well recognized that zeolites are widely used as catalysts for a broad range of hydrocarbon transformation. Although the major applications of zeolites as catalysts, especially in the Petroleum Industry, are based on acid form zeolites several investigators have discovered that specific catalytic properties are shown by transition metal ion exchanged zeolites toward reactions generally catalyzed by the parent metal ions in solution. In this paragraph some recent aspects of the catalytic properties of "immobilized transition metal ions" in zeolites will be given. It is worthwhile to recall that most of the work which will be described deals with faujasite-type zeolites, these materials appearing the most suitable carrier for "Immobilizing" soluble catalysts because of the relatively large space of their cavities.

3-1- Oligomerisation, addition, cyclodimerisation of unsatured hydrocarbons

The homogeneous catalytic formation of imines by addition of

primary aliphatic amines to acetylenes in the presence of zinc acetate needs high pressure (62). Zn(II)-exchanged Y zeolite was found
active for the addition of methylamine to methylacetylene at atmospheric pressure. N-isopropylidene methylamine resulted following the
reaction (63)

$$CH_3NH_2 \;+\; CH_3-C \equiv CH \;\rightarrow\; (CH_3)_2C = NCH_3$$

However Ni, Pd, Pt and Cu(I) exchanged zeolites (isoelectronic with
Zn(II)) were found inactive in this reaction. The lack of activity
was attributed to a rapid reduction of the transition metal ions.
Thus it is clear that when using these materials for reactions catalyzed by ions in solution, it is important to avoid the reduction
of the exchanged transition metal ions into their metallic forms.

Benzene formation from the trimerisation of acetylene was catalyzed
by Ni^{2+}-NaY zeolite. The reaction was found to occur within the large cavities of the zeolite (64). The dimerisation of ethylene over
rhodium exchanged zeolite has been investigated 565). It was found
that the rate of formation of n-butenes depended on the temperature
of activation of RhY, the maximum rate being reached when RhY was
activated around 400°C. Furthermore it was found that the dimerisation activity was lowered by the addition of pyridine or CO which
are known to interact with the Rh cations, while an appropriate
amount of HCl increased the reaction rate. Xps measurements indicated the existence of monovalent Rh(I) on the active Rh-Y samples.
It was concluded that Rh(I) in NaY was the active sites for the di-
merization of ethylene. The effect of HCl is similar to that encountered in homogeneous systems for which it has been shown that
the addition of HCl to Rh(I) complexes activates the catalyst (66).

Copper-exchanged zeolites have shown interesting catalytic properties for the cyclodimerisation of butadiene to vinylcyclohexene (67).
Nickel (O) complexes in solution catalyze the cyclodimerisation of
butadiene into 4-vinylcyclohexene (68). Thus it appeared that it is
Cu(I) which is isoelectronic with Ni(O) which is the active site for
the cyclodimensation of butadiene by Cu(II)-Y zeolites Cu(I)-Y were
further investigated for the cyclodimerisation of butadiene (69).
Cu(I)-Y samples were prepared either by direct exchange with cuprous
solution or by reduction of Cu(II)-Y by CO following the procedure
given in (70), monovalent-copper containing zeolites exhibited a
high selectivity (90 %) for the formation of 4-vinylcyclohexene.
However Cu(I)-Y samples obtained from CO reduction showed a rapid
deactivation due to the formation and deposition of polymer butadiene on the zeolite surface. The active sites for polymerisation
were Brönsted/Lewis centres generated during reduction of Cu(II)-Y
to Cu(I)-Y. When the acid sites were neutralized by ammonia the catalyst deactivation during the cyclodimerisation of butadiene was
lowered. Cu(I)-Y prepared by direct exchange with cuprous solution
was found substantially more stable due to the absence of acid sites.

In conclusion, it was stated that the difference between Cu(I)-Y and CO-reduced Cu(II)-Y is due to the acidity generated by CO-reduction. The role of the zeolite framework is to stabilize monovalent copper ions, which allowed the oxidative coupling of two butadiene molecules.

3-2- Oxidation reactions. Transition metal ions exchanges Y zeolites were found active for selective oxidation of hydrocarbons. In the oxidation of ethylene into acetaldehyde (Wacker process) which in the liquid phase is catalyzed by a solution containing Cu(II) and Pd(II) ions was found to occur in the gas-solid reaction using Pd(II), Cu(II)-exchanged Y zeolite (71). It should be noted that the yield of acetaldehyde decreased at temperatures higher than 115°C as the result of reduction of Pd(II) and Cu(II) into metal forms.

Cu(II)-Y zeolite was also investigated in the vapor phase oxidation of benzyl alcohol at a temperature range of 300-390°C (72). The active sites for the oxidation of benzyl alcohol into benzaldehyde were Cu(II) in the zeolite framework. The conversion of benzylalcohol increased abruptly beyong 30 % Cu(II) ion exchange, which indicated that Cu(II) ions were, below 30 % exchange level in hidden sites. Beyong 30 % exchange Cu(II) ions were located within the zeolite supercages and thus accessible to the reactants. Cobalt(II)-exchanged NaY zeolite was also found active and selective in the oxidation of benzyl alcohol into benzaldehyde (73). Co-NaY was found much more selective than Cu-NaY. As in the case of Cu-NaY, the yield of benzaldehyde increased abruptly beyong about 20 % exchange. The effect of amine addition on the benzaldehyde yield demonstrated the similarity of the behevior of Co(II) in solution and within the zeolite framework. The addition of pyridine or piperidine would pull out of the sodalite cages Co(II) and thus increases the number of accessible Co(II) ions, which would lead to an increase of the benzaldehyde yield. However when ethylene diamine was adsorbed, the yield of benzaldehyde decreased. This was attributed to the formation of $Co(III)(en)_2O_2^-$ complex in the large cavities which appeared relatively stable. In contrast Co(II) in the presence of pyridine or ammonia forms with O_2 a dimeric Cobalt-oxygen adduct such as $L_xCo(II)-O_2-Co(II)L_x$ which was considered as the precursor for dissociation of O_2 molecule. The formation of Co-O is thus facilitated in the presence of pyridine resulting in the increase of the benzaldehyde yield.

The oxidation of cyclohexene over molybdenum zeolites in the liquid-phase was found to occur with relatively high selectively toward epoxidation (74) at 50 % conversion the selectivity for cyclohexane oxide was about 50 %. It was concluded that at low Mo content the cyclohexene epoxidation was initiated by a radical mechanism, the formation of cyclohexenyl hydroperoxide being the limiting step Zeolites containing both molybdenum and cobalt exhibited activities and selectivities in cyclohexene oxidation comparable to homogeneous catalysts such as $Co(acac)_2/MoO_2(acac)_2$.

3-3- Carbonylation of methanol

The carbonylation of methanol into acetic acid or methylacetate was developed by Monsanto in the liquid phase using rhodium based catalysts, in the presence of an iodide promotor (74). Rhodium exchanged zeolites were found active and selective for the vapor phase carbonylation of methanol (75). Similarly iridium exchanged zeolite showed interesting activity in this reaction. The reaction was carried out at 150-180°C and at atmospheric pressure. The kinetic studies revealed identical rate expression both in liquid phase and in vapor phase that is : first order with respect to CH_3I promotor and (Rh) concentration and zero order with respect to CH_3OH and CO. The various steps of the reaction were investigated by infrared and the results parallel those obtained with the homogeneous rhodium catalysts. In the presence of CO Rh(III) ions were reduced to Rh(I) which coordinated two CO molecules to give $Rh(I)(CO)_2$ active sites. These species added CH_3I through an oxidative addition with the subsequent formation of $[CH_3-Rh(III)(CO)_2(I)]$ complexes. This complex was relatively unstable and rearranged rapidly into a rhodium acetyl complex $[(CH_3CO)Rh(CO)(I)]$. Methanol or water reacted easily with the acetyl adduct and methyl acetate or acetic acid were evolved. The role of the zeolite appeared to be as a carrier for better stabilization and for better dispersion of Rh(I) species.

These few examples are sufficient to show that a large number of reactions catalyzed by soluble metal complexes in liquid phase can be carried out in the vapor phase, by zeolites exchanged with the analogous metal ions. In general the zeolite matrix affords the highest metal ion dispersion in comparison with other supports and affords high stabilisation for cations in low oxidation state. In addition in several cases, on zeolites the reaction can be carried out at much lower pressures than those required by other homogeneous catalysts in liquid phase. The applicability of zeolites containing transition metal ions is far from being exhausted, and it is clear that several new applications for producing chemicals will be found in the near futur.

392

REFERENCES

1. J. A. Rabo ed. Zeolite Chemistry and Catalysis, ACS monograph, 171 (1976).

2. J. H. Lunsford, Catalysis Review, 12, 137 (1975.

3. C. Naccache and Y. Ben Taarit, Acta Physica et Chemica, Hungaria 24, 23 (1978).

4. J. H. Lunsford, "Molecular Sieves II" (J. R. Katzer ed.) 40, 39 (1977).

5. G. K. Boreskov and Kh. M. Minachev "Application of zeolites in Catalysis", Akademiai Kiado, Budapest (1979).

6. C. Naccache and Y. Ben Taarit, Pure and Appl. Chem., 52, 2175, (1980).

7. Y. Ben Taarit and M. Che, in "Catalysis by zeolites" (B. Imelik et al. eds), Elsevier Co, Amsterdam (1980), p. 167.

8. P. Gelin, F. Lefebvre, B. Elleuch, C. Naccache and Y. Ben Taarit, Intra zeolitic symposium, Meeting of the American Chemical Soc., Kansas City Missouri, A.C.S. Symposium Series, 218 (1983), p. 455.

9. A. Cremers, Molecular Sieves II (J.R. Katzer ed.), A.C.S. Symposium Series, 40, 179 (1977).

10. Hsu Rhu Reng and Yu Guo Chen, Proc. Fifth Int. Conference on Zeolites (ed. L.V. Rees) Heyden London (1980), p. 231.

11. K.G. Ione, P. N. Kuznetsov, V. N. Romannikov, Application of zeolites to catalysis (eds G. K. Boreskov, Kh. M. Minachev) Akad. Kiodo, Budapest (1979), p. 87.

12. V. F. Anufrienko et al., Application of zeolites to catalysis (eds G.K. Boreskov, Kh. M. Minachev) Akad. Kiodo, Budapest, (1979), p. 109.

13. A. M. Rubinstein, A. A. Slinkin, M. I. Loktev, E. A. Fedorovskaya, H. Bremer, F. Vogt, Z. Anorg. Alg. Chem. 423, 164 (1976).

14. A. H. Badran, J. Dwyer, N. P. Evmerides and J. A. Manford, Inorg. Chim. Acta, 21, 61, (1977).

15. J. R. Pearce, D. E. Sherwood, M. B. Hall and J. H. Lunsford, J. Phys. Chem., 84, 3215 (1980).

16. J. R. Pearce, W. J. Mortier and J. B. Uytterhoeven, J. Chem. Soc. Faraday Trans I, 77, 937 (1981).

17. J. A. Rabo, V. Schomaker and P. E. Pickert, Proceedings 3rd Int. Congr. Catal., North Holland, Amsterdam, p. 1264 (1965).

18. P. Gallezot, Catal. Rev. Sc. Eng., 20, 121 (1979).

19. C. Naccache, J. F. Dutel and M. Che, J. Catal. 29, 179 (1973).

20. P. Gallezot and G. Bergeret, Metal microstructures in zeolites (eds P. A. Jacobs, N. I. Jaeger, P. Jiru, G. Schylz-Ekloff), Studies in Surface Sciences and Catalysis, Elsevier Amsterdam, 12, p. 167 (1982).

21. B. Christensen and M. Scurrel, J.C.S. Faraday I, 73, 2036, (1977).

     Y. Okamoto, N. Ishida, T. Imanaka, S. Teranishi, J. Catal., 58, 82, (1979).

22. R. Shannon et al., to be published.

23. C. Naccache, Y. Ben Taarit, M. Boudart, A.C.S. Symp. Ser., 40, 156 (1972).

24. P. Gelin, M. Dufaux, C. Naccache, "Catalysis by zeolites (B. Imelik et al. eds) Elsevier Amsterdam, p. 261 (1980).

25. J. H. Lunsford "Metal microstructures in zeolites" (P. A. Jacobs et al. eds) Elsevier Amsterdam, p. 1 (1982).

26. R. M. Barrer, Nature, 164, 112 (1949).

27. T. M. Troup, A. Clearfield, J. Phys. Chem. 74, 2578, (1970).

28. P. S. E. Dai, J. Lunsford, J. Catal., 64, 173, (1980).

29. J. Scherrer, D. Fort, J. Catal., 71, 111 (1981).

30. P. Gallezot, G. Coudurier, M. Primet, B. Imelik, Molecular Sieves (J. Katzer ed.), A.C.S. Symposium Series 40, p. 144 (1977).

31. D. Ballivet-Tkatchenko, G. Coudurier, Nguyen Duc Chau,
    Metal microstructures in zeolites (P. A. Jacobs et al. eds)
    Elsevier Amsterdam, p. 123 (1982).

32. Th. Bein, P. A. Jacobs, F. Schmidt, Metal Microstructures
    in zeolites (P. A. Jacobs et al. eds) Elsevier Amsterdam
    p. 111 (1982).

33. K. M. Minachev, G. V. Antoshin, E. S. Shpiro, T. A. Navruzov,
    Isv. Akad. Nauk. SSR, Ser. Khim 2131 (1973).

34. K. H. Minachev, G. V. Antoshin, E. S. Shpiro, Y. A. Yusifov,
    Proc. 6th Inter. Congress. Catal., London 1976 (G.C. Bond,
    F.C. Thompkins eds), vol. $\underline{2}$, p. 621, Chemical Soc. (1977).

35. M. Primet, J. C. Vedrine, C. Naccache, J. Mol. Catal., $\underline{4}$,
    411, (1978).

36. S. Lars, T. Anderson, M. S. Scurrell, J. Catal., $\underline{71}$, 233
    (1981).

37. M. Niwa, J. H. Lunsford, J. Catal., $\underline{75}$, 302 (1982).

38. J. W. Ward in Zeolite Chemistry and Catalysis (J.A. Rabo ed.)
    A.C.S. Monograph $\underline{171}$, p. 118 (1976).

39. J. Morice, L. V. C. Rees, Trans. Farad. Soc. $\underline{64}$, 1388 (1977).

40. R. L. Garten, W. N. Delgass, M. Boudart, J. Catal. $\underline{18}$, 90
    (1970).

41. P. H. Kasai, R. J. Bishop, J. Phys. Chem., $\underline{81}$, 1527 (1977).

42. C. Naccache, Y. Ben Taarit, Acta Physica et Chemica Hungaria
    $\underline{24}$, 23 (1978).

43. J. J. Verdonck, R. A. Schoonheydt, P. Jacobs, J. Phys. Chem.
    $\underline{85}$, 2393, 1981.

44. C. P. Madhusudhan, M. D. Patil, M. L. Good Inorg. Chem.
    $\underline{18}$, 2383 (1979).

45. M. Goldwasser, A. Abou Kais, J. F. Dutel, C. Naccache,
    to be published.

46. R. F. Howe, J. H. Lunsford, J. Phys. Chem., $\underline{79}$, 1836, (1975).

47. Y. Ben Taarit, J. C. Vedrine, J. F. Dutel, C. Naccache.
    J. Magn. Resonance

48. P. H. Kasai and R. J. Bishop, Jr., J. Phys. Chem., 77, 2308, (1973).

49. C. Naccache and Y. Ben Taarit, J. Chem. Soc. Farad. Trans. I 69, 1475 (1973).

50. J. W. Jermyn, T. J. Johnson, E. F. Vansant and J. H. Lunsford, J. Phys. Chem., 77, 2964 (1977).

51. J. W. Ward, in "Zeolite Chemistry and Catalysis" (J. A. Rabo ed.) A.C.S. Monograph 171, p. 118 (1976).

52. P. Gelin, G. Coudurier, Y. Ben Taarit, C. Naccache, J. Catal., 70, 32, (1981).

53. E. Montovani, N. Palladino and A. Zanobi, J. Mol. Catal. 3, 285, (1977).

54. P. Gelin, Y. Ben Taarit, C. Naccache, J. Catal. 59, 357, (1979).

55. J. R. Anderson, Structure of Metallic catalysts, Academic Press London (1978).

56. T. A. Dorling, B. W. J. Lynch and R. L. Moss, J. Catal. 20, 190 (1971).

57. J. A. Rabo, V. Schomaker and P. E. Pickert, Proc. Int. Congress Catal. 3rd, 2, 1264 (1964).

58. R. A. Dalla Betta and M. Boudart, Proc. Int. Congress Catal., 5th North Holland Pub. 2, 1329 (1973).

59. P. Gallezot, J. Datka, J. Massardier, M. Primet and B. Imelik, Proc. Intern. Cong. Catal. 6th, 696 (1976).

60. C. Naccache, N. Kaufherr, M. Dufaux, J. Bandiera and B. Imelik, A.C.S. Symposium Series 40, Molecular Sieves-II (ed. Katzer, J.R.) 538 (1977).

61. P. Gallezot, Catal. Rev. Sci. Eng., 20, 121 (1979).

62. C. W. Kruse and R. F. Kleinschmidt, J. Am. Chem. Soc., 83, 213 (1961).

63. R. S. Neale, L. Elek and R. E. Malz Jr., J. Catal., 27, 432 (1972).

64. P. Pichat, J. C. Vedrine, P. Gallezot and B. Imelik, J. Catal., 32, 790 (1974).

65. Y. Okamoto, N. Ishida, T. Imanaka and S. Teranishi, J. Catal. 58, 82 (1979).

66. R. Cramer, J. Am. Chem. Soc. 87, 4717 (1965).

67. i) U.S. Patent 3, 444, 253 (1969) and 3, 497, 462 (1970) to Union Carbide Corp.

    ii) H. Reimlinger, U. Krüerke and E. Ruiter, Chem. Ber. 103, 2317 (1970).

68. P. Heimbach, P. W. Jolly and G. Wilke, advan. Organometal. Chem., 8, 29 (1970).

69. I.E. Maxwell, R. S. Downing and S. A. J. Van Laugen, J. Catal. 61, 485 (1980).

70. C. Naccache and Y. Ben Taarit, J. Catal. 41, 412 (1976).

71. H. Arai, T. Yamashiro, T. Kubo and H. Tominaga, Bull. Japan Petroleum Institute 18, 39 (1976).

72. S. Tsuruya, Y. Okamoto and T. Kuwada, J. Catal., 50, 52 (1979).

73. S. Tsuruya, H. Miyamoto, T. Sakae and M. Masai, J. Catal. 64, 260 (1980).

74. J. F. Roth, J. H. Craddock, A. Hershman, P. E. Paulik, Chem. Tech., 600 (1971).

75. B. K. Nefedov, N. S. Seergerva and L. L. Krasnova, Izv Akad. Nauk., SSSR, Khim, 614 (1977).

    B. Christensen and M. S. Scurrell, J. Chem. Soc. Farad. Trans., 74, 2313 (1978).

    N. Takakashi, Y. Orikasa and T. Yashima, J. Catal. 59, 61 (1979).

    P. Gelin, Y. Ben Taarit and C. Naccache, VII International Congress on Catalysis, Tokyo (1980).

ZEOLITE BIFUNCTIONAL CATALYSIS

M. Guisnet and G. Perot

Laboratoire Associé au CNRS - Catalyse Organique
Université de Poitiers, France

Bifunctional zeolite catalysts, generally metal-loaded acid zeolites are employed in numerous processes in petroleum refining and in petrochemical industries (1,2) : hydrocracking, selectoforming, dewaxing, hydroisomerization of C_5-C_6 alkanes, hydroisomerization of C_8 aromatics... The catalysts used in these processes present two types of sites : metallic sites whose main function is to hydrogenate and to dehydrogenate and acid sites whose main function is to crack or to isomerize.

Because of the progressive evolution in the supply towards heavier crudes and in the demand towards light products, hydrocracking, the most important process involving bifunctional catalysis, is destined to have an important development. Compared to catalytic cracking, it has the advantage of being more flexible indeed improvements in processes and in catalysts have made it possible to convert a large variety of feedstocks (from naphtha to heavy gasoils) and to obtain a large variety of high quality products. Hydrocracking requires catalysts with a high acidity counterbalanced by a high hydrogenation activity, which implies the use of bifunctional zeolite catalysts. The cracking function is provided by a Y zeolite in which the sodium ions have been replaced by hydrogen, by rare earth or by divalent cations and the hydrogenating function is provided by noble or non-noble metals (1,3). In marked contrast to amorphous silica-alumina catalysts, zeolite catalysts can operate in the presence of substantial amounts of ammonia and other basic nitrogen compounds. This greater ability of the zeolites to tolerate basic compounds can be attributed to their greater number of acid sites. Moreover this greater acidity enhances the resistance of the hydrogenating function to poisoning by sulfur

compounds and hydrocracking catalysts show a high stability with
regards to sulfur poisoning.

Selectoforming and Mobil Dewaxing (MDDW (4)) processes take
advantage of the shape-selective properties of zeolites. Selectoforming allows the selective hydrocracking of n-paraffins in a
gasoline reformate. In this process, a bifunctional catalyst with
a non-noble metal as hydrogenating component and a small pore
zeolite (T zeolite) as a cracking component is employed. The MDDW
process uses a Pd or Ni exchanged ZSM-5 zeolite (an intermediate
pore-size zeolite) which cracks preferentially the n-paraffins
particularly those with higher boiling temperatures.

Hydroisomerization of light naphtha (C_5-C_6 alkanes) is carried
out in the Hysomer process (Shell) on a noble metal highly dispersed
on a large pore zeolite with a high acidity. Bifunctional catalysts
containing mordenite or ZSM-5 zeolite can also be used for the
isomerization of C_8 aromatic cuts.

Numerous parameters govern the activity and the selectivity
of bifunctional zeolite catalysts. This is namely the case for the
characteristics of the hydrogenating and acid functions as well as
for their "balance". To show the influence of these parameters,
examples will be selected from the most typical reactions of the
industrial processes that is the isomerization and the cracking of
n-alkanes. Naturally, the characteristics of bifunctional zeolite
catalysts depend largely on the preparation procedure and particularly on the way the hydrogenating component is introduced.
However the preparation of metal-loaded catalysts,, described
elsewhere (5,6) will not be discussed here.

1. MECHANISMS OF ALKANE TRANSFORMATION ON BIFUNCTIONAL CATALYSTS

1.1 Bifunctional Catalysis

1.1.1 Generalities. In bifunctional catalysis, reactions occur in
successive steps involving two different types of sites (7). As an
example, the conventional bifunctional process of isomerization of
n-hexane into methylpentanes found on platinum-silica alumina is
shown in Fig. 1. This catalyst presents two types of sites :
　　i) platinum sites whose function is to dehydrogenate n-hexane
into n-hexenes (reaction 1) and to hydrogenate methylpentenes into
methylpentanes (reaction 5) ;
　　ii) acid sites whose function is to isomerize hexenes into
methylpentenes (reaction 3).
　　Beside these chemical steps, a bifunctional process requires
diffusion steps of the intermediate species. In this case, olefin
intermediates diffuse from the metallic to the acid sites (step 2)
and from the acid to the metallic sites (step 4).

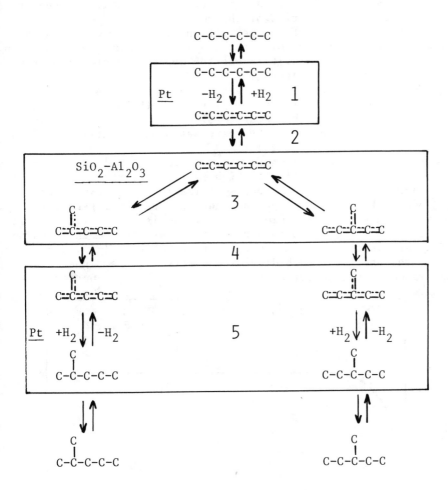

Fig. 1. Bifunctional process of n-hexane isomerization on
platinum-silica alumina.

The existence of this bifunctional process is now well
established :
i) although highly unfavoured thermodynamically under the
usual operating conditions, the intermediate olefins were detected
by GLC or by mass spectrometry (8,10). Moreover, the skeletal isomerization of olefins (reaction 3 in Fig. 1.) is known to occur very
readily on acid catalysts (7,8,11) ;
ii) the participation in the reaction of both acid and hydrogenating centers was clearly demonstrated by using physical mixtures

of an acid catalyst and of a metal deposited on an inert carrier
(7,12-14) : the activities of the mixtures were definitely greater
than the sum of the activities of the components ;
    iii) the change in the isomerization activities of bifunc-
tional catalysts (differing by their platinum content) as a function
of their hydrogenation activities was the change expected from the
multistep bifunctional process (15) :
    - for low hydrogenation activities the limiting step is the
n-hexane dehydrogenation (reaction 1) or the methylpentene hydrogenation (reaction 5) on the metallic sites ; under these conditions, the isomerization activity increases proportionally to the
hydrogenation activity (Fig. 2.) ;
    - for high hydrogenation activities, reactions 1 and 5 become
very fast compared to reaction 3 which is then the rate-limiting
step ; under these conditions the isomerization activity of the
bifunctional catalysts no longer depends on their hydrogenation
activity but only on their acidity. Thus, the isomerization activity
of bifunctional catalysts with a given acid carrier remains constant
beyond a certain value of the hydrogenation activity (Fig. 2.).
Moreover, the greater the acidity of the carrier, the higher the
maximum value of the isomerization activity of the bifunctional
catalyst.

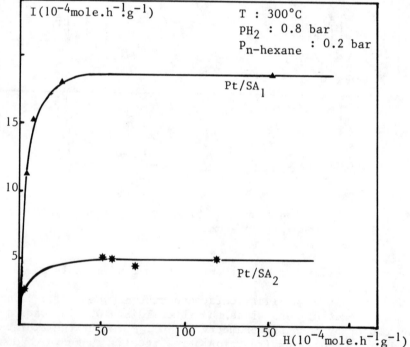

Fig. 2.   n-Hexane isomerization on platinum-silica alumina (Pt/SA).
          Influence of the hydrogenating activity (H) and of the
          acidity (SA_1 is more acid than SA_2) of the catalysts on
          their isomerization activities (I) (from reference 12).

1.2 Mechanisms of the reactions occurring on bifunctional catalysts.

Isomerization and cracking of alkanes can take place by the bifunctional process but can also be catalyzed independently by the acid or by the metallic sites.

1.2.1 Acid - catalyzed reactions. The mechanism of alkane isomerization and cracking on acid sites is shown in Fig. 3. Carbocations, involved as intermediates, are formed either by alkane adsorption on Brönsted or Lewis acid sites :

$$RH + H^+ \rightleftharpoons R^+ + H_2$$

$$RH + L \rightleftharpoons R^+ + LH$$

or by hydride transfer from the alkane to a preadsorbed carbocation :

$$RH + R^{+\prime} \rightleftharpoons R^+ + R'H$$

1.2.2 Metal - catalyzed reactions. Two mechanisms have been proposed to account for alkane isomerization on metals, particularly on platinum : the bond-shift mechanism attributed to the formation of α α γ-triadsorbed intermediates bonded to two adjacent metal atoms (16,17) and the cyclic mechanism which involves the formation of cyclopentanic intermediates and their scission (18,19). On large platinum crystallites, isomerization occurs mainly by the bond-shift mechanism whereas on small platinum crystallites, the cyclic mechanism is privileged (20).

The alkane cracking on metal sites (hydrogenolysis) involves the scission of highly dehydrogenated species adsorbed by two adjacent carbon atoms on adjacent metal sites (21). Contrary to the cracking reactions by acid or bifunctional catalysis, hydrogenolysis produces methane and ethane.

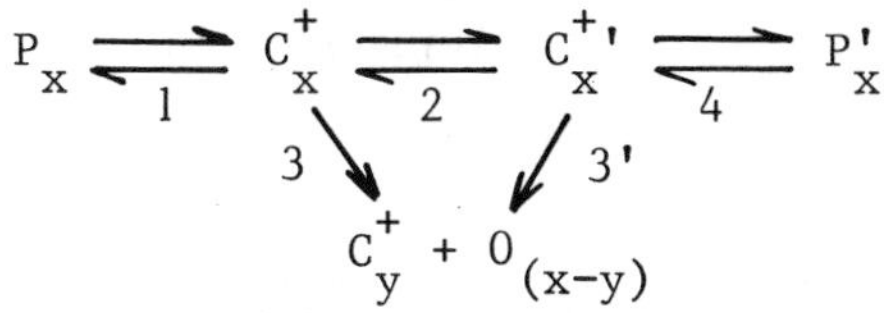

$$P_x \underset{1}{\rightleftharpoons} C_x^+ \underset{2}{\rightleftharpoons} C_x^{+\prime} \underset{4}{\rightleftharpoons} P_x'$$

$$3 \searrow \qquad \swarrow 3'$$

$$C_y^+ + O_{(x-y)}$$

Fig. 3. Isomerization and cracking of paraffins on acid catalysts. P : paraffin ; O : olefin ; C^+ : carbocation ; x,y : number of carbon atoms.

402

1.2.3 Reactions involved in the bifunctional process. This bifunctional process of alkane isomerization has already been developed in the n-hexane example (Fig. 1.). The cracking reactions occur by scission of olefin intermediates. The Horiuti Polanyi mechanism is commonly accepted. to explain olefin hydrogenation (reaction 5) or alkane dehydrogenation (reaction 1). The cracking and the isomerization of intermediate olefins on acid sites involve carbenium ions formed by olefin adsorption on protonic sites (Fig. 4.). The rearrangement (step 2) and the cracking (steps 3, 3') of carbocations are the limiting steps of isomerization and cracking respectively, since the carbocation formation and desorption are very rapid.

It is widely admitted that skeletal rearrangements of carbocations without change in the chain length (termed type A (22)) proceed via alkyl-shift :

whereas rearrangements with change in the chain length (type B) proceed via protonated cyclopropane intermediates (23,24) :

This process avoids the formation of the very unstable primary carbocations which would be involved in an alkyl-shift branching mechanism :

For the sake of conciseness, the protonated cyclopropanes will be pictured as face-protonated cyclopropanes (a). Such an entity seems however to have little physical reality compared to alkyl-bridged (b) or edge-protonated cyclopropanes (c) (25) :

a   b   c

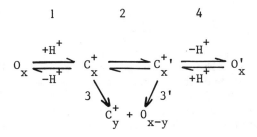

Fig. 4. Isomerization and cracking of olefins on acid catalysts.
O : olefin ; C^+ : carbocation ; x,y : number of carbon
atoms.

The β-scission of a carbocation leads to an olefin and a carbocation :

$$C\text{-}C\text{-}C\text{-}C\text{-}C \longrightarrow C\text{-}C^+ + C\text{=}C\text{-}C$$

A primary carbocation is obtained by β-scission of a linear carbocation :

$$C\text{-}C\text{-}C\text{-}C\text{-}C\text{-}C\text{-}C \longrightarrow C\text{-}C\text{=}C + \overset{+}{C}\text{-}C\text{-}C\text{-}C$$

Consequently, this type of scission (step 3, Fig. 4. if O_x is a
linear olefin) will be very slow ; a linear carbocation will only
be isomerized (type B rearrangement) and the cracking will occur
by scission of carbocations with mono and multibranched skeletons
(step 3'), leading to secondary or tertiary carbocations :

$$C\text{-}C\text{-}C\text{-}C\text{-}C\text{-}C \longrightarrow C\text{-}\overset{+}{C}\text{-}C + C\text{=}C\text{-}C\text{-}C$$

$$C\text{-}C\text{-}C\text{-}C\text{-}C \longrightarrow C\text{-}\overset{+}{C}\text{-}C + C\text{=}C\text{-}C$$

This is the reason why the cracking of n-alkanes follows their
isomerization.

2. ACTIVITY OF BIFUNCTIONAL ZEOLITE CATALYSTS

2.1 Influence of the hydrogenating activity

On pure acid zeolites, pentanes and hexanes isomerize and crack whereas long-chain alkanes undergo only cracking. The deactivation is rapid at low hydrogen pressure but becomes very slow at high pressure (26). The isomerization to cracking rate ratio depends not only on the zeolite but also on operating conditions (27) : thus, for n-hexane transformation on H mordenite at atmospheric pressure, this ratio is practically nul at 400°C but is equal to 0.2 at 250°C ; it increases with catalyst deactivation by coke deposit (27,28) and with hydrogen pressure (26). However the initial isomerization rate decreases when hydrogen pressure increases (29).

When a hydrogenating component is added to the zeolite, its activity, its stability and its isomerization selectivity generally increase. Fig. 5. shows, for n-hexane isomerization and cracking the changes of the activities of Pt-HY catalysts (measured after an aging period) as a function of their metallic surface areas (30). These catalysts, with a platinum content ranging from 0 to 17.7 wt %, were prepared by exchange of a stabilized Y zeolite : the platinum crystallites were between 2 to 5 nm depending on the samples. The isomerization activity increases very rapidly at low metallic surface areas, then remains practically constant above 0.5 m^2g^{-1} (Fig. 5.). This type of curve, expected in the case of a bifunctional process (see Fig. 2.), was found for the same reaction with other catalysts :Pt-LaY (31), Pt-H mordenite (32). It is also the case for the isomerization and the cracking of long-chain n-alkanes (up to nC_{16}) with physical mixtures of mordenite and platinum deposited on inert alumina, provided that the activities after aging be taken into consideration (13). Both isomerization and cracking activities per gram of mordenite first increase proportionally to the hydrogenation activity and then reach a plateau (see as an example in Fig. 6. for n-octane transformation) ; for all the alkanes, a_{HI}, the value of the hydrogenation activity required to obtain the plateau for isomerization is greater than a_{HC} which is the value required to obtain the plateau for cracking (Fig. 6.). If one considers the activities before deactivation, the plateau is reached for cracking but not for isomerization (27). The same observation can be made for n-heptane transformations on mixtures with Y zeolite. On platinum-alumina, ZSM-5 mixtures, the plateau is not obtained either for isomerization nor for cracking of n-heptane. This means that the hydrogenation activity is not sufficient to "balance" the high initial acidity of these zeolites and particularly that of ZSM-5 zeolite. The same conclusion was reached by Jacobs et al. for n-decane transformation on 0.5 wt % Pt-ZSM-5 (33). Physical mixtures of mordenite and NiMo sulfides deposited on

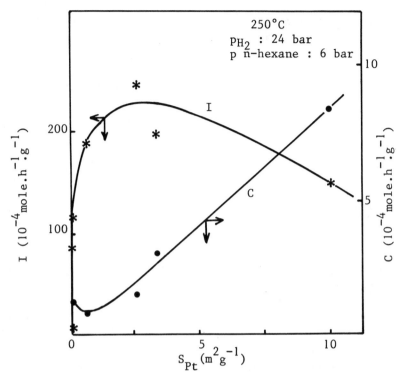

Fig. 5. Rates of n-hexane isomerization (I) and cracking (C) against the metal surface area of the samples (S_{Pt}) (from reference 30).

alumina were also used. In n-decane transformation at 400°C under 30 bar hydrogen pressure, a synergistic effect was observed, which accounts for the bifunctional catalysis. However the poor hydrogenation activity of NiMo sulfides was not sufficient to "balance" the high acidity of the mordenite and thus the isomerization and cracking activities per gram of mordenite increased with the NiMo sulfide content without reaching a plateau (34). It can be noted that on all these mixtures of a zeolite and a hydrogenating component, the bifunctional character of isomerization and cracking is clearly shown by the fact that their activities (considered before or after deactivation) are much greater than the sum of the activities of both components.

For the isomerization of C_5-C_6 alkanes on noble metal-loaded acid zeolites, various authors contest the existence of a bifunctional process. They suggest that the increase in isomerization activity caused by the introduction of a hydrogenating component in the mordenite would only be due to the increase in the

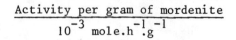

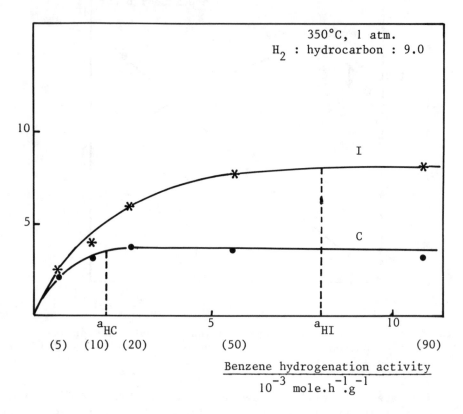

Fig. 6. n-Octane transformation. Effect of platinum-alumina
content (numbers in parentheses) or hydrogenation activity
on the hydroisomerization (I) and hydrocracking (C) activities of physically mixed catalysts (from reference 13).

catalyst's stability ; the only role of the hydrogenating component
would be to keep the surface of the mordenite free of coke deposit.
Actually, when working under conditions where the pure mordenite
does not deactivate, i.e. under high hydrogen pressure, the addition
of the hydrogenating component has only a slight positive effect on
the isomerization activity. In certain cases, even a decrease in
this activity has been observed (35). The positive effect is also
very weak at low hydrogen pressure, when the initial activities are
taken into consideration : Chicks et al. (28) found that the
increase in isomerization activity occurs essentially at the expense

of the cracking activity ; they explain the action of the hydrogenating component by a modification in the acid properties of the
mordenite. For all the above authors, isomerization on bifunctional
mordenite catalysts therefore occurs by the acid mechanism described
in Fig. 3. However, a combination of acid and bifunctional mechanism
has also been proposed (26).

Again, if the bifunctional mechanism allows one to explain the
cracking of long-chain alkanes (with more than 6 carbon atoms), it
is not the case for the cracking of light alkanes (32,36). The
cracking steps of light carbenium ions involving unstable carbocations are very slow. Thus the β-cracking of a secondary carbocation
nC_6^+ always leads to a primary carbocation :

$$C-\overset{+}{C}-C-C-C-C \longrightarrow C-C=C \;+\; \overset{+}{C}-C-C$$

In the same way, the cracking of the carbocations iC_6^+ with a
monobranched or a bibranched skeleton involves, in the most favorable case, two secondary carbocations :

$$C-C-\overset{+}{C}-\overset{+}{C}-C \longrightarrow C-\overset{+}{C}-C \;+\; C=C-C$$
$$\underset{C}{|}$$

and could be slow compared to the isomer desorption and to the
alkane scission on metallic sites (hydrogenolysis). It is effectively what is observed in n-hexane cracking on the series of Pt-HY
and Pt-H mordenite catalysts (32) : the distribution of the cracking
products is the one expected from a simple type hydrogenolysis
reaction, namely an important formation of methane and ethane as
well as a $(C_1 + C_2)/(C_4 + C_5)$ molar ratio of about 1. Moreover,
the greater the metal area the greater the activity (Fig. 5.).

2.2   Influence of the zeolite characteristics

As expected from a bifunctional process (as well as from an
acid-catalyzed reaction), the isomerization and cracking rates
depend strongly on the zeolite acidity. Thus, the smaller the
sodium content of the Y zeolite, the greater the activity for
n-pentane isomerization of Pd-HY catalysts will be ; the effect is
especially pronounced at low sodium contents. Thus a decrease from
0.27 to 0.02 wt % Na_2O enables a reduction of 50°C in reaction temperature for a 30 % n-pentane conversion (26).

The participation of Brönsted acid centers in n-decane hydrocracking is demonstrated by the positive effect of water on the
activity of a Pd-Re X catalyst. Indeed, this positive effect can
be attributed to the generation of protonic sites via rare earth
cation hydrolysis. On the other hand, with platinum on a non-
stabilized HY zeolite a negative effect due to a decrease of the

acid strength by hydrating protons is observed (37).

The effect of the SiO_2/Al_2O_3 ratio has also been determined in n-pentane isomerization on bifunctional mordenite catalysts : a maximum activity was observed for a ratio of about 16 (26) ; this optimum value would be the result of two antagonistic effects of the dealumination : one positive, due to the elimination of obstructions in the mordenite channels, the other negative, due to the decrease in the number of acid centers (1).

Because the isomerization activity depends so highly on Na removal, comparing zeolites with different structures is difficult. However, the comparison between low sodium Pd-H mordenite and low sodium Pd-HY shows that both materials have about the same activity for n-hexane isomerization (26).

2.3 Influence of poisons

The effect of sulfur poisoning was studied on a Pd-Ca Y catalyst by Rabo et al. (38) using an n-pentane feed in which sulfur concentrations (as n-butyl mercaptan) were varied from zero to 3000 ppm. Sulfur concentrations of 6 ppm had no effect on catalyst activity ; this observation can be readily explained by the bifunctional mechanism if the hydrogenating activity of the catalyst remains high enough to maintain a pseudo-equilibrium between paraffins and olefins and to keep as limiting step the skeletal olefin isomerization on acid sites. Above this concentration, sulfur acted as a temporary poison : it decreased the isomerization activity but this activity was restored on reversion to a clean feed.

Modifications in activity and in selectivity caused by poisoning with dimethyl disulfide (220 ppm) and n-butylamine (800 ppm) of a Pt-HY zeolite with 6 wt % of platinum were also those expected from the bifunctional mechanism (30). Since sulfur poisons reduce the active platinum area, the poisoned catalyst will act like a catalyst with a smaller platinum content. This was actually observed : the poisoned catalyst had the isomerization selectivity and the activity of a catalyst with a 0.09 to 0.5 wt % platinum content. A basic poison reduces the number of the acid sites ; since with the catalyst used, the limiting step was the skeletal olefin isomerization on acid sites, poisoning by n-butylamine must provoke a decrease in activity with no modification in selectivity. This is effectively what was observed.

The addition of dimethyldisulfide (1 %) to n-decane caused a significant decrease in both the isomerization (I) and cracking (C) activities of a mixture of H mordenite and platinum-alumina. As expected from the well known I/C decrease with decreasing hydrogenation activity, I was much more affected than C. By reversion to pure n-decane, the catalyst recovered half of its original isomerization activity but only 10 % of its original cracking activity and I/C was 4.5 times greater than on the fresh catalyst. Since a poisoning of the metal should result in a lowering of I/C, it would

appear that the irreversible poisoning effect is due to a modification of the mordenite rather than to a poisoning of the metal (39).

Catalyst deactivation and coke formation were examined on
mixtures of platinum-alumina and zeolite. Nearly all the coke was
found to be deposited on the zeolite (14) ; therefore the deactivation can be attributed to a decrease in the acidity of the bifunctional catalyst. The decrease in cracking activity as a function
of time on stream was always greater than the decrease in isomerization activity. To explain this phenomenon one could propose that
isomerization and cracking occur on different catalytic centers,
the cracking sites being more readily poisoned by coke than the
isomerization sites (14). However, this phenomenon can also be
explained by the fact that cracking is consecutive to isomerization.
Indeed the comparison of deactivated catalysts with fresh catalysts
shows that poisoning by coke results in a decrease in zeolite
concentration without any change in the isomerization to cracking
ratio. The fact that poisoning by coke does not cause any segrega-
tion among the catalytic sites is in favor of a single type of sites
for isomerization and cracking (40).
Moreover, it was found that the amount of alkane transformed
into coke during one experiment was proportional to the total
amount of cracked alkane whereas no clear relationship existed
between coking and isomerization activities. This indicates that
coke formation probably results from a secondary transformation of
olefins produced by the cracking reaction (40,41).

3. SELECTIVITY OF BIFUNCTIONAL ZEOLITE CATALYSTS

3.1 n-Hexane isomerization

3.1.1 Influence of hydrogenating activity. The selectivity of
n-hexane transformation on pure zeolites has been described by
numerous authors. The kinetic model represented in Fig. 7. allows
us to account for the n-hexane isomerization on H mordenite or on
HY zeolite (32,35) : n-hexane (nC_6) leads directly to a thermodynamic equilibrium mixture of methylpentanes (MP) and 2,3-dimethyl-

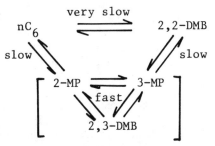

Fig. 7.   Kinetic model of n-hexane isomerization an acid zeolites.

butane (2,3-DMB) and to a small quantity of 2,2-dimethylbutane
(2,2-DMB). This is well explained by the mechanism in Fig. 3. in
which the limiting step is step 1 that is the formation of secondary nC_6^+ carbocations. This step is slower than the carbocation
isomerization (step 2) and slower than the desorption of MP or
2,3-DMB (step 4) which involves tertiary carbocations. The 2,2-DMB
formation is slow because the carbocations corresponding to this
alkane are either primary or secondary.

2,3-DMB formation rate decreases when the metallic surface area
of Pt-HY or Pt-H mordenite catalysts increases (30,32) : thus, the
2,3-DMB content in the mixture of MP and 2,3-DMB which is close to
its thermodynamic value on pure mordenite, decreases down to 40 %
of this value for the high platinum content mordenite catalysts.
The decrease is more pronounced in the case of Pt-HY catalysts :
the 2,3-DMB content drops to 10 % of its thermodynamic value when
the platinum content is equal to 0.5 %. This selectivity is between
the selectivity expected of an acid mechanism and that of a bifunctional process in which the limiting step is the isomerization of
the intermediate olefins (step 3 in Fig. 1.). Indeed, with this
bifunctional process, the methylpentanes must be the only primary
products of n-hexane isomerization : the methylpentenes formed by
n-hexene rearrangements on the acid sites are immediately hydrogenated and cannot isomerize into dimethylbutenes.

To explain the direct formation of 2,3-DMB, at least two
proposals can be made (32) :
  - an acid and a bifunctional mechanism would both participate
simultaneously in the isomerization. On the portion of the catalyst carrying platinum crystallites, nC_6 would isomerize into MP
through the bifunctional mechanism. MP would rapidly isomerize
into 2,3-DMB on acid sites distant from platinum crystallites.
This would imply that the part of catalyst with platinum and that
without platinum work independently.
  - a more probable explanation is that the migration of the
intermediate olefins from one metal site to another metal site is
slower than their isomerization on acid sites. In this case,
n-hexenes have the possibility of reacting successively on several
acid sites before being hydrogenated and therefore 2,3-DMB will
appear as a primary product of nC_6 isomerization. The largest direct
2,3-DMB formation on Pt-H mordenite would be due i) to the greater
reactivity of olefins on the stronger acid sites of mordenite
ii) and/or to the diffusional limitations in the one-dimensional
porous structure of mordenite which are greater than in the three-
dimensional structure of Y zeolite (32).

3.1.2 Influence of the zeolite porous structure. On ZSM-5 bifunctional catalysts, the direct formation of 2,3-DMB is much less
pronounced than on Y bifunctional catalysts and the formation of
2,2-DMB does not occur (27). This can be explained by limitations
to the diffusion of the olefins with a 2,3- or especially with a
2,2-DMB skeleton. Moreover, the formation of 2-MP is more favoured

with ZSM-5 bifunctional catalysts than with Y catalysts. A possible explanation for this is that the environment of the ZSM-5 acid sites is such that it privileges scission a of the protonated cyclopropane intermediate in comparison to scission b :

$$
\begin{array}{l}
\underset{\text{C-C---C-C-C}}{\overset{\displaystyle b \; \overset{\text{C}}{\triangle} \; a}{}} \quad \overset{a}{\Longrightarrow} \quad \overset{\text{C}}{\underset{}{}} \;\; C\text{-}C\text{-}\overset{+}{C}\text{-}C\text{-}C \\[4mm]
\qquad\qquad\qquad \overset{b}{\searrow} \quad \overset{\text{C}}{\underset{}{}} \;\; C\text{-}\overset{+}{C}\text{-}C\text{-}C\text{-}C
\end{array}
$$

On bifunctional catalysts with a small pore size zeolite as acid component (erionite...), n-alkanes only undergo cracking ; the formation of branched alkanes does not occur. Indeed, only linear compounds can reach the acid sites and be desorbed from the porous structure. This characteristic is used to eliminate linear alkanes (which have a low octane number) from naphthas without transforming the other components, especially the branched alkanes (selectoforming).

3.1.3 Influence of reaction temperature. On a 0.6 wt % Pt-H mordenite it has been shown that temperature is a determining factor in the isomerization selectivity (42) :
    - at 250°C, 2,3-DMB content in the mixture of MP and 2,3-DMB formed by n-hexane isomerization is very close to its thermodynamic value as was the case with pure zeolites.
    - at 400°C, 2,3-DMB formation is very slow and the selectivity is practically that expected from the conventional bifunctional mechanism.

In conclusion, the conventional bifunctional mechanism or the acid mechanism or their combination allow one to explain the selectivity of n-hexane isomerization on bifunctional zeolite catalysts. However, it can be noted that another mechanism, invoking cyclohexane-type bimolecular intermediates, was proposed by Bolton and Lanewala to explain the particular selectivity of a Pt-Re NH$_4$ Y catalyst (43).

3.2 Long-chain alkane isomerization and cracking

3.2.1 "Ideal bifunctional catalysis. A detailed analysis of the hydroisomerization and hydrocracking of long-chain alkanes has been reported by Weitkamp (36,44-49). The reactions were carried out on a 0.5 wt % Pt-Ca Y catalyst at temperatures ranging from 200 to 300°C, with a total pressure of about 40 bar and an H$_2$/alkane ratio of about 20. In these conditions, isomerization and cracking occur by the conventional bifunctional process in which the limiting steps are the rearrangement or the cracking of carbenium ion intermediates ; according to Weitkamp's terminology the bifunctional catalysis is then "ideal" (44).

Various conversion rates were obtained by varying the contact time or more generally the reaction temperature. As an example, the change of the isomerization and cracking conversion of n-tridecane versus the reaction temperature is shown in Fig. 8. With all alkanes the hydroisomerization is the only reaction observed at low conversion (up to 40 % for n-tridecane, Fig. 8.). The hydroisomerization conversion passes through a maximum which is due to the consumption of branched isomers by hydrocracking.

The isomer distribution is markedly dependent on the conversion rate. The branching occurs by consecutive reactions (49) : at low conversion monobranched alkanes (methyl and also ethyl, propyl...) are formed almost exclusively (at least 95 %) ; by increasing the conversion rate, the content of bibranched alkanes increases. The distribution of monobranched isomers was examined in detail (49). At low conversion the 2-methyl isomers were generally formed at a lower rate than the 3-methyl isomers. The branching mechanism via

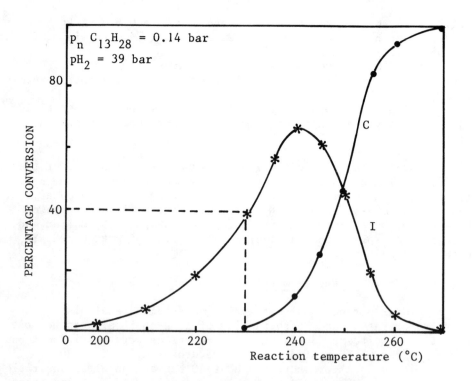

Fig. 8. Hydroisomerization (I) and hydrocracking (C) of n-tridecane on Pt-Ca Y zeolite (from reference 49).

protonated cyclopropanes (type A rearrangement) can explain this observation. However, it does not explain the formation, observed even at low conversion rate, of ethyl, propyl... monobranched isomers. To account for their formation, type A rearrangements of methyl branched carbocations, which are known to occur faster than type B rearrangements (24,47) were proposed (49). Thus the formation of ethylpentane from n-heptane should involve the following carbocation rearrangements :

$$CH_3-\overset{+}{C}HCH_2-CH_2CH_2CH_2CH_3 \xrightarrow[slow]{\triangle \overset{+}{H}} CH_3-\overset{+}{C}H-\overset{CH_3}{\underset{|}{C}H}-CH_2-CH_2-CH_3$$

$$\xrightarrow[very\ fast]{2 \curvearrowright H} CH_3-CH_2-\overset{CH_3}{\underset{|}{C}}\overset{+}{C}H-CH_2-CH_3 \longrightarrow CH_3-CH_2-\overset{\overset{CH_3}{/}}{\underset{|}{C}H+}-CH_2-CH_3$$

Even at high conversion rates (up to 70-90 %) the cracking product distribution was fully symmetrical and the sum of fragments amounted to 200 moles per 100 moles of alkane cracked, indicating pure primary cracking. Methane and ethane were not formed, which rules out hydrogenolysis on platinum. There was a significant amount of branched alkanes (essentially monobranched) in the cracking product (36). The iso to normal alkane ratio was generally higher than its thermodynamic value which demonstrates that the branched alkanes are primary products of the cracking reaction and do not result from a secondary isomerization of normal alkanes (50). The cracking product distribution is actually the one expected from the cracking of monomethylbranched tertiary carbocations provided that one assumes that these cations are equally reactive and that their relative concentrations are represented by the relative concentrations of methyl isomers (36). However this cracking reaction requires the highly endothermic (55-60 kcal/mole) β-scission of a tertiary cation to form a primary cation (22).

$$ex.\ : \quad C-\overset{+}{C}-C-C-C-R \longrightarrow C-\overset{C}{\underset{|}{C}}=C \ + \ \overset{+}{C}-C-R$$

Such a reaction is most unlikely. Yet, β-scission could be concerted with hydride shift so that a secondary carbocation would be produced directly (22) :

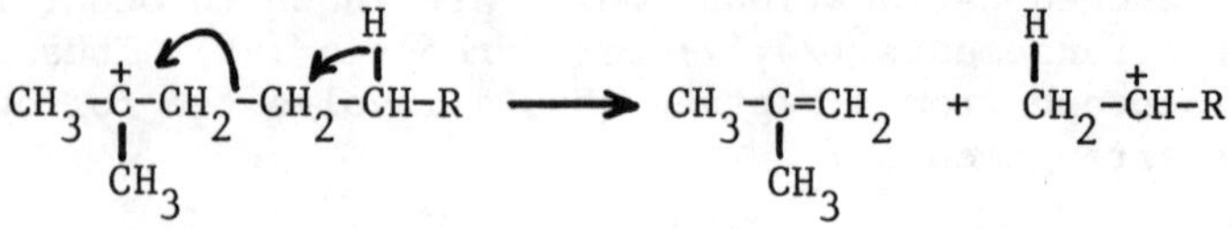

$$CH_3-\overset{+}{\underset{\underset{CH_3}{|}}{C}}-CH_2-CH_2-\overset{\overset{H}{|}}{CH}-R \longrightarrow CH_3-\underset{\underset{CH_3}{|}}{C}=CH_2 \quad + \quad \overset{\overset{H}{|}}{CH_2}-\overset{+}{CH}-R$$

In any case, this mechanism cannot explain the high iso to normal ratio in the cracked products, particularly in butanes (50), and the absence of cracking up to a 40 % isomerization of methylnonanes (47). To explain this, cracking of carbocations with a bibranched skeleton was proposed (50,51). This route involving secondary and tertiary carbocations is energetically favorable :

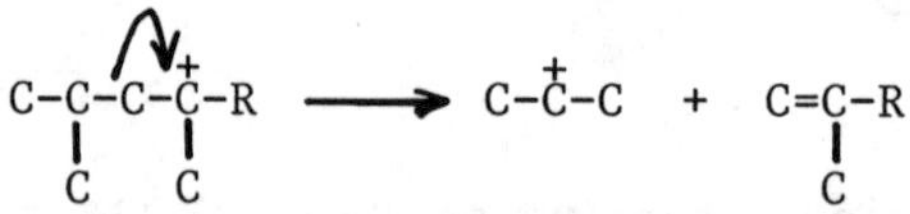

$$C-C-\overset{\overset{+}{\curvearrowleft}}{C}-\overset{+}{C}-R \longrightarrow C-\overset{+}{C}-C \quad + \quad C=\underset{\underset{C}{|}}{C}-R$$

$$\underset{C}{|} \quad \underset{C}{|}$$

In conclusion, it can be said that n-alkane transformation by "ideal" bifunctional catalysis occurs via the successive reactions shown in the rake-type scheme of Fig. 9. The transformations of adsorbed carbocations are the limiting steps, their desorption into alkanes via the olefins being very rapid. Consequently, monobranched isomers can be obtained with a high selectivity, practically in

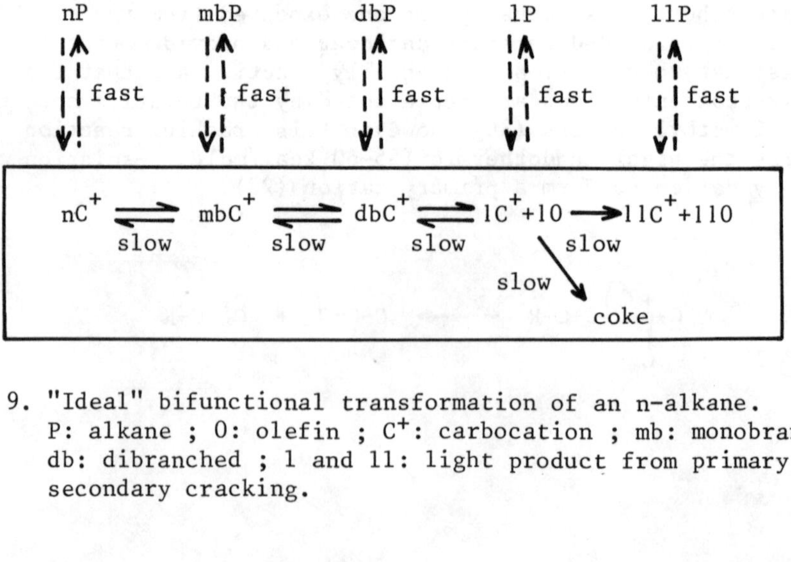

Fig. 9. "Ideal" bifunctional transformation of an n-alkane.
P: alkane ; O: olefin ; C+: carbocation ; mb: monobranched ;
db: dibranched ; 1 and 11: light product from primary and
secondary cracking.

thermodynamic equilibrium with the n-alkane ; with a longer contact
time, the thermodynamic mixture of mono and bibranched alkanes could
be selectively formed ; with an even longer contact time primary
cracking products would be obtained. Secondary cracking and coking
reactions will occur only for extremely long contact time. To
obtain "ideal" bifunctional catalysis, the bifunctional catalyst
must be such that all the steps involved in the alkane desorption
from carbocations be very rapid i.e.

    a) the desorption of olefins from carbocations,
    b) the diffusion of olefins from the acid to the metallic sites,
    c) the olefin hydrogenation on the metallic sites.

3.2.2 Deviations from the "ideal" bifunctional catalysis. Step *a*
is commonly considered to be more rapid than the carbocation
rearrangement or its cleavage, but steps *b* and *c* can become the
limiting steps of the alkane transformation. This is the case for
step *b* when the distance between hydrogenating sites, i.e. the
diffusional path of intermediate olefins, is too long and consequently comprises too many active acid sites ; therefore, olefins
with a monobranched skeleton formed on an active acid site, can
adsorb on other acid sites and successively lead to olefins with
a dibranched skeleton, cracking products and coke. Under these
conditions, dibranched alkane isomers and cracking products will be
primary products of n-alkane transformation. The same behaviour will
be observed if the olefin hydrogenation is the limiting step. This
occurs when the hydrogenating activity is too small in relation to
the activity of the acid sites. Then bifunctional catalysis will be
"ideal" only if the catalysts have many highly active hydrogenating
sites well dispersed among the acid sites. In the limit case the
diffusional path of the intermediate olefin (between two hydrogenating active sites) will comprise only one acid site of such
strength that it will allow only one olefin transformation during
one sojourn.

    As already noted, dibranched isomers are formed as primary
products of n-hexane isomerization on Pt-HY and above all on Pt-H
mordenite catalysts. Diffusional limitations in the olefin migration have been invoked to explain the non-"ideal" behaviour of
these catalysts. In the same way dibranched isomers but also
cracking products are found as primary products of the n-heptane
transformation on physical mixtures of platinum-alumina and Y or
ZSM-5 zeolites (40). The isomerization/cracking rate ratio (I/C)
increases and the percentage of bibranched alkanes decreases when
the platinum-alumina content increases. However, even with a
platinum-alumina content equal to 90 %, the distance between hydrogenating sites is not small enough and the hydrogenating activity
high enough to obtain a selective transformation of n-heptane into
its monobranched isomers. Diffusional limitations in the narrow
porous structure of the ZSM-5 zeolite account for the very low
value of I/C found with the platinum-alumina, ZSM-5 mixtures (40).

On 0.5 wt % Pt-ZSM-5 catalyst almost no hydroisomerization of n-decane is found (52) ; moreover a secondary cracking is observed at relatively low conversion rates ($\simeq$ 20 %). This can be due to an "imbalance" of the acid and the hydrogenating functions (the hydrogenating activity would be too small to counterbalance the high activity of the very strong ZSM-5 acid sites) as well as to configurational limitations. Another explanation would be that the strength of ZSM-5 acid sites would be such that several consecutive transformations could occur during the same sojourn of olefins on a zeolite site.

3.2.3 Influence of the zeolite porous structure. The isomerization and cracking selectivities of bifunctional catalysts with Y or with ZSM-5 zeolite were significantly different (40,52,53). Thus in n-heptane isomerization, the 2-methylhexane/3-methylhexane molar ratio is higher with ZSM-5 than with Y catalysts and 2,3- and 2,4-dimethylpentanes are formed in definitely smaller quantities ; the 2,2- and 3,3-dimethylpentanes were formed with Y but not with ZSM-5 catalysts. These selectivities are very similar to those observed in n-hexane isomerization and therefore can be explained in the same way. On all the catalysts, cracking gives mainly propane and butane in approximately equimolar amounts. Curiously, the iso/n-butane molar ratio is higher with ZSM-5 than with Y catalysts whereas it was much smaller in n-octane and n-decane cracking. Since the formation of bibranched heptanes is very unfavoured on ZSM-5 catalysts, the following cracking reaction is proposed to explain the high iso/n-butane ratio :

$$CH_3-\overset{+}{\underset{CH_3}{C}}-CH_2-CH_2-\overset{\overset{H}{|}}{CH}-CH_3 \longrightarrow CH_3-\underset{CH_3}{C}=CH_2 + CH_3-\overset{+}{CH}-CH_3$$

However the authors (40) do not exclude another possibility : the preferential cracking of carbenium ions with a 2,4- or especially with a 2,2-dimethylpentane skeleton, whose desorptions from the ZSM-5 porous structure are inhibited.

β-scission of tertiary monobranched decyl cations and even of linear secondary decyl cations were first proposed by Jacobs et al. (52) to explain the products of n-decane cracking on Pt-ZSM-5. However, in a more recent publication (53) another proposal was preferred : cracking would imply two reactions :

  i) bibranched carbenium ion cracking
  ii) monobranched secondary carbenium ion cracking producing secondary carbenium ions such as

$$C-\underset{C}{C}-C-\overset{+}{C}-R \longrightarrow C-\overset{+}{C}-C + C=C-R$$

The distribution of methylnonanes formed from n-decane isomerization on bifunctional zeolite catalysts depends very much on the porous structure of the zeolite (53). With Y zeolite, the distribution approaches thermodynamic equilibrium at high conversion whereas with ZSM-5 zeolite, 2-methylnonane is much favoured. Curiously, with ZSM-11, another pentasil zeolite, a different methylnonane distribution is obtained. These differences between ZSM-5 and ZSM-11 would be mainly the results of transition-state shape selectivity (53).

CONCLUSION

The activity, the stability and the selectivity of bifunctional zeolite catalysts are clearly governed by the characteristics of their acid and hydrogenating sites. As was shown for alkane isomerization and cracking, the highest activity would be obtained if all the acid sites were very active and working at their maximum. For this to occur, the acid sites must be sufficiently supplied with olefinic intermediates which requires numerous and well distributed active hydrogenating sites. In the "ideal" case, the diffusional path of olefins (between two hydrogenating sites) will contain only one active acid site. Here the catalyst will also be the most selective one in the series of consecutive reactions : thus, when transforming n-alkanes, it will give the best yield of monobranched isomers and the most selective isomerization. On such a catalyst, coke formation will be very slow and consequently the stability very great. Zeolites are perfectly adapted to the preparation of "ideal" bifunctional catalysts : their acid sites are numerous and highly active ; moreover, a high dispersion and a high activity of hydrogenating sites can be obtained when introducing the metal by an ion-exchange process.

ACKNOWLEDGMENTS

The authors thank the following publishers for having released their copyrights on the following figures. Academic Press (Fig. 5.), Heyden (Fig. 6.), American Chemical Society (Fig. 8.).

REFERENCES

1. Bolton A.P., Zeolite Chemistry and Catalysis, J.A. Rabo, Ed., ACS Monograph 171 (American Chemical Society, Washington, D.C., 1976) pp. 714-779.
2. Minachev Kh.M. and Isakov Ya.I., Zeolite Chemistry and Catalysis, J.A. Rabo, Ed., ACS Monograph 171 (American Chemical Society, Washington, D.C., 1976) pp. 552-611.

418

3. Marcilly C., Franck J.P. in Catalysis by Zeolites, B. Imelik et al., Eds., Studies in Surface Science and Catalysis 5, (Elsevier Scientific Publishing Company, Amsterdam, Oxford, New York, 1980) pp. 93-104.
4. Gorring R.L., Mobil Oil Corp., U.S. Patent 3, 894, 938, 1975. Gorring R.L. and Smith R.L., Mobil Oil Corporation, U.S. Patent, 4, 153, 540, 1979.
5. Uytterhoeven J.B., Proceedings of the Symposium on Zeolites, Szeged (1978) Acta Physica and Chemica 24 (1978), 53.
6. Gallezot P., Catalysis Reviews Science and Engineering, 20 (1979) 121.
7. Weisz P.B., Advances in Catalysis, 13 (1962) 137.
8. Weisz P.B., Swegler E.W., Science 126 (1957) 31.
9. Silvestri A.J., Naro P.A., Smith R.L., J. Catal. 14 (1969) 386.
10. Weitkamp J., Schulz H., J. Catal. 29 (1973) 361.
11. Mills G.A., Heinemann H., Milliken T.H. and Oblad A.G., Ind. Eng. Chemistry 45 (1953) 134.
12. Chevalier F., Thèse Poitiers, 1979.
13. Perot G., Montes A. and Guisnet M., Gueguen C. and Bousquet J., Proceedings of the 5th International Conference on Zeolites, L.V. Rees Ed. (Heyden, London, Philadelphia, Rheine, 1980) pp. 640-648.
14. Perot G., Montes A., Hilaireau P., Chevalier F. and Guisnet M., Catalyst Deactivation, B. Delmon and G.F. Froment Eds., Studies in Surface Science and Catalysis, 6 (Elsevier Scientific Publishing Company, Amsterdam, Oxford, New York, 1980) pp. 431-438.
15. Chevalier F., Guisnet M. and Maurel R., C.R. Acad. Sc. Paris 282 (1976) 3.
16. Anderson J.R. and Avery N.R., J. Catal. 7 (1967) 315.
17. Anderson J.R., Mac Donald R.J. and Shimoyama Y., J. Catal. 20 (1971) 147.
18. Barron Y., Maire G., Cornet D. and Gault F.G., J. Catal. 2 (1963) 152.
19. Corolleur C., Tomanova D. and Gault F.G., J. Catal. 24 (1972) 401.
20. Dartigues J.M., Chambellan A. and Gault F.G., J. Amer. Chem. Soc. 98 (1976) 856.
21. Sinfelt J.H., Advances in Catalysis 23 (1973) 91.
22. Poutsma M.L., Zeolite Chemistry and Catalysis, J.A. Rabo, Ed., ACS Monograph 171 (American Chemical Society, Washington, D.C., 1976) pp. 437-528.
23. Brouwer D.M., Rec. Trav. Chim. 87 (1968) 1435.
24. Chevalier F., Guisnet M. and Maurel R., Proceedings 6th Int. Congr. Catal. G.C. Bond, B.P. Wells F.C. Tompkins Eds. Vol. 1 (The Chem. Soc., London, 1977) pp. 478-487.
25. Olah G.A., J. Amer. Chem. Soc. 94 (1972) 808.
26. Kouwenhoven H.W., Molecular Sieves, W.M. Meier and J.B. Uytterhoeven, Eds., Adv. Chem. Ser. (American Chemical Society, Washington, 1973) pp. 529-539.

27. Hilaireau P., Perot G. and Guisnet M. to be published.
28. Chick D.J., Katzer J.R. and Gates B.C., Molecular Sieves II, J.R. Katzer Ed., ACS Symposium Series 40 (American Chemical Society, Washington, 1977) pp. 519-527.
29. Minachev Kh.M., Garanin V., Isakova T., Kharlamov V. and Bogomolov, Molecular Sieves, Zeolites II. Adv. Chem. Ser. 102 (American Chemical Society, Washington, 1971) pp. 441-450.
30. Ribeiro F., Marcilly C. and Guisnet M., J. Catal. 78 (1982) 267.
31. Lanewala M.A., Pickert P.E., Bolton A.P., J. Catal. 9 (1967) 95.
32. Ribeiro F., Marcilly C. and Guisnet M., J. Catal. 78 (1982) 275.
33. Jacobs P.A. and Uytterhoeven J.B., Steijns H. and Froment G., Weitkamp J., Proceedings of the 5th International Conference on Zeolites, L.V. Rees Ed. (Heyden, London, Philadelphia, Rheine, 1980) pp. 607-615.
34. Romero M., Thèse Poitiers 1981.
35. Beecher R. and Voorhies A., Jr, Ind. Eng. Chem., Product. Research and Development 8 (1969) 3607.
36. Weitkamp J., Amer. Chem. Soc., Symp. Ser. (J.W. Ward Ed.) 20 (1975) 489.
37. Yan T.Y., J. Catal. 25 (1972) 733.
38. Rabo J.A., Pickert P.E., Mays R.L., Ind. Eng. Chem. 53 (1961) 733.
39. Romero M., Perot G., Gueguen C. and Guisnet M., Applied Catalysis 1 (1981) 273.
40. Perot G., Hilaireau P. and Guisnet M., 6th International Conference on Zeolites, Reno, 1983 (Submitted).
41. Montes A., Perot G. and Guisnet M., React. Kinet. Catal. Lett. 13 (1980) 77.
42. Montes A., Perot G. and Guisnet M., to be published.
43. Bolton A.P., Lanewala M.A., J. Catal. 18 (1970) 1.
44. Pichler H., Schulz H., Reitemeyer H.O. and Weitkamp J., Erdöl, Kohle - Erdgas Petrochem. 25 (1972) 494.
45. Schulz H., Weitkamp J., Ind. Eng. Chem., Prod. Res. Dev. 11 (1972) 46.
46. Schulz H., Weitkamp J. and Erberth H., Proc. 5th Intern. Congr. Catal., J.W. Hightower, Ed., Vol 2 (Miami Beach, Amsterdam, 1973) p. 1229.
47. Weitkamp J. and Farag H., Proceedings of the Symposium on Zeolites, Szeged, 1978.
48. Weitkamp J., Erdöl, Kohle Erdgas Petrochem. 31 (1978) 13.
49. Weitkamp J., Ind. Eng. Chem., Prod. Res. Dev. 21 (1982) 550.
50. Steijns M. and Froment G., Jacobs P. and Uytterhoeven J., Weitkamp J. Ind. Eng. Chem., Prod. Res. Dev. 20 (1981) 657.
51. Steijns M. and Froment G.F., Ind. Eng. Chem., Prod. Res. Dev. 20 (1981) 660.

420

52. Jacobs P.A. and Uytterhoeven J.B, Steijns M. and Froment G.F.,
     Weitkamp J., Proceedings of the 5th International Conference
     on Zeolites, L.V. Rees Ed. (Heyden, London, Philadelphia,
     Rheine, 1980) pp. 607-615.
53. Jacobs P.A., Martens J.A., Weitkamp J. and Beyer H.K.,
     Faraday Discussions of the Chemical Society 72 (1981) 353.

PART IV

INDUSTRIAL APPLICATIONS

DESIGN ASPECTS OF CATALYTIC REACTORS AND ADSORBERS

A.Rodrigues,C.Costa,R.Ferreira,J.Loureiro and S.Azevedo

Department of Chemical Engineering
University of Porto
4099 Porto Codex, Portugal

1. INTRODUCTION

This lecture is intended to provide the fundamentals for the design of catalytic reactors and adsorbers,that is,the equipment necessary for the "mise-en-oeuvre" of zeolites as catalysts and adsorbents.

Industrial applications of zeolites have been listed by several authors (Breck |1|,Lee |2|,Heineman |3|,Meuqel |4|,etc.); Tables I and II summarize catalytic and adsorption processes,respectively, which use zeolites.

A reactor (or adsorber) is an ensemble of particles (catalyst or adsorbent); it is then worthwhile to look first at the structure of the particle and the mechanisms of mass (heat) transfer at the particle level.

A zeolite pellet can be considered |8| as having two pore systems: micropore or adsorbing pores of zeolite crystals and transport pores,i.e.,interstices between contacting crystals. This biporous structure is sketched in Fiqure 1 and is similar,for mathematical purposes,to other bidisperse materials,such as macro-reticular resins,although in zeolites micropores have a very uniform size in the range of 3 to 10 angstroms.

For pellets of diameter $d_p=0.16$ cm in packed beds with porosity 0.32 some typical values for design are:

intracrystal voids   0.191
intercrystal voids   0.227

424

TABLE I - Some catalytic processes using zeolites |3,4|

- Cracking (10-40% rare earth exchanged H-Y Zeolite dispersed
  in a matrix of silica alumina or clay)
- Hydrocracking (large pore Y zeolites with 0.5% Pt)
- Isomerization of C_5 and C_6 paraffinic hydrocarbons (large
  pore mordenite with Pd)
- Shape selective hydrocracking
- Isomerization of aromatics
- Methanol to gasoline (ZSM-5)

TABLE II - Some adsorption processes using zeolites |4|

Separations of
- n-paraffins from iso-paraffins (Molex |5|;N-IsElf |6|)
- aromatics,e.g.,p-xylene from a mixture of xylene isomers
  and ethylbenzene (Parex |5|)
- olefins from paraffins
- air |7|

Epuration of
- synthesis gas
- streams containing HCl, SO_x, NO_x (Pura-Siv-N)
- steam-cracker effluent containing CO_2
- LPG (sweetening)

Dehydration of
- cracking gases for ethylene production
- olefins
- hydrogen
- solvents

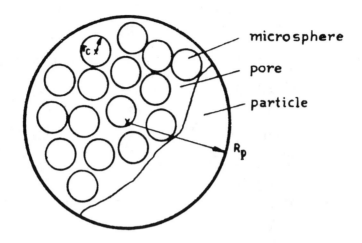

Figure 1 - Bidisperse structure of a zeolite pellet

interpellet voids	0.32
solid portion of crystals	0.199
solid portion of binder	0.063
ρ_a (pellet)	1.12 g/cm^3
ρ_b	0.76 g/cm^3
$\bar{c}_p$ (40-260 $^\circ$C)	0.22 cal/g$^\circ$C
crystal size of the order of μm	

Either mass transfer through transport pores or diffusion in the zeolite crystals can be the controlling rate process. Garg and Ruthven |9| state that the relative importance of macropore and micropore resistances can be measured by the parameter Ω,

$$\Omega = \frac{3\omega(1-\varepsilon_p)}{\varepsilon_p} \frac{D_c}{D_p} \left(\frac{R_p}{r_c} \right)^2 \frac{dq}{dc} \qquad |1|$$

where ω is the volume fraction of the pellet occupied by zeolite crystals, ε_p is the pellet porosity, D_c and D_p are the diffusivities in the zeolite crystals and in macropores, respectively, R_p and r_c are the radii of pellet and crystal, respectively and dq/dc is the slope of the adsorption equilibrium isotherm. If $\Omega < 1$ micropore diffusion is the controlling step while for $\Omega > 100$ macropore diffusion is the limiting one.

Macropore diffusion has been found |10| to be the controlling mechanism for liquids in molecular sieve pellets (Davison 525); for a tracer C_6H_{12} in a feed containing C_6H_6, Lee and Ruthven found $D_p = 0.29 \times 10^{-5}$ cm²/s.

Diffusion measurements in zeolites have been made by numerous workers; Ma and Lee [11] reported values for diffusion of n-butane in X zeolite pellets of $R_p = 0.23$ cm and $r_c = 1.08 \times 10^{-4}$ cm equal to $D_p = 0.046$ cm²/s and $D_c = 5.1 \times 10^{-14}$ cm²/s. Ruthven and Doetsch [12] dealing with n-C_7H_{16}, C_6H_{12} and $C_6H_5CH_3$ in 13X zeolite found diffusivities in the range 10^{-8}-10^{-9} cm²/s while Doelle and Riekert [13] for the system n-butane/NaX zeolite obtained a diffusivity of 2×10^{-7} cm²/s at 300 K (large crystal zeolite with $d_c = 80$ μm so $\tau_d = 8$ s). Kumar et al [14] using chromatographic techniques studied diffusion of i-C_4H_{10} in 4A zeolite obtaining $D_p = 0.021$ cm²/s (molecule too large to penetrate the zeolite crystal) while for He in N_2 carrier they obtained $D_c = 1.7 \times 10^{-10}$ cm²/s. For 4A zeolite pellets ($R_p = 0.39$ cm; $r_c = 1.74$ μm) they also reported reciprocal values of the diffusion time constants $(1/\tau_d)$ at 306 K for CH_4, Ar and CO, respectively 4×10^{-4} s^{-1}, 10^{-2} s^{-1} and 1.7×10^{-3} s^{-1}.

The subject of diffusion in zeolites is a very exciting one since, as pointed out by Weisz [15], "the field of shape selective catalysis relies on the different diffusivities of molecules through spaces of near-molecular dimensions". Classic diffusion regimes are well known:

- Knudsen diffusion, when pore size < molecular mean free path and thus diffusivity ∝ (pore diameter)$^{-1}$

- ordinary diffusion, when pore size > molecular mean free path and then diffusivity = (constriction factor)x(ordinary diffusion coefficient)

Now a new regime appears called the "configurational regime" where the diffusion is affected by the size and configuration of the molecules (Figure 2).

In view of the bidisperse nature of these catalyst/adsorbents, mass conservation equations for a volume element of the pellet should be correctly formulated and as a result the concept of catalyst effectiveness factor should be extended.

Let us recall the definition of catalyst effectiveness factor for a homogeneous particle of volume V_p. If (c_s, T_s) are the concentration and temperature at the surface the effectiveness factor, η,

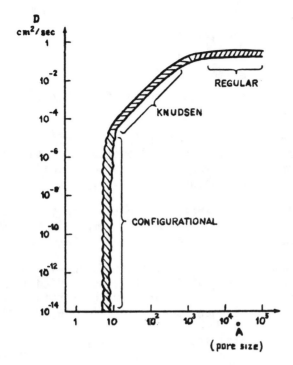

Figure 2 - Diffusional regimes (after Weisz |15|)

is defined as the ratio between the observed rate, $r_{obs} = \dfrac{1}{V_p} \iiint\limits_{V_p} r(V)dV$

and the intrinsic rate, $r_{int}(c_s, T_s)$ calculated at the surface conditions. For irreversible n^{th} order reaction the steady state mass balance in the particle volume element, in isothermal operation, is:

$$\frac{d^2f}{dx^2} + \frac{\alpha-1}{x} \frac{df}{dx} - \phi^2 f^n = 0 \qquad\qquad |2|$$

where $f = c/c_s$ is the reduced reactant concentration inside the catalyst, $x = z/\ell$ is the reduced space coordinate for the catalyst, α is the shape factor (1 for slab, 2 for cylinder, 3 for sphere) and $\phi = \ell \sqrt{k c_s^{n-1}/D_e}$ is the Thiele modulus (ℓ - characteristic dimension: half-thickness of the slab or particle radius; D_e - effective diffusivity; k - kinetic constant). The concentration profile is easily obtained for first order reactions if we take into account the boundary conditions, $x = 1, f = 1$ and $x = 0, df/dx = 0$; the effectiveness factor η is then:

slab $\qquad \eta = \dfrac{th\phi}{\phi}$ $\qquad$; diffusional regime (ϕ high) $\eta \simeq \dfrac{1}{\phi}$

cylinder $\qquad \eta = \dfrac{2}{\phi}\dfrac{I_1(\phi)}{I_0(\phi)}$; $\qquad\qquad\qquad\qquad \eta \simeq \dfrac{2}{\phi}$

sphere $\qquad \eta = \dfrac{3}{\phi}\left(\dfrac{1}{th\phi} - \dfrac{1}{\phi}\right)$; $\qquad\qquad\qquad \eta \simeq \dfrac{3}{\phi}$

with I_1, I_0 modified Bessel functions of first kind, first and zero orders respectively.

For zero order reactions concentration profiles will reach, in some cases, zero values at a point x^* inside the catalyst where also $df/dx = 0$. Solution of the model equations will lead to the effectiveness factor $\eta = 1 - x^{*\alpha}$ shown in Figure 3.

Thiele modulus values below which the catalyst is operating in the chemical regime are $\sqrt{2\,\alpha}$ while the asymptotic expression of the effectiveness factor for high ϕ, is simply $\alpha\sqrt{2}/\phi$.

The importance of the effectiveness factor is obvious since it determines the real quantity of catalyst to be used in order to reach a certain degree of conversion; moreover the working regime of the catalyst can be crucial when looking for high selectivities.

For bidisperse catalysts Ihm et al |16| proposed an overall effectiveness factor which is calculated from the knowledge of micro and macroeffectiveness factors. These authors considered a pellet as made of microspheres with pores among them; moreover a fraction γ of the active sites is at the microsphere surface or pore walls.

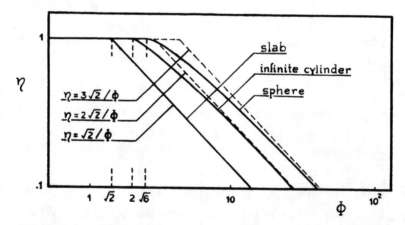

Figure 3 - Effectiveness factor η versus ϕ for zero order reactions.

The governing steady state equations are then:

pore space

$$\frac{d^2 f_a}{dx_a^2} + \frac{2}{x_a}\frac{df_a}{dx_a} = \gamma\phi_a^2 f_a^n + 3(1-\gamma)\left.\frac{\phi_a^2}{\phi_i^2}\left(\frac{df_i}{dx_i}\right)\right|_{x_a,x_i=1} \qquad |3a|$$

$$x_a = 0 \qquad \frac{df_a}{dx_a} = 0 \qquad\qquad\qquad |3b|$$

$$x_a = 1 \qquad f_a = 1 \qquad\qquad\qquad\qquad |3c|$$

with $\phi_a = R_p\sqrt{kc_s^{n-1}/(V_a D_a \epsilon)}$, $f_a = c_a/c_s$, $x_a = R/R_p$ and V_a- volume of a pellet.

microsphere

$$\frac{d^2 f_i}{dx_i^2} + \frac{2}{x_i}\frac{df_i}{dx_i} = \phi_i^2 f_i^n \qquad\qquad |4a|$$

$$x_i = 0 \qquad \frac{df_i}{dx_i} = 0 \qquad\qquad\qquad |4b|$$

$$x_i = 1 \qquad f_i = f_a(x_a) \qquad\qquad\qquad |4c|$$

with $\phi_i = r_c\sqrt{(1-\gamma)kc_s^{n-1}/(n'V_i D_i)}$, $f_i = c_i/c_s$, $x_i = r/r_c$, V_i- volume of a microsphere and n'- number of microspheres in the pellet.

It is obvious that the effectiveness factor for a microsphere is

$$\eta_i = \frac{\text{diffusional flux at the microparticle surface}}{\text{intrinsic rate in the microsphere at the surface conditions}}$$

or

$$\eta_i = \left.\frac{3}{\phi_i^2 f_a^n}\left(\frac{df_i}{dx_i}\right)\right|_{x_a,x_i=1} \qquad |5|$$

The average microeffectiveness factor $\bar{\eta}_i$ is defined as the ratio between the diffusion flux into all the microspheres of the pellet and the intrinsic rate in all the microspheres at the surface concentration, c_a ; then

$$\bar{\eta}_i = \frac{\int_0^1 x_a^2 f_a^n \eta_i dx_a}{\int_0^1 x_a^2 f_a^n dx_a} \qquad |6|$$

430

On the other hand the overall effectiveness factor, η_{ov}, is the ratio between the diffusional flux at the pellet surface, $x_a = 1$, and the intrinsic rate in the whole pellet, i.e., pore space and microspheres, supposed at the surface conditions, c_s; then

$$\eta_{ov} = \frac{3}{\phi_a^2} \left(\frac{df_a}{dx_a}\right)\Bigg|_{x_a=1} \qquad |7|$$

If we introduce η_a defined as:

$$\eta_a = \frac{\text{diffusional flux at the pellet surface}}{\begin{bmatrix}\text{intrinsic rate in} \\ \text{the pores at the} \\ \text{concentration } c_s\end{bmatrix} + \begin{bmatrix}\text{diffusional flux in all} \\ \text{microspheres with sur-} \\ \text{face concentration } c_s\end{bmatrix}}$$

we can show that

$$\eta_a = \frac{3}{\phi_a^2(\gamma+(1-\gamma)\overline{\eta}_i)}\left(\frac{df_a}{dx_a}\right)_{x_a=1} \qquad |8|$$

and finally

$$\eta_{ov} = \eta_a(\gamma+(1-\gamma)\overline{\eta}_i) \qquad |9|$$

Equation $|9|$ contains some limiting cases:
a) Reaction takes place only in the microspheres; $\gamma = 0$

$$\eta_{ov} = \overline{\eta}_i \eta_a \qquad |9a|$$

b) Reaction only in the pore walls; $\gamma = 1$

$$\eta_{ov} = \eta_a \qquad |9b|$$

In the case of zero-order reaction the treatment becomes more complex since the concentration profiles, for the macrosphere and/or the microsphere can reach zero values at certain radial positions.

In Figure 4 effectiveness factors η_a, η_{ov} and η_i are plotted as functions of ϕ_i for second order reaction at given values of ϕ_a and γ.

2. FUNDAMENTALS OF CATALYTIC REACTION ENGINEERING

In this section we will discuss some aspects of design and operation of catalytic reactors. First of all it should be stressed

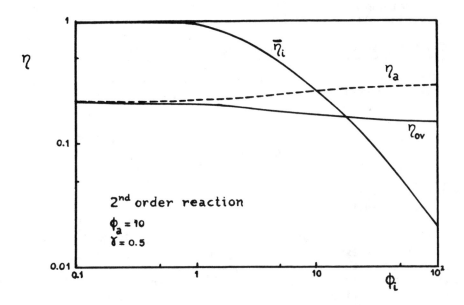

Figure 4 - η_{ov}, η_a, $\overline{\eta}_i$ versus ϕ_i.

that the choice of the type of reactor is the first step to be done; then preliminary estimates of concentration and temperature gradients, both in the film and inside the particle, are needed in order to guide the use of a suitable model.

For homogeneous catalysts we have |17|:

external concentration gradient, $\overline{\Delta c}_e = \dfrac{c_{Ab} - c_{As}}{c_{Ab}} = Ca = \dfrac{r_{obs} \ell}{k_f c_{Ab}}$

internal concentration gradient, $\overline{\Delta c}_i = \dfrac{c_{As}}{c_{Ab}} = 1 - Ca$

maximum internal temperature rise, $\Delta T_i^* = \dfrac{T_{max} - T_s}{T_s} = \beta = \dfrac{(-\Delta H) D_e c_{As}}{\lambda_e T_s}$

maximum external temperature rise, $\overline{\Delta T}_e^* = \dfrac{T_{s,max} - T_b}{T_b} = \overline{\beta}b =$

$= \dfrac{(-\Delta H) D_e c_{Ab}}{\lambda_e T_b} \dfrac{Bi_m}{Bi_h}$

These calculations are useful particularly when we are involved with the determination of reaction kinetics in the laboratory and we try to avoid falsification of kinetics due to diffusion intrusion.

2.1 - Tubular Fixed Bed Catalytic Reactors (TFBCR)

Historically this is one of the oldest arrangements for conducting gas-solid or liquid-solid chemical reactions on an industrial scale.

A TFBCR is simply an assembly of uniformly sized particles (viz. the catalyst) which are randomly arranged and which are held firmly in position within a tube or pipe (viz. the reactor). Intimate contact is achieved between the particles and the reactant fluid as the latter flows in a random manner between, around and, in the case of porous catalysts, into the particles.

Examples of fixed bed catalytic processes employed in basic chemical industries are steam reforming, carbon monoxide oxidation and methanation, the synthesis of ammonia, sulphuric acid and methanol. Catalytic reforming, isomerization and polymerization processes are utilised in the petroleum refining industry. The oxidation of olefins and aromatics, the synthesis of vinyl acetate, the production of styrene and other dehydrogenation processes are examples in the petrochemical industry.

2.1.1 - Theoretical Aspects on Modelling. The general problem may be placed in perspective by viewing the reactor in terms of long and short range gradients along the spacial coordinates. Gradients of species concentrations and fluid-solid temperatures which prevail throughout the geometric confines of the reactor are termed inter-particulate. Those which may persist in the fluid immediately surrounding the catalyst particles are defined as interphase gradients whilst those within the solid porous catalytic phase are defined as intraphase gradients.

The transport characteristics inside the particle are distinct from those outside and consequently a realistic model for a TFBCR should consider the single particle case and then construct from this the overall reactor model in which the conservation equations include transfer between the phases in contact.

A model resulting from this approach is clearly a complex learning model with such computational costs as to preclude its use for optimisation and control purposes. Hence predictive simplified models are continually sought both at particle and overall reactor levels.

The mathematical treatment of the behavior of porous catalyst pellets rests on relations which link the fluxes of the reacting species to gradients in composition, pressure and temperature. A variety of flux relations of varying degrees of complexity is available in the literature |18|.

It is quite apparent that the point of interest lies in the calculation of the global rate for a catalyst particle and where this leads to a rate of reaction per unit volume of reactor bed when the analysis is extended to the whole TFBCR.

The method of expressing this global rate in a catalyst pellet as its intrinsic rate under surface conditions multiplied by an effectiveness factor is now well established.

In the developments which follow it is assumed that such global rates can be obtained in that way.

The focus of attention will then be on the form of conservation models for the integral reactor, i.e., on how they account for inter-particle gradients of species concentrations in the fluid, inter-particle gradients of fluid and solid temperatures and also for the interphase transport of mass and heat.

The more complex structures recognize the existence of both phases and form the group of the heterogeneous models.

Resulting from the simplification of these latter an alternative group of essentially predictive structures has emerged, viz. the pseudo-homogeneous models. In these the assumptions are made that the resistance to interphase transport of mass and heat is negligible that the physical properties of the fluid vary only slightly across a particle diameter and that the fluid film around a catalyst par-ticle is small - such that each particle with its surrouding boun-dary is regarded as a point source within a homogeneous field.

The two most common forms of deterministic description of the physical and chemical processes occurring in a TFBCR are the Fickian models (a convenient general name for those models based on Fick's and Fourier's laws for mass and thermal dispersion, respectively) and the cell models of Deans and Lapidus |19| where interstices between packing elements are idealized as perfectly stirred mixers which give rise to dispersion-type behavior.

The classification suggested by Froment |20| for Fickian models has been extended by Gros and Bugarel |21| to include all models and is presented in Table III.

One of the reasons for the wider use of Fickian models is possibly the fact that previous mass and heat transfer experiments have been analysed almost exclusively on the basis of such models

TABLE III - Classification of Fickian and Cell Models for Tubular Fixed Bed Catalytic Reactors

	Fickian Models				Cell Models			
	Pseudo-homogeneous		Heterogeneous		Pseudo-homogeneous		Heterogeneous	
	Characteristic	Code	Characteristic	Code	Characteristic	Code	Characteristic	Code
One-dimensional	Plug flow	PH1	PH1+interphase transport	H1				
			H1+intraparticle transport	HI1				
	PH1+axial dispersion	PH2	H1+axial dispersion	H2	Cells in series	C2	C2+interphase transport	CH2
			HI1+axial dispersion	HI2				
	PH1+radial dispersion	PH3	H1+radial dispersion	H3				
			HI1+radial dispersion	HI3				
Two-dimensional	PH2+radial dispersion	PH4	H2+radial dispersion	H4	Two-dimensional arrays of cells	C4	C4+interphase transport	CH4
			HI2+radial dispersion	HI4				

and consequently a large amount of compatible parameter values are available.

The other reason is concerned with numerical procedure. The cell model was originally introduced when it appeared that the solution of large number of algebraic equations for the steady-state or ordinary differential equations for the unsteady-state was simpler than the solution of the partial differential equations required by the dispersion models. Indeed, even today, the simple implementation of methods of solution of non-linear algebraic equations are generally more cumbersome.

During the past three years some works have been published which question the validity of the Fickian approach |22,23|.

It is indeed apparent that at present no second-order continuous model satisfies all the requirements which the experimental and theoretical analysis in packed beds suggest.

At the present state of the art, however, the Fickian models are a feasible and generally satisfactory approach for design purposes.

2.1.2 - Fickian Model Analysis. In this section a heterogeneous, a hybrid and a pseudo-homogeneous models for the steady-state of TFBCRs will be examined. The aim is to develop essentially predictive structures for design purposes.

It is obvious from the literature that radial dispersion of heat and mass in cooled-wall reactors is of considerable significance and therefore has to be included throughout.

Criteria which may be applied to determine the extent of the influence of thermal and mass axial mixing (Young and Finlayson |24|, Mears |25|) show that in general such influence is minimal for conditions of industrial practice (flow relations and length/ particle diameter ratio). But for adiabatic regimes such axial mixing phenomena are usually neglected as the survey of thirty-two experimental works presented by Feyo de Azevedo |26| confirms.

Following the classification of Table III the models to be developed are then of the types H3 or HI3 and PH3.

The Heterogeneous Model

The continuity equations for the key reacting component A and the energy equations are seen to constitute a set of parabolic partial differential equations coupled with a non-linear algebraic equation, viz.

Fluid Phase (C,T without subscript)

$$u \frac{\partial C}{\partial z'} = D_r \nabla^2_{r'} C - a_v k_c (C - C_p^S) \qquad |10a|$$

$$u \rho c_p \frac{\partial T}{\partial z'} = k_{rf} \nabla^2_{r'} T - a_v h (T - T_p^S) \qquad |10b|$$

Solid Phase (subscript p for pellet variable)

$$a_v k_c (C - C_p^S) = (1-\varepsilon) \eta r_A \qquad |10c|$$

$$a_v h (T_p^S - T) = (-\Delta H)(1-\varepsilon) \eta r_A + k_{rp} \nabla^2_{r'} T_p \qquad |10d|$$

The initial and boundary conditions are

for $z' = 0$; $\qquad\qquad C = C_o$ and $T = T_o(r')$ $\qquad$ |10e|

for $0 < z' \leq L$ and $r' = 0$; $\frac{\partial C}{\partial r'} = 0$ and $\frac{\partial T_p}{\partial r'} = \frac{\partial T}{\partial r'} = 0$ $\qquad$ |10f|

for $0 < z' \leq L$ and $r' = R$; $\frac{\partial C}{\partial r'} = 0$ $\qquad$ |10g|

$$h_{wf}(T_w - T) = k_{rf} \frac{\partial T}{\partial r'} \qquad |10h|$$

and $\qquad h_{wp}(T_w - T_p) = k_{rp} \frac{\partial T_p}{\partial r'} \qquad |10i|$

where the prime denotes variables with dimensions and a_v represents the interphase transfer area per unit reactor volume.

Some further remarks are due:

i - The continuity of thermal transport through the solid phase is taken into account as suggested by De Wasch and Froment |27| by including a dispersion term in equation |10d|.

ii - A major difference between this heterogeneous model and the hybrid and pseudo-homogeneous models is in the concept of transport coefficients involved - in the former coefficients for each phase (k_{rf}, k_{rp}, h_{wp} and h_{wf}) are considered whilst in the latter effective parameters (k_r, h_w) are involved.

iii - The meaning of the interphase coefficient h should also be clarified:
The interparticular radiation and conduction in the heterogeneous

model are accounted for within the effective conductivity of the solid and in principle only the convective contribution h_c should be considered for the film coefficient h_{fp}, i.e.,

$$h = (h_{fp})_0 = h_c \qquad |10j|$$

If the assumption of pellet isothermality (or near isothermality) is made then the coefficient h is a quantity interpolated between the film coefficient h_{fp} and the particle conductivity as given by equation $|10\ell|$:

$$\frac{1}{h} = \frac{1}{h_{fp}} + \frac{1}{\beta} \frac{d_p}{k_p} \qquad |10\ell|$$

The above set of equations $|10a|$ to $|10i|$ may be written in dimensionless form using the transformations:

$$r = \frac{r'}{R} \qquad z = \frac{z'}{L}$$

$$X = \frac{C}{C_0} \qquad X_p = \frac{C_p}{C_0}$$

$$\theta = \frac{T}{T_0} \qquad \theta_p = \frac{T_p}{T_0}$$

thus giving the following relationships:

$$\frac{\partial \theta}{\partial z} = b_1 \nabla_r^2 \theta - a_1(\theta - \theta_p^S) \qquad |11a|$$

$$\frac{\partial X}{\partial z} = b_2 \nabla_r^2 X - a_2(X - X_p^S) \qquad |11b|$$

$$a_1(\theta_p^S - \theta) = c_1(1-\varepsilon)nr_A + d_1 \nabla_r^2 \theta_p \qquad |11c|$$

$$a_2(X - X_p^S) = c_2(1-\varepsilon)nr_A \qquad |11d|$$

with the initial and boundary conditions as:

for $z = 0$; $X = 1$ and $\theta = \theta_0(r)$ $\qquad |11e|$

for $0 < z \leq 1$ and $r = 0$; $\frac{\partial X}{\partial r} = 0$ and $\frac{\partial \theta}{\partial r} = \frac{\partial \theta_p}{\partial r} = 0$ $\qquad |11f|$

for $0 < z \leq 1$ and $r = 1$; $\frac{\partial \theta}{\partial r} = - Bi_f(\theta - \theta_w)$ $\qquad |11g|$

438

$$\frac{\partial \theta_p}{\partial r} = - Bi_p(\theta_p - \theta_w)$$ |11h|

and $$\frac{\partial X}{\partial r} = 0$$ |11i|

where

$$a_1 = \frac{a_v h L}{u \rho c_p} \quad , \quad b_1 = \frac{k_{rf}}{u \rho c_p d_p} \frac{d_p L}{R^2} = \frac{1}{Pe_{hr}^f} \frac{d_p L}{R^2}$$

$$a_2 = \frac{a_v k_c L}{u} \quad , \quad b_2 = \frac{D_r}{u d_p} \frac{d_p L}{R^2} = \frac{1}{Pe_{mr}} \frac{d_p L}{R^2}$$

$$c_1 = \frac{(-\Delta H)L}{T_o u \rho c_p} \quad , \quad d_1 = \frac{k_{rp}}{u \rho c_p d_p} \frac{d_p L}{R^2}$$

$$c_2 = \frac{L}{C_o u}$$

$$Bi_f = \frac{h_{wf} R}{k_{rf}} \quad , \quad Bi_p = \frac{h_{wp} R}{k_{rp}}$$

The Hybrid Model

This model distinguishes between conditions in gas and solid phases but makes use of the effective transport concept for a pseudo-homogeneous medium.

A similar treatment to that carried out in the previous section allows the following equations to be established:

$$\frac{\partial \theta}{\partial z} = f_1 \nabla_r^2 \theta - e_1(\theta - \theta_p^s)$$ |12a|

$$\frac{\partial X}{\partial z} = f_2 \nabla_r^2 X - e_2(X - X_p^s)$$ |12b|

$$e_1(\theta_p^s - \theta) = g_1(1 - \varepsilon) n r_A$$ |12c|

$$e_2(X_p^s - X) = g_2(1 - \varepsilon) n r_A$$ |12d|

with initial and boundary conditions:

for z = 0 $\qquad$; X = 1 and $\theta = \theta_o(r)$ $\qquad$ |12e|

for $0 < z \leq 1$ and r = 0 ; $\dfrac{\partial X}{\partial r} = 0$ and $\dfrac{\partial \theta}{\partial r} = 0$ $\qquad$ |12f|

for $0 < z \leq 1$ and r = 1 ; $\dfrac{\partial X}{\partial r} = 0$ $\qquad$ |12g|

$\qquad$ and $\qquad$ $\dfrac{\partial \theta}{\partial r} = - Bi(\theta - \theta_w)$ $\qquad$ |12h|

and the following quantities written as:

$$e_1 = \frac{a_v h L}{u \rho c_p} \,,\ f_1 = \frac{k_r}{u \rho c_p d_p} \frac{d_p L}{R^2} = \frac{1}{Pe_{hr}} \frac{d_p L}{R^2}$$

$$e_2 = \frac{a_v k_c L}{u} \,,\ f_2 = \frac{D_r}{u d_p} \frac{d_p L}{R^2} = \frac{1}{Pe_{mr}} \frac{d_p L}{R^2}$$

$$g_1 = \frac{(-\Delta H) L}{T_o u \rho c_p} \,,\ g_2 = \frac{L}{C_o u}$$

$$\text{and} \qquad Bi = \frac{h_w R}{k_r}$$

The heat transfer coefficient h contained within the parameter e_1 is again a form interpolated between the true film coefficient and the particle effective conductivity. However, in this model the pellet temperature is obtained by a heat balance where the heat removal is described solely by the surface heat transfer coefficient h. Since the total heat removal from each pellet must be accounted for in this balance, it is necessary to include the interphase radiative and conductive modes of transfer.

Thus the film coefficient is now:

$$h_{fp} = h_c + h_r + h_p \qquad\qquad |12i|$$

where h_r and h_p represent the radiative and the conductive contributions, respectively, and again:

$$\frac{1}{h} = \frac{1}{h_{fp}} + \frac{1}{\beta} \frac{d_p}{k_p} \qquad\qquad |10\ell|$$

The reader is referred to Table IV for the details of the calculation of h_{fp}.

The Pseudo-Homogeneous Model

In this more simplified approach the interphase gradients are considered negligible. It is further assumed that physical properties of the fluid vary only slightly across the particle diameter and that the fluid film around a catalyst particle is small compared to the dimensions of the particle and consequently each particle with its surrounding gas layer can be regarded as a point source in a homogeneous fluid.

The model equations are:

$$\frac{\partial \theta}{\partial z} = \alpha_1 \nabla_r^2 \theta + \beta_1 (1 - \varepsilon) \eta \, r_A \qquad |13a|$$

$$\frac{\partial X}{\partial z} = \alpha_2 \nabla_r^2 X - \beta_2 (1 - \varepsilon) \eta \, r_A \qquad |13b|$$

with initial and boundary conditions:

for $z = 0$; $X = 1$ and $\theta = \theta_o(r)$ $|13c|$

for $0 < z \le 1$ and $r = 0$; $\dfrac{\partial X}{\partial r} = 0$ and $\dfrac{\partial \theta}{\partial r} = 0$ $|13d|$

for $0 < z \le 1$ and $r = 1$; $\dfrac{\partial X}{\partial r} = 0$ $|13e|$

       and $\dfrac{\partial \theta}{\partial r} = -\, Bi \, (\theta - \theta_w)$ $|13f|$

where

$$\alpha_1 = \frac{1}{Pe_{hr}} \frac{d_p L}{R^2} \quad , \quad \beta_1 = \frac{(-\Delta H) L}{u \rho c_p T_o}$$

$$\alpha_2 = \frac{1}{Pe_{mr}} \frac{d_p L}{R^2} \quad , \quad \beta_2 = \frac{L}{C_o u}$$

$$\text{and } Bi = \frac{h_w R}{k_r}$$

2.1.3 - Details of the solution to the models. A close inspection of each model described above reveals the kind of problems that it is necessary to investigate if a meaningful description of a parti- cular reactor is to be obtained.

One particular group of problems can be related to the mode of operation of the reactor. Two details too often unduly neglected

involve the radial temperature profile of the feed at the reactor
inlet and the mathematical description of the axial variation of
the reactor wall temperature.

The other group concerns mainly the fundamental problems of
the transport of heat in the packed bed and the kinetics which
occur in the catalyst pellet.

In spite of the fact that most fundamental objections to con-
tinuum dispersion models concern the mass transport mechanisms
chosen, the major problems of analysis lie in the determination of
the heat transport parameters (leaving here aside the kinetic problem which is very much system-dependent). The development of
structures relating the "effective parameters" of the hybrid and
pseudo-homogeneous models to the more elementary parameters of the
heterogeneous model has been considered a correct procedure for
some time - not only as a form of improving the understanding of
those elementary mechanisms but also as a way of developing more
meaningful predictive relationships |28|.

The representation of the pure thermal behavior of packed beds
by the steady-state models given before has been extensively
examined by Feyo de Azevedo |26,29| who has established sufficient
conditions for equivalence in high temperature regimes where
radiative heat transfer can not be neglected. Such conditions are
presented in terms of correlations between the 'effective' and the
'phase' parameters.

A resumé of such correlations is presented in Table IV.

Some of the parameters which characterize the basic transport
processes have been identified as only being poorly known and consequently their estimates from the literature lack reliability,viz.

   i - The turbulent limit Peclet group, $Pe_r(\infty)$.

  ii - The fluid/wall heat transfer coefficient $(h_{wf})_o$, given by
       equation |14b|, particularly for the range of particle
       Reynolds number below 100.

 iii - The solid/wall heat transfer coefficient $(h_{wp})_o$ expressed
       in the corresponding Biot group $(Bi_p)_o$.

This lack of reliability and consequent scatter in such parameters appears as related to the scatter which is also generally
observed in the literature concerning the effective transport
parameters $(k_r)_o$, $(h_w)_o$.

It is clear that no definite theory has yet been advanced to explain the transport of heat in packed beds. The best example of this is possibly the transport of heat through the reactor walls so far characterized by the 'convenient' wall heat transfer coefficients.

By correlating experimentally determined values of the Peclet and Biot groups $(Pe_{hr})_o$ and $(Bi)_o$ employing the approaches presented in Table IV it is possible to obtain estimates for those poorly known parameters which are both physically and statistically acceptable.

By employing such estimated values an equivalence is found between the pseudo-homogeneous, hybrid and heterogeneous models (as given in section 2.1.2) in the description of a packed bed thermal behavior in absence of reaction under conditions where the conductive, the convective and the radiative contributions are all significant.

TABLE IV - Synthesis of Correlations for the Estimation of Transport Parameters in Packed Beds |26,29|

Note - Parameters or dimensionless groups which are presented contained within brackets and with the subscript o characterize the transport phenomena in absence of radiation, i.e., including the conductive and convective modes only.

All correlations are in SI units.

---

1. General Form for Heat Transport Relationships

    i - Fluid-phase Peclet Group for Heat Transfer (in absence of radiation), $(Pe_{hr})_o$

$$\frac{1}{(Pe_{hr}^f)_o} = \frac{(k_{rf})_o}{Gc_p d_p} = \frac{1}{Pe_r(\infty)} + \frac{\varepsilon}{\tau_m RePr}$$
    |14a|

    with $\tau_m = 1.29$ (Gunn |30|).

    For $Pe_r(\infty)$ see sections 2 and 3 of the Table.

    ii - Fluid/wall Nusselt Group (in absence of radiation), $(Nu_{wf})_o$

$$(Nu_{wf})_o = \frac{(h_{wf})_o d_p}{k_g} = a\ Pr^{1/3}\ Re^b$$
    |14b|

    For a,b see section 3 of the Table.

TABLE IV - (continuation)

iii - Solid/wall Biot Group, Bi_p

$$Bi_p = (Bi_p)_o \qquad |14c|$$

For $(Bi_p)_o$ see section 3 of the Table.

iv - The Fluid/solid Heat Transfer Coefficient, h

$$\frac{1}{h} = \frac{1}{h_{fp}} + \frac{1}{\beta} \frac{d_p}{k_p} \quad \text{or} \quad \frac{1}{h} = \frac{1}{(h_{fp})_o} + \frac{1}{\beta} \frac{d_p}{k_p} \qquad |14d|$$

where $\beta = 10, 8, 6$ for spheres, cylinders and slabs, respectively.

The total film coefficient is given by:

$$h_{fp} = h_c + h_r + h_p \qquad |14e|$$

a) Convective contribution (Gunn |31|)

$$\frac{h_c d_p}{k_g} = (7 - 10\,\varepsilon + 5\,\varepsilon^2)(1 + 0.7\,Re^{0.2}\,Pr^{1/3}) +$$
$$+ (1.33 - 2.4\,\varepsilon + 1.2\,\varepsilon^2)\,Re^{0.7}\,Pr^{1/3} \qquad |14f|$$

for $0.35 < \varepsilon < 1$, $Re_p \le 10^5$

b) Conductive and radiative contributions

b_1)- Argo and Smith |32|

$$h_p = \frac{k_p^*(2k_p + h_{fp}\,d_p)}{d_p k_p} \qquad |14g|$$

with $\log k_p^* = -1.52 + 7.49 \times 10^{-3} \frac{k_p}{\varepsilon}$

$$h_r = \frac{k_{rad}(2k_p + h_{fp}\,d_p)}{d_p k_p} \qquad |14h|$$

with k_{rad} defined in vii.

TABLE IV - (continuation)

Which gives, combining equations $|14e|$ to $|14h|$:

$$h_{fp} = (h_c + \frac{2k_p^*}{d_p} + \frac{2k_{rad}}{d_p})/(1 - \frac{k_p^*}{k_p} - \frac{k_{rad}}{k_p}) \qquad |14i|$$

or, in absence of radiation:

$$(h_{fp})_o = (h_c + \frac{2k_p^*}{d_p})/(1 - \frac{k_p^*}{k_p}) \qquad |14j|$$

b_2)- Adderley $|33|$

$$h_r = k_{rad}/(2\,d_p) \qquad |14\ell|$$

$$h_p = k_p^*/(2\,d_p) \qquad |14m|$$

and use equation $|14e|$ for h_{fp}.

In the absence of radiation:

$$(h_{fp})_o = h_p + h_c \qquad |14n|$$

usually $\quad h_p \ll h_c$

v - Effective Biot Group (in absence of radiation), $(Bi)_o$
(Dixon and Cresswell $|28|$)

$$(Bi)_o = \frac{(h_w)_o R}{(k_r)_o} = \frac{R}{d_p} \frac{(Nu_{wf})_o (Pe_{hr}^f)_o}{Re\ Pr} \qquad |14o|$$

vi - Effective Radial Peclet Group for Heat Transfer (in absence of radiation), $(Pe_{hr})_o$, $|28|$

$$\frac{1}{(Pe_{hr})_o} = \frac{(k_r)_o}{Gc_p d_p} = \frac{1}{Pe_r(\infty)} + \frac{1}{RePr}\left[\frac{\varepsilon}{\tau_m} + \frac{(k_{rp})_o}{k_g} \frac{\dfrac{(Bi)_o}{(Bi)_o + 4}}{\dfrac{8}{(N_s)_o} + \dfrac{(Bi_p)_o + 4}{(Bi_p)_o}}\right] \qquad |14p|$$

TABLE IV - (continuation)

where $(N_s)_o = a_v R^2 h/(k_{rp})_o$

with $(k_{rp})_o$ obtained from the Zehner and Schlunder equation, $|34|$:

$$\frac{(k_{rp})_o}{k_g} = (1 - \varepsilon)^{1/2} \frac{k_r^{o'}}{k_g} \qquad |14q|$$

with $$\frac{k_r^{o'}}{k_g} = \frac{2}{1 - \frac{k_g B}{k_p}} \left[\frac{(1 - \frac{k_g}{k_p}) B}{(1 - \frac{k_g}{k_p})^2} \ln(\frac{k_p}{B k_g}) - \frac{B+1}{2} - \frac{B-1}{1 - \frac{k_g B}{k_p}} \right] \qquad |14r|$$

where $B = c \left(\frac{1 - \varepsilon}{\varepsilon}\right)^{10/9}$

and $c = 1.25$ for spheres

or 1.4 for crushed particles.

The coefficient h is calculated from equation $|14d|$ using $(h_{fp})_o$.

vii - Effective Radial Conductivity, k_r

$$k_r = (k_r)_o + k_{rad} \qquad |14s|$$

with $k_{rad} = 4 \psi_r \sigma d_p T^3$

$$\psi_r = \frac{2}{\frac{2}{e} - 0.264}$$

for the emissivity e see section 3 of the Table.

viii - Apparent Wall Heat Transfer Coefficient, h_w

$$h_w = (h_w)_o + h_{rad} \qquad |14t|$$

with $(h_w)_o = (Bi)_o (k_r)_o/R$, $h_{rad} = 4 \frac{e}{2 - e} \sigma T^3$

TABLE IV - (continuation)

ix - Radiative Contribution to the Heterogeneous Model
Parameters

a) Fluid-phase thermal conductivity, k_{rf}

$$k_{rf} = (k_{rf})_o + \varepsilon \, k_{rad} \qquad\qquad |14u|$$

b) Fluid/wall heat transfer coefficient, h_{wf}

$$h_{wf} = (h_{wf})_o + \varepsilon \, h_{rad} \qquad\qquad |14v|$$

c) Effective radial solid thermal conductivity, k_{rp} (Zehner and Schlunder |34|)

$$\frac{k_{rp}}{k_g} = (1 - \varepsilon)^{1/2} \left[\frac{k_r^{o'}}{k_g} + \frac{k_{rad} \, k_p}{k_g k_p + k_g k_{rad}} \right] \qquad |14x|$$

2. <u>Radial Mass Dispersion</u> (Gunn |30|)

$$\frac{1}{Pe_{mr}} = \frac{1}{Pe_r(\infty)} + \frac{\varepsilon}{\tau_m \, Re \, Sc} \qquad\qquad |14y|$$

with $\quad Pe_r(\infty) = A \left[1 + 19.4 \, \left(\frac{d_p}{D}\right)^2 \right]$

for the value of $Pe_r(\infty)$ see section 3 of this Table.

3. <u>Values Proposed for the Parameters</u> (Feyo de Azevedo |26|)

3.1 - Experimental conditions

i - Packing:
Commercial V_2O_5 catalyst for sulphur trioxide synthesis

Average particle diameter	$d_p = 0.0022$ m
Ratio tube/particle diameter	$D/d_p = 21.6$
Bed porosity	$\varepsilon = 0.31$

TABLE IV - (continuation)

---

  ii - Fluid-air

 iii - Ratio $k_p/k_g \simeq 7.5$

  iv - Ratio $(k_{rp})_o/k_g \simeq 3.75$

   v - Range of experimental work

      $26.9 \leq Re \leq 76.2$

      Temperature up to 702 K

3.2 - Estimated values for the parameters:

   i - $Pe_r(\infty) = 12$

  ii - $(Bi_p)_o = 11.3$

 iii - Constants a,b in the fluid/wall Nusselt group
      correlation (equation |14b|)

      $a = 0.280$

      $b = 0.617$

  iv - Emissivity $e = 0.675$

2.2 - Radial Flow Fixed Bed Reactors (RFBR)

    Cylindrical radial fixed bed reactors are found in a number of industrial situations and their study has been tackled by several authors |35,36,37|.

    A typical RFBR consists of a hollow cylindrical vessel packed with catalyst and is sketched in Figure 5; it is claimed that an advantage of the RFBR over axial flow units is the high flow rate surface area per volume of catalyst leading to narrow, low pressure drop beds.

    Ponzi and Kaye |36| considered a plug flow model and reactions with no change in the number of moles and analysed the influence of gas maldistribution on the conversion (for first and second order reactions) and selectivity (for series and parallel schemes) in the case of both isothermal and adiabatic operation.

    They considered that the superficial velocity at the inner

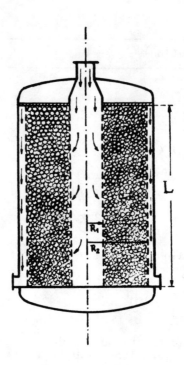

Figure 5 - Radial flow fixed bed reactor (outward flow)

basket $V_r(y)$ is related to the axial velocity in the central pipe, W_0 by $V_r(y) = -\dfrac{R_1 W_0}{2L} \dfrac{du}{dy}$ where $y = z/L$ and $u = W/W_0$. If there is no flow maldistribution $\dfrac{du}{dy} = -1$; otherwise the degree of maldistribution can be simulated by the residence time distribution of the individual elements. The fraction of gas entering the bed between axial positions y and $y + dy$ is $-\dfrac{du}{dy} dy = E(t_r)dt_r$, the residence time being $t_r(y) = \dfrac{c_0 dV}{dF}$ with $dV = \pi(R_2^2 - R_1^2)dL$ and F the feed molar flow; the mean residence time is $\bar{t}_r = \{(R_2^2 - R_1^2) L/(R_1^2 W_0)\}$.

Balakotaiah and Luss |37| extensively studied the influence of the flow direction on conversion for isothermal single reactions, $A \rightarrow (1 + \nu) B$ taking into account dispersion phenomenon.

For outward flow the mass balance equation for reactant A and associated boundary conditions are:

$$\frac{d}{dr}\left(-D_r \frac{dc_A}{dr}\right) + \frac{d}{dr}\left(u\, r\, c_A\right) + r\, g(c_A) = 0 \qquad |15a|$$

$$u(c_{Af} - c_A) = -D\,\frac{dc_A}{dr} \quad , \quad r = R_1 \qquad |15b|$$

$$\frac{dc_A}{dr} = 0 \qquad\qquad , \quad r = R_2 \qquad |15c|$$

In this equations u is the molar average velocity at radial

position r , i.e., $u = \dfrac{N(R_1)(1+\nu)+\nu a D\,\dfrac{dc_A}{dr}}{a(c_{Af}+\nu c_A)}$ with N - total molar flow

rate, y -mole fraction of A and a - area normal to flow direction.

For Re > 1, we have $Bo = \dfrac{u\, d_p}{D} \approx 2$ and then:

$$\frac{1}{Pe}\, y'' - y' - \frac{Da\ x}{1+\nu}\, f(y)\ (1+\nu y - \nu y' / Pe)^2 = 0 \qquad |16a|$$

$$\frac{y'}{Pe} = y - 1 \quad , \quad x = x_1 \qquad |16b|$$

$$y' = 0 \qquad\qquad , \quad x = 1 \qquad |16c|$$

where
$$Da = \frac{R_2^2\ g(c_{Af})}{u_1\ R_1\ c_{Af}}$$

$$Pe = \frac{u_2\ R_2}{D_2}$$

$$f(y) = g(y\, c_{Af}) / g(c_{Af})$$

$$x = r/R_2$$

For inward flow we get:

$$\frac{\omega''}{Pe} + \omega' - \frac{Da^* \ x}{1+\nu}\, f(\omega)\ \left(1 + \nu\omega + \frac{\nu\ \omega'}{Pe}\right)^2 = 0 \qquad |17a|$$

$$\omega' = 0 \quad , \quad x = x_1 \qquad |17b|$$

$$\frac{\omega'}{Pe} = 1 - \omega \ , \ x = 1 \qquad |17c|$$

with $\quad Da^* = \dfrac{R_2^2 \; g(c_{Af})}{u_2 R_2 c_{Af}}$

At $\quad Re < 1$, $D =$ const. and then:

- outward flow

$$\frac{(1+\nu y)\,(x\,y')' - \nu\,x\,(y')^2}{Pe^*} - (1+\nu)y' - Da\,x\,f(y)(1+\nu y)^2 = 0 \quad |18a|$$

$$\frac{y'}{Pe} = y - 1 \qquad , \; x = x_1 \qquad\qquad\qquad\qquad\qquad |18b|$$

$$y' = 0 \qquad\qquad , \; x = 1 \qquad\qquad\qquad\qquad\qquad\quad |18c|$$

with $\quad Pe^* = \dfrac{u_1\,R_1}{D}$

- inward flow

$$\frac{(1+\nu\omega)(x\,\omega')' - \nu\,x\,(\omega')^2}{Pe} + (1+\nu)\omega' - Da\,x\,f(\omega)\,(1+\nu\omega)^2 = 0 \quad |19a|$$

$$\omega' = 0 \qquad\qquad , \; x = x_1 \qquad\qquad\qquad\qquad\qquad |19b|$$

$$\frac{\omega'}{Pe} = 1 - \omega \qquad , \; x = 1 \qquad\qquad\qquad\qquad\qquad |19c|$$

The conclusions reached by the authors are:

i) in absence of dispersion, $Pe \to \infty$, the conversion is independent on the flow direction

ii) for zero order reaction the conversion is also independent
on the flow direction

iii) for constant Bo or constant dispersion and first order
reaction with no change in the number of moles, conversion is
independent on flow direction.

iv) for $\nu = 0$ and high Pe, outward flow gives a higher conversion
for all convex kinetic laws $(n > 1)$ and lower conversions for $0 < n < 1$.

In fact when Bo or D are constants the dispersion term at each point
is proportional to x/Pe and increases monotonically with x; thus in
the outward flow mixing occurs later than in the inward flow and,
for a convex rate equation, a delay in the mixing improves the
conversion.

v) for $\nu \neq 0$ the conversion depends on flow direction except for zero order reactions.

An industrial example of RFBR is found in the production of p-xylene from xylene isomerization with simultaneous disproportionation reactions; also catalyst deactivation by coke occurs.

An excellent study of determination of kinetic and diffusivity parameters for this system, using an amorphous silica-alumina catalyst, has been recently published by Orr, Cresswell and Edwards |38|. The catalyst has a surface area (BET) of 264 m^2/g with internal porosity of $\varepsilon_p = 0.5$, apparent density $\rho_a = 1.085$ g/cm^3 and a mean pore radius of 40 Å. The authors suggest a reaction scheme o-xylene $\rightleftharpoons$ m-xylene $\rightleftharpoons$ p-xylene and final parameters (at 723 K) were estimated:

Isomerization $\quad k_1 = 8.92 \times 10^{-6}$ mol/(Kg cat x s) (kPa)

$\qquad\qquad\qquad k_2 = 1.21 \times 10^{-5}$ mol/(Kg cat x s) (kPa)

$\qquad\qquad\qquad E_1 = E_2 = 111.4$ KJ/mole

Coking $\qquad\qquad k_c = 7.93 \times 10^{-9}$ (kPa)$^{-2}$ s^{-1}

$\qquad\qquad\qquad E_c = -34.1 \qquad$ KJ/mole

Disproportionation (o-xylene + o-xylene $\rightarrow$ products; m-xylene + + m-xylene $\rightarrow$ products)

$\qquad\qquad\qquad k_6 = 1.29 \times 10^{-7}$ mole/(Kg cat x s) (kPa)^2

$\qquad\qquad\qquad k_7 = 2.37 \times 10^{-8}$ mole/(Kg cat x s) (kPa)^2

$\qquad\qquad\qquad E_6 = E_7 = 97.6 \qquad$ KJ/mole

Some industrial units use zeolites as catalyst. Typical operating conditions are: T = 400-450°C, product equilibrium compositions: m-xylene/xylenes = 52%, p-xylene/xylenes = 24.5%, o-xylene/xylenes = 23.5%, ethylbenzene (EB)/C$_8$ aromatics = 8% . The reactions taking place are: m-xylene $\rightleftharpoons$ o-xylene $\rightleftharpoons$ p-xylene and EB $\rightleftharpoons$ m-xylene. The second procedes as EB + 3 H$_2$ $\rightleftharpoons$ ECH (ethylcyclohexane), ECH $\rightleftharpoons$ DMCH (dimethylcyclohexane) and DMCH $\rightleftharpoons$ m-xylene + 3 H$_2$.

452

3. DESIGN OF FIXED BED ADSORBERS

Fixed bed adsorption is an unsteady state process and then its behavior is governed by the propagation of heat and concentration waves through the column. The factors which influence wave travelling can be grouped in equilibrium, hydrodynamic and kinetic factors.

A general picture of adsorption in fixed beds is provided by Figure 6, where it is shown the progress of a front (which keeps its form - stationary front) through the bed.

Under certain ideal conditions (isothermal operation, no diffusional resistances, plug flow of the fluid phase and favorable equilibrium) a step change in concentration at the input will propagate as a shock or discontinuity as shown in Figure 7.

These figures help us in introducing some useful terminology: U is the flowrate, c_o and c_e the inlet and outlet concentrations, respectively; t_{Bp} the breakthrough time, t_{st} is the stoechiometric time and t_f the time needed to completely saturate the bed.

It is obvious that in the ideal situation described in Figure 7

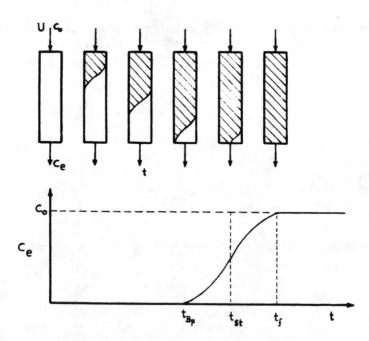

Figure 6 - Propagation of a stationary front through a fixed bed adsorber and breakthrough curve c_e vs. time.

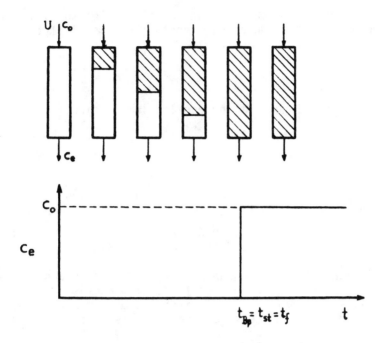

Figure 7 - Propagation of a shock in a packed bed adsorber

we introduced in the column during a time t_{st} a certain amount of solute $(= U c_o t_{st})$ which was retained in the voids of the bed $(= \varepsilon c_o v)$ and sorbed in the adsorbent $\{=(1 - \varepsilon) Q v\}$, where v is the bed volume, ε the porosity and Q the adsorbed solute concentration , in equilibrium with c_o, referred to the solid volume. This equality leads to $t_{st} = \tau (1 + \xi_m)$ where $\tau := \varepsilon v / U$ is the space time and $\xi_m = (1-\varepsilon) Q / (\varepsilon c_o)$ is the mass capacity factor. The adsorbent volume needed to operate the bed during a time t_{st} is then $v_s = (1-\varepsilon) U t_{st} / \{\varepsilon (1+\xi_m)\}$. However since dispersive effects are always present some designers just suggest the use of a higher volume $v_s' = s_f v_s$, s_f being a safety factor |39|.

This empirical approach for designing fixed bed adsorbers can be replaced by a semi-empirical approach which uses the concept of mass transfer zone (MTZ). Coming back to Figure 6 we can see that, at a given time $\underline{t}$, the column has three different regions: near the entry it is saturated, near the outlet no solute is yet adsorbed and in between those regions mass transfer is occurring between fluid and solid. We can then define, more or less arbitrarily, the MTZ as the distance between positions in the column where $c_2 = 0.99 c_o$ and $c_1 = 0.01 c_o$.

454

Considering the state of the column (of length L) at the time t_{Bp} in Figure 6, the unused bed length (LUB) is simply:

$$LUB = L - z_{st} \qquad |20|$$

and

$$L = LUB + LES \qquad |21|$$

where LES is the length of the equilibrium section.

This leads to:

$$LUB = L(1 - t_{Bp}/t_{st}) \qquad |22|$$

From an experimental breakthrough curve we are able to calculate LUB by measuring t_{Bp} and t_{st}; different runs at various operating conditions (flowrate, particle diameter, etc.) will show how LUB depends on those factors and then improve design.

But a satisfactory design will rely on understanding wave propagation which will enable us to develop models for fixed bed adsorbers. In the next section principles of isothermal fixed bed behavior and models of such adsorbers are reviewed.

3.1 - Isothermal Fixed Bed Adsorption

3.1.1 - Single component case. The prediction of breakthrough curves in this case is quite simple at least if we assume the equilibrium model |38|. Assuming plug flow of the fluid phase and combining the mass balance equation with the adsorption equilibrium isotherm, $q^* = f(c^*)$, we get:

$$u_c \equiv \left(\frac{\partial z}{\partial t}\right)_c = \frac{u_i}{1 + \frac{1-\varepsilon}{\varepsilon} f'(c)} \qquad |23|$$

Thus, the velocity of propagation of a given concentration c is inversely proportional to the slope of the isotherm. For unfavorable isotherms, $f''(c) > 0$, the wave is dispersive in nature; in the particular case of an isotherm with constant separation factor $K < 1$,

$$y = \frac{Kx}{1 - x + Kx} \qquad |24|$$

where y and x are reduced solid and fluid concentrations, respectively, we have:

$$x = \frac{\sqrt{K/T} - 1}{K - 1} \qquad |25|$$

with $T = \dfrac{c_o(V - \varepsilon v)}{(1 - \varepsilon) Q v}$ (throughput parameter) and V - volume passed through the column.

For favorable isotherms, $f''(c) < 0$, we will get a compressive wave which implies that a shock or discontinuity is formed travelling with a velocity u_s:

$$u_s = \frac{u_i}{1 + \dfrac{1-\varepsilon}{\varepsilon}\dfrac{\Delta q}{\Delta c}} \qquad |26|$$

the changes in concentration being calculated between the feed state (c_o, q_o) and the presaturation state of the bed {for a clean bed: $(0, 0)$}. It can be shown that even when axial dispersion is significant the stationary front will move at a velocity equal to u_s.

3.1.2 - Multicomponent adsorption - Equilibrium Model. Let us consider adsorption of n components contained in a feed at concentrations $c_{i,in}$ (i = 1,2,...,n) the bed being presaturated with solid concentrations q_{io} (i = 1,2,...,n). We assume affinities ordered as $a_1 > a_2 > \cdots > a_i > \cdots > a_n$ and Langmuir equilibrium relationships:

$$q_j = \frac{a_j c_j}{1 + \sum_{i=1}^{n} b_i c_i} \qquad j = 1,2,\ldots,n \qquad |27|$$

The calculation of concentration profiles in the bed can be made according to the theory developed, for instance, by Helfferich and Klein |40| for stoechiometric ion exchange as well as for adsorption processes. Recently Tien et al |41| illustrated this theory using an example.

We will briefly describe the steps to design concentration profiles:

i) conversion of an n-component adsorption system into the equivalent n + 1 component, stoechiometric ion exchange system, by the introduction of a fictitious component (n + 1) with mole fraction:

$$x_i = c_i \frac{b_i R}{a_i - R} \qquad i = 1,2,\ldots,n \qquad |28|$$

$$y_i = q_i \frac{b_i}{a_i - R} \qquad i = 1,2,\ldots,n \qquad |29|$$

456

and

$$x_{n+1} = 1 - \sum_{i=1}^{n} x_i \qquad |30|$$

$$y_{n+1} = 1 - \sum_{i=1}^{n} y_i \qquad |31|$$

with

$$\frac{y_j}{x_j} = \frac{1}{\sum\limits_{i=1}^{n} \alpha_{ij} x_i} \qquad |32|$$

$$\alpha_{ij} = \frac{a_i}{a_j} \qquad i = 1,2,\ldots,n \qquad |33|$$

$$\alpha_{i,n+1} = \frac{a_i}{R} \qquad |34|$$

where R is the transformation factor.

ii) The h-transformation leads to:

$$\sum_{i=1}^{n+1} \frac{x_i}{h - \alpha_{1i}} = 0 \quad \text{or} \quad \sum_{i=1}^{n+1} x_i \prod_{\substack{j=1 \\ \neq i}}^{n+1} (h - \alpha_{1j}) = 0 \qquad |35|$$

If x_i are known, the h values are found from the roots of the above equation.

iii) Calculating h-profiles versus $\frac{1}{T}$ where $T = \dfrac{(V-\epsilon v) \sum\limits_{i=1}^{n} c_{i,in}}{(1-\epsilon)v \sum\limits_{i=1}^{n} q_{i,in}}$

there are n+1 plateaus where the h_i are constants; changes in h_i take place between plateaus (gradual or abrupt transitions).

Let $h_{k\ell}$ represent the h value for species k at the ℓ plateau; we have:

$$h_{k1} = h_{k2} = h_{k3} = \cdots = h_{kk}$$

$$h_{k,k+1} = h_{k,k+2} = \cdots = h_{k,n+1}$$

There is only one change in value in the h-profile for any species; thus knowing the feed condition and the presaturation of the bed, the first plateau of h_i vs. $1/T$ corresponds to the feed condition and the last to the presaturation condition, i.e.,

$$c_{i,in} = c_{i1} \qquad i = 1,2,\cdots,n$$

$$q_{io} = q_{i,n+1} \qquad i = 1,2,\cdots,n$$

The values of q_{i1} and $c_{i,n+1}$ are calculated through equation $|27|$. The nature of change of the value h_k is determined by

$$\bar{R}_{k,k+1} = \frac{h_{k,n+1}}{h_{k1}} \; ;$$

if $\bar{R}_{k,k+1} < 1$ the transition is abrupt and if $\bar{R}_{k,k+1} > 1$ it is gradual. The changes occur at

$$\bar{T}_{k,k+1} = (h_{k,k+1} h_{k1} P_k)^{-1} \text{ with } P_k = \prod_{\substack{i=1 \\ \neq k}}^{n} h_{ik} \Big/ \prod_{i=1}^{n+1} \alpha_{1i} \; .$$

iv) conversion of h-profiles into concentration profiles using:

$$x_{jk} = \frac{\prod_{i=1}^{n} (h_{ik} - \alpha_{1j})}{\prod_{\substack{i=1 \\ \neq j}}^{n+1} (\alpha_{1i} - \alpha_{1j})} \qquad \begin{array}{l} j = 1,2,\cdots,n+1 \\ k = 1,2,\cdots,n+1 \end{array} \qquad |36|$$

$$y_{jk} = \frac{\prod_{i=1}^{n} (\frac{1}{h_{ik}} - \alpha_{j1})}{\prod_{\substack{i=1 \\ \neq j}}^{n+1} (\alpha_{i1} - \alpha_{j1})} \qquad \begin{array}{l} j = 1,2,\cdots,n+1 \\ k = 1,2,\cdots,n+1 \end{array} \qquad |37|$$

From these mole fractions we obtain:

$$c_i = \frac{a_i - R}{b_i R} x_i \qquad\qquad |38|$$

458

$$q_i = y_i \; \frac{a_i - R}{b_i} \tag{39}$$

Tien et al |41| claimed that different choices of R lead to different transition bounds although the h values are not changed; they then try to prove which R value is correct. This is irrelevant since as has been pointed out by Helfferich and Klein $1/\overline{T}_{k,k+1} \propto R$ and in fact whatever R is, the concentration versus distance profile is unique (and this is the final goal).

Modelling of multicomponent fixed bed adsorbers involving diffusional resistances has been done by several researchers |42,43, 44| although using simplified kinetic laws. Numerical techniques used in solving model equations are either finite differences or collocation methods. Recently, Sereno |45| solved a single pore diffusion model using moving finite elements.

An important industrial application of selective adsorption is the Parex process for separation of p-xylene from a mixture containing xylene isomers and ethylbenzene. In recent publications |46,47,48| an italian research group got some information on equilibrium adsorption of xylenes in a Y-zeolite adsorbent as well as breakthrough curves.

In Figures 8 and 9 we present two breakthrough curves from the references mentioned above and the responses predicted by the equilibrium model.

The experimental conditions were:

Fixed bed characteristics:

$L = 39$ cm

$ID = 2.04$ cm

$R_p = 0.065$ cm

$\varepsilon = 0.42$

$\varepsilon_p = 0.2$

$\rho_t = 1.4$ g/cm^3

$m_s = 83$ g

flowrate $= 3.9$ cm^3/min

Initial conditions of the bed:

    Case 1 - Figure 8 - 100% vol n-octane
    Case 2 - Figure 9 -   5% vol m-xylene
                          5% vol p-xylene
                         90% vol n-octane

Feed compositions (% vol):

    Case 1 - Figure 8 - 5% m-xylene
                         5% p-xylene
                        90% n-octane

    Case 2 - Figure 9 - 12.5% toluene
                        87.5% n-octane

The histories predicted by the equilibrium theory are derived below taking into account the equilibrium data for this system:

	T = 20 °C			T = 57 °C		
	b_j	a_j	Q_j	b_j	a_j	Q_j
m-xylene	6	11.76	1.96	4.2	8.232	1.96
o-xylene	8	15.68	1.96	4.7	9.212	1.96
p-xylene	36	70.56	1.96	24.0	47.04	1.96
ethylbenzene	21	41.16	1.96	12.0	23.52	1.96
toluene	25	49	1.96	8.0	15.68	1.96

Note: b_j in liter of solution/mole of solute
      a_j in liter of solution/liter of solid
      Q_j in mole of solute/liter of solid

Detailed calculations of theoretical histories from multi-component equilibrium theory (57 °C)

Case 1

Feed composition (% vol) : 5% p-xylene   (component 1)
                           5% m-xylene   (component 2)
                          90% n-octane   (solvent)

Then: $c_{11} = 0.4029$ mole/ℓ solution = 0.1692 mole/ℓ of bed

$c_{21} = 0.4156$ mole/ℓ solution = 0.1746 mole/ℓ of bed

Equilibrium data:

$b_1 = 24$ ℓ solution/mole = 57.143 ℓ of bed/mole

$b_2 = 4.2$ ℓ solution/mole = 10 ℓ of bed/mole

$Q = 1.75 \times 10^{-3}$ mole/g solid = 1.96 mole/ℓ solid = = 1.1368 mole/ℓ of bed

$a_1 = 64.96$ (dimensionless)

$a_2 = 11.368$ (dimensionless)

Then the solid composition in equilibrium with feed conditions is:

$$q_{11} = \frac{a_1 c_{11}}{1 + b_1 c_{11} + b_2 c_{21}} = 0.8853 \text{ mole/}\ell \text{ of bed}$$

$$q_{21} = \frac{a_2 c_{21}}{1 + b_1 c_{11} + b_2 c_{21}} = 0.1599 \text{ mole/}\ell \text{ of bed}$$

Choosing $R = 1$, the selectivity coefficients $\alpha_{ij} = a_i/a_j$ are:

$\alpha_{11} = 1$ $\alpha_{12} = 5.7143$ $\alpha_{13} = 64.96$

and the composition (corresponding to the feed state) of the $n+1$ component equivalent stoechiometric system is:

$x_{11} = 0.1512$ $x_{12} = 0.1684$ $x_{13} = 0.6804$

$y_{11} = 0.7909$ $y_{12} = 0.1542$ $y_{13} = 0.0549$

The composition of the initial state of the bed is:

$x_{13} = 0$ $x_{23} = 0$ $x_{33} = 1$

$y_{13} = 0$ $y_{23} = 0$ $y_{33} = 1$

Using Eq.|35|, we obtain:

$h^2 - 26.362 h + 70.9529 = 0$

and

$h_{11} = 3.0427$ $h_{21} = h_{22} = 23.3193$ (1st plateau)

Also for the third plateau we get:

$h_{13} = h_{12} = 1$ $\qquad$ $h_{23} = 5.7143$ $\qquad\qquad$ (3rd plateau)

Since $\bar{R}_{12} = h_{13}/h_{11} < 1$ and $\bar{R}_{23} = h_{23}/h_{21} < 1$ the transitions are both abrupt.

To calculate the position of the transitions we first need the values of

$$P_k = \prod_{\substack{i=1 \\ \neq k}}^{n} h_{ik} \prod_{i=1}^{n+1} \alpha_{i1} \text{ ,i.e.,}$$

$P_1 = 0.06282$ $\qquad$ $P_2 = 2.694 \times 10^{-3}$

Then $\bar{T}_{k,k+1} = (h_{k,n+1} \, h_{k1} \, P_k)^{-1}$ leads to:

$\bar{T}_{1.2} = 5.2316$ $\qquad$ $\bar{T}_{2.3} = 2.7857$

The velocity of the transition is given by:

$$u_{transition} = \frac{u_i}{1 + \bar{T}R}$$

Since $u_i = 2.841$ cm/min we get

$u_{1.2} = 0.4559$ cm/min $\qquad\qquad$ $u_{2.3} = 0.7505$ cm/min

and finally:

$t_{1.2} = 85.545$ min $\qquad\qquad$ $t_{2.3} = 51.968$ min

The concentrations at the second plateau can be now obtained:

$$x_{12} = \frac{(h_{12} - \alpha_{11})(h_{22} - \alpha_{11})}{(\alpha_{12} - \alpha_{11})(\alpha_{13} - \alpha_{11})} = 0$$

$$x_{22} = \frac{(h_{12} - \alpha_{12})(h_{22} - \alpha_{12})}{(\alpha_{11} - \alpha_{12})(\alpha_{13} - \alpha_{12})} = 0.2972$$

or $\qquad$ $c_{22} = x_{22} \dfrac{a_2 - R}{b_2 R} = 0.3081$ mole/ℓ of bed $= 0.7335$ mole/ℓ solution $=$

$$= 1.765 \, c_o$$

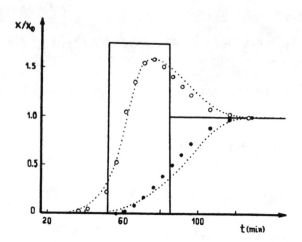

Figure 8 - Breakthrough curves for case 1. Experimental points |47|: (o) m-xylene; (•) p-xylene. Calculated curves: (···) from |47|; (——) equilibrium theory (this work).

Case 2

In this case we have:

j	component	initial condition % vol.	feed composition % vol.
1	p-xylene	5%	-
2	toluene	-	12.5%
3	m-xylene	5%	-
-	n-octane	90%	87.5%

Feed state

$c_{11} = 0$, $c_{21} = 1.177$ mole/ℓ solution = 0.4943 mole/ℓ bed, $c_{31} = 0$

Initial state of the bed

$c_{14} = 0.4029$ mole/ℓ solution = 0.1692 mole/ℓ bed , $c_{24} = 0$

$c_{34} = 0.4156$ mole/ℓ solution = 0.1746 mole/ℓ bed

From equilibrium data

$b_1 = 24$ ℓ solution/mole $= 57.143$ ℓ bed/mole
$b_2 = 8$ ℓ solution/mole $= 19.048$ ℓ bed/mole
$b_3 = 4.2$ ℓ solution/mole $= 10$ ℓ bed/mole
$Q = 1.1368$ mole/ℓ bed

Then $\quad a_1 = b_1 Q = 64.96$
$\qquad a_2 = 21.654$, $a_3 = 11.368$

Solid phase concentrations in equilibrium with feed state and initial state of the bed are, respectively:

$q_{11} = a_1 c_{11}/(1 + b_1 c_{11} + b_2 c_{21} + b_3 c_{31}) = 0$
$q_{21} = 1.0277$ mole/ℓ bed , $q_{31} = 0$

and

$q_{14} = 0.8853$ mole/ℓ bed , $q_{24} = 0$, $q_{34} = 0.1599$ mole/ℓ bed

If we take $R = 1$, then

$\alpha_{11} = 1$, $\alpha_{12} = 3$, $\alpha_{13} = 5.7143$, $\alpha_{14} = a_1/R = 64.96$

The composition for the feed and presaturation states of the equivalent (n+1)-component stoichiometric system are, respectively:

$x_{11} = 0$, $x_{21} = 0.4559$, $x_{31} = 0$, $x_{41} = 0.5441$
$y_{11} = 0$, $y_{21} = 0.9478$, $y_{31} = 0$, $y_{41} = 0.0522$

and

$x_{14} = 0.1512$, $x_{24} = 0$, $x_{34} = 0.1684$, $x_{44} = 0.6804$
$y_{14} = 0.7909$, $y_{24} = 0$, $y_{34} = 0.1542$, $y_{44} = 0.0548$

Calculation of h's:

- 1st plateau

$0 + 0.4559(h-\alpha_{11})(h-\alpha_{13})(h-\alpha_{14}) + 0 + 0.5441(h-\alpha_{11})(h-\alpha_{12})(h-\alpha_{13}) = 0$

the roots being $\quad h_{11} = 1$
$\qquad\qquad\qquad h_{21} = h_{22} = 5.7143$
$\qquad\qquad\qquad h_{31} = h_{32} = h_{33} = 31.2476$

464

- 4th plateau

$0.1512(h-\alpha_{12})(h-\alpha_{13})(h-\alpha_{14}) + 0 + 0.1684(h-\alpha_{11})(h-\alpha_{12})(h-\alpha_{14}) +$
$+ 0.6804(h-\alpha_{11})(h-\alpha_{12})(h-\alpha_{13}) = 0$

the roots being $\qquad h_{14} = h_{13} = h_{12} = 3$
$$h_{24} = h_{23} = 3.0427$$
$$h_{34} = 23.3193$$

Nature of the transitions:

Since $\bar{R}_{12} = h_{14}/h_{11} > 1$, the first transition is gradual whilst
$\bar{R}_{23} = h_{24}/h_{21} < 1$ and $\bar{R}_{34} = h_{34}/h_{31} < 1$ are abrupt.

Location of the transitions:

a) $\bar{T}_{2.3} = (h_{24}\ h_{21}\ P_2)^{-1}$

$P_2 = h_{12}h_{32}(\alpha_{11}\alpha_{21}\alpha_{31}\alpha_{41}) = 3 \times 31.2476 \times \dfrac{1}{3 \times 5.7143 \times 64.96} =$
$$= 0.08418$$

$\bar{T}_{2.3} = \dfrac{1}{P_2 \times 3.0427 \times 5.7413} = 0.68324$

b) $\bar{T}_{3.4} = (h_{34}\ h_{31}\ P_3)^{-1}$

$P_3 = h_{13}h_{23}(\alpha_{11}\alpha_{21}\alpha_{31}\alpha_{41}) = 3 \times 3.0427 \times \dfrac{1}{3 \times 5.7143 \times 64.96} =$
$$= 8.1969 \times 10^{-3}$$

$\bar{T}_{3.4} = \dfrac{1}{P_3 \times 23.3193 \times 31.2476} = 0.16742$

c) Diffuse boundary 1.2:

- For the side of zone (1)

$\bar{T}_{1.1'} = (h_{11}^2\ P_1)^{-1}$

$P_1 = h_{21}h_{31}(\alpha_{11}\alpha_{21}\alpha_{31}\alpha_{41}) = 0.16034$

$\bar{T}_{1.1'} = 6.23664$

- For the side of zone (2)

$\bar{T}_{2'2} = (h_{14}^2\ P_1)^{-1} = 0.69296$

Times at which transitions are located (outlet of the bed):

a) transition 1.2

   side (1) $u_{1.1'} = 0.393$ cm/min $t_{1.1'} = 99.34$ min

   side (2) $u_{2'.2} = 1.678$ cm/min $t_{2'.2} = 23.24$ min

b) transition 2.3 $u_{2.3} = 1.688$ cm/min $t_{2.3} = 23.11$ min

c) transition 3.4 $u_{3.4} = 2.434$ cm/min $t_{3.4} = 16.03$ min

Calculation of compositions (mole fractions - Eq. |36|)

- 2nd plateau

$$x_{12} = \frac{(h_{12} - \alpha_{11})(h_{22} - \alpha_{11})(h_{32} - \alpha_{11})}{(\alpha_{12} - \alpha_{11})(\alpha_{13} - \alpha_{11})(\alpha_{14} - \alpha_{11})} = 0.4729$$

and then (from Eq. |38|):

   $c_{12} = 0.5293$ mole/ℓ bed = 1.2603 mole/ℓ solution

for components 2 and 3: $x_{22} = 0$; $c_{22} = 0$

$x_{32} = 0$; $c_{32} = 0$

This is a very small plateau since we are near a watershed point ($h_{24} = 3.0427 \simeq h_{14} = \alpha_{12} = 3$).

- 3rd plateau

   $x_{13} = 0.2049$ $c_{13} = 0.2294$ mole/ℓ bed = 0.5461 mole/ℓ solution

   $x_{23} = 0$ $c_{23} = 0$

   $x_{33} = 0.2442$ $c_{33} = 0.2532$ mole/ℓ bed = 0.6028 mole/ℓ solution

For the diffuse boundary, we have for components 1 and 2:

$$x_j = \frac{(u_k/P_k)^{1/2} - \alpha_{1j}}{h_{k1} - \alpha_{1j}} x_{jk} \qquad |A1|$$

$$y_j = \frac{(P_k/u_k)^{1/2} - \alpha_{j1}}{1/h_{k1} - \alpha_{j1}} y_{jk} \qquad |A2|$$

when $x_{kk} = y_{kk} = 0$ and $h_{k1} = \alpha_{1k}$ the species k appears in the diffuse boundary; then:

$$x_k = \frac{(u_k/P_k)^{1/2} - \alpha_{1k}}{h_{k,k+1} - \alpha_{1k}} \; x_{k,k+1} \qquad \qquad |B1|$$

$$y_k = \frac{(P_k/u_k)^{1/2} - \alpha_{k1}}{1/h_{k,k+1} - \alpha_{k1}} \; y_{k,k+1} \qquad \qquad |B2|$$

this is the case for species 1 in our case; so we use Eq.|B1|:

$\bar{T}_1 = 1/u_1 = 5$	$x_1 = 0.0276$	$t = 82.37$ min
$\bar{T}_1 = 3.5$	$x_1 = 0.0792$	$t = 61.77$ min
$\bar{T}_1 = 2$	$x_1 = 0.1811$	$t = 41.18$ min
$\bar{T}_1 = 1$	$x_1 = 0.3540$	$t = 27.46$ min
$\bar{T}_1 = .6962$	$x_1 = 0.4729$	$t = 23.24$ min

For species 2 we use Eq.|A1|; in the same points we obtain:

$\bar{T}_1 = 5$	$x_2 = 0.4293$	$t = 82.37$ min
3.5	0.3796	61.77
2	0.2813	41.18
1	0.1146	27.46
0.6926	0	23.24

In Figure 9 we compare these predictions with the experimental and calculated results obtained by Santacesaria et al |47|.

3.2 - Non Isothermal Adsorption

The basic theoretical framework for the understanding of non isothermal, and particularly, adiabatic adsorption has been developed by Rhee et al |49,50,51,52|, Pan and Basmadjian |53,54,55| and Basmadjian et al |56,57,58,59| for single and multicomponent cases. A review of the area has been presented by Sweed |60|.

Practical systems have been considered in drying |61,62,63| and for the separation of gas mixtures of n-pentane from iso-pentane by modulation of feed temperature |64,65|. Methods for the obtention of adsorption equilibrium data |66,67.| and kinetic laws |68,69,70| have been implemented, specially using single particle technique and microbalance apparatus |71,72,73,74|; we should emphasize the work done by Ma and his group on ZSM-5 zeolite and silicalite. Modelling of such processes is object of a number of papers |75,76, 77| using a staged approach and simplified hypothesis. A mechanistic

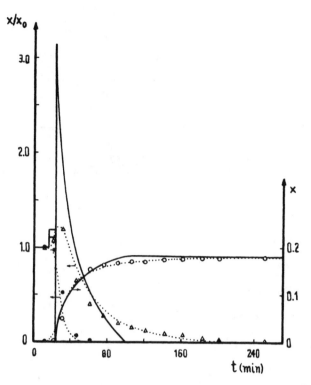

Figure 9 - Breakthrough curves for case 2. Experimental points
|47|: (●) m-xylene; (▲) p-xylene; (○) toluene.
Calculated curves: (···) from |47|; (——) equilibrium
theory (this work).

explanation of nonisothermal sorption has been recently published
by Helfferich |78|.

   3.2.1 - Non isothermal sorption in a perfectly mixed sorber.
Let us consider the system sketched in Figure 10, where T is the
temperature, F the total mass flux ($F_i = U\rho_i$; i for active species,
I for inerts), V the mixture volume and $\dot{q} = \widetilde{U} A (T_o - T)$ the heat flux
removed by the cooling medium. After introducing dimensionless
quantities, mass and heat balances take the form |79|:

$$\frac{1}{1 + \xi_m} \frac{dx_i}{d\theta} + \frac{\xi_m}{1 + \xi_m} \frac{dy_i}{d\theta} = 1 - x_i \qquad |40|$$

$$\frac{1 + \xi_h}{1 + \xi_m} \frac{d\widehat{T}}{d\theta} - \frac{\xi_m \beta}{1 + \xi_m} \frac{dy_i}{d\theta} = (1 - \widehat{T})(1 + \alpha) \qquad |41|$$

468

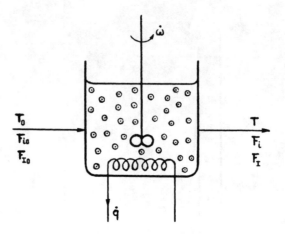

Figure 10 - The nonisothermal perfectly mixed sorber

with $x_i = \rho_i/\rho_o$, $y_i = q_i/Q$, $\hat{T} = T/T_o$, $\theta = t/t_{st}$.

The model parameters are:

ξ_m , mass capacity factor

$$\xi_m = \frac{1-\varepsilon}{\varepsilon} \frac{Q}{\rho_o}$$

ξ_h , heat capacity factor

$$\xi_h = \frac{1-\varepsilon}{\varepsilon} \frac{\rho_s \, cp_s}{\underset{i+I}{\Sigma} \rho_i \, cp_i}$$

β , thermicity factor

$$\beta = \frac{(-\Delta H)\, \rho_o}{(\underset{i+I}{\Sigma} \rho_i \, cp_i)\, T_o}$$

α , heat transfer number

$$\alpha = \frac{\tilde{U}\,A}{\dot{\Gamma}\, p}$$

When developing the model equations one should note that:
a) specific enthalpy at temperature T for species i
$$h_i(T) = h_{io} + cp_i\,(T - T_o)$$
b) specific enthalpy of the adsorbent
$$h_s(T) = h_{so} + cp_s\,(T - T_o)$$

c) specific enthalpy of adsorbed species i

$$h_i^*(T) = h_{io} + cp_i (T - T_o) + \Delta H$$

where $\Delta H < 0$ (-ΔH is the heat of adsorption in cal/g)

The assumptions made were:

i) $\Gamma_p = \sum_{i+I} v_f \rho_i cp_i = const$ (Γ_p in cal K^{-1})

ii) $\dot\Gamma_p = \sum_{i+I} F_i cp_i = const$ ($\dot\Gamma_p$ in cal $s^{-1} K^{-1}$)

iii) $v_s cp_i q_i \ll \Gamma_p$

iv) $cp_i (T - T_o) \ll (-\Delta H)$

v) $T_c = T_o$ (T_c - cooling medium temperature)

For adiabatic systems $\alpha = 0$. Assuming no mass and heat transfer resistances between fluid and solid phases, the adsorption equilibrium at the interface is governed by

$$y_i = \frac{K(T) x_i}{1 + x_i \{K(T) - 1\}}$$ $|42|$

where $K(T) = K(T_o) \exp\{-\gamma(1/\overline{T} - 1)\}$ with $\gamma = -\Delta H/(R T_o)$

The dynamic behavior of the stirred cell was then studied through computer simulation; we just present in Figures 11,12 and 13 the influence of ξ_h, ξ_m and α, respectively, on the histories of concentration and temperature.

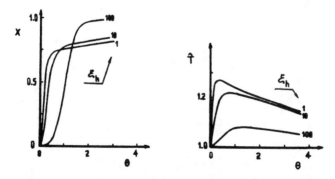

Figure 11 - Effect of the thermal capacity factor, ξ_h, on the response of the perfectly mixed sorber. $K(T_o) = 100$, $\beta = 1$, $\xi_m = 10$, $\alpha = 0$.

470

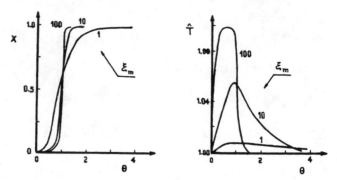

Figure 12 - Effect of the mass capacity factor, ξ_m, on the
response of the perfectly mixed sorber.
$K(T_o) = 100$, $\beta = 0.1$, $\xi_h = 10$, $\alpha = 0$.

As can be seen from Figure 11 (or from Figure 12), the removal
of a component by adsorption is negativelly affected by decreasing
ξ_h (or ξ_m) for adiabatic systems.

3.2.2 - Non isothermal sorption in a cascade of stirred cells.
Sorption columns are often simulated by a series of J stirred cells,
J being then a measure of the mass axial dispersion.

In Figures 14 and 15, travelling waves of concentration and
temperature are shown for J = 30; we can see the histories at the
exit of the 5th, 10th, 20th and 30th sorbers.

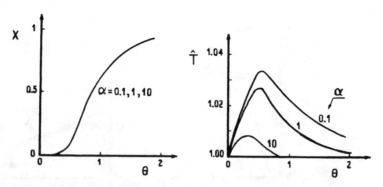

Figure 13 - Influence of the heat transfer number, α, on
the response of the perfectly mixed sorber.
$T_o = 300$ K, $\beta = 0.1$, $K(T_o) = 100$, $\xi_m = \xi_h = 1$.

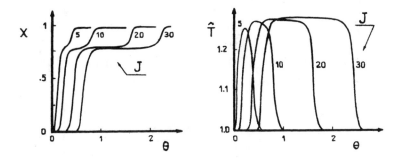

Figure 14 - Response of a cascade of adiabatic sorbers.
$T_o = 300$ K, $K(T_o) = 100$, $\xi_m = \xi_h = \beta = 1$, $\alpha = 0$,
$(-\Delta H) = 200$ cal/g.

For adiabatic systems (Figure 14) and $T_o = 300$ K, $K(T_o) = 100$, $\xi_m = \xi_h = \beta = 1$, $(-\Delta H) = 200$ cal/g, the concentration history shows a plateau between two transitions; similarly the temperature history has a plateau at the same place.

If heat is removed from the system, the plateau is deformed and it will disappear for isothermal systems (Figure 15).

3.2.3 - Simplified analysis of nonisothermal sorption. Some understanding of the behavior of the nonisothermal fixed bed adsorber can be obtained through a simplified analysis involving linearization of the model equations and use of Laplace transform technique.

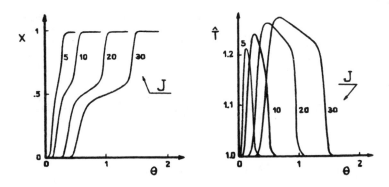

Figure 15 - Response of a cascade of nonadiabatic sorbers.
$T_o = 300$ K, $K(T_o) = 100$, $(-\Delta H) = 200$ cal/g,
$\xi_m = \xi_h = \beta = \alpha = 1$.

For adiabatic stirred tanks and linear isotherms

$$q_i = K(T) c_i = K_o \exp(-\frac{\Delta H}{RT}) c_i \qquad |43|$$

linearization around (c_{i_o}, T_o) leads to the model equations, in the Laplace domain and matrix form:

$$\begin{vmatrix} 1+s & -\gamma \dfrac{\xi_m}{1+\xi_m} s \\[3ex] -\beta \dfrac{\xi_m}{1+\xi_m} s & 1 + \dfrac{1+\xi_h+\gamma\beta\xi_m}{1+\xi_m} s \end{vmatrix} \begin{Bmatrix} \bar{x} \\[3ex] \tilde{T} \end{Bmatrix} = \begin{Bmatrix} \dfrac{1}{s} \\[3ex] 0 \end{Bmatrix} \qquad |44|$$

where $\tilde{T} = (T - T_o)/T_o$.

The system behaves as a second order one with:

$$\tau^* = \frac{\sqrt{1+\xi_h+\xi_m(1+\gamma\beta+\xi_h)}}{1+\xi_m} \qquad |45|$$

$$\zeta = \frac{2+\xi_m(1+\gamma\beta)+\xi_h}{2\sqrt{1+\xi_h+\xi_m(1+\gamma\beta+\xi_h)}} \qquad |46|$$

For $\zeta > 1$

$$x(\theta) = 1 - \frac{\exp(-\zeta \frac{\theta}{\tau^*})}{2\tau^*\sqrt{\zeta^2-1}} \{ (1-\tau^*(\zeta-\sqrt{\zeta^2-1})) \exp(\frac{\theta}{\tau^*}\sqrt{\zeta^2-1}) +$$

$$+ (\tau^*(\zeta+\sqrt{\zeta^2-1})-1) \exp(-\frac{\theta\sqrt{\zeta^2-1}}{\tau^*}) \} \qquad |47|$$

$$\tilde{T}(\theta) = \frac{\xi_m\beta}{(1+\xi_m)} \frac{\exp(-\zeta \frac{\theta}{\tau^*})}{2\tau^*\sqrt{\zeta^2-1}} \{ \exp(\frac{\theta}{\tau^*}\sqrt{\zeta^2-1}) - \exp(-\frac{\theta}{\tau^*}\sqrt{\zeta^2-1}) \} \qquad |48|$$

The maximum temperature is:

$$\tilde{T}_{max} = \frac{\beta \xi_m}{\tau^*(1+\xi_m)} \ (\zeta - \sqrt{\zeta^2-1})^{(\zeta/\sqrt{\zeta^2-1})} \qquad |49|$$

and occurs at:

$$\theta_{max} = \frac{\tau^*}{2\sqrt{\zeta^2-1}} \ \ln \frac{\zeta+\sqrt{\zeta^2-1}}{\zeta-\sqrt{\zeta^2-1}} \qquad |50|$$

For isothermal systems, $\beta = \gamma = 0$, and we get:

$$\tau^* = \sqrt{(1+\xi_h)/(1+\xi_m)} \qquad |45a|$$

$$\zeta = \frac{1}{2\tau^*} \ (1+\tau^{*2}) \qquad |46a|$$

and $x(\theta) = 1 - e^{-\theta}$ $\qquad\qquad |47a|$

For adiabatic fixed bed adsorbers with plug flow for the fluid phase we have the model equations in the Laplace domain:

$$\frac{d\bar{x}}{dz} + s\bar{x} - \frac{\xi_m\gamma}{1+\xi_m} \ s\ \bar{\bar{T}} = 0 \qquad |51|$$

$$\frac{d\bar{\bar{T}}}{dz} + \frac{1+\xi_h+\gamma\beta\xi_m}{1+\xi_m} \ s\ \bar{\bar{T}} - \frac{\xi_m\beta}{1+\xi_m} \ s\ \bar{x} = 0 \qquad |52|$$

or, using τ^* and ζ:

$$\frac{d^2\bar{x}}{dz^2} + 2\zeta\tau^*s \frac{d\bar{x}}{dz} + \tau^{*2}s^2 \bar{x} = 0 \qquad |53|$$

$$z = 0: \ \bar{x} = 1/s, \ \ \bar{\bar{T}} = 0$$

Finally, the histories of concentration and temperature are:

$$x(1,\theta) = \frac{\tau^*(\zeta+\sqrt{\zeta^2-1})-1}{2\tau^*\sqrt{\zeta^2-1}} \ H\{\theta-\tau^*(\zeta-\sqrt{\zeta^2-1})\} + \qquad |54|$$

$$+ \frac{1+\tau^*(\sqrt{\zeta^2-1}-\zeta)}{2\tau^*\sqrt{\zeta^2-1}} \ H\{\theta-\tau^*(\zeta+\sqrt{\zeta^2-1})\}$$

474

$$\tilde{T}(1,\theta) = \beta \frac{\xi_m}{1+\xi_m} \frac{1}{2\tau^*\sqrt{\zeta^2-1}} \{H(\theta-\tau^*(\zeta-\sqrt{\zeta^2-1}))- \qquad |55|$$

$$-H(\theta-\tau^*(\zeta+\sqrt{\zeta^2-1}))\}$$

Several situations can then occur as shown in Figure 16, depending on the relative values of ξ_h and ξ_m and/or $\gamma\beta$.

Some of these cases have been experimentally confirmed, as reported in recent work |80,81,82|, concerning adsorption, on 5A-zeolite, of ethane + helium and n-butane + helium mixtures.

4. NONLINEAR ADSORPTION COUPLED WITH CHEMICAL REACTION

In a number of situations,from which chromatographic reactor is an example, nonlinear adsorption is coupled with reaction. Analysis of the transient behavior of the column becomes then very complex; for irreversible and reversible reactions coupled with multicomponent Langmuir adsorption, results have been presented by Rodrigues et al |83|.

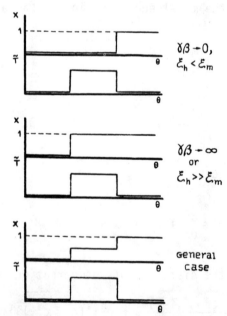

Figure 16 - Possible concentration and temperature
histories in nonisothermal sorption.

The point we want to briefly discuss here is the validity of usual methods for getting equilibrium and kinetic parameters from experiments. Often chromatographic techniques are used and, from the retention time of the peak and the ratio between the area under the peak and the injected quantity, the kinetic constant is obtained, provided the reaction is irreversible, first order and the isotherm is linear.

If the adsorption isotherm is nonlinear, this method fails. In order to have some insight on how nonlinear adsorption and reaction are coupled, we studied this phenomenon in a perfectly mixed reactor.

For a nth order irreversible reaction $A \overset{k}{\to} B$ coupled with non linear adsorption of component A alone, we have mass balance equations for A and B:

$$U c_{Ao} = U c_A + \varepsilon v \frac{dc_A}{dt} + (1-\varepsilon) v \frac{dq_A}{dt} + (1-\varepsilon) v r \qquad |56|$$

$$0 = U c_B + \varepsilon v \frac{dc_B}{dt} - (1-\varepsilon) v r \qquad |57|$$

for a feed containing only A. Assuming a Langmuir adsorption

isotherm $q_A = \dfrac{QK' c_A}{1+K' c_A}$ and $r = k c_A^n$ and introducing $x = c/c^o$,

$\theta = t/\tau$, $y_A = q_A/q^o$ (c^o is the maximum solute concentration in the fluid phase, for a step change at the inlet in the case of adsorption alone, or for an impulse in the case of reaction alone) we get $|84|$:

$$x_{Ao} = x_A + \{1 + \frac{\xi K}{(1+(K-1)x_A)^2}\} \frac{dx_A}{d\theta} + N_r x_A^n \qquad |58|$$

$$0 = x_B + \frac{dx_B}{d\theta} - N_r x_A^n \qquad |59|$$

with $\quad K = 1 + K' c^o$

$\qquad \xi_m = (1-\varepsilon) q^o/(\varepsilon c^o)$

$\qquad N_r = (1-\varepsilon) k\tau (c^o)^{n-1}/\varepsilon$

476

For a step change in concentration we obtain:

i) zero order reaction

$$\theta = \{1 + \frac{\xi K}{(1+(K-1)(1-N_r))^2}\} \ln\left|\frac{N_r - 1}{(N_r-1)+x_A}\right| + \qquad |60a|$$

$$+ \frac{\xi K}{1+(K-1)(1-N_r)} \{\frac{1}{1+(K-1)(1-N_r)} \ln|1+(K-1)x_A| + \frac{(K-1)x_A}{1+(K-1)x_A}\}$$

ii) first order reaction

$$\theta = -\{\xi K \frac{N_r+1}{(N_r+K)^2} + \frac{1}{N_r+1}\} \ln|1-(N_r+1)x_A| + \qquad |60b|$$

$$+ \xi K \frac{K-1}{N_r+K} \frac{x_A}{1+(K-1)x_A} + \xi K \frac{N_r+1}{(N_r+K)^2} \ln|1+(K-1)x_A|$$

For a Dirac impulse we obtain:

i) zero order reaction

$$\theta = \{1 + \frac{\xi K}{(N_r(K-1)-1)^2}\} \ln\frac{N_r+x_A(0^+)}{N_r+x_A} - \frac{\xi K}{(N_r(K-1)-1)^2} \times \qquad |61a|$$

$$\times \ln\left|\frac{1+(K-1)x_A(0^+)}{1+(K-1)x_A}\right| - \frac{\xi K}{N_r(K-1)-1} \{\frac{1}{1+(K-1)x_A(0^+)} - \frac{1}{1+(K-1)x_A}\}$$

ii) first order reaction

$$(1+N_r)\theta = \xi K \ln\left|\frac{1+(K-1)x_A}{1+(K-1)x_A(0^+)}\right| + (1+\xi K)\ln\frac{x_A(0^+)}{x_A} - \qquad |61b|$$

$$- \xi K \frac{(K-1)(x_A(0^+)-x_A)}{(1+(K-1)x_A(0^+))(1+(K-1)x_A)}$$

with $x_A(0^+)$ defined by equation $|62|$:

$$x_A(0^+) = \frac{-(2+K(\xi-1))+\sqrt{4\xi K+K^2(1-\xi)^2}}{2(K-1)} \qquad |62|$$

and shown in Figure 17.

In Figure 18 we show the response of a perfectly mixed tank to step and Dirac changes in concentration for zero order reaction and Langmuir adsorption.

Figure 19 shows the response of a perfectly mixed adsorptive reactor for first order reaction.

The determination of parameters can be easily made for first order reactions:

a) From a tracer experiment we get ε.

b) By comparing impulsional responses without and with reaction we see that at a given outlet concentration, c_{A1}, the corresponding times are, respectively, θ_1 and $(1+N_r)\theta_1$ from which N_r can be obtained; also from the area under the curve in the presence of reaction, we get area $= c^o/(1+N_r)$.

c) From the impulsional response and taking into account that

$$(1+N_r)\int_0^\theta c_A \, d\theta = c^o - \{c_A + \frac{1-\varepsilon}{\varepsilon} f(c_A)\}$$

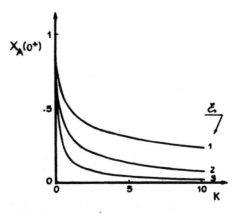

Figure 17 - $x_A(0^+)$ versus ξ and K.

478

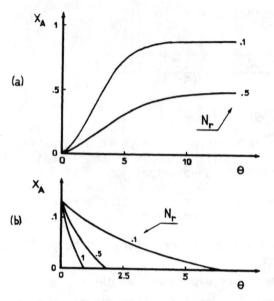

Figure 18 - Response of a perfectly mixed adsorptive reactor
to: a) step change in inlet concentration;
b) Dirac impulse. (zero order reaction, $\xi=2$, K=5)

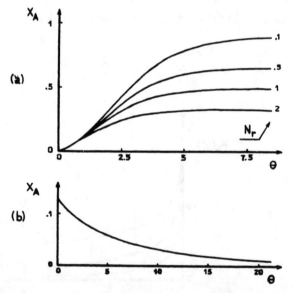

Figure 19 - Response of a perfectly mixed adsorptive reactor
to: a) step change in inlet concentration;
b) Dirac impulse. (first order reaction, $\xi=2$, K=10)

we get for each c_A the corresponding $f(c_A)$ and then the equilibrium isotherm.

It is also interesting to notice that, in the case of Dirac response, the ratios (R_{1n}) between the areas under the responses in the case of reaction alone and adsorption alone are:

$$R_{1n} = \frac{\left(\begin{array}{c}\text{area under the response}\\ \text{for reaction (order n)}\end{array}\right)}{\left(\begin{array}{c}\text{area under the}\\ \text{response for adsorption}\end{array}\right)}$$

zero order reaction, $n = 0$
$$R_{10} = 1 - N_r \ln \frac{1+N_r}{N_r}$$

first order reaction, $n = 1$
$$R_{11} = \frac{1}{1+N_r}$$

second order reaction, $n = 2$
$$R_{12} = \frac{1}{N_r} \ln(1+N_r)$$

On the other hand, the ratios (R_{2n}) between the areas under the Dirac responses for reaction coupled with linear adsorption and adsorption alone are:

$$R_{2n} = \frac{\left(\begin{array}{c}\text{area under the Dirac}\\ \text{response for reaction +}\\ \text{+ linear adsorption}\end{array}\right)}{\left(\begin{array}{c}\text{area under the Dirac}\\ \text{response for adsorption}\end{array}\right)}$$

zero order reaction, $n = 0$
$$R_{20} = (1+\xi)N_r \ln \frac{(1+\xi)N_r}{(1+\xi)N_r+1} + 1$$

first order reaction, $n = 1$
$$R_{21} = \frac{1}{1+N_r}$$

second order reaction, $n = 2$
$$R_{22} = \frac{1+\xi}{N_r} \ln(1 + \frac{N_r}{1+\xi})$$

It can be seen that we obtain R_{2n} from R_{1n} if in the expressions giving R_{1n} we replace N_r by $N_r(1+\xi)^{1-n}$.

The ratios $\tilde{R}_{2n}$ for the case of Langmuir type adsorption isotherms are:

$n = 0$
$$\tilde{R}_{20} = x_A(0^+) - \left\{1 + \frac{\xi K}{(N_r(K-1)-1)^2}\right\} N_r \ln \frac{N_r + x_A(0^+)}{N_r} -$$

$$- \frac{\xi K x_A(0^+)}{(N_r(K-1)-1)(1+(K-1)x_A(0^+))} + \frac{\xi K N_r}{(N_r(K-1)-1)^2} \ln|1+(K-1)x_A(0^+)|$$

$n = 1$
$$\tilde{R}_{21} = \frac{1}{1+N_r}$$

$n = 2$
$$\tilde{R}_{22} = \frac{1}{N_r} \left[1 + \frac{\xi K(\frac{N_r}{K-1})^2}{(\frac{N_r}{K-1}-1)^2}\right] \ln(1+N_r x_A(0^+)) -$$

$$- \frac{\xi K x_A(0^+)}{(\frac{N_r}{K-1}-1)(1+(K-1)x_A(0^+))} - \frac{\xi K N_r}{(K-1)^2(\frac{N_r}{K-1}-1)^2} \ln|1+(K-1)x_A(0^+)|$$

This kind of analysis can be useful when carrying out experiments in microreactors in view of the determination of kinetic parameters.

NOMENCLATURE

a_v external particle surface area per unit reactor volume

A   reaction (adsorption) component

b   ratio between the mass and the heat Biot numbers

Bi   Biot number

Bi_f fluid/wall Biot number

Bi_p solid/wall Biot number

Bi_h heat Biot number

Bi_m mass Biot number

Bo   Bodenstein number

c_p specific heat

c_i concentration of component i in the fluid phase

C_i concentration of component i

C_{pi} concentration of component i inside the pellet

c_s reactant concentration at the surface of the pellet

d_p particle diameter

D   reactor diameter, diffusivity

Da   Damköhler number

D_c diffusivity in zeolite crystlas

D_e effective diffusivity

D_p diffusivity in macropores

D_r radial effective mass diffusivity

e   emissivity

E   activation energy

f   reduced reactant concentration inside the catalyst (c/c_s)

F   mass flux

h   interpolated heat transfer cofficient for film surrounding a particle

h_c convection heat transfer coefficient for film surrounding a particle

h_c, h_r, h_p convective, radiative and conductive contributions for heat transfer coefficient for film surrounding a particle

h_r radiation contribution for the film heat transfer coefficient

h_{rad} radiative contribution to the apparent wall heat transfer coefficient

h_w apparent wall heat transfer coefficient

h_{wf} wall heat transfer coefficient for the fluid

h_{wp} wall heat transfer coefficient for the solid phase

$(-\Delta H)$ heat of reaction (adsorption)

I_o, I_1 modified Bessel functions of the first kind, zero and first orders respectively

k kinetic constant

k_c gas phase mass transport coefficient referred to unit interfacial area

k_f film mass transfer coefficient

k_g molecular thermal conductivity of the fluid

k_p pellet effective thermal conductivity

k_r radial effective thermal conductivity

k_{rad} radiative contribution to the static radial thermal conductivity

k_{rf} radial thermal conductivity of the fluid

k_r^o defined by equation $|14r|$

k_{rp} radial thermal conductivity of the solid

K constant separation factor for adsorption isotherm

L reactor (column) length

n' number of microspheres in the pellet

N_r reaction number

N_s interphase heat transfer group

Nu_{fp} fluid/solid Nusselt number

Nu_{wf} fluid/wall Nusselt number

P pressure

Pe_{ha}^f fluid phase axial Peclet number

Pe_{hr} radial Peclet number for heat transfer

Pe_{hr}^r radial fluid Peclet number for heat transfer

Pe_{mr} radial Peclet number for mass transfer

Pe_{hz}, Pe_{ma} axial Peclet number for mass transfer

$Pe_r(\infty)$ turbulent limit Peclet number

q adsorbed solid concentration

Q   adsorbed solid concentration in equilibrium with c_o

r   dimensionless radial coordinate

r'   radial coordinate

r_A intrinsic rate of disappearance of component A per unit volume of pellet

r_c zeolite crystal radius

R   reactor radius

Re   particle Reynolds number

R_p catalyst pellet radius

s   Laplace transform parameter

Sc   Schmidt number

t   time

t_{Bp} breakthrough time

t_{st} stoechiometric time

T   temperature

T_o reference temperature

T_p temperature inside the pellet

T_s temperature at the surface of the pellet

T_w wall temperature

$\hat{T}, \tilde{T}$ reduced temperature

u   superficial gas velocity

u_i intersticial velocity

U   flowrate

$\tilde{U}$ overall heat transfer coefficient

v   reactor (adsorber) volume

V_p, V_a pellet volume

V_i volume of a microsphere

x   reduced fluid phase concentration

$\overline{X}$ radially averaged ratio, C/C_o

484

Greek letters

α shape factor

α heat transfer number

β Prater thermicity factor

$\bar{\beta}$ Prater thermicity factor in the bulk conditions

γ fraction of active sites at the microsphere surface or at pore walls

γ Arrehnius number $(-\Delta H/RT_o)$

ϵ reactor (adsorber) voidage

ϵ_p pellet porosity

η effectiveness factor

η_a effectiveness factor, defined by equation $|8|$

η_i effectiveness factor for a microsphere

$\bar{\eta}_i$ average microeffectiveness factor

η_{ov} overall effectiveness factor

θ dimensionless time

λ_e effective thermal conductivity

ρ_a, ρ_s apparent density of support

ρ_b bulk density (density of the catalyst bed)

ρ_t true density

τ space time

τ^* time constant (2nd order dynamic system)

ξ_h heat capacity factor

ξ_m mass capacity factor

ω volume fraction of the pellet occupied by zeolite crystals

Ω parameter, defined by equation $|1|$. Measures the relative importance between macropore and micropore mass transfer resistances

ζ damping factor (2nd order dynamic system)

REFERENCES

(1) Breck,D., Zeolite Molecular Sieves: Structure, Chemistry and
     Use (New York: John Wiley & Sons, 1973)

(2) Lee,M., Novel Separation with Molecular Sieve Adsorption, in
     N. Li,ed., Recent Developments in Separation Science, vol. 1,
     (New Jersey: CRC Press,1972), pp. 75-112.

(3) Heineman,H., Cat.Rev.Sci.Eng. 23 (1981) 315

(4) Meugel,M., Informations Chimie 227 (1982) 99

(5) Rosset,A.,Neuzil,R. and D. Broughton, Industrial Applications
     of Preparative Chromatography, in A. Rodrigues and D.Tondeur,
     eds., Percolation Processes: Theory and Applications (Alphen
     den Rijn: Sijthoff & Noordhoff, 1981), pp. 249-281.

(6) Jacob,P., NATO ASI "Zeolites: Science and Technology", Lisbon
     (1983)

(7) Kotsis,L.,Argyelan,J.,Szolcsanyi,P. and K.Patak, React.Kinet.
     Cat.Lett. 18 (1981) 149

(8) Dubinin,M.,Erashko,I.,Ulin,V.,Voloschuk,A. and P.Zolotarev,
     Diffusion Processes in Biporous Adsorbents and Catalysts, in
     G.Boreskov and Kh.Minachev,eds., Application of Zeolites in
     Catalysis (Budapest: Akademiaikiado, 1979), pg. 161

(9) Garg,D. and D.Ruthven, Chem.Eng.Sci. 28 (1973) 791

(10) Lee,L. and D.Ruthven, IEC Fund. 16 (1977) 290

(11) Ma,Y. and T.Lee, IEC Fund. 16 (1977) 44

(12) Ruthven,D. and I.Doetsch, AIChE J. 22 (1976) 882

(13) Doelle,H. and L.Riekert, in J.Katzer,ed., Molecular Sieves II,
     ACS Symp.Ser. 40 (1977) 401

(14) Kumar,R.,Duncan,R. and D.Ruthven, Can.J.Chem.Eng. 60 (1982)493

(15) Weisz,P., Chem.Tech. 3 (1973) 498

(16) Ihm,S.K.,Suh,S and I.Oh, 2nd World Congress in Chemical
     Engineering, Montreal, (1981)

486

(17) Rodrigues,A., Scientific Basis for the Design of Two Phase
     Catalytic Reactors, in A.Rodrigues, N.Sweed and J.Calo,eds.,
     Multiphase Chemical Reactors (The Hague: M.Nijhoff, 1982),
     pp. 65-133.

(18) Aris,R., The Mathematical Theory of Dispersion and Reaction
     in Permeable Catalysts (Oxford: Clarendon Press, 1975)

(19) Deans,H.A. and L.Lapidus, AIChE J. 6 (1960) 656

(20) Froment,G.F., Adv.Chem.Ser. 109 (1972) 1

(21) Gros,J.B. and R.Bugarel, The Chem.Eng.J. 13 (1977) 165

(22) Sunderasan,S.,Amundson,N.R. and R.Aris, AIChE J. 26 (1980) 529

(23) Hlavacek,V. and P.Rompay, Chem.Eng.Sci. 36 (1981) 1587

(24) Young,L.C. and B.C.Finlayson, Ind.Eng.Chem. Fundam. 12(1973)412

(25) Mears,D.E., Ind.Eng.Chem. Fundam. 15 (1976) 20

(26) Feyo de Azevedo,S., "Modelling and Operation of a Tubular
     Fixed Bed Catalytic Reactor", Ph.D. Thesis, University of
     Wales (Swansea), U.K. (1982)

(27) De Wasch,A.P. and.G.F.Froment, Chem.Eng.Sci. 26 (1971) 629

(28) Dixon,A.G. and Cresswell,D.L., AIChE J. 17 (1979) 247

(29) Feyo de Azevedo,S. and A.P.Wardle, Heat Transport in Packed
     Bed - A Model Equivalence Study, to be published.

(30) Gunn,D.J., Trans.Inst.Chem.Eng. 47 (1969) T341

(31) Gunn,D.J., Int.J. of Heat and Mass Transfer 21 (1978) 467

(32) Argo,W.B. and J.M.Smith, Chem.Eng.Progr. 49 (1953) 443

(33) Adderley,C.I.,"The Dynamics and Stability of Fixed Bed
     Catalytic Reactors", Ph.D. Thesis, University of Leeds, U.K.
     (1973)

(34) Zehner,P. and E.P.Schlünder, Chem.Ing.Tech. 42 (1970) 933

(35) Calo,J., ACS Symposium Series 65 (1978) 530

(36) Ponzi,P. and L.Kaye, AIChE J. 25 (1979) 100

(37) Balakotaiah,V. and D.Luss, AIChE J. 27 (1981) 442

(38) Orr,N.,Cresswell,D. and D.Edwards, Ind.Eng.Chem.Proc.Des.Dev. 22 (1983) 135

(39) Rodrigues,A., Modelling of Percolation Processes, in A.Rodrigues and D.Tondeur,eds., Percolation Processes: Theory and Applications (Alphen den Rijn: Sijthoff & Noordhoff, 1981)

(40) Helfferich,F. and G.Klein, Multicomponent Chromatography (New York: M. Dekker, 1970)

(41) Tien,C.,Hsieh,J. and R.Turian, AIChE J. 22 (1976) 498

(42) Hsieh,J.,Turian,R. and C.Tien, AIChE J. 23 (1977) 263

(43) Morbidelli,M.,Servida,A.,Storti,G. and S.Carra, Ind.Eng.Chem. Fundam. 21 (1982) 123

(44) Wong,Y. and J.Niedzwieck, "A Simplified Model for Multicomponent Fixed Bed Adsorption", AIChE Meeting, Orlando (1982)

(45) Sereno,C., Internal Report, Technical University of Denmark, Lyngby (1983)

(46) Santacesaria,E.,Morbidelli,M.,Danise,P.,Mercenari,M. and S. Carra, Ind.Eng.Chem.Proc.Des.Dev. 21 (1982) 440

(47) Santacesaria,E.,Morbidelli,M.,Servida,A.,Storti,G. and S.Carra, Ind.Eng.Chem.Proc.Des.Dev. 21 (1982) 446

(48) Carra,S.,Santacesaria,E.,Morbidelli,M.,Storti,G. and D.Geiosa, Ind.Eng.Chem.Proc.Des.Dev. 21 (1982) 451

(49) Rhee,H. and N.R.Amundson, The Chem.Eng. J. 1 (1970) 241

(50) Rhee,H.,Heerdt,E. and N.R.Amundson, The Chem.Eng. J. 1(1970)279

(51) Rhee,H.,Heerdt,E. and N.R.Amundson, The Chem.Eng.J. 3(1972)22

(52) Rhee,H. and N.R.Amundson, The Chem.Eng. J. 3 (1972) 121

(53) Pan,C. and P.Basmadjian, Chem.Eng.Sci. 22 (1967) 285

(54) Pan,C. and P.Basmadjian, Chem.Eng.Sci. 25 (1970) 1653

(55) Pan,C. and P.Basmadjian, Chem.Eng.Sci. 26 (1971) 45

488

(56) Basmadjian,P.,Ha,K. and C.Pan, Ind.Eng.Chem.Proc.Des.Dev. 14 (1975) 328

(57) Basmadjian,P.,Ha,K. and D.Proulx, Ind.Eng.Chem.Proc.Des.Dev. 14 (1975) 340

(58) Basmadjian,P., Ind.Eng.Chem.Proc.Des.Dev. 19 (1980) 137

(59) Basmadjian,P., AIChE J. 26 (1980) 625

(60) Sweed,N., "Nonisothermal and Nonequilibrium Fixed Bed Sorption", in A.Rodrigues and D.Tondeur,eds., Percolation Processes: Theory and Applications (Alphen den Rijn: Sijthoff & Noordhoff, 1981)

(61) Chi,C. and D.Wasen, AIChE J. 16 (1970) 23

(62) Marcussen,L. and C.Vinding, Chem.Eng.Sci. 37 (1982) 311

(63) Marcussen,L., Chem.Eng.Sci. 37 (1982) 299

(64) Jacob,P. and D.Tondeur, 5th Scandinavian Congress of Chemical Engineering (1980)

(65) Jacob,P. and D.Tondeur, Sep.Sci. and Tech. 15 (1980) 1563

(66) Ruthven,D.,Lee,L. and M.Yucal, AIChE J. 16 (1980) 16

(67) Carleton,F.,Kershenbaum,L. and W.Wakeham, Chem.Eng.Sci. 33 (1978) 1239

(68) Brunovska,A.,Hlavacek,V.,Ilavsky,J. and J.Valtyvi, Chem.Eng. Sci. 33 (1978) 1385

(69) Brunovska,A.,Hlavacek,V.,Ilavsky,J. and J.Valtyvi, Chem.Eng. Sci. 35 (1980) 757

(70) Rudobashta,S. and A.Planovskii, Zhurnal Prikladnoi Khimii, 50 (1976) 804

(71) Ilavsky,J.,Brunovska,A. and V.Hlavacek, Chem.Eng.Sci. 35 (1980) 2475

(72) Brunovska,A.,Ilavsky,J. and V.Hlavacek, Chem.Eng.Sci. 36 (1981) 123

(73) Wu,P.,Debebe,A. and Y.Ma, AIChE Meeting, Orlando (1982)

(74) Kmiotek,S.,Wu,P. and Y.Ma, submitted for publication.

(75) Carter,J., Trans.Inst.Chem.Eng. 46 (1968) 1213

(76) Ikeda,K., Chem.Eng.Sci., 34 (1979) 1941

(77) Ozil,P. and L.Bonnetain, Chem.Eng.Sci. 33 (1978) 1233

(78) Helfferich,F., Journal of Chemical Education 59 (1982) 646

(79) Rodrigues,A.,Almeida,F.,Machado,I.,Costa,C. and J.Loureiro, Chempor 81, Braga, Portugal (1981)

(80) Sircar,S.,Kumar,R. and K.Anselmo, Ind.Eng.Chem.Proc.Des.Dev. 22 (1983) 10

(81) Sircar,S. and R.Kumar, Ind.Eng.Chem.Proc.Des.Dev. (1983a) in press.

(82) Sircar,S. and R.Kumar, Ind.Eng.Chem.Proc.Des.Dev. (1983b) in press.

(83) Loureiro,J.,Costa,C. and A.Rodrigues, submitted to The Chem. Eng. J.

(84) Loureiro,J., Thesis in preparation (Univ. Porto,Portugal)

ENGINEERING ASPECTS OF CATALYTIC CRACKING

H. de LASA

Chemical Engineering Department
Faculty of Engineering Science
The University of Western Ontario
London, Ontario, Canada   N6A 5B9

The technology. State of the art and expected progress.

The technology of fluidized bed catalytic cracking (FCC) has shown a remarkable change in the last 30 years. The conventional FCC process, intensively applied during the 50's, being basically the combination of two dense fluidized beds (the reactor and the regenerator) and two transport lines, may be considered as the *first generation* of catalytic crackers. The heat required for the endothermic cracking reactions was supplied by the exothermal coke combustion. From an overall view point the industrial process was operated under conditions close to the thermal equilibrium where the silica-alumina catalyst was transferring the heat from the hot regions (regenerator) to the cold regions (reactor) and vice-versa. (Fig. 1).

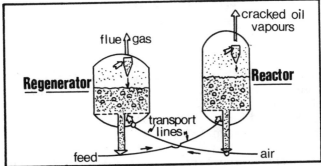

Fig. 1   Schematic Description of the First
          Generation of Crude Oil Catalytic
          Cracking Process

However, the commercial introduction of the microcrystalline silica-alumina catalysts (zeolites) at the beginning of the 60's, indicated the need of handling the overall process of the catalytic cracking in a new system where the remarkable activity and selectivity of zeolites (1) were fully used. This situation generated the interest in a new FCC concept, the *second generation* of catalytic crackers, where the catalytic reaction was conducted in a shorter period of time (2),(3),(4). In fact, the cracking reactions were started in the transport line, conveying the catalyst to the reactor (3-4s) and they were proceeded in a shallow dense fluidized bed (3-4s) reducing in this way the undesirable over-cracking phenomena (3), (Fig. 2). This overcracking, resulting from the secondary reactions and normally generating excessive amounts of coke and gases, was too dominant in deep fluidized beds constituted by the so active zeolites (5).

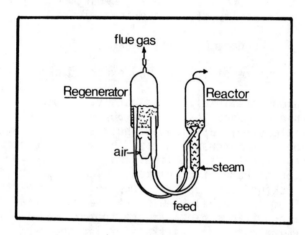

Fig. 2   Schematic Description of the Second Generation of Crude Oil Catalytic Cracking Process

Consequently, the second generation of transported-shallow bed FCCs represent an important effort to take full advantage of zeolitic catalysts minimizing at the same time these undesirable cracking transformations. A typical example of this process is the design patented by Texaco (6).

However, during the late 60's it was stressed the importance of a full use of the specific activity of the new microcrystalline silica-alumina catalysts (zeolites) recently introduced in the market (6),(7). Some difficulties were nevertheless forecasted (2) for the use of these materials having a dominant microporous structure (2-7A), (9), much smaller than the amorphous silica-alumina (30-70A) (10). In favour of zeolites a 100-10000 times improved activity, in relation with the amorphous silica-alumina, was claimed (1). This fact coupled with an improved hydrogen transfer capability, presumably a consequence of hydrogen mobility between cracked molecules, was indicating an important potential of zeolitic materials for increasing conversions to gasoline with low coke yields (9),(11),(12). Besides that, the growing refinery problem resulting from the necessity of processing crudes of different source and quality, showed that an accurate control of reaction times between the catalyst and the hydrocarbons was a must for FCC crackers. This approach was a very promising concept for the FCC units. In this way was born the *third generation* of crackers based on a reactor which was, in fact, a transport line unit (2),(13),(8),(14). This concept was developed in several patents, case of Exxon in the Flexicracking unit, Gulf in Gulf FCC unit, Kellogg in the Orthoflow and the Heavy oil processes, Standard Oil in the Ultracat technology, UOP in the riser cracker units (2), (15),(16),(17),(18),(19),(20).

The peculiar characteristics of FCC transported reactors (21) provided better control of gasoline yields (60-65%) improved feedstock conversions (75-80%) and smaller coke yields (coke/feed:3-6%). For instance, the capability of zeolitic catalysts for cracking naphtenic and parafinic hydrocarbons avoiding at the same time the slow decomposition of the aromatic molecules, normally generating high coke yields, was a significative progress for the FCC technology (4). It must be stressed at the same time that the introduction of the riser reactors in the *third generation* of FCC units was still combined with dense fluidized bed regenerators. However, the high thermal resistance of the zeolites allowing high regeneration temperatures, 740°C instead of 630°C without a noticeable particle deactivation or sintering (16),(22) gave fast coke combustion rates and low CRC (coke on regenerated catalyst) levels. In this way, CRC of the order of 0.05 - 0.2% were obtained (23),(24), (25),(4),(26). These low CRC levels seem to be an important condition for reactivating the zeolitic sites on the catalysts and for achieving a maximum use of zeolite surface (27). Other advantages like a better overall plant thermal balance (14), a reduced recycled oil streams (8) were also claimed for the riser cracking reactors.

Finally, more recent developments in the manufacturing of the cracking catalyst have been oriented to increase the zeolite performance. For instance, it has been analysed in which way the type of zeolite affects the conversion and the selectivity (28),(29), how the resistance to the attribution or the R.O.N. yields on gasoline could be improved (30),(2), how the hydrocarbon molecule size affects the zeolite performance (1),(9).

In spite of all this progress, during the mid 70s the FCC units were challenged again. The energy crisis resulting from the first Arab embargo showed that new modifications or changes were required to face the new situation. In fact, new problems needed to be considered for efficiently processing the heavy crudes (14). The bottom of the barrel or heavy crudes, fractions that could not have been economically cracked before the 1975 crisis, were more and more considered for this purpose (31). This situation continues to be the present challenge for the refineries. However, another problem appeared together with the ones already described. Particularly the growing tendency to mix a bigger recycled or heavy crude fraction in the cracker feed (26), increased the catalyst contamination with nickel and vanadium metals (33). These metals once deposited on the zeolitic matrix favour the dehydrogenation reactions giving more hydrogen and more coke (33),(14),(34),(35), (36),(37),(38). At the same time metals such as nickel and vanadium reduce the active sites and the specific active surface per unit weight decreasing the overall catalytic activity and selectivity to produce gasoline. Some authors claimed that nickel is around four times as bad as vanadium at the same concentrations (39),(36),(38). As a consequence and for having a standardized basis of reference an "effective metal concentration" on the catalyst is defined. This effective parameter is four times the nickel level plus the vanadium concentration (36),(32). It was also pointed out, however, that nickel and vanadium activity would probably be related to other factors like thermal activity history (40), (38),(41). The refineries would have then, with the heavy crudes, the potential problems resulting from additional coke formation and smaller catalyst activity (14).

In fact, typical coke yields for heavy crudes have given coke concentrations values as high as 8-16%. (The basis for these yields is the fresh feed), (31),(14). A possible way of tackling the problem is either the use of antimony passivators of nickel and vanadium as suggested by Phillips Petroleum (42),(39) or the employment or the new families of zeolitic catalyst, the Redsicat, the GRZ and the Filtrol (24),(34). These catalysts would allow the cracking process to perform with significant nickel and vanadium levels, permitting at the same time adequate operation of the FCC units with relatively low coke and gas yields. It is expected that these catalysts would provide high crude oil conversions, with effective $(4N_i + V)$ concentrations as high as 5000 ppm, conditions where a standard cracking catalyst would have shown a noticeable activity decay (24). We may visualize then that we assist today to the development of a significant technical effort towards the production of new FCC catalysts required for the processing of heavy crudes.

At the same time, the operation of the FCC units has been contested again by the new environmental regulations. The problem of how to reduce the CO emissions, transforming CO in CO_2, and gaining in the same operation the CO heat of combustion has been one of the major refinery concerns during the 70s, (44),(45),(43). A first approach involved the use of CO burners, located in the plant after the regenerator. These burners transformed the CO gases producing at the same time steam for the cracking plant (44). However, a more advanced concept seeks a direct CO to CO_2 conversion in the regenerator itself. With this purpose catalytic materials named *combustors* were developed (23),(45).

Three basic types of combustors have been suggested: - a platinum group metal (Pt, Pd, Ir) deposited in small concentrations on the cracking catalyst, - a solid promotor mixed to a cracking catalyst without additives (0.1-1%) - a liquid additive injected into the regenerator, (42). Unfortunately, the specific references about the chemicals promoting these effects are very incomplete. For the case of the platinum group metal it would seem that the metal is incorporated to the zeolitic matrix in such a way that would be only contacted by CO and O_2 but not by the bigger hydro-carbon molecules. In this manner the CO transformation would be promoted and the adverse dehydrogenation reactions catalyzed by the same metal controlled (45). These combustors, which normally limit the CO emissions to levels below 500 ppm (maximum allowed CO environmental emission concentration) (46) have shown to be a more appropriate technology than the CO external boilers (23),(45). In fact, through the advanced CO transformation method significant capital gains have been claimed as well as important reductions of CO concentrations in the dilute regenerator phase (250-500 ppm) (43),(35),(14). This low CO concentration in the entrainment phase region is possibly a very important factor for eliminating the troublesome CO oxidation in the lean regenerator phase (postburning) (45). In this way, the uncontrolled temperature increase in the upper part of the regenerator, consequence of CO postburning and frequent cause of catalyst sintering and cyclone damages, is avoided.

However, in spite of this progress no technical information is, at least to our knowledge, available to predict the behaviour of the catalytic combustors in front of heavy oils with high nickel and vanadium contents. This is, in our opinion, a research area where significant technological improvement must be achieved in the next few years to really make of the FCC a technological break-through.

KINETICS OF CATALYTIC CRACKING

When a gas oil is catalytically cracked, an ample diversity of chemicals, going from hydrogen and methane to coke forming poly-meric materials deposited on the catalyst, are generated. The most popular kinetic models used for the process description are basically based on a simplified representation involving three characteristic groups of reactants and products:

A_1: gas oil; A_2: gasoline(C_5^+ - 210°C); A_3: butanes, light, gases and coke

This certainly leads to the conception of the catalytic cracking of the gas oil as follows (48),(49),(50),(51),(52),(53).

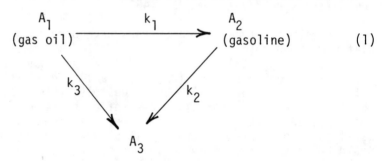

$$\begin{array}{ccc} A_1 & \xrightarrow{\ \ k_1\ \ } & A_2 \\ \text{(gas oil)} & & \text{(gasoline)} \end{array} \qquad (1)$$

(butanes, light gases and coke)

It is important to point out that the catalytic process may be interfered to some extent by the thermal cracking reactions which simultaneously occur. Corrections have been suggested to the gas oil conversion to discount the noncatalytic effect (54),(55),(56). These corrections are around 2% at 490°C (51).

Consequently, the scheme presented above with equation [1] may be a usual approach to predict both the overall gas oil conversion and the selective transformation of gas oil in gasoline. Certainly the second aspect is a key factor for an efficient operation of the FCC units. These two parameters have been estimated by several kinetic models. The most frequently adopted kinetic representation (52),(57),(58),(50),(51),(59), considers that the overall gas oil transformation is a second order decomposition and the further gasoline cracking is a first order reaction.

$$r_1 = k_0 \phi_1 (t_c) y_1^2 \quad \text{(gas oil cracking)} \qquad [2]$$

$$r_2 = k_2 \phi_2 (t_c) y_2 \quad \text{(gasoline cracking)} \qquad [3]$$

This difference between the gas oil and the gasoline cracking reaction orders may be explained as follows:

- The gasoline is a mixture of hydrocarbons with boiling points varying in a limited range. In this sense it is normal to expect that the gasoline will closely behave like a pure hydrocarbon showing a reaction order equal to one (51).

- Conversely, the gas oil is a far more complex mixture with an ample diversity of cracking reaction rates. In this respect the gas oil cracking is the summation of a large number of reactions acting in parallel with very different kinetic constants. The consequence is an overall reaction order larger than one (60). These facts have been justified both intuitively and theoretically (57), (61), (62).

Another kinetic scheme (63), (49) based on a similar idea to equation [2], that use instead of weight fractions the more appropriate molar fractions, introduces a parameter named crude oil refractoriness. This parameter lumps in a single factor the cracking feedstock and catalyst characteristics:

$$r_1' = k_0'\phi(t_c) \, y_1'^{W+1} p^{W+1} \quad \text{(cracking of crude oil)}[4]$$

The parameter W is certainly different from zero for conventional crudes which normally have boiling points varying in a wide temperature range (around 280-550°C). The interesting aspect of this model is the condition of a W parameter oscillating between 0 and 3.3 and showing some temperature dependence (64), (55), (65), (66), (49), (67), (62). The lower W values corresponded to experiments with a variety of cracking catalysts where some diffusional limitations were suspected. The higher W were considered for LaY zeolitic catalysts where diffusional limitations could probably be neglected (28). In any case it would seem that for zeolites cracking heavy feedstocks the catalyst operation with important diffusional limitations is highly possible (9).

At the same time on equations (2), (3) and (4) it may be noticed that two time dependent parameters are included. Both $\phi_1(t_c)$ and $\phi_2(t_c)$ introduce in these expressions the catalytic activity decay factors. It may also be observed that both $\phi_1(t_c)$ and $\phi_2(t_c)$ are functions of t_c, the *reaction time* or the *catalyst on stream time*. This time is a key variable when modeling a heterogeneous reactive system like the FCC process. The cause of this activity decay is mainly related to the deposition or adsorption of the polyaromatic compounds which polymerize on the catalyst surface forming the coke. (59), (61), (68), (29). This polymerization appears activated by the olefins capability of abstracting hybride ions from naphtenic and aromatic hydrocarbons (9), (56), (69) which produces a highly condensed aromatic solid residue (70). In spite of the general acceptation of these facts some arguments were presented (56), (71) indicating the difficulties of formulating a simple relationship between catalyst activity and coke concentration.

Moreover and because it is normally assumed that the same type of active sites will crack both gas oil and gasoline molecules, an identical set of $\phi_1(t_c)$ and $\phi_2(t_c)$ have been proposed (58). Under these conditions equations [2] and [4] may be written as follows:

$$r_1' = k_0'\phi(t_c) \cdot y_1'^{W+1} \cdot p^{W+1} \quad\quad\quad\quad\quad\quad [5]$$

$$r_2' = k_2'\phi(t_c) \cdot y_2' \cdot p^{W+1} \quad\quad\quad\quad\quad\quad\quad [6]$$

Different types of $\phi(t_c)$ functions have been proposed to represent both the catalyst activity decay and the coke formation. In particular, two of these mathematical forms are the most popular ones. One of them considers that the catalyst activity decay may be represented with a power law, $\phi(t_c) = a\, t_c^{-n}$(72), (60), (51), (73). The other one approaches the $\phi(t_c)$ function using a decreasing exponential $\phi(t_c) = b\, \exp(-\alpha t_c)$,(59), (52). Some argue-ments like a finite catalyst activity when $t \to 0$ (59) and the possible relationship of coke formation with an irreversible adsorption mechanism would support the exponential decay law (74).

It must however, be indicated that some efforts were oriented to modify the power law correcting essentially the anomaly when $t_c \to 0$, $\phi(t_c) = a(1 + G\, t_c)^{-n}$ with $G = (m-1)K$ and $n = 1/(1-m)$ (61), (75), (64), (63), (65), (76), (77), (78). Moreover, this equation has the additional advantage of allowing the derivation of an exponential law when m=1. In other words the exponential law is a particular case of the modified power law (63). This modified power law expression is also supported by experimental evidence which showed n and m values changing between 0.6-30 and 1.03-2.57, respectively (64), (66). Then the observed m values, depending on the couple catalyst feedstock and on the operating conditions, match the exponential behaviour, m = 1 only in a few cases which certainly showed the convenience of a more general $\phi = a(1+G\, t_c)^{-n}$ expression.

It is important to mention, however, that both exponential or modified power laws contain a major simplification, the dependence of hydrocarbon concentration on $\phi(t_c)$ is ignored. This hypothesis does not always seem appropriate for fixed bed catalytic cracking laboratory units (79), (20). A modified exponential law has been proposed (81) as a more suitable $\phi(t_c)$ relationship.

At the same time other effects, having consequences on catalyst activity, were described. For instance the action of nitrogenated bases on silica-alumina and zeolitic catalysts received particular attention. These compounds, coming along with the feedstock stream, would decrease the acidity of the catalyst matrix reducing in this way its activity and selectivity (82). Nevertheless, the approp-riate way of including this effect in the kinetic model is, at the present time, quite unclear.

We will consider now the interest of equations [5] and [6] for defining the gas oil conversion and gasoline formation rates. In fact, once those expressions are derived, it is possible to estimate the selective transformation of gas oil in gasoline.

At the same time because gas oil conversion tests were frequently performed in laboratory scale fixed bed reactors, some interesting observations were developed for those systems. It has been realized, for instance, that in laboratory scale units t_v, the residence time of the hydrocarbon mixture, is normally much smaller than the catalyst on stream time $(t_v \ll t_c)$ (58). In fact t_c is in the order of a few minutes when in opposite t_v is only in the order of 1-2 seconds.

When $t_v \ll t_c$, the gas oil molecules travel through the reactor with such a high speed, in relation with the rate of activity decay, that the bed may appear to them as having a uniform activity (50), (57). Under these conditions the following equations may be proposed to describe the behaviour of the cracking reactor.

$$-\frac{dA_1}{dt_v} = \phi(t_c)k_0{}'y_1{}^{,W+1}p^{W+1} = \phi(t_c)(k_1{}' + k_3{}')y_1{}^{,W+1}p^{W+1} \qquad [7]$$

$$\frac{dA_2}{dt_v} = \phi(t_c)(S\ k_1{}'y_1{}^{,W+1}p^{W+1} - k_2{}'y_2{}'P) \qquad [8]$$

If the equation [8] is divided by equation [7] the differential change of y_2 with respect to y_1 may be derived. It must be mentioned however, that equation [9] considers constant total pressure and neglects the change of the molecular weight of the mixture.

$$-\frac{dy_2}{dy_1} = \frac{k_1}{k_0} - \frac{k_2}{k_0}\frac{y_2}{y_1{}^{W+1}} \qquad [9]$$

An integration of equation [9] allows us to derive the selective conversion of gas oil in gasoline (58), (50).

$$y_2 = (k_1 k_2/k_0{}^2)\ \exp\ (-\frac{k_2}{k_0}\ y_1)\ [\frac{k_2}{k_0}\ \exp\ (\frac{k_2}{k_0})\ -$$

$$y_1\frac{k_0}{K_2}\ \exp\ (\frac{k_2}{k_0}\cdot\frac{1}{y_1})\ +\ E\ (\frac{k_2}{k_0}\cdot\frac{1}{y_1})\ -\ E(\frac{k_2}{k_0})\ ] \qquad [10]$$

with $E(x) = \int_{-\infty}^{x}\frac{e^u}{u}\ du$ = exponential integral

It is important to stress that the integrated form of equation [9] provides an expression for estimating the *instantaneous* gas oil conversion in gasoline. This expression, allowing the prediction of a maximum gasoline yield for a given set of kinetic parameters, is certainly different from the measured *mean* selectivity values $(\overline{y_2}/(1-\overline{y_1})$ obtained when fixed bed runs are employed for testing the adequacy of cracking kinetic models. These $\overline{y_2}/(1-\overline{y_1})$ selectivity values are avarages performed during the t_c total time on stream period.

Another interesting question is given by the fact that equation [9] or its integrated form allows to visualize that the instantaneous selectivity is not dependent on $\phi(t_c)$, catalyst activity decay function, none of the following factors: t_c, t_v or catalyst to gas oil ratio (48), (54). It must nevertheless be stressed that $y_2 \neq f(\phi(t_c)$ behaviour results of $\phi_1(t_c) = \phi_2(t_c)$ postulate, a common hypothesis of difficult verification.

Conversely, it has been shown that the average selectivity, a frequent result of fixed bed laboratory units, depends on $\phi(t_c)$, t_c, t_v and catalyst/gas oil relationship (48), (50). This has stimulated researchers to design catalytic cracking experiments in such a way that the key instantaneous selectivity parameter could be estimated. The importance of this instantaneous parameter may also be understood in the light of its interest to define the performance of industrial FCC units (48).

A most valuable way of finding this instantaneous selectivity function, $y_2 = f(y_1)$ is through the determination of an evolvent curve which limits all the possible average selectivity functions (52). This instantaneous y_2 curve is, at the same time, an optimum selectivity curve, giving the maximum possible yield for a certain couple crude feedstock-catalyst type. For that reason it has been named the *optimum performance envelope* (OPE) (49), (63), (76).

Another aspect of high interest for the process of crude oil cracking is related to the way the different crude oil types, available as potential refinery feedstocks, affect the kinetic scheme already described. It has been verified that, in general terms, this scheme is still valid. Only changes in the kinetic constants, k_0, k_1, k_2 and the α activity decay constant may be expected (50). This analysis was performed considering three typical groups or classes of gas oil constituents:-aromatic compounds, α, parafinic compounds, p, naphthenic compounds, n, (51).

It was possible to observe through these studies that crude oils with a high parafinic and naphtenic hydrocarbon content present higher k_0 and k_1 and smaller k_2 and α values (83). A generalization of this concept, including a factor weighting the aromatic to naphtenic fraction in k_0 and k_1 definition was proposed (84).

These facts showed the high crackability of the p and n compounds and their small tendency to form coke (15). The reverse is also right. The hydrocarbon mixtures with a high aromatic fraction showed smaller k_0 and k_1 and higher α and k_2 values. In a word the crudes with a significant fraction of a compounds will have a higher refractivity to be cracked and a stronger tendency to form coke (51), (50), (61), (15).

Finally, more advanced models have been proposed adding more specifications to the parafinic, naphtenic and aromatic basic distinction. For instance, the hydrocarbons were classed in light and heavy, pure aromatic compunds or branced to parafinic or naphtenic groups (61). This last specification provides a more accurate description of feedstock cracking behaviour. For example, this kind of model is compatible with observations which indicate that aromatic rings have a special capability to activate cracking on side branches (85). However, in counterpart with these approaches, the simplicity of kinetic representations, as the ones given by equations [2] and [3] is partially lost.

To complete the modeling of catalytic cracking the dependence of reaction rates with temperature was also extensively analysed. Typical energies of activation recommended for the 480-540°C temperature range (61) may be classified as follows: - for the reactions of gasoline and coke formation from p and n hydrocarbons, E = 5-9 Kcal/mol (51) - for the reactions of gasoline and coke formation from α hydrocarbons, E = 14=18 Kcal/mol - for the transformation of gasoline and coke, E = 20 Kcal/mol. These values are substantially different from the 50 Kcal/mol levels suggested for the gas oil cracking (66), (49), (67), (69) and for the gasoline formation reaction (69).

Again the possible influence of internal diffusional controls on the reaction rates may explain this difference. This view is confirmed by the fact that the difference of the energy of activation appears coupled with a variation on the overall cracking reaction orders (49), (67), (69).

KINETIC OF REGENERATION

The combustion of the coke deposited on the matrix of a cracking catalyst particle is a process resembling the burning of graphite (86). Some arguments like the presence in the coke of turbostratic layers similar to the graphite structures were advanced (87). It is also important to point out that some similarities on the overall kinetic behaviour of coke burning has further supported this view (86).

However, coke presents a hydrogen fraction intimately associated to the carbon that cannot be neglected. This hydrogen fraction largely depends on the effectiveness of the hydrocarbon stripping operation at the end of the cracking reactor. In fact, the catalyst is stripped with steam at the exit of the cracker to remove the hydrocarbons remaining in the porous matrix. This explains the coke formula reported in the literature, CH_n with n ranging between 0.4 to 1 (88), (89), (86). Even for the first coke fractions burned some authors propose $n = 2(88)$. It is important to stress that the uncertainty in the hydrogen content definition has concrete consequences on the prediction of the heat of coke combustion, key parameter for modeling an industrial regenerator (47). A direct estimation of hydrogen in coke is then highly advisable for an appropriate regenerator modeling (90).

In any case, a reasonable average of coke composition seems to be $CH_{0.7}$ (90). Under those conditions the couple of stoichiometric equations describing CO and CO_2 formation may be written as follows (90).

$$CH_{0.7} + 0.675 \ O_2 \xrightarrow{\ k_4\ } CO + 0.35 \ H_2O \qquad [11]$$

$$CH_{0.7} + 1.175 \ O_2 \xrightarrow{\ k_4\ } CO_2 + 0.35 \ H_2O \qquad [12]$$

A prevailing approach to represent the coke combustion is based on the assumption that the rate of coke disappearance is a first order law with respect to both oxygen and coke concentrations (86), (91).

$$r_c = (k_4 + k_5)C_c \ y_3 \ P \qquad [13]$$

In this sense this kinetic model considers that the coke burning rate may be seen as a graphite combustion rate, giving no significant weight to the coke heterogeneities. This equation has proved to be adequate for 60 μm particles and coke concentrations in the order of 6% or smaller than 6% for both amorphous silica-alumina (86) and zeolitic catalysts (91). It is important to mention, however, that this equation is only appropriate to describe the rate of combustion of carbonaceous deposits if the coke has been well stripped of volatile hydrocarbons. Otherwise equation [13] may be questionable (92). A second order equation describing a combination of vapourization and oxidation during the early stages of catalyst regeneration may be more appropriate.

It has also been shown that 60 μm cracking particles with a coke concentration above 6% level experienced some diffusional limitations derived of a non-uniform coke distribution in the internal surface of the catalyst (86). It seems then, highly advisable to limit the coke deposit below 6% in order to have a coke combustion process dominated by the intrinsec reaction rate.

This also may contribute to an appropriate modeling of the regeneration process (90), (93).

Another interesting property of equation [13] is given by the fact that this relationship is still valid for cokes formed with different types of feedstocks on differenct cracking catalysts (86), (91). Some significant variations were, however, observed on the specific values assigned to the kinetic parameters for different catalysts. For instance, energies of activation of 35-41 Kcal/mol were reported for coke burning on silica-alumina (86), (89), (94) (88). These values are very close to the typical energies of activation obtained by burning graphite surfaces (86). Nevertheless, coke combustion runs in a similar temperature range (420-630°C) but using a series of zeolitic catalysts, gave energies of activation in the order of 26 Kcal/mol (91). This might suggest diffusional controls of the overall combustion coke process in those zeolitic materials.

Another important question when describing the coke transformation is given by the relative extension of CO_2 and CO formation reactions. This relative extension is weighted in equation [13] by the kinetic parameters k_4 and k_5 which are functions of $\sigma = CO_2/CO$. The CO_2/CO ratio is also a key parameter for simulating an industrial scale regenerator. To estimate the CO_2/CO relationship an exponential function, temperature dependent only, which results from the peculiar characteristics of equating two rates involving the same reaction orders was derived (95). This CO_2/CO equation predicts a CO_2/CO ratio declinding with temperature. However, practical experience shows that this relationship, being correct below 600-610°C, starts being deficient above the 600-610°C level (96). Then the CO_2/CO relationship grows with temperature and becomes dependent on the cracking catalyst characteristics. To overcome the difficulties of CO_2/CO prediction different authors advise to develop a correlation for the particular catalyst being used (96), (90), (93). Nevertheless, the complex task of estimating CO_2/CO ratios recently experienced a considerable simplification with the introduction of combustors which shift the coke transformation to an almost complet CO conversion in CO_2. Under those conditions ($\sigma \to \infty$) the kinetic modeling of coke regeneration depends mainly on the adequate choice of the k kinetic parameter involved in equation [13] and on the appropriate specification of the hydrogen content in the coke formula.

SIMULATION AND MODELLING OF THE INDUSTRIAL UNITS

When considering this topic a basic distinction between the different FCC representations may observed: - multivariable search approaches involving normally power series with fitting parameters without a particular physical meaning - phenomenological models including differential and algebraic equations with

504

with parameters having clear physical sense.

The multivariable models were proposed in few contributions (97),(98),(99) for searching the optimum FCC operating conditions. In this type of representation a major question is the number of key operating variables and their interactions to be considered (97). For instance it was claimed (98) that a two variable polynomial model involving reactor temperature and mass flow velocity of the feedstock may be a useful tool for predicting the coke formation and optimum gasoline yields. Nevertheless, further contributions indicated that at least a four variable polynomial may be required (99).

The phenomenological approaches are certainly far more popular that the multivariable models. It seems that broad agreement exists concerning the strategy to adopt for further progress in this field. More advances could be expected following the phenomenological route for both design and simulation of the FCC units. In this respect different contributions consider the reactor and the regenerator as individual entities. Those studies search appropriate descriptions of each unit through a careful consideration of the kinetics, the fluid dynamic, the mass transfer and the heat transfer phenomena. For instance, the researchers concerned with the cracking reactor proposed phenomenological models for dense fluidized bed crackers (100),(57),(101) and later on considered representations for the more advanced riser units. Then the quick transition from dense phase fluidized beds to transported beds was a clear trend on the process of modelling the cracking reactor. In opposite the regenerators models has been mainly centered on dense fluidized beds. The advances on this section of the FCC plant tried to increase the understanding about the factors controlling regenerator performance such as: jet region and entrainment phase (104),(93),(90),(105),(106),(107),(108),(44).

When the cracking reactor is designed using the more advanced cracking riser technology in addition to the basic description of the cracking process given by equation [1] the following hypotheses are adopted: - adiabatic reactor, acceptable simplification considering the heat losses, below 5%, measured in a commercial unit (102), - particle and fluid circulating at the same velocity, quite reasonable approach still in curved riser units (109),(110) taking into account the 60μm typical average particle size, - temperature and hydrocarbon concentrations constant in the radial direction, due to the high degree of turbulence in the suspension, - piston flow model, adequate hypothesis (111) if D/L ratio is small enough. This kind of riser reactor representation seems to predict fairly well both conversions and selectivities of a 30m length commercial unit (102).

However, an important assumption, used frequently in the riser cracker models (112), lumping the significant volume expansion and the changing quality of the uncoverted fee in a second reaction order (57) need further verification. A more suitable approach correcting the fluid-particle velocity in riser reactors was recently proposed by us (113).

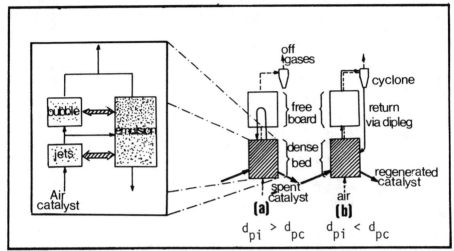

Fig. 3 Schematic of the fluidized dense bed including the jet region, the bubbles and the emulsion phase.

Fig.4 Schematic diagram of the fluidized bed including the freeboard region.

When the regenerator is in turn simulated using the more advanced fluidized bed approaches (Fig. 3 and Fig. 4) the equation [13] is coupled to a model including the following hypothesis: - a two phase model is appropriate for describing the gas circulation in the bed - bubbles are represented as DSTR, - the emulsion phase is described as a CSTR, considering the intense solid mixing in the bed (93) - the piston flow model is adopted for the entrainment phase (104) - the particles ejected at the bed surface are related to the bubble wake characteristics (114) - the particle trajectories in the freeboard region may be predicted neglecting particle - particle interactions (115),(114). This type of model requires the evaluation of a series of fluid dynamic parameters such as bubble size, jet penetration, specific particle trajectories, jet velocities and bubble velocities. Care must be taken to appropriately evaluate these parameters in order to reduce the uncertainty of coke conversion estimations. It is important to mention that these representations have strongly contributed to establish the relative influence of different regions in the regenerator.

Finally some efforts have been addressed to combine the reactor and the regenerator models in the same way they are linked in the plant (116),(117),(118),(112). This kind of analysis allows: - to take into account the important interactions between the two units - to design the auxiliary equipment required in the plant (117) - to predict important effects such as multiplicity of steady state operating conditions (119). We believe, however, that much more progress is required at this stage in the combined reactor-regenerator simulation in order to achieve a reliable model of the overall FCC process.

Nomenclature

a	constant related to the power law catalyst decay function
A_1, A_2, A_3	symbols representing gas oil, gasoline and butane-light gases-coke molecules, respectively
b	constant related to exponential catalyst decay function
C_c	coke concentration (g coke/g catalyst)
D	diameter of the riser reactor (cm)
E	energy of activation (Kcal/mol)
G	aging parameter (1/s) (function of temperature)
k	overall kinetic regeneration constant (Kmol gas/Kmol O_2.atm.s)
k_0, k_1, k_2, k_3	kinetic constants involved in cracking reactions
k_4, k_5	$k/(1+\sigma)$ and $\sigma k(1+\sigma)$ kinetic constants for the regenerative reaction (Kmol O_2.atm.s)
K	deactivation constant (function of temperature)
L	total length of the riser (cm)
n, m	constants of the activity decay function
P	total pressure (atm)
r_1, r_2, r_c	reaction rates for the gas oil consumption, gasoline production and coke consumption respectively (g converted/g cat.s)
S	number of moles of gasoline formed per mol of gas oil converted
t_c, t_v	catalyst on stream time and vapour residence time (s)
W	refractoriness
y_1, y_2, y_3	weight fractions of gas oil, gasoline and oxygen respectively
$\bar{y}$	time averaged weight fraction

Greek Letters

α	activity decay constant (1/s)
σ	(CO)/(CO)
ϕ	activity decay function

Superscripts

$'$	molar property

1. Miale, J.C., N.Y. Chen and P.B. Weisz, "Catalysis by
    Crystalline Alumino Silicates IV Attainable Catalytic Cracking
    Rate Constants and Superacidity". J. Catal., 6, 278 (1966).
2. Blazek, J.J., "Gains from FCC Revival Evident Now", Oil and
    Gas Journal, Oct., 65 (1973)
3. Strother, C.W., Vermillion, W.L. and A.J. Conner, "FCC getting
    boost from all-riser cracking". Oil and Gas Journal, May,
    103 (1972).
4. Whittington, E.L., Murphy, J.R. and I.H. Lutz. "Striking
    advances show up in modern FCC design". Oil and Gas Journal
    Oct., 49 (1972).
5. Parasakos, J.A., Shah, Y.T., McKinney, J.D. and N.L. Carr.
    "A Kinematic Model for Catalytic Cracking in a Transfer line
    Reactor". Ind. Engng. Chem. Process Design Dev. 15; 165 (1976)
6. Bunn, D.P., Gruenke, G.F., Jones, H.B., Luessenthop, D.C. and
    D.J. Youngblood. "Texaco's Fluid Catalytic Cracking Process".
    Chem. Eng. Progress, 65; 6, 88 (1969).
7. Gussov, S., Higginson, G.W. and I.A. Schwint. "New FCC
    Catalyst score high commercially". Oil and Gas Journal 70,
    25, 71 (1972).
8. Montgomery, J.A., "Recycle Rates Reflect FCC Advances" Oil
    and Gas Journal 70, 50, 81 (1972).
9. Gates, B.C., Katzer, J.R. and G.C.A. Schuit, "Chemistry of
    Catalytic Processes" McGraw-Hill Inc. (1979).
10. Johnson, M.F.L., Kreger, W.E. and H. Erickson. "Gas Oil
    Cracking by Silica-Alumina Bead Catalysts". Ind. Engng.
    Chem. 49; 283 (1957).
11. Plank, C.J., Rosinski, E.J. and W.P. Hawthorne, "Acidic
    Crystalline Aluminosilicates". Ind. Engng. Chem. Prod. Res.
    Develop. 3; 165 (1964).
12. Nace, D.M., "Catalytic Cracking over Crystalline Aluminosili-
    cates II. Application of Microreactor Technique to the
    Investigation of Structural Effects of Hydrocarbon Reactants".
    Ind. Engng. Chem. Prod. Res. Develop. 8; 31 (1969).
13. Strother, C.W., Vermillion, W.L. and A.J. Conner "Riser
    cracking gives advantages". Hydrocarbon Processing May, 89
    (1972).
14. Hemler, C.L. and W.L. Vermillion. "New Jobs for FCC", Oil
    and Gas Journal. 71, 45, 88 (1973).
15. Murcia, A.A., Soudek, M., Quinn, G.P. and G.J. D'Souza,
    "Add Flexibility to FCC's". Hydrocarbon Processing Sept.,
    131 (1979).
16. Murphy, J.R. and M. Soudek. "Modern FCC units incorporate
    many design advances". Oil and Gas Journal Jan. 71 (1977)
17. Finneran, J.A., Murphy, J.R. and E.L. Whittington. "Heavy-
    Oil cracking boost distillates". Oil and Gas Journal,
    Jan. 53 (1974).

508 <u>References</u>:

18. Pierce, W.L., Souter, R., Kaufman, T.G. and D.F. Ryan, "Innovations in Flexicracking", Hydrocarbon Processing, May 92 (1972).
19. Bryson, M.C. and G.P. Huling, "Gulf explores riser cracking", Hydrocarbon Processing May, 85 (1972).
20. Bryson, M.C., Hulling, G.P., Glausser, W.E. and C.F. Braun, "New Gulf FCC process in five units", Oil and Gas Journal, May, 97 (1972).
21. Aalund, L.R. "Custom Building a Riser Cracker", Oil and Gas Journal 72, 42, 105 (1974).
22. Chester, A.W. and W.A. Stover, "Steam Deactivation Kinetics of Zeolitic Cracking Catalysts", Ind. Engng. Chem. Prod. Res. Develop. 16; 285 (1977).
23. Rheaume, L., Ritter, R.E., Blazek, J.J. and J.A. Montgomery, "New FCC catalysts cut energy and increase activity", Oil and Gas Journal, May, 103 (1976).
24. Magee, J.S., Ritter, R.E. and L. Rheaume, "A Look at FCC Catalyst Advances". Hydrocarbon processing, Sep., 123 (1979).
25. de Lasa, H.I., Errazu, A., Barreiro, E. and S. Solioz, "Analysis of fluidized bed catalytic cracking regenerators models in an industrial scale unit". Can. J. Chem. Engrg. 54; 549, (1981).
26. Wollaston, E.G., Haflin, W.J., Ford, W.D. and G.J. D'Souza, "What influences catalytic cracking", Hydrocarbon Process Sep., 93 (1975).
27. Ritter, R.E., "Tests make case for coke-free regenerated FCC catalyst", Oil and Gas Journal, Sep., 41 (1975).
28. Yeh, J., and B.W. Wojciechowski, "Comparison of Catalytic Cracking on LaX and LaY Catalysts", Can. Journal of Chem. Engng. 56; 599 (1978).
29. Appleby, W.G., Gibson, J.W. and G.M. Good, "Coke Formation in Catalytic Cracking", Ind.Engng.Chem. Process Design Develop, 1: 102 (1962).
30. Yeh, J. and B.W. Wojciechowski, "Comparison of the Product Distribution in the Catalytic Cracking of Dewaxed Natural Distillate over Lanthanum-Exchanged X and Y type Zeolites", Can. Journal of Chem. Engng. 57; 292 (1979).
31. Rudder, J.K., "Up Octanes and fuel Oil in FCC", Hydrocarbon Process, 57; 207 (1978).
32. Murphy, J.R., Whittington, E.L., and C.P. Chang, "Review ways to upgrade resids", Hydrocarbon Processing, Sept. (1979).
33. Edison, R.R., Siemssen, J.O. and G.P. Masologites, "Crude and Residua can be cat-cracker feeds", Oil and Gas Journal, Dec. 55 (1976).
34. Masagutov, R.M., Danilona, R.A. and G.A. Berg, "Dry demetallization of a poisoned silica-alumina catalyst". Int.Chem. Engng. 10; 368 (1970).
35. Edelman, A.M., Lipuma, C.R. and F.G. Turpin, "Developments in Thermal and Catalytic Cracking Processes for Heavy Feeds, 10th World Petroleum Congress, Bucharest, Rumania (1979).

References: 509

36. Habib, E.T., Owen, H., Synder, P.W., Streed, C.W. and
     P.B. Venuto, "Artificially Metals-Poisoned Fluid Catalysts.
     Performance in Pilot Plant Cracking of Hydrotreated Resid".
     Ind. Eng. Chem., Prod. Res. Div., 16, 4, 291 (1977).
37. Cimbalo, R.N., Foster, R.L. and S.J. Wachtel, "Deposited
     Metals Poison FCC catalyst", Oil and Gas Journal, 70, 20, 120,
     (1976).
38. Mills, G.A., "Aging of Cracking catalysts. Loss of
     Selectivity", Ind. Engng. Chem. 42, 182, (1950).
39. Connor, J.E., Rothrock, J.J., Birkheimer, E.R., and L.N. Leum,
     "Fluid Cracking Catalyst Contamination. Some Fundamental
     Aspects of Metal Contamination". Ind.Engng. Chem. 49, 276
     (1957).
40. Dale, G.H. and D.L. McKay, "Passivate Metals in FCC feeds",
     Hydrocarbon Processing, 56, Sept., 97 (1977).
41. Meisenheimer, R.G., "A Mechanism for the Deactivation of
     Trace Metal Contaminants on Cracking Catalysts". J. of
     Catalysis, 1: 356 (1962).
42. Rothrock, J.J., Birkhimer, E.R., and L.N. Leum, "Fluid Crack-
     ing Catalyst Contamination. Development of a Contamination
     Test". Ind. Engng. Chem., 49; 272 (1957).
43. Davis, J.C., "FCC Units get crack catalysts". Chem. Engng.
     84, 12, 77 (1977).
44. Ford, W., Reineman, R.C., Vasalos, I.A. and R.J. Fahrig,
     "Operation Cat Crackers for Maximum Profit". Chem. Engng.
     Prog. 73 (4), 92, (1977).
45. Prescott, J.H., "FCC Regeneration Routes Boots Yields, Cut
     Energy". Chem. Engng. Sep., 64 (1974).
46. Hartzell, F.D., and A.W. Chester, "CO burn promotor produces
     multiple FCC benefits". Oil and Gas Journal, 77, 16, 33,
     (1979).
47. Ewell, R.B. and G. Gadner, "Design cat crackers by computer",
     Hydrocarbon Processing, April, 125 (1978).
48. Weekman, V.W. and D.M. Nace, "Kinetics of Catalytic Cracking
     Selectivity in Fixed, Moving and Fluid Bed Reactors", AIChE
     Journal, 16; 397 (1970).
49. Pachovsky, R.A., and B.W. Wojciechowski, "Temperature Effects
     of Gasoline Selectivity in the Cracking of a Neutral
     Distillate:, J. of Catalysis, 37; 368 (1975).
50. Gross, B., Nace, D.M., and S.E. Sterling, "Application of a
     Kinetic Model for Comparison of Catalytic Cracking in Fixed
     Bed Microreactor and a Fluidized Dense Bed". Ind. Engng.
     Chem. Process Design Dev. 13; 199 (1974).
51. Nace, D.M. "Catalytic Cracking over Crystalline Aluminosili-
     cates, I. Instantaneous rate measurements for hexadecane
     cracking". Ind.Eng. Chem. Prod. Res. Develop. 8; 24 (1969).
52. Weekman, V.W., "Kinetics and Dynamics of Catalytic Cracking
     Selectivity in Fixed Beds". Ind.Engng. Chem. Process Design
     Develop. 8; 385 (1969).

53. Campbell, D.R., and D.W. Wojciechowski, "Theoretical Patterns of Selectivity in Aging Catalysts with Special Reference to the Catalytic Cracking of Petroleum", Can. J. Chem. Engng. 47, 413 (1969).

54. John, T.M. and B.W. Wojciechowski, "On Identifying the Primary and Secondary Products of the Catalytic Cracking of Neutral Distillates", J. Catalysis 37; 240, (1975).

55. Pachovsky, R.A., and B.W. Wojciechowski, "Theoretical Interpretation of Gas Oil Conversion Data on X-Sieve Catalyst", Can. J. Chem. Engng. 49; 365 (1971).

56. John, T.M., Pachovsky, R.A., and B.W. Wojciechowski, "Coke and Deactivation in Cracking Catalyst". Advances in Chemistry Series 133, 422 (1974).

57. Weekman, V.W., "A Model of Catalytic Cracking Conversion in Fixed, Moving and Fluid-Bed Reactors". Ing. Engng. Chem. Process Design Dev., 7; 90 (1968).

58. Weekman, V.W., and D.M. Nace, "Kinetics of Catalytic Cracking Selectivity in Fixed, Moving and Fluid Bed Reactors", A.I.Ch.E Journal, 16, 397, (1970).

59. Gustafson, W.R., "Evaluation Procedure for Cracking Catalysts". Ind. Eng. Chem. Process Design Develop., 11; 507 (1972).

60. Blanding, F.H., "Reaction Rates in Catalytic Cracking of Petroleum", Ind. Engng. Chem. 45; 1186 (1953).

61. Jacob, S.M., Gross, B., Voltz, S.E., and V.W. Weekman, "A Lumping and Reaction Scheme for Catalytic Cracking". AIChE Journal 22; 701 (1976).

62. Kemp, R.R.D., and B.W. Wojciechowski, "The Kinetic of Mixed Reactions", Ind. Engng. Chem. Fundam. 13; 332 (1974).

63. Pachovsky, R.A., John, T.J., and B.W. Wojciechowski, "Theoretical Interpretation of Gas Oil Selectivity Data on X-Sieve Catalyst". AIChE Journal 19; 802 (1973).

64. Pachovsky, R.A., Best, D.A., and B.W. Wojciechowski, "Applications of the time-on-stream theory of Catalyst Decay", Ind. Eng. Chem., Process Des. Develop. 12; 254 (1973).

65. Pachovsky, R.A., and B.W. Wojciechowski, "Effects of Diffusion Resistance on Gasoline Selectivity in Catalytic Cracking", AIChE. Journal 19; 1121 (1973).

66. Pachovsky, R.A. and B.W. Wojciechowski, "Effects of Charge Stock Composition on the Kinetic Parameters in Catalytic Cracking", Can. J. Chem. Engng., 53; 308 (1975).

67. Pachovsky, R.A. and B.W. Wojciechowski, "Temperature Effects on Conversion in the Catalytic Cracking of a Dewaxed Neutral Distillate". J. of Catalysis 37; 120 (1975).

68. Venuto, P.B., Hamilton, L.A. and P.S. Landis, "Organic Reactions Catalyzed by Crystalline Alumino Silicates II. Alkylation Reactions. Mechanistic and Aging Considerations". J. of Catalysis 5, 484 (1966)

69. John, T.M., and B.W. Wojciechowski, "Effect of Reaction Temperature on Product Distribution in the Catalytic Cracking of Neutral Distillates", J. Catalysis 37; 348 (1975)

70. Eberly, P.E., Kimberlie, C.N., Miller, W.H., and H.V. Drushel, "Coke Formation on Silica-Alumina Cracking Catalysts" Inc. Engng. Chem. Proc. Design Develop, 5;193 (1966).

71. Andrews, J.M., "Cracking Characteristics of Catalytic Cracking Units", Ind. Eng. Chem. 51; 507 (1959).

72. Voorhies, A., "Carbon Formation in Catalytic Cracking", Ind. Engng. Chem. 37; 318 (1945).

73. Ruderhausen, C.G. and C.C. Watson, "Variables affecting activity of molybdena-alumina hydroforming catalyst in aromatization of cyclohexane", Chem. Eng. Sci., 3; 110 (1954).

74. Tan, C.H. and O.M. Fuller, "A Model of Fouling in Zeolite Catalyst". Can. J. Chem. Engng., 48; 174 (1970).

75. Wojciechowski, B.W., 'A Theoretical Treatment of Catalyst Decay". Can. J. Chem. Engng. 46; 48 (1968).

76. Campbell, D.R., and B.W. Wojciechowski, "Theoretical Patterns of Selectivity in Aging Catalysts with special reference to the Catalytic Cracking of Petroleum". Can. J. Chem. Engng. 47; 413 (1969).

77. Campbell, D.R., and B.W. Wojciechowski, "Selectivity of Aging Catalysts in Static, Moving and Fluidized Bed Reactors", Can. J. Chem. Engng. 48; 224 (1970).

78. Campbell, D.R. and B.W. Wojciechowski, "The Catalytic Cracking of Cumene on Aging Catalysts II. An Experimental Study". J. of Catalysis, 23; 307 (1971).

79. Froment, G.F., and K.M. Bischoff, "Non-Steady state behaviour of fixed bed catalytic reactors due to catalyst fouling", Chem. Engng. Sci. 16; 189 (1961).

80. Froment, G.F. and K.B. Bischoff, "Kinetic data and product distribution from fixed bed catalytic reactors subject to catalyst fouling", Chem. Engng. Sci. 17; 105 (1962).

81. Corella, J., Asua, J.M., and J. Bilbao, "Kinetic of the Deactivation of a 10% Cu-0.5% Cr_2O_3 asbestos catalyst for benzyl alcohol dehydrogenation". Chem. Eng. Science, 35; 1447 (1980).

82. Voltz, S.E., Nace, D.M., Jacob, S.M. and V.W. Weekman, "Application of a Kinetic Model for Catalytic Cracking III Some Effects of Nitrogen Poisoning and Recycle", Ind. Engng. Chem., Process Design Develop. 11; 261 (1972).

83. Nace, D.M., Voltz, S.E. and V.W. Weekman, "Application of a Kinetic Model for Catalytic Cracking. Effects of Charge Stocks". Ind. Engng. Chem. Processes Design and Develop 10; 530 (1971).

84. Voltz, S.E., Nace, D.M., and V.W. Weekman, "Application of a Kinetic Model for Catalytic Cracking. Some Correlations of Rate Constants", Ind. Engng. Chem. Process Design Develop. 10; 538 (1971).

512

85. Thomas, C.L., "Chemistry of Cracking Catalysts", Ind. Engng. Chem., 41; 2564 (1949).

86. Weisz, P.B., and R.D. Goodwin, "Combustion of Carbonaceous Deposits Within Porous Catalyst Particles", II Intrinsic Burning Rate. J. of Catalysis, 6: 227 (1966).

87. Eberly, P.E., Kimberlie, C.N., Miller, W.H. and H.V. Drushel, "Coke formation on Silica-Alumina Cracking Catalysts". Ind. Engng. Chem. Proc. Design Develop. 5; 193 (1966).

88. Massoth, F.E., "Oxidation of Coked Silica Alumina Catalysts". Ind. Engng. Chem. Proc. Design and Develop 6, 2 (1967).

89. Pansing, W.F., "Regeneration of Fluidized Bed Cracking Catalysts". AIChE Journal, 2; 71 (1956).

90. de Lasa, H.I., Errazu, A., Barreiro, E., and S. Solioz, "Analysis of fluidized bed catalytic cracking regenerators models in an industrial scale unit". Can. J. Chem. Engng. 54; 549 (1981)

91. Hano, T., Nakashio, F., and K. Kusonoki, "The Burning Rate of Coke Deposited on Zeolytic Catalyst", J. Chem. Engng. Japan, 8; 127 (1975).

92. Metcalfe, T.B., "Kinetics of Coke Combustion on Catalyst Regeneration". Brit. Chem. Engng. 12; 388 (1967).

93. Errazu, A.F., de Lasa, H.I., and F. Sarti, "A Fluidized Bed Catalytic Cracking Regenerator Model: Grid Effects". Can. J. Chem. Engng. 57; 191 (1979).

94. Johnson, M.F.L., and H.G. Maryland,"Carbon Burning Rates of Cracking Catalyst in the Fluidized State". Ind. Engng. Chem., 47; 127 (1955).

95. Arthur, J.R., "Reactions Between Carbon and Oxygen", Trans. Faraday Soc. 47; 164 (1951).

96. Weisz, P.B., "Combustion of Carbonaceous Deposits Within Porous Catalyst Particles. III. The CO_2/CO Product Ratio" J. of Catalysis 6, 425 (1966).

97. Fiero, W.J., and P.E. Kelly, "To optimize the FCC Unit", Hydrocarbon Process, 56, 9, 117 (1977).

98. Shumskii, V.M., "Experimental determination of the steady-state characteristics of the fluidized-bed catalytic cracking process". Int. Chem. Eng. 9; 508 (1969).

99. Shumskii, V.M., "Determination of a static model of catalytic cracking". Int. Chem. Eng. 11; 64 (1971).

100. Tigrel, A.Z. and D.L. Pyle, "A model for a fluidized bec catalytic cracker", Chem. Engng. Sci. 26; 133 (1977).

101. Kato, K., Inomata, M., Onoda, K., and M. Yamagishi, "Process Design for Packed Fluidized-Bed Catalytic Reactors" Int. Chem. Eng. 19; 96 (1979).

102. Shah, Y.T., Huling, G.P., Parasakos, J.A. and J.D. McKinney "A Kinematic Model for an Adiabatic Transfer Line Catalytic Reactor". Ind. Engng. Chem., Process Design Dev. 16;89 (1977).

103. Parasakos, J.A., Shah, Y.T., McKinney, J.D., and N.L. Carr, "A Kinematic Model for Catalytic Cracking in a Transfer Line Reactor", Ind. Engng. Chem. Process Design Dev. 15; 165 (1976)

104. de Lasa, H.I., and J.R. Grace, "The Influence of the Freeboard Region in a Fluidized Bed Catalytic Cracking Regenerator", AIChE Journal, 25; 984 (1979).
105. de Lasa, H.I. and A.F. Errazu, "Ignition of a Fluidized Bed Catalytic Cracking Regenerator. Freeboard Region Influence", Proceedings of 1980 International Fluidization Conference, ed. J.R. Grace and J.M. Matsen, Plenum Publishing Corporation, New York (1980).
106. de Lasa, H.I., Errazu, A., Barreiro, E. and S. Solioz, "Analysis of Fluidized Bed Catalytic Cracking Regenerator Models in a Revamped Unit", American Chemical Society Meeting, Las Vegas (1980)
107. de Lasa, H.I., "Simulation of and Industrial Scale Regenerator. Influence of the Different Constitutive Fluidized Bed Regions". Proceedings World Chem. Engng. 54; 549 (1981).
108. Grace, J.R., and H.I. de Lasa, "Reaction Near the Grid in Fluidized Beds". AIChE Journal, 24; 364 (1978).
109. de Lasa, H.I., Errazu, A., Porras, J., and E. Barreiro, "Influence of the pneumatic transport line in the simulation of a Fluidized Bed Catalytic Cracking Regenerator". Lat. J. of Chem. Engng. 11, 139 (1981).
110. Errazu, A.F., Porras, J.A., and H.I. de Lasa, "Modelling a Catalytic Cracking Regenerator. Influence of the Pneumatic Transported Riser", X Jornadas de Ingenieria Quimica. Santa Fe, Argentina (1978).
111. de Lasa, H.I. and G. Gau, "Influence des Agrgats sur le Rendement d,un reacteur a Transport Pneumatique", Chem. Engng. Sci. 28; 1875 (1973).
112. Wollaston, E.G., Haflin, W.J., Ford, W.D., and G.J. D'Souza, "FCC Model Valuable Operating Tool", Oil and Gas Journal Sep., 87 (1975).
113. de Lasa, H.I. and L.K. Mok, "Entrained Coal Gasifiers; Modeling the Particle Acceleration". AIChE Meeting, Philadelphia (1980). Can. J. Chem. Engng. 56; 658 (1981).
114. George, S.E., and J.R. Grace, "Entrainment of Particles from Aggregative Fluidized Beds", AIChE Symp. Ser., 74, 176, 67 (1978)
115. Do, H.T., Grace, J.R., and R. Clift, "Particle Ejection and Entrainment from Fluidized Beds", Powder Technology, 6; 195 (1972)
116. Seko, H., Tone, S., and T. Otake, "Consideration of the Treatment of coke distribution in a fluid catalytic cracker", J. Chem. Engng., Japan, 10; 493 (1977).
117. Ewell, R.B., and G. Gadner, "Design cat crackers by computer", Hydrocarbon Processing, April, 125 (1978).
118. Corella, J., Bilbao, R. and J. Delgado Puche "Modelo Macrocinetico del Process de Craqueo Catalitco del Gas Oil en Lecho Fluidizado (FCC) en Estado Estacionario" Ingenieria Quimica, October (1980).

514

119. Elnashaie, S.S.E.H. and I.M. El-Hennawi, "Multiplicity
     of the Steady State in Fluidized Bed Reactors - 4. Fluid
     Catalytic Cracking (FCC) Chem. Engng. Sci. 34; 1113 (1979)

CONVERSION OF METHANOL TO GASOLINE OVER ZEOLITE CATALYSTS
I.   REACTION MECHANISMS

Eric G. Derouane

Mobil Research and Development Corporation
Central Research Division
P. O. Box 1025
Princeton, New Jersey   08540   U.S.A.

1.   INTRODUCTION

The conversion of methanol into hydrocarbons occurs over a large variety of acidic catalysts. These can be non-zeolitic compounds such as phosphoric acid, heteropolyacids, and silica-aluminas (1-5) or more commonly zeolites (6-11). The use of zeolite ZSM-5 as acid catalyst (8,12-25) is particularly attractive as it offers a new and viable route for the direct production of high-grade gasoline from methanol synthesized from coal or natural gas resources. Zeolite-based ZSM-5 catalysts are the key to Mobil's Methanol-to-Gasoline (MTG) process for which a commercial plant is presently being built in New Zealand. The MTG process has unique advantages:

(a) Hydrocarbons are produced in a rather narrow compositional range; little methane and no hydrocarbons larger than C_{11} are formed.

(b) A high conversion of methanol can be combined with a high selectivity for aromatics and isoparaffins of high octane value.

(c) ZSM-5 based catalysts are characterized by a very low aging rate.

This paper reviews and discusses the major mechanistic aspects of the MTG reaction and delineates some of the factors which confer unusual catalytic properties to zeolite ZSM-5. An accepted overall reaction scheme consists in the rapid equilibration of methanol with dimethylether (DME), the production of light olefins from these reactants, and the conversion of the light olefins into higher molecular weight products by repeated alkylation with methanol or dimethylether (DME), followed by cracking and classical conjunct polymerization (8,17,18,26). The following discussion will be

focused essentially on the still controversial mechanism by which
the first C-C bond is formed, the autocatalytic nature of the
methanol conversion, and the molecular shape-selective effects
which explain the yield of a high octane product.

2. METHANOL CONVERSION MECHANISM - THE ROUTE TO LIGHT OLEFINS

The most intriguing question associated with the MTG conversion
is the reaction pathway which accounts for the first C-C bond formation from methanol or DME. Figures 1 and 2 summarize the essential
steps which characterize some of the mechanisms proposed at this
date.

The hypothesis of carbene intermediacy (Figure 1A) was proposed
by Chang and Silvestri (8), carbene intermediates being produced by
α-elimination of water from methanol due to the cooperative actions
of acidic and basic sites or methylene transfer occurring between
two R-O-CH_3 species. The latter two processes have, however, a low
probability because of the low active site density in ZSM-5 or the
constraints imposed on the bimolecular transition complex explaining
the methylene transfer. It appears nevertheless that carbene-type
C_1 intermediates might play a role as suggested by ^{13}C-isotope
labeling studies of the conversion of methanol in the presence of
added propane (27). Abnormally low iso/n-butane ratios are explained
by a carbene-like species insertion into a s-p^3 C-H bond (of propane), this suggestion being supported by the ^{13}C isotope distribution in the C_4-products (27) and the observation of shape-selectivity
effects which moderate the isomerization of isobutane into n-butane
over HZSM-5 (28).

Kaeding et al. (16) suggested an intermolecular reaction
between a surface dimethyloxonium species and the (slightly) negatively charged methyl group of an incoming DME molecule, aided by
a basic site on the catalyst surface (see Figure 1B). However, the
basic properties of ZSM-5 have not been recognized at this time and
basic sites, if present, would be very weak as a result of the
strong catalyst acidity.

Van den Berg et al. (29,30) have suggested a mechanism which
transfers the known chemistry of tetramethylammonium cations to
trimethyloxonium cations. As shown in Figure 1C, the key step is
the intramolecular, Stevens-like (31), rearrangement of a trimethyloxonium ion formed by the bimolecular addition of a methanol molecule
to a dimethyloxonium cation. Questions about this reaction pathway
relate again to the availability and role of basic sites on the
zeolite surface. In addition, this type of rearrangement has not
been observed, at least at present, in more classical reaction
conditions such as homogeneous transformations in the presence of
a strong base. Nevertheless, this mechanism appears most attractive

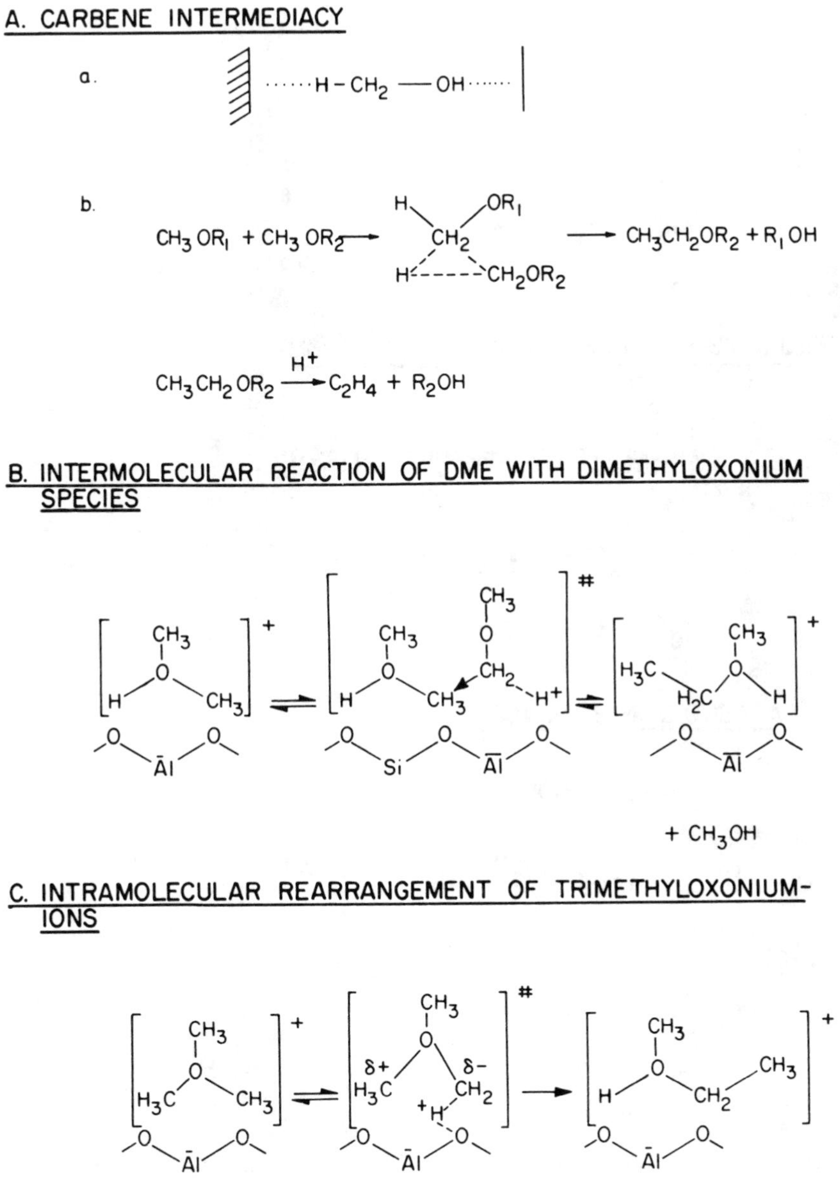

A. CARBENE INTERMEDIACY
a.
H - CH2 — OH
b.
CH3 OR1 + CH3 OR2
H
OR1
CH2
H
CH2OR2
CH3CH2OR2 + R1 OH
H+
CH3 CH2 OR2
C2H4 + R2OH
B. INTERMOLECULAR REACTION OF DME WITH DIMETHYLOXONIUM SPECIES
CH3
O
H
CH3
Al
Si
CH2
H+
#
H3C
H2C
+ CH3OH
C. INTRAMOLECULAR REARRANGEMENT OF TRIMETHYLOXONIUM-IONS
δ+
δ-

Figure 1. First C-C bond formation from methanol or DME.

D. METHYLCARBONIUM ION ATTACK ON METHANOL OR DME

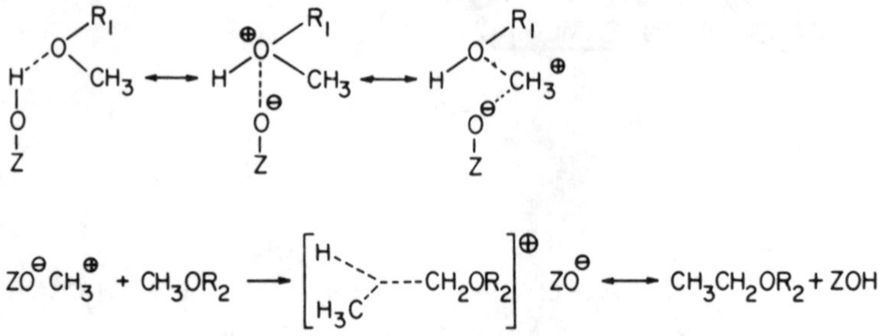

E. OXONIUM ATTACK ON METHANOL OR DME

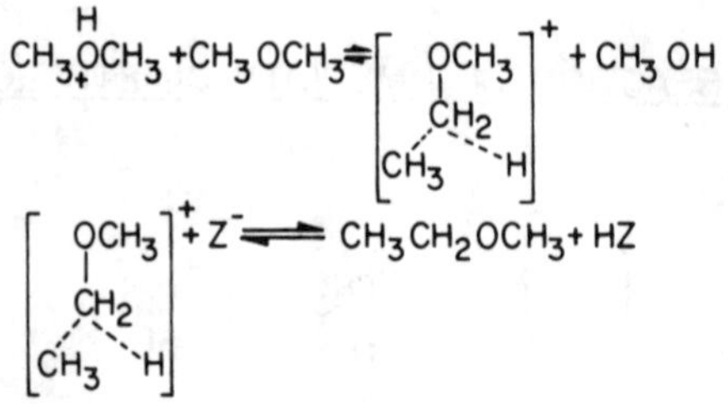

F. CARBENOID INTERMEDIATES

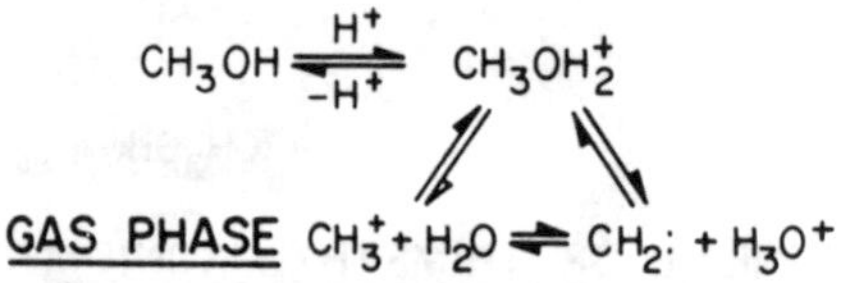

GAS PHASE

ZEOLITE

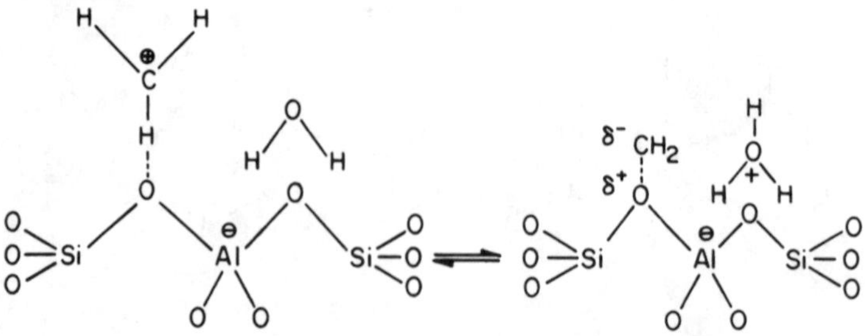

Figure 2. First C-C bond formation from methanol or DME (continued)

and deserves additional attention as it explains satisfactorily
changes in the light olefinic products distribution as a function
of reaction conditions (temperature, methanol/H_2O ratio, etc.).

CH_3^+ species appear as unlikely intermediates as one would
then logically expect the formation of large amounts of methane by
hydride abstraction. This led Zatorski et al. (9) to postulate the
occurrence of a radical mechanism (which has received no support so
far) and Ono et al. (22) to envisage the attack of methylcarbonium
ions on the C-H bond of methanol or DME via the formation of penta-
coordinated carbon-species of Olah's type (32) (see Figure 2D), in
a scheme resembling that originally proposed by Pearson (1). Such
a mechanism implies that HZSM-5 possesses superacidic properties for
which no evidence is presently available; it also ignores the recent
observation (33) that the acid strength of the sites active for the
methanol or DME conversion is intermediate between those required to
isomerize ortho-xylene and disproportionate toluene.

In an effort to ascertain the inter- vs. intramolecular nature
of the DME conversion, Van den Berg (30) compared the trialkyl-
ammonium vs. the carboxonium route to ethylene. These pathways are
schematized in Figure 3. The initial step in route A is the pro-
tonation of DME while in route B a hydride is initially abstracted.
Kinetic experiments, thermodynamic, and theoretical considerations
favor route A and indicate that the formation of the oxonium ylide
(reaction A2) is probably rate determining.

This scheme has recently obtained some support from labeling
experiments which confirm that the conversion of DME occurs via an
intermolecular process and point out the important role played by
ether intermediates in the methanol conversion (23).

Consideration of the former mechanisms indicate that two
essential points remain controversial, i.e., (i) the possible role
played by carbenoid species and (ii) the nature of the initial olefinic product(s). Semi-empirical quantum mechanical calculations by
Beran and Jiru (36) have suggested that methanol molecules, in the
presence of a strong electric field such as it may exist in zeolites,
can be transformed into CH_2O and CH_2 species. Hence, it is not un-
realistic to consider the scheme shown in Figure 2F which suggests
that an equilibrium could exist between a carbonium-type and a
carbenoid species, that is between a surface methoxy group and a
surface ylide. The internal energy variation corresponding to the
deprotonation of CH_3+ to CH_2 (carbene) in the gas phase is of ca.
39 kcal mol^{-1} (38); it is likely to be smaller for the same transformation occurring on the zeolite surface because of the electrophilic character of both the carbene and hydroxonium entities. It
is conceivable that such an equilibrium could be shifted in one or
the other direction depending on the actual consumption of the

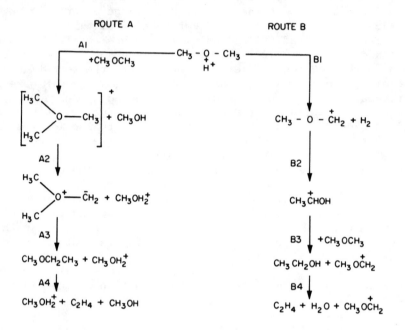

Figure 3. Trialkyloxonium (A) vs. carboxonium route (B) for C-C bond formation (from reference 30).

C_1-species, methyl carbonium ions being effective in alkylation reactions and carbenoid-moieties acting in carbene insertion processes or the formation of light olefins (21).

Reacting methanol at low conversion can, in principle, give a hint into the nature of the primary olefinic products. Propylene has most often been suggested as the primary olefinic product (3, 23,29,30,33,37,39). Recently, Haag et al. (24) indicated that this conclusion could be affected by diffusion disguise, i.e., when the precursor species to ethylene (sometimes referred to as a surface C_2 entity (17)) leads preferentially to sequential reaction products (intermediates) rather than to desorbed ethylene. This analysis meets the suggestion that the conversion of methanol occurs via a rake-type mechanism as depicted in Figure 4 in which ether intermediates play an essential role (23,40). It is now felt that ethylene is a major olefinic product when diffusion effects are reduced (24).

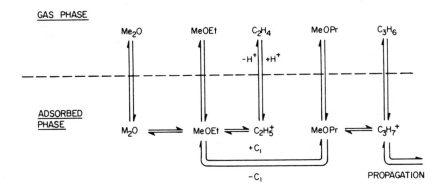

Figure 4   Rake-type mechanism for the conversion of methanol and dimethylether into hydrocarbons (adapted from reference 40).

Studies on the conversion of ^{13}C labeled methanol by HZSM-5 in the presence of propylene or 1-hexene by Dessau and LaPierre (41) have, however, indicated that ethylene was essentially produced by the cracking of higher olefins. It should, therefore, be considered as a primary methanol reaction product only during the initial initiation phases leading to C_{3+} olefins. The selectivity to the formation of light olefins is also determined by the structure of the zeolite used as catalyst, in particular its pore size. Cormerais et al. (33) have investigated the conversion of DME over H-erionite, HZSM-5 and H-Y zeolites. They found that smaller pore sizes led to higher yields in ethylene and propylene while the formation of iso-butane and C_{5+} hydrocarbons was strongly reduced as illustrated in Figure 5.

3.   THE AUTOCATALYTIC CONVERSION OF METHANOL

In the MTG conversion, the consumption of methanol rapidly increases as the hydrocarbon product concentration (olefins and aromatics) increases. The autocatalytic nature of the methanol conversion has been discussed by various authors (42-44), the analysis proposed by Chang (44) predicting, in particular, conversion and selectivities over a wide range of pressures. Most of the results were obtained for HZSM-5 catalysts although their interpretation is probably not restricted to this particular zeolite.

522

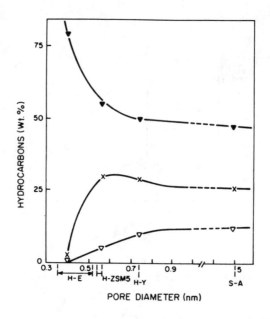

Figure 5. Influence of pore aperture on the distribution of hydrocarbon products in the zeolite-catalyzed methanol/DME conversion:
▼, ethylene + propylene; ✕, C_5+ hydrocarbons; ▽, isobutane (from
reference 33, with permission from Butterworth & Co., Pub.)

Early in-situ ^{13}C-NMR studies by Derouane et al. (15) suggested
the occurrence of preferential reaction pathways by which methanol
was consumed in alkylation reactions of olefinic and aromatic products. They were confirmed recently by complementary investigations
of the reaction of methanol in the presence of ethylene using [13]C-
labeled reactants (45). Figure 6 shows ^{13}C-NMR spectra of such
reaction mixtures at various temperatures and conversions with either
labeled methanol or ethylene. Clearly, methanol is essentially used
up in aliphatic chain growth or aromatic ring alkylation. The
ethylene conversion, which is inhibited by the presence of methanol
at temperatures below 200°C, yields ultimately higher olefinic and
aromatic products.

In support of the autocatalytic conversion of methanol or DME,
Van den Berg (30) observed that two routes are operative in the conversion of DME. At low temperature, trialkyloxonium ions are mostly
formed via condensation reactions of methanol and DME while above
280°C reaction of dimethylether or methanol with alkyl cations, i.e.,
alkylation processes, become predominant thereby offering a more
direct and efficient route for the conversion of the oxygenated

reactants. Perot et al. (23) have compared the reactivities of ethylene and propylene, both in the presence and absence of DME. At 350°C, propylene alone is about 20 times more reactive than ethylene as reported qualitatively earlier (46). A mixed ethylene-DME feed has essentially the same reactivity as DME, as also concluded by Gilson (45), while that of a mixed propylene feed exceeds the sum of the individual reactant reactivities by nearly 50%, an observation which also explains in part the autocatalytic methanol-DME conversion.

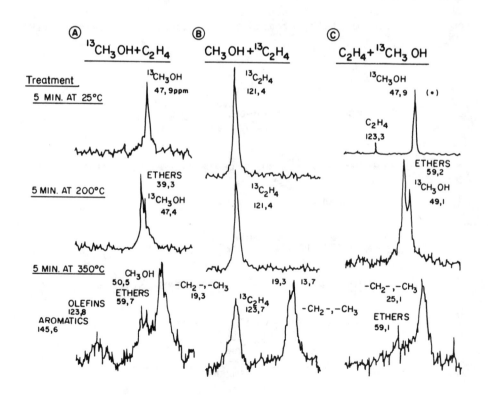

Figure 6. In-situ ^{13}C-NMR spectra of labeled methanol-ethylene mixtures reacted over HZSM-5 (from reference 45).

4. THE FORMATION OF AROMATIC COMPOUNDS

It has been well established that C_{2-5} olefins could act as precursors of aromatic compounds (17,18). Reaction pathways are essentially analogous to those describing conjunct polymerization (47,48), that is, the conversion of olefins (by oligomerization and/or cyclization) into naphthenes and the dehydrogenation of naphthenes into aromatics by hydrogen transfer reactions of the type:

1 Naphthene (C_nH_{2n}) + 3 Olefins (C_nH_{2n})

1 Aromatic (C_nH_{2n-6}) + 3 Paraffins (C_nH_{2n+2})

This scheme predicts a stoichiometric paraffinic/aromatic ratio of 3 which is effectively observed in a normally run MTG conversion.

Figure 7 schematizes the aromatization pathway of olefins, using propylene as a model reactant. UV-spectroscopy allows the identification of the carbocations intermediates and shows that cyclopentenyl carbocations are first formed. Cyclohexenyl species, aromatics, and (poly)alkylaromatics then appear successively (18).

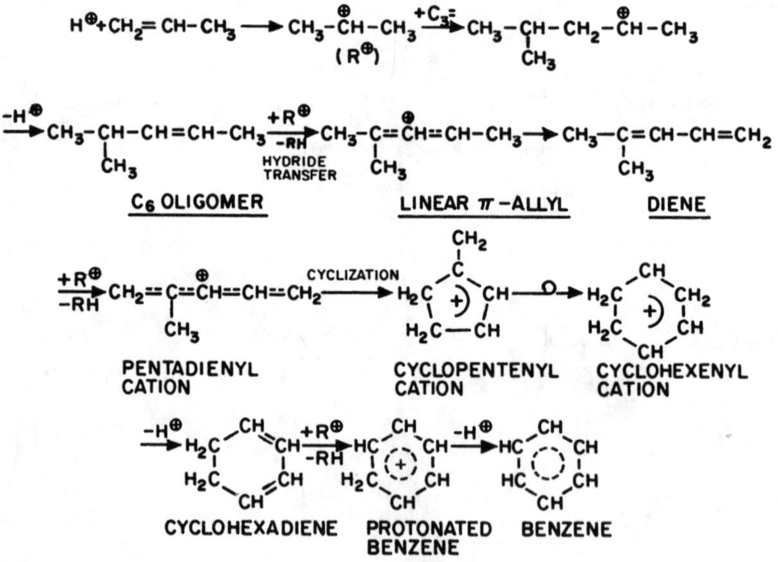

Figure 7. Typical intermediates in the conversion of olefins to naphthenes and aromatics (from reference 18, with permission from Elsevier Pub. Co.).

Restricted transition-state shape-selectivity has been claimed as an essential factor to explain the high iso/normal paraffins ratio in the aliphatic product and the cut-off at C_{10-11} in the aromatics distribution (49,50). C_{4-6} olefins are obtained by oligomerization and/or alkylation of the primary olefins and by cracking of higher olefinic products. Oligomerization (C_{3-5}) and dehydrocyclization (C_{6-10}) of the olefins occur at the channel intersections of HZSM-5 where molecular shape-selectivity constraints act on the formation of intermediate complex structures (50). Steric inhibition, limiting the accommodation of isoaliphatics at these intersections (51) and precluding the accommodation of bimolecular transition complexes of more than ten carbon atoms (49), appears as a decisive parameter.

Haag et al. (24) have shown that the distribution of aromatic compounds resulted from the combination of two factors: the methanol alkylation of lower aromatics which is very effective at low methanol conversion and diffusion limitations (product selectivity) which stem from the comparable sizes of the zeolite pores and of the methylaromatics. Evidence for such diffusion constraints was gained from the multiple labeling of the polymethylbenzenes produced in the unlabeled toluene-^{13}C-methanol reaction (25).

Molecular traffic control (52,53,54) which directs the flows of (small) reactant and (larger) product molecules in the zeolite channels was mentioned as a likely reason for the absence of major counterdiffusion limitations. Some support seems to have been gained for its existence in comparative studies of the methanol conversion and of the para-xylene alkylation by methanol over HZSM-5 and HZSM-11 catalysts (20). These data need, however, to be further assessed.

As discussed elsewhere (50,52), alkylaromatics cannot be converted further in ZSM-5 (by dehydrocyclization and subsequent alkylation). It explains its low coking activity and its unusually high stability as a methanol conversion catalyst.

5. CONCLUSIONS

The conversion of methanol-to-gasoline-type hydrocarbons occurs via a well-established network of sequential and parallel reaction steps. Among those, the formation of the primary olefins, i.e., of the first C-C bond, still remains controversial. An attractive, usually well-accepted, mechanism postulates trimethyloxonium ions as intermediates while new evidence was recently presented to support the hypothesis of carbenoid intermediacy. In any event, this reaction step is slow and it is followed by competitive and very fast subsequent reactions. Ether-type species are found to play an important role in the autocatalytic conversion of methanol.

The latter is also readily consumed in aromatic alkylation reacti
ons. The formation of aromatics occurs via classical reaction pathways which have received ample support.

Restricted transition-state shape-selectivity is found to
play a role in the oligomerization/dehydrocyclization of olefins
into naphthenes and aromatics while diffusion/desorption disguises
clearly affect the distribution of aromatic compounds.

REFERENCES

1. D.E. Pearson, J. Chem. Soc. Chem. Comm., (1974) 397.
2. L. Kim, M.M. Wald, and S.G. Brandenberger, J. Org. Chem.,
   43 (1978) 3432.
3. B.J. Ahn, J. Armando, G. Perot, and M. Guisnet,
   C.R. Acad. Sci. Paris, 288 C (1979) 245.
4. T. Baba, J. Sakai, and Y. Ono, Bull. Chem. Soc. Jpn., 55
   (1982) 2657.
5. T. Baba, J. Sakai, H. Watanabe, and Y. Ono,
   Bull. Chem. Soc. Jpn., 55 (1982) 2555.
6. P.B. Venuto and P.S. Landis, Advan. Catal.-Relat. Subj.,
   18 (1968) 259.
7. K.V. Topchieva, A.A. Kubasov, and T.V. Dao,
   Vestn. Mosk. Univ. Khim., 37 (1972) 620; Khimyia, 27 (1972)
   628.
8. C.D. Chang and A.J. Silvestri, J. Catal., 47 (1977) 249.
9. W. Zatorsky and S. Krzyzanowski, Acta Phys. Chem., 24 (1978)
   347.
10. M.S. Spencer and T.V. Whittam, Acta Phys. Chem., 24 (1978)
    307.
11. M.S. Spencer and T.V. Whittam, J. Molec. Catal., 17 (1982)
    271.
12. S.L. Meisel, J.P. McCullough, C.P. Lechthaler, and
    P.B. Weisz, Chem. Tech., 6 (1976) 86.
13. S.E. Voltz and J.J. Wise, "Development Studies on Conversion
    of Methanol and Related Oxygenates to·Gasoline", Final Re-
    port, DOE Contract No. E (49-18)-1773 (1976).
14. A.Y. Kam and W. Lee, "Fluid-Bed Process Studies on Selective
    Conversion of Methanol and Related Oxygenates to Gasoline",
    Final Report, DOE Contract No. Ex-76-C-01-2490 (1978).
15. E.G. Derouane, J.B.Nagy, P. Dejaifve, J.H.C. Van Hooff,
    B.P. Spekman, J.C. Vedrine, and C. Naccache, J. Catal., 53
    (1978) 40.
16. W.W. Kaeding and S.A. Butter, J. Catal., 61 (1980) 155.
17. P. Dejaifve, J.C. Vedrine, V. Bolis, and E.G. Derouane,
    J. Catal., 63 (1980) 331.

18. J.C. Vedrine, P. Dejaifve, E.D. Garbowski, and
    E.G. Derouane, in "Catalysis by Zeolites", Stud. Surf. Sci.
    Catal., Vol. 4, B. Imelik et al., eds. (Elsevier Sci. Pub. Co.,
    Amsterdam, The Netherlands, 1980), p. 29.
19. P. Dejaifve, A. Auroux, P.C. Gravelle, J.C. Vedrine,
    Z. Gabelica, and E.G. Derouane, J. Catal., 70 (1981) 123.
20. E.G. Derouane, P. Dejaifve, Z. Gabelica, and J.C. Vedrine,
    Faraday Disc. Chem. Soc., 72 (1981) 331.
21. E.G. Derouane, P. Dejaifve, J.P. Gilson, J.C. Vedrine, and
    V. Ducarme, in "Abstracts 7th North American Meeting of the
    Catalysis Society", Boston, Massachussetts, October 11-15
    (1981).
22. Y. Ono and T. Mori, J. Chem. Soc. Faraday Trans. I, 77 (1981)
    2209.
23. G. Perot, F.X. Cormerais, and M.Guisnet,
    J. Molec. Catal., 17 (1982) 255.
24. W.O. Haag, R.M. Lago, and P.G. Rodewald,
    J. Molec. Catal., 17 (1982) 161.
25. R.M. Dessau and R.B. LaPierre, J. Catal., 78 (1982) 136.
26. V.N. Ipatieff and H. Pines, Ind. Eng. Chem., 28 (1936) 684.
27. C.D. Chang and C.T. Chu, J. Catal., 74 (1982) 203.
28. P. Hilaireau, C. Bearez, F. Chevalier, G. Perot, and
    M. Guisnet, Zeolites, 2 (1982) 69.
29. J.P. Van den Berg, J.P. Wolthuizen, and J.H.C. Van Hooff,
    in "Proc. 5th Int. Conf. Zeolites", L.V.C. Rees, ed.
    (Heyden and Sons, London, 1980); p. 649.
30. J.P. Van den Berg, Ph. D. Thesis, Technical University of
    Leiden, 1981.
31. T.S. Stevens and W.E. Watts, in "Selected Molecular
    Rearrangements" (Van Nostrand Reinhold Pub. Co., London,
    1973).
32. G.A. Olah, G. Klopman, and R.H. Schlosberg,
    J. Amer. Chem. Soc., 91 (1969) 3261.
33. F.X. Cormerais, Y.S. Chen, M. Kern, N.S. Gnep, G. Perot,
    and M. Guisnet, J. Chem. Research (S), (1981) 290.
34. D. Kagi, J. Catal., 69 (1981) 242.
35. C.D. Chang, J. Catal., 69 (1981) 244.
36. S. Beran and P. Jiru, React. Kinet. Catal. Lett., 9 (1978)401.
37. S. Ceckiewicz, J. Coll. Interface Sci., 90 (1982) 183.
38. J.G. Fripiat and E.G. Derouane, unpublished results.
39. C.D. Chang, W.H. Lang, and R.L. Smith, J. Catal., 56 (1979)
    169.
40. F.X. Cormerais, G. Perot, F. Chevalier, and M. Guisnet,
    J. Chem. Research (S), (1980) 362.
41. R.M. Dessau and R.B. Lapierre, J. Catal., 78 (1982) 136.
42. N.Y. Chen and W.J. Reagan, J. Catal., 59 (1979) 123.
43. Y. Ono, E. Imai, and T. Mori, Zeit. Phys. Chem., N.F.,115
    (1979) 99.

44. C.D. Chang, Chem. Eng. Sci., 35 (1980) 619.
45. J.P. Gilson, Ph. D. Thesis, University of Namur, 1982.
46. J.C. Vedrine, P. Dejaifve, C. Naccache, and
    E.G. Derouane, Proceedings VIIth Intern. Congr. Catal.,
    Tokyo (1980).
47. V.N. Ipatieff and H. Pines, Ind. Eng. Chem., 28 (1936) 684.
48. M.L. Poutsma, in "Zeolite Chemistry and Catalysis",
    J.A. Rabo, ed. (American Chemical Society, Washington, D.C.,
    (1976), p. 488.
49. E.G. Derouane and J.C. Vedrine, J. Molec. Catal., 8 (1980)
    479.
50. E.G. Derouane, in "Catalysis by Zeolites",
    Stud. Surf. Sci. Catal., Vol. 4, B. Imelik et al., eds.
    (Elsevier Sci. Pub. Co., Amsterdam, The Netherlands, 1980),
    p. 5.
51. J. Valyon, J. Mihalyfi, H.K. Beyer, and P.A. Jacobs, in
    "Proceedings Workshop on Adsorption", Berlin, 1979; 1
    (1979) 134.
52. See lecture by E.G. Derouane on "Molecular Shape-Catalysis
    by Zeolites", this volume.
53. E.G. Derouane and Z. Gabelica, J. Catal., 65 (1980) 486.
54. E.G. Derouane, Z. Gabelica, and P.A. Jacobs, J. Catal., 70
    (1981) 238.

CONVERSION OF METHANOL OVER ZEOLITE CATALYSTS
II INDUSTRIAL PROCESSESS

Zelimir Gabelica

Facultés Universitaires de Namur
Département de Chimie, Laboratoire de Catalyse
Rue de Bruxelles, 61, B-5000 Namur, Belgium

1. INTRODUCTION

In view of the latest critical energy situation in the Western world, extensive research is being conducted to reconsider the potentialities of other non-petroleum materials to become important sources of fuels. In particular, considerable efforts are expanded to develop and improve the various existing technologies for the conversion of coal, heavy petroleum residues, tar sands, shale and biomass, to gasoline. The Bergius (1) and Fischer-Tropsch (2) processes are among the best known that are currently used to produce gasoline from coal.

In the Bergius process (coal liquefaction), a synthetic crude is produced by slurrying finely divided coal with recycle oil, in the presence of iron catalyst, then by hydrogenating the mixture at high pressure and temperature:

The product mixture resulting from this liquefaction process requires extensive and costly hydrogenative upgrading to provide high quality fuels.

Such processes were used extensively in Germany during world war II but no more Bergius plants exist today.

Other coal liquefaction-derived processes, such as Synthoil (a variation of the Fischer-Tropsch process, producing a mixture of

gasoline-diesel) are now being developped and large pilot plants
are in preparation or under construction. However, these processes
still yield hydrogen-deficient products which are not desirable for
a high grade gasoline.

The classical Fischer-Tropsch approach consists in producing
motor fuel from coal. The latter is firstly gasified to synthesis
gas ($CO + H_2$), which is catalytically converted into products.
These consist of a wide spectrum of hydrocarbons (C_1 to C_{40}, or
higher) and oxygenates covering a broad range of alcohols, ketones,
acids and esters:

Again, products need to be adequately separated and upgraded to
produce a high quality gasoline. Such a processing includes costly
hydrogenation, reforming, (hydro)cracking or dewaxing procedures.
Nevertheless, these technologies are now well established and the
by-products are used as raw materials in a variety of industrial,
extremely important, secondary processes (3).
Plants producing gasoline from coal using the Fischer-Tropsch principle are being working currently in South-Africa.

Methanol, a potential motor fuel, is one among the products
readily formed in large amounts, from coal or synthesis gas, by
existing technologies (4,5). It can be either used directly as a
fuel in automotive engines (6),or be blended with gasoline (7,8).
Although methanol presents some minor advantages over gasoline,
such as reduced emissions, significant difficulties are encountered
when it is used alone as fuel. For example, low temperature starting
is difficult. Methanol also remains highly corrosive to many materials used in classical engines so that modified systems need to be
developped. Other safety and toxicity problems specific to methanol,
used either alone or as a component to gasoline, are encountered,
some of them being quite costly to solve adequately. Methanol's
high sensitivity to water, its corrosiveness, its toxicity, its low
volumetric energy content and its unusual volatility, are some of
these disadvantages. A methanol-gasoline mixture enhances the fuel-
overall octane quality but significant problems related to performance and reliability must be solved.

The formation of hydrocarbons from methanol was first reported over classical NaX zeolites (10,11) and, since then, over a
large variety of acidic zeolitic or non-zeolitic catalysts (12).

A novel process for the straightforward conversion of methanol to gasoline over ZSM-5 zeolite catalyst has been recently
developed by Mobil Oil Co. (9,13). It is commonly referred to as

" Methanol-to-Gasoline " (MTG) process.

Crude methanol, such as produced from synthesis gas using the current commercially available methanol technology (5), is rapidly and quantitatively converted into olefines, (cyclo)paraffins and (substituted)aromatics boiling in the gasoline range:

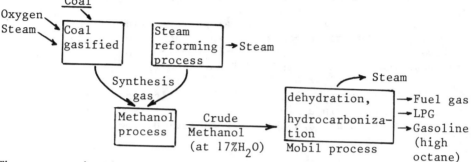

The process is highly selective and can be schematized by the following overall reactions:

$$\left[2CH_3OH \underset{+H_2O}{\overset{-H_2O}{\rightleftharpoons}} CH_3OCH_3\right] \xrightarrow{-H_2O} \underset{\text{olefins}}{C_2 - C_5} \longleftarrow \begin{array}{l}\text{iso(paraffins)}\\ \text{cycloparaffins}\\ \text{aromatics}\end{array}$$

The MTG process is distinct from the former conventional discoveries in that:

a) Essentially isoparaffins and monocyclic methyl-substituted aromatic hydrocarbons are produced
b) A sharp cutoff at C_{10} is observed in the hydrocarbon product distribution
c) Deactivation by coke deposition is slow.

The major mechanistic considerations about MTG reactions are reviewed and discussed in the part I of this work (12).

The present part II summarizes some of the most important technological informations on general industrial processes concerning the MTG conversions. For more details, the reader is referred to technical reparts (20,21,23,28), reviews (1-9,13,14,16,22,27,41, 42) or related publications in current (11,12,15,18,29,30,34-39) or patent (10,17,24-26,31-33,40,43) literature.

2. METHANOL CONVERSION OVER H-ZSM-5 : THE MOBIL PROCESS

Crude methanol (or selected oxygenates) obtained from coal or natural gas by the above sketched classical routes may be converted to gasoline on various (shape selective) zeolitic catalysts, among which those of H-ZSM-5 type are currently preferred.

The active form of ZSM-5 cayalyst is obtained by heating the initially synthesized precursor in inert atmosphere at 550-600° C, followed by exchange with diluted mineral acids or ammonium salts and recalcination in air at 570° C for 24 h (15).

The conversion proceeds by the schematic route depicted in the introduction part. In a single pass, methanol is converted at 99 % to a mixture of hydrocarbons, the composition of which is shown in table 1.

Table 1

Typical hydrocarbon distribution in the MTG process over H-ZSM-5 (fixed-heat reactor, VHSV=0.5h^{-1}, p=205 psig and inlet temperature=700° F (from ref 9).

Products	wt %
Methane, ethane, ethene	1.5
Propane	5.6
Isobutane	9.0
n-butane	2.9
Propylene and butenes	4.7
C_5^+ non-aromatics	49.0
Aromatics	27.3

Essentially no hydrocarbons are produced above C_{11}, which corresponds to the end point of conventional gasoline. Some 25 % of the hydrocarbons are gases, among which less than 2 % is methane or ethane.

This unique narrow range of product molecular weights is consistant with the constrained structure of the zeolite. Hydrocarbons higher than C_{11} cannot escape from the catalyst; only subsequent isomerization and rearrangements reactions allow them to escape with a lower melecular weight. This is a convenient advantage for gasoline manufacture, since no further distillation step to remove heavy ends is required.

About 75 % of the hydrocarbons produced are in the C_5^+ gasoline fraction. A significant amount of additional gasoline can be produced by alkylating C3 and C4 olefins with isobutane. The resulting mixture is rich in isoparaffins and aromatics, with an unleaded Research Octane Nomber (RON) of 90 to 100. This quality is far superior to that stemming from a classical Fischer-Tropsch procedure (table 2).

Table 2

Comparison of MTG conversion on ZSM-5 with Fischer-Tropsch technology (from ref 16)

Product	Fischer-Tropsch		MTG
	Fixed bed	Fluidized bed	
Light gases ($C_1 + C_2$)	11	23	2
LPG($C_3 + C_4$)	11	29	22
Gasoline ($C_5 - C_{11}$)	25	34	76
Fuel Oil (C_{11})	51	5	–
Oxygenates	2	9	–
	100	100	100
Octane of gasoline (clear)	75		95

The highest molecular weight member of the alkyl aromatics series which can escape from the catalyst, is durene (1,2,4,5-tetramethylbenzene). This compound, while having a high octane blending number, melts at 80° C and may potentially give carburator problems at high concentrations. Although its appearence in the product mixture is marginal, its production rate must be controlled by using various parameters. For example, lower durene yields (as well as a higher resistance of the catalyst to coke deposition) are achieved by embedding the active phase in a binder that is preferably an Al-free material (17). This improvement proved particularly useful for the MTG conversion conducted at superatmospheric pressures, which enhance the formation of polymethylbenzenes, among which durene (18) (see 4.2, table 6).

A comparison of ZSM-5 with other zeolites for the MTG conversion is shown in table 3.

534

Table 3

Comparison of ZSM-5 with other catalysts in the MTG conversion
(from ref 16)

	Montmoril-lonite	Narrow pore Zeolite(5Å)	ZSM-5 (5.6Å)	Large pore Zeolite(9-10Å)
% conversion of oxygenates	0.4	11.0	100	18.8
HC distribution (wt %)				
C_1-C_4	100	100	23.7	93.8
C5 aliphatics	–	–	48.9	3.9
Aromatics	–	–	27.4	2.3

3. PROCESS DEVELOPMENT

A major problem in running the methanol conversion is the disposal and/or removal of the heat of reaction (about 406 kcal (13,2
kcal) released per kg (mole) of converted CH3OH), which could lead
to an adiabatic temperature rise as high as 600° C. Various types
of reactors have been described which enable the best use of the
exothermic heat.

3.1. Two stage fixed bed reactor (14,19-23)

The schematic Mobil fixed-bed pilot plant is shown in fig. 1.

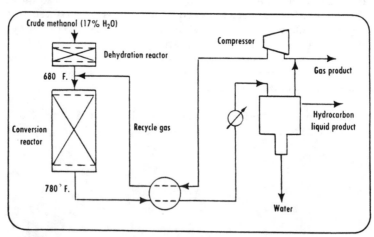

Figure 1. Schematic of Mobil fixed-bed pilot plant

In the first dehydration reactor, methanol is converted on a conventional dehydration catalyst such as γ-Al$_2$O$_3$ to an equilibrium mixture H$_2$O - CH$_3$OCH$_3$ - CH$_3$OH. These products, which leave the reactor at 360° C, are diluted with recycle gas and pass through the second reactor containing the ZSM-5 based catalyst, where it is converted into hydrocarbons. The reactor effluent is condensed and the water separated from the liquid hydrocarbon. The purification of the latter can be achieved by conventional separation methods. A rather high recycle ratio is necessary to reduce the adiabatic temperature rise to 50 - 110° C under typical operatory conditions. Methanol is converted to hydrocarbons (44 wt %) and water (56 wt %) in nearly stoechiometrical amounts. Small amounts of CO , CO$_2$ and coke are formed as by-products. Only the catalyst in the second reactor requires periodic oxidative regeneration to maintain catalyst activity (21,23).
Fixed-bed pilot plant studies were completed and their test results reported with respect to long-term catalyst aging (23).

A variant of this system consists in controlling the highly nexothermic olefin forming reaction step,which accounts for about 40 % of the reaction heat. In such a process, the reactor tubes in the conversion reactor are adjusted in number, size and length as to perform the adequate exchange capacity. A facility is provided to upgrade the recycle gas (C3 - C5 olefins) together with the product stream coming from the dehydration reactor. The complete scheme of such a "Tube Heat Exchanger Reactor" is detailed in the patent literature (24).

Typical process conditions and yields are shown in table 4.

Table 4

Yields from methanol in fixed-bed reactor system (28)

Temperature, °C	
Inlet	360
Outlet	415
Pressure, kPa	2170
Recycle Ratio (mole)	9.0
Space Velocity (WHSV)	2.0

Yields, Wt % of Methanol Charged

Methanol + Ether	0.0
Hydrocarbons	43.4
Water	56.0
CO, CO$_2$	0.4
Coke, Other	0.2
	100.0

Table 4 (continued)
Hydrocarbon Products, Wt %

Light Gas	1.4
Propane	5.5
Propylene	0.2
i-Butane	8.6
n-Butane	3.3
Butenes	1.1
C5 + Gasoline	79.9
	100.0
Gasoline	85.0
LPG	13.6
Fuel Gas	1.4
	100.0

It appears that the hydrocarbon product is primarily gasoline. Its yield, on the basis of the total hydrocarbons produced, is 85 % Wt. Butenes and propene are alkylated with isobutane, the amount of the latter being limited by the yield of the former. Excess isobutane is put into LPG (yield=13.6 % Wt). The remaining 1.4 % Wt is light gas which could be used as fuel gas.

3.2. Fluid bed reactors

Methanol may also be converted in fluid-bed reactors in a single step (21,23,25,26). This provides means for continuous regeneration of the catalyst, so that steady state operation at nearly constant catalytic activity can be reached (compared to the cyclic operation of a fixed-bed). With a fluid bed, the heat of reaction is removed by using heat exchange coils within the bed. Heat transfer is so fast that danger of local overheating is minimal. By operating at nearly isothermal conditions, with significantly lower recycle than the fixed bed system, higher gasoline yields are obtained.
The scheme of a small scale fluid bed reactor is shown in fig. 2. The reactor is pre-heated by an electrical resistance furnace. During operation, charge stock and N_2 (carrier gas) are pumped through a preheater coil where the charge is vaporized. The vapor then passes through the distributor and conversion occurs within the dense fluid bed containing four baffles to minimize by-passing. Gaseous products are carried out of the reactor and the mixed H_2O-hydrocarbon product collected after condensation. The gases are analyzed by gas chromatography as described elsewhere (14,29).
The product distribution obtained from a fluid bed pilot unit is given in table 5.

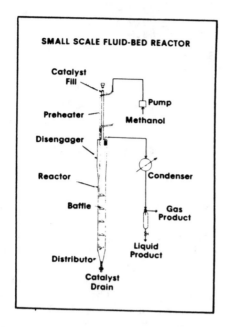

Figure 2 Schematic of fluidized dense bed pilot plant

Table 5

Yields from Methanol in 4 B/D Fluid-Bed Pilot Unit (28)

Average Bed Temperature, ° C	413
Pressure, kPa	275
Space Velocity (WHSV)	1.0

Yields, Wt % of Methanol Charged

Methanol + Ether	0.2
Hydrocarbons	43.5
Water	56.0
CO, CO_2	0.1
Coke, Other	0.2
	100.0

Hydrocarbon Product, Wt %

Light Gas	5.6
Propane	5.9
Propylene	5.0
i-Butane	14.5
n-Butane	1.7
Butenes	7.3

Table 5 (continued)

C$_5$ + Gasoline	60.0
	100.0
Gasoline (including Alkylate)	88.0
LPG	6.4
Fuel Gas	5.6

Fluid bed produces more light olefins and less C$_5^+$ gasoline, with respect to the fixed bed system. However, when propylene and butenes are alkylated with isobutane, the final gasoline yield is superior. The principal reasons for the preferred choice of a fluid bed process are the following :
a) The superior heat transfer characteristics of the latter which would simplify the heat removal.
b) The fluid-bed, when coupled with alkylations, gives a higher gasoline selectivity.
c) Catalyst activity in a fluid-bed reactor can be controlled at the optimum gasoline selectivity level by continuous regeneration.

An improved fluidized bed was recently described (32). The catalyst, stored in the upper region of the bed at elevated temperature, is transferred to a heat exchanger. The latter is advantageously used to preheat the reactors. That means for example that liquid methanol is charged and vaporized before being fed in the mixing zone.

4.   EFFECT OF PROCESS VARIABLES (14,27)

4.1. Temperature

The effect of temperature on the methanol conversion was discussed by Chang et al. (30). The yields of C$_5^+$ aliphatics and aromatics show a maximum in the 370-440 ° C temperature range, at atmospheric pressure. Above 350° C, the reaction rate increases by a factor of 1.5 for each 28° C (21,23) and the production of durene decreases at higher temperatures. Finally the steady state operation of the process is greatly improved by feeding methanol preheated to a selected temperature close to its boiling point, so that the heat needed for its vaporization and rise to reaction temperature is nearly equal to the isothermic heat of conversion (31).

4.2. Pressure

Increasing pressure results in higher methanol conversion, increased gasoline yield and durene formation (9,18,21,23,33)(table6).

Table 6

Effect of pressure on the MTG conversion (700° F, VHSV=1.5h^{-1}, conversion 98-99 % , Fixed bed)(9)

Pressure	0	59	169	353	720
C/R[a]	0.375	0.415	0.450	0.463	0.561
% durene in HC	3.2	4.4	9.4	9.5	20.5

[a] C/R =(C in aromatic side chains)/(C in aromatic rings)

In addition, the catalyst activity is shown to be inhibited at higher pressures by either the adsorption of reactants or products on its surface or by mass transport(diffusion)limitations. Chang et al. also noted an increase in secondary alkylation reaction at higher pressure (9,33). The effect of varying pressure is, as a whole, to change the selective rates of the dehydration and conjunct polymerization steps in the reaction sequences. Increasing pressure increases the overlap of the two reactions and promotes the formation of higher (polymethyl) aromatics, such as the unwanted durene.These effects are shown in fig. 3. More recently, Chang (34) has proposed a kinetic model which describes the methanol conversion path and predicts conversion and selectivities over a wide pressure range (0.04-50 atm).

4.3. Space velocity

Equilibrium between water, dimethylether and methanol is reached rapidly (fig. 3 - pressure 1 atm). Light olefins appear next and aromatics are formed together with aliphatics,as they are related to hydrogen transfer reactions. These effects are discussed in more detail elsewhere (12).

4.4. Water partial pressure

The nature of the product distribution can also be altered by varying the water partial pressure at the inlet of the conversion reactor. In agreement with proposed mechanisms (12), an increase in H_2O partial pressure essentially decreases the selectivity to aromatics (33) and leads to more light olefins by levelling the acidity of the catalyst and by competing for the strong acid sites,which are responsible for hydrogen transfer reactions. Intracrystalline steam distillation effects could also accelerate the desorption of the less volatile products from the zeolite particles (35). Durene yields can also be minimized by recycling it to the inlet of the conversion reactor,where isomerizations and dealkylations are favoured (23).

540

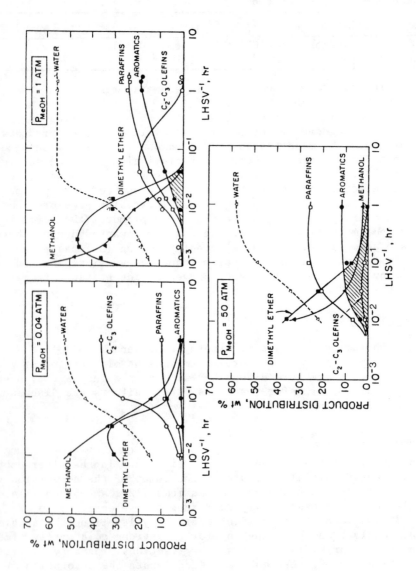

Figure 3. Pressure and space velocity effects on the MTG reactions (18).

5. AGING AND REGENERATION OF THE REACTOR SYSTEMS

Short terms aging tests (21,23) have indicated that the methanol conversion decreases during each cycle (fixed bed). The fraction of light hydrocarbons and the olefins to paraffins ratio decrease while the aromatics concentration increases in parallel. Similarly, long terms aging tests have shown that the durene and coke
formation is reduced. These effects may be understood by considering that aging (deactivation by deposition of carbonaceous residues) will eventually enhance the apparent shape selective properties of the zeolite. For H-ZSM-5, coking essentially occurs at the
external surface (36). Pore mouth restrictions will partially occur
(37) and aromatics such as p-xylene will be produced with an increasing yield (36,38,39).

Although ZSM-5-based catalysts show an unusually high resistance to coking (36), they nevertheless have to be regenerated
either periodically (fixed bed) or continuously (fluid bed).

Fixed bed regeneration can be achieved by controlled treatment with air or steam at 340-480° C (21,23). The heat due to the
exothermic reaction can be more homogeneously redistributed by
mixing the active zeolite with an inert phase, the mixture moving
downwards to the regeneration zone (26).

Fluid bed conversion-regeneration units require the use of
special technologies which are described in detail in the patent
literature (25). A simple, fast and very effective way to regenerate an aged catalyst has been disclosed (40). It is called "se-
lictivation" and may be considered as a pre-coking process of a
fresh catalyst. The latter is activated in a hydrogen-toluene atmosphere. The resulting coking induces pore mouth restrictions in
ZSM-5,which maintain an increased molecular shape selective activity of the zeolite and avoids the need for further controlled coke-
redeposition pretreatment.

6. FUTURE DEVELOPMENTS OF THE MTG PROCESS

The actual use of the Mobil process to make gasoline will depend on competitive economies with other synthetic fuel processes.
Technical and economical comparisons have been made at the end of
the last decade (41), on the basis of the Mobil pilot plants information, with the commercially available SASOL-type-Fischer-
Tropsch (F-T) technology. Table 2 shows that the Mobil process
produces 70 % more gasoline than the F-T route. It has a higher
thermal efficiency, and also requires less investment so that the
gasoline cost is cheaper by 14-31 %. A recent modification of the

MTG process may launch a new era in the F-T technology (42). This Mobil variant is labelled MFT (Mobil-Fischer-Tropsch). The basic idea is to pass the total effluent from a F-T reactor into a second reactor containing ZSM-5. The products include 70 % gasoline with an octane number of 92 and about 20 % LPG and light gas. The MFT process eliminates the heavy conventional products by cracking.

Another modification of the basic MTG process consists in maximizing olefin production. A modification will yield up to 60 % of C_2 - C_4 olefins in a pilot plant and even greater yields have been produced in the laboratory. One version of this process is being piloted in South Africa.

Another variant is the MOGD process, which produces diesel fuel (80 %) with gasoline (20 %) as a side product, from mixed propylene and butylene as feedstock (42).

It thus seems that the Mobil process, either as the basic MTG conversion or in the form of its various modifications, becomes economically attractive, mainly in view of recent price increases of crude oil, in areas where coal and/or natural gas are readily available. The first commercial installation of the MTG process is expected to come on stream in the mid-1980s in New Zealand. It will convert natural gas (instead of coal) to synthesis gas, from which the methanol will be made. This circumstance is peculiar to New Zealand which has a huge amount of natural gas available in its offshore gasfields. The plant will use a fixed-bed reactor system but Mobil is now being developping a fluid-bed variant for the same purpose. The fluid-bed reactor is considered more appropriate for the very huge installations and is being scaled up for commercial operation. By 1985, the New Zealand MTG plant will begin to draw 140 million cu.ft. of gas per day with, as the net result expected, about 14,000 bbl per day of high-quality motor fuel.

7. REFERENCES

1. "Encyclopedia of Chemical Technology" (Kirk Othmer, Ed.) Vols 4 and 13, Wiley, New York, 1972
2. H.H. Storch, N. Golumbic and R.B. Anderson, "The Fischer-Tropsch and Related Synthesis", Wiley, New York, 1951
3. For example: H. Perry, Scientific American 230, 19 (1974)
4. "Methanol Technology and Economics" (C.A. Danner, Ed.), Chem. Eng. Progr. Symp. Ser. N° 98, 66 (1970)
5. "Methanol Technology and Application in Motor Fuels" (J.K. Paul, Ed.), Noyer Data Corp., Park Ridge, (N.J.), 1978, Chap2,pp107-165
6. For example: G.A. Mills and B.A. Harney, Chemtech. 4, 26 (1974)
7. "Methanol/Gasoline Blends", chap 5,pp 241-315 (ref 5)
8. T.B. Reed and R.M. Lerner, Science 1982, 1299 (1973)
9. C.D. Chang, J.C.W. Kuo, W.H.Lang, S.M. Jacob, J.J.Wise and A.J. Silvestai, Ind. Eng. Chem. Process Des. Dev. vol 17, 255 (1978)
10. W.J. Mattox, U.S. Pat. 3,036,134 (1982)
11. E.I. Heiba and P.S. Landis J. Catal 3,471 (1964)
12. E.G. Derouane, in"Zeolites, Science and Technology", Lisbon, 1983 (this meeting), part I of the present work, refs (1) to (11)
13. S.L. Meisel, J.P. McCullough, C.H. Lechthaler and P.B. Weisz Chemtech. 6, 86 (1976)
14. see ref (5), p 375-420
15. D.H. Olson, W.O. Haag and R.M. Lago J. Catal. 61, 390 (1980)
16. H. Heinemann,"Proc. Vth Ibero-American Symp. Catalysis"(M. Farinha Portela and C.M. Pulido, Eds.) Lisboa, Portugal, 1978, Vol I,p 10
17. C.D. Chang and W.H. Lang, U.S. Pat. 4,013,732 (1977)
18. C.D. Chang, W.H. Lang and R.L. Smith, J. Catal 56, 169 (1979)
19. S.Yurchak, G.E. Voltz and J.P. Warner, Ind. Eng. Chem.Process Des. Dev. 18, 527 (1979)
20. N. Daviduk, J. Mazuik and J.J. Wise,"Proc 11th Intersociety Energy Conversion", vol. I (1976)
21. J.E. Voltz and J.J. Wise "Development Studies on Conversion of Methanol and Related Oxides to Gasoline" Final Report, ERDA Contract n° E(49-18) -1773,Nov. 1976
22. J.J. Wise and A.J. Silvestri, Oil and Gas J. (1976)
23. J.J. Wise and J.E. Woltz, Synth. Fuels Proc. Res. Dig. ORNL-FE-1, pp 51-63, Nov. 1977
24. C.D. Chang and J.J. Grover, U.S. Pat 4,058,576 (1977)
25. H. Owen and P.B. Venuto, U.S. Pat 4,046,825 (1977);ibid 4,071,573 (1978)
26. N.Y. Chen, U.S. Pat 4,118,431 (1978)
27. D. Liederman, S.M. Jacob, J.E. Voltz and J.J. Wise, Ind. Eng. Chem. Process Des. Dev. 17, 340 (1978)
28. J.E. Penick, S.L. Meisel, W. Lee and A.J. Silvestri "Alcohol Fuels Conference", Sydney, Australia, Aug. 1978
29. J.M. Stockinger, J. Chromatogr. Sci. 15, 198 (1977)
30. C.D. Chang and A.J. Silvestri J. Catal. 47, 249 (1977)
31. C.D. Chang, A.J. Silvestri and J.C. Zahner, U.S. Pat. 4,044,061 (1977)

544

32. W. Lee and S. Yurchak, U.S. Pat. 4,197,418 (1980)
33. C.D. Chang, A.J. Silvestri and R.L. Smith, U.S. Pat. 3,894,103 (1975); ibid, 3,928,483 (1975)
34. C.D. Chang, Chem. Eng. Sci. 35, 619 (1980)
35. E.G. Derouane and J.C. Vedrine J.Molec. Catal. 8, 479 (1980)
36. P. Dejaifve, A. Auroux, P.C. Gravelle, J.C. Vedrine, Z. Gabelica and E.G. Derouane, J. Catal. 70, 123 (1981)
37. Z. Gabelica, J.P. Gilson, G. Debras and E.G. Derouane, in "Thermal Analysis, Proc. 7th Int. Conf. Thermal Anal.", (B. Miller, Ed.), vol II, Wiley-Heyden, New York, 1982, pp 1203
38. N.Y. Chen, W.W. Keading and F.G. Dwyer, J. Amer. Chem. Soc. 101, 6783 (1979)
39. E.G. Derouane, P. Dejaifve, Z. Gabelica and J.C. Vedrine Faraday Disc. Chem. Soc., 72, 331 (1981)
40. W.O. Haag, W.W. Keading, D.H. Olson and P.D. Rodenwald, Eur. Pat. Appl. 9,894 (1979)
41. W. Lee, J. Mazuik and C. Portail, "Informations Chimie N° 188," April 1979, pp 165-73.
42. J. Maggin, Chem. Eng. News, Dec. 1982, pp 9-15
43. ref (5), "Gasoline from Methanol Patent Technology", pp 421-464.

USE OF PLATINUM HY ZEOLITE AND PLATINUM H MORDENITE IN THE HYDROISOMERIZATION OF N-HEXANE

F. Ramôa Ribeiro

Grupo de Estudos de Catálise Heterogénea

Instituto Superior Técnico

Av.Rovisco Pais, 1096 Lisboa Codex, Portugal

1. INTRODUCTION

Petroleum cuts containing C_5/C_6 paraffins have a research octane number of about 70 due to the high content of monobranched paraffins. To upgrade the octane number of the light gasoline fractions, consisting predominantly of C_5/C_6 paraffins, antiknock additives,e.g. tetraethyllead and tetramethyllead were used, but at present the new environmental regulations defines very low permissible concentration levels for automative pollutants, particulary for the lead compounds (1).

As a consequence, both industrial and academic research efforts have been made to increase the octane number of the fractions C_5/C_6 paraffins, in order to compensate for the removal of the lead antiknock additives, through the isomerization of normal paraffins into branched paraffins with higher octane-level.

Thermodynamically, the isomerization of C_5/C_6 paraffins is reversible and slightly exothermic (1-5 Kcal/ /mole) and consequently low temperatures are suitable for obtaining maximum yield of branched paraffins, as shown in Figure 1 (2).

Liquid acid catalysts, such as those of the Friedel Crafts type, present a good activity at 85^oC, and the superacid $Sb F_5 - HF$ (3) at 20^oC, but they are extremely corrosive and easy to contaminate.

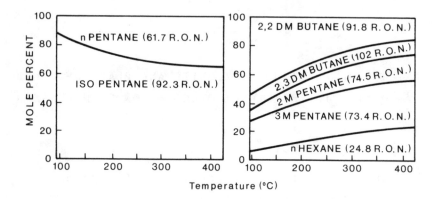

Figure 1. Equilibrium distributions and the research octane number for the pentanes and hexanes (ref.2)

During World War II, due to the great need of high-octane aviation fuel, some isomerization plants were built, using aluminum chloride as a catalyst. The process was expensive and the disposal of the sludge formed was a severe problem.

In the 1950's these processes were replaced by the use of bifunctional catalysts composed of a dispersed noble metal on a classic support such as chloride alumina or silica-alumina (4,5) and several plants (6,7,8) using this type of catalyst were built.

The process using chloride alumina operates at lower temperature but requires careful feed pretreatment for removal of deactivating substances such as water, sulphur and benzene.

The process using silica-alumina as support operates at 400°C, too high a temperature, from the thermodynamical point of view, which limits the conversion and favours the hydrocarbon side reactions (cracking).

Consequently, the need to discover other catalysts, active at low temperatures and not easily contaminable, was very urgent. It was also important to operate at a temperature sufficiently high to obtain good conversions, close to equilibrium, with high yields of dimethylbutanes.

Since the 1960's many research workers (9, 10,
11, 12) have been reporting the high activity for isomerization of zeolitic catalysts, mainly zeolite Y, and
mordenite, loaded with a noble metal to stabilize the
activity.

The development of highly active zeolitic catalysts has resulted in a hydroisomerization process,
known as the Hysomer Process (13) launched by the Shell
Oil Company. The first comercial unit was built at the
La Spezia refinary in Italy and at present there are
about ten units in operation throughout the world.

The catalyst, manufactured by Union Carbide, is
a platinum exchanged acid mordenite of very low sodium
content, presenting a much better resistance to water,
sulphur and aromatic compounds over amorphous catalysts.
In fact the catalyst remains active with sulphur and
water levels in the feedstock of 10 and 50 p.p.m. respectively, and concentrations of aromatics up to 2 wt%.

The design unit is shown schematically in Figure 2 (14).

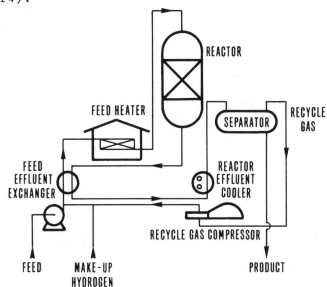

Figure 2. Shell Hysomer process (ref. 14)

This process converts low octane pentane and he
xane streams (RON ~ 70) to higher octane products
(RON ~ 83) containing isopentane, methylpentanes and
dimethylbutanes. It is not used on C_7 or higher paraffins because undesired reactions (cracking) occur on
these acid zeolitic catalysts.

The process operates at temperatures of 250°C and hydrogen pressure of 10-25 bars.

Table I (14) gives the typical properties of C_5/C_6 feed and isomerized products.

TABLE I (ref. 14)

Products from hydroisomerization of C_5/C_6 tops ex.

Middle East Crude at 260°C

	Feed	Isomerizate
R.O.N. of C_5^+	67.5	79.2
Composition (wt.%)		
Methane	–	0.18
Ethane	–	0.18
Propane	–	1.06
Isobutane	–	1.60
n-Butane	0.57	1.10
Isopentane	17.43	31.33
n-Pentane	28.12	15.07
2,2 - Dimethylbutane	0.44	9.34
2,3 - Dimethylbutane	1.97	4.06
2 - Methylpentane	12.43	14.39
3 - Methylpentane	9.72	9.91
n-Hexane	21.15	8.35
Cyclopentane	1.75	1.30
Methylcyclopentane	3.26	1.50
Cyclohexane	0.88	0.63
Benzene	1.11	–
C_7^+	1.17	–

These data show that the hydroisomerization of the light gasoline fractions C_5/C_6 substantially increase the research octane number.

For the full understanding of the hydroisomerization of n-paraffins on Pt-zeolites there is a need to clarify several important aspects such as the preparation conditions, the influence of the reduction temperature of platinum on the activity, the mechanism and kinetics of the reaction.

We will present a comparison between the Pt H Y and Pt H mordenite using as test reaction the hydroisomerization of n-hexane.

2. CATALYSTS

The hydroisomerization of n-paraffins use bifunctional catalysts - platinum acid zeolites - with a hydrogenating and an acidic function.

2.1 Acidic Function

Acidic function arises from the hydrogen form of zeolite.

Rabo and his associates (9) found the sodium form of zeolite Y as support for the platinum, inactive in the n-hexane isomerization and have shown that the decationization increased the activity so that the catalyst could be used at lower temperatures.

Kouwenhoven (20) has shown that reducing the sodium content to very low values, in a catalyst of Pt HY for the hydroisomerization of n-pentane, the activity is increased considerably.

However, a completely sodium free zeolite HY (19) presents lower stability and, in order to obtain high stability and consequently high activity, some very low concentrations of sodium are required.

Figure 3 (15) compares the extent of sodium removed by ammonium ion exchange for the zeolite Y (Union Carbide, 9.9 wt.% Na) and the mordenite (Norton Zeolon 100, 6.0 wt.% Na), and it is clear that it is more difficult to remove the residual sodium of the type Y, because it presents some cations located in small β-cages (16). However, in agreement with the literature (17, 18, 19) we found that a further exchange with ammonium ions after steam calcination of zeolite Y reduced the sodium content to 0.3 wt.%.

After three exchanges with ammonium ions at 20°C, the mordenite presented a very low sodium content (0.016 wt.%), because the cations are located in the main channels and they are easily removed.

The study of the acidity by infrared spectroscopy (pyridine desorption at increasing temperatures) has shown a strong Brönsted acidity more marked for the H mordenite than for HY (54).

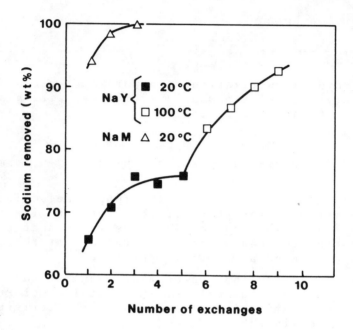

Figure 3. Comparison of the extent of decationization on Y zeolite and mordenite

2.2 Hydrogenating function

2.2.1 Metal impregnation

The hydrogenating function is provided by noble metals, particulary platinum. It is very important to employ techniques of metal impregnation in order to obtain good metal dispersion and a homogeneous macroscopic distribution of metal into the zeolite.

To achieve this last purpose we developed a technique, called competitive cation exchange (21,22).

It is well known that in order to obtain a good dispersion of the metal we must promote a strong interaction between the metal and the support, but this condition is not sufficient.

A classical ion exchange of metallic ions in a small quantity, in comparison with the number of exchangeable centers of the support, leads to a heterogeneous macroscopic distribution of the metal over the surface, due to the exchange rate being higher than the diffusion rate and to an important affinity of the platinum ions for the zeolite Y, as shown in Figure 4 (15).

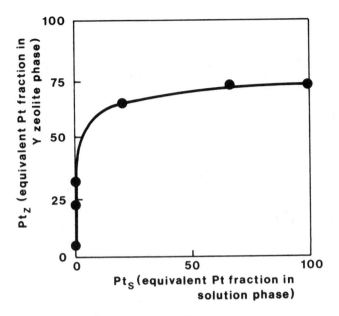

Figure 4. $NH_4Y / \left[Pt(NH_3)_4{}^{2+} \right]$ exchange isotherm at $25^{\circ}C$ and 0.1 N

Under such conditions the metallic ions fix themselves at the exchange centers on the surface and, because the concentration of those ions in the solution is low, there is no diffusion of the platinum inside the grain of catalyst, and they stay confined to a thin layer over the periphery.

For a zeolite under the $NH_4{}^+$ form the ion exchange reaction, for the platinum under the cationic form is

$$\left[Pt(NH_3)_4{}^{2+} \right]_S + \left[NH_4{}^+ \right]_Z \rightleftarrows \left[Pt(NH_3)_4{}^{2+} \right]_Z + \left[NH_4{}^+ \right]_S$$

S - Solution
Z - Zeolite

The technique of ion exchange with competition (23,24) enables a homogeneous macroscopic distribution to be obtained. It consists of introducing an excess of ions for competition ($NH_4{}^+$ ions) which displace the above equilibrium towards the left thus increasing the concentration of metallic ions in the solution and consequently increasing the rate of diffusion and migration of the metal towards the inside region of the support.

This effect, referred to as competition, reflects itself on an experimental curve, which represents the fraction of metal in equilibrium in the solution versus the competition.

The competition is defined as $\dfrac{N_o + N_z}{n\,M_o^{n+}}$

where

N_o - number of competing gram ion equivalents added to the solution phase

N_z - number of competing gram ion equivalents initially present in solid phase

$n\,M_o^{n+}$ - number of metal equivalents present in solid-solution system

$n\,M_s^{n+}$ - number of metal equivalents in solution phase

Upon rearranging the equilibrium constant equation (24) of ion exchange reaction we obtain

$$\frac{x}{\alpha} = \left(\frac{1}{\alpha} - 1 + \frac{Y}{\alpha}\right)\left[\left(\frac{C\,y/\alpha}{1-y/\alpha}\right)^{1/2} + 1\right]$$

where

$$x = \frac{N_o + N_z}{N_z},\quad y = \frac{n\,M_s^{n+}}{N_z},\quad \alpha = \frac{n\,M_o^{n+}}{N_z},$$

$$C = \left(\frac{K_a}{\varepsilon}\right)\frac{V_S}{V_Z}$$

K_a - equilibrium constant for the ion exchange reaction

ε - quantity which groups the activity coefficients together

V_S - volume of the solution

V_Z - volume of the zeolite

The competition curves for the zeolites NH_4Y and MH_4M and platinum are shown in Figure 5.

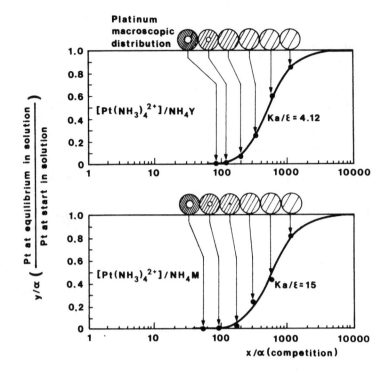

Figure 5. Competition curves for the zeolites NH_4Y
and NH_4M (ref. 24) (Conditions: $T = 20°C$,
$pH = 7$, $\dfrac{V \text{ Solution}}{P \text{ } NH_4Y} = \dfrac{40}{3}$ cm^3 g^{-1},
$C_{\left[Pt(NH_3)_4\right]^{2+}} = 356$ mg 1^{-1})

The curves were obtained using relation $\dfrac{Y}{\alpha} = f\left(\dfrac{x}{\alpha}\right)$
(24) and the mean values of $\dfrac{K_a}{\varepsilon}$ calculated from the experimental data, and it is found that the experimental in
formation shows good agreement with theoretical predicti
ons.

The importance and pratical use of this technique
of competition lies in the fact that the competition
curve enables one to know which is the minimum quantity
of competing ions to be added to the solution, in order
to have metal uniformly distributed over the support
with simultaneously the least possible fraction of the
metal in the solution. For the zeolites NH_4Y and NH_4M
the optimum value of the competition from the industrial point of view should lie between 200 and 300.

The Pt - zeolites, prepared using the competitive

ion exchange, present a greater stability and activity, as compared with identical catalysts, in which the platinum was deposited using a classical ion exchange.

Studies described elsewhere (25) confirmed the above conclusion. Two PtHY zeolites with the same platinum content (about 0.5 wt.% Pt) have been prepared in different conditions: in one of them (catalyst I) the platinum was deposited using the competitive ion exchange and in the other (catalyst II) using a classical ion exchange. Catalyst II presents an heterogeneous platinum distribution, with an important content of metal on the external surface.

The performances of the two PtHY catalysts were studied using the hydroisomerization of n-hexane under a total pressure of 30 bars and $H_2/n\ C_6 = 4$ (mole ratio), as the test reaction. As shown in Table II, the activity obtained at 230-280°C using semidifferential conditions, is 2.5 times higher for catalyst I, as compared with that of catalyst II, and both present very low values of selectivity for the dimethylbutanes.

Table II

Comparison between two PtHY catalysts ($\sim$ 0.5 wt% Pt) one being prepared by competitive ion exchange (cat.I), and the other by classical ion exchange (cat.II)

$T\ (°C)$	$10^4\ a_I$ (mole $h^{-1}\ g^{-1}$ of PtHY)	
	I	II
230	39	–
250	180	70
260	317	140
270	613	280
280	–	540

The difference of activity for catalysts I and II may be interpreted as follows: for catalyst I (homogeneous platinum distribution) all the surface actuates according to a classical bifunctional metal-acid mechanism, whereas for catalyst II only the external surface actuates in this way and its internal surface without platinum actuates through an acid mechanism with low activity due to a faster deactivation.

2.2.2 Thermal treatments

The production of PtHY and PtHM zeolites includes several thermal treatments, whose experimental conditions have a great influence on their stability and catalytic properties.

2.2.2.1 Preparation of the HY and HM

The calcination of NH_4Y and NH_4M not only removes the water vapour and ammonia, leading to the HY and HM forms, but also tends to produce important structural changes depending on temperature, time and steam partial pressure conditions.

A study described in references 26 and 27 shows the influence of the temperature and composition of the calcination atmosphere on the structure of a very low sodium zeolite. The structure of the HY zeolite quickly collapses at 450°C when calcined under dry air, but the presence of very little water vapour delays the destruction, as shown in Figure 6

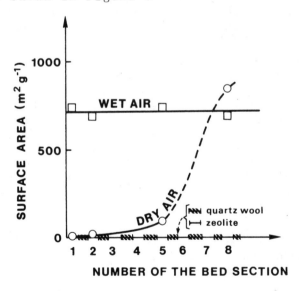

Figure 6. Evolution of the surface area as a function of the number of the powder bed section.

The presence of steam at high temperatures during calcination contributes to stabilize the structure, decreasing the value of the unit cell parameter down to 24.4 $\overset{\circ}{A}$ (24.65 $\overset{\circ}{A}$ for Na Y) and increasing the SiO_2/Al_2O_3 ratio. This thermal treatment also modifies the acidic properties, decreasing the Brönsted acidity (28).

A literature (17, 18, 29-35) survey on this subject indicates several techniques to obtain the more or less stabilized structures.

John Ward (35) describes, in a U.S. patent, the sequence of operations to prepare a "stabilized" zeolite Y, involving three steps: partial ion exchange of sodium with ammonium, calcination in the presence of steam above 500°C, followed by an ammonium ion exchange, which reduces soda levels down to 1 wt.%.

The stabilization of the zeolite Y structure in the presence of steam is the consequence of two simultaneous processes: extraction of Al atoms and their replacement by Si atoms in positions that are not "strategic". The calcination under very dry air accelerates the first process and leads to the destruction of the structure (28).

The presence of steam at high temperature (> 400°C) during calcination favours the second process, which must have a higher activation energy, and leads to the stabilization of the H Y structure.

The ammonium form of mordenite in the presence of steam at high temperatures, presents (as the NH_4Y does) a greater stability as compared with that of mordenite calcinated under dry air. Thus the values of the surface areas of two H mordenites calcinated under wet air and dry air, after recalcination at 950°C are 235 m^2/g and 9 m^2/g respectively (15). These results show that the steam calcination of NH_4M leads to a stabilization of the structure. However, the high stability of this other mordenite does not require such particular conditions for calcination.

2.2.2.2 Calcination of platinum amino complex-
- containing NH_4 zeolites

The calcination conditions of the $[Pt(NH_3)_4 \, NH_4$ zeolite] must be well chosen in order to obtain the best metallic phase dispersion, without damaging the zeolitic structure.

For zeolite Y, Boudart (36) has shown that the reduction of $[Pt(NH_3)_4 \, Ca \, Y]$ with hydrogen at 350°C causes the formation of a very labile $[Pt(NH_3)_4 H_2]$ complex and consequently leads to the metal sintering. In order to obtain highly dispersed platinum in zeolite Y, it is essential to carry out a staged calcination in dry air (37, 38).

Our studies (39) have shown that the presence of high contents of platinum contributes to the damaging of the structure during the calcination when the zeolite Y is not previously stabilized.

The calcination of $[Pt(NH_3)_4 NH_4$ mordenite$]$ must be made under a stream of air, using staged heating (40, 41). Contrary to what happens with the zeolite Y, the structure of the mordenite (39), either stabilized or not not, always remains well organized after calcination of the $[Pt(NH_3)_4 NH_4 M]$.

2.3.3 Influence of the reduction temperature on the catalytic behaviour of PtH mordenite

The reduction conditions of the bifunctional catalyst, platinum H zeolite, have a marked influence on their activity and selectivity.

Studies reported in a recent publication (42) have shown the influence of the reduction temperature of a series of platinum H mordenite (PtHM) catalysts with platinum contents varying from 0 to 10.2 wt %, on the n-hexane isomerization and cracking under 24 bars hydrogen pressure.

Figures 7 and 8 (42) compare the isomerization and cracking rates of n-hexane for two series of PtHM catalysts reduced at 450 and 500°C. It is clear that the reduction at 500°C increases their isomerization activity and decreases their hydrogenolysis activity. The reduction temperature has no influence on the selectivities for the isomerization products and does not affect the metallic phase dispersion.

We have shown (43) that for the zeolitic catalysts, with high content of platinum, the formation of light cracking products results mainly from the hydrogenolysis of n-hexane on metal sites and, for low content of platinum, results from the hydrocracking (bifunctional reaction). The reduction treatment at 500°C, having a greater influence on the catalysts with higher content of platinum, affects, therefore, essentially the hydrogenolysis sites.

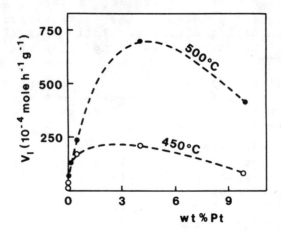

Figure 7. Influence of catalyst reduction temperature on the isomerization rate of n-
-hexane at 260°C on a series of PtHM
with several platinum contents

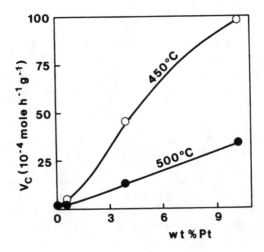

Figure 8. Influence of catalyst reduction temperature on the cracking rate of n-hexane
at 260°C on a series of PtHM with several platinum contents

The decrease of the hydrogenolysis activity after
the pretreatment under hydrogen at high temperatures,
as proposed by Froment et al. (44, 45), seems to be cau
sed by a strong chemisorption of H_2 on platinum.

The increase of isomerization activity can be explained by the fact a decrease of coke formation on the zeolitic catalysts reduced at 500°C has been shown to be associate with a weaker hydrogenolysis activity than when the reduction was carried out at 450°C. Our earlier studies (46) on PtHM catalysts with high platinum content have shown a simultaneous increase in hydrogenolysis activity and in coke formation.

An important conclusion to be drawn is that the PtHM catalysts reduced at 500°C are much more selective for the isomerization of n-hexane than those reduced at 450°C.

3. Kinetics and mechanism of n-hexane hydroisomerization on Pt H zeolites

The hydroisomerization of n-hexane on metal-loaded zeolites has been the subject of numerous studies (9, 10, 20, 47, 53), particulary its kinetics and mechanism.

The main products of the reaction are 2,2-dimethylbutane, 2,3-dimethylbutane, 2-methyl-pentane and 3-methyl-pentane. Side reactions, such as hydrocracking, occur on those bifunctional catalysts and lead to light alkanes from C_1 to C_5, which we must minimize to increase the selectivities for the products of isomerization.

3.1 Effect of platinum content on the stability and activity of Pt H zeolite

The study of the aging period for Pt H Y zeolites (Figure 9) at 300°C, under a total pressure of 30 bars and a molar ratio $\frac{H_2}{n\,C_6} = 4$, shows clearly the effect of platinum content on the isomerizing activity and stability of zeolitic catalysts.

The stability of the conversion level during the operating time increases with the increasing of platinum content.

The incorporation of platinum onto the hydrogen Y zeolite increases the isomerization, the maximum occurring for low platinum contents. We can assume that the platinum prevents the catalyst deactivation by promoting the hydrogenation of coke precursors.

For higher platinum contents there is a stability of conversion level during the operating working time but the isomerization decreases due to an increase of hydrocracking and/or hydrogenolysis of n-hexane.

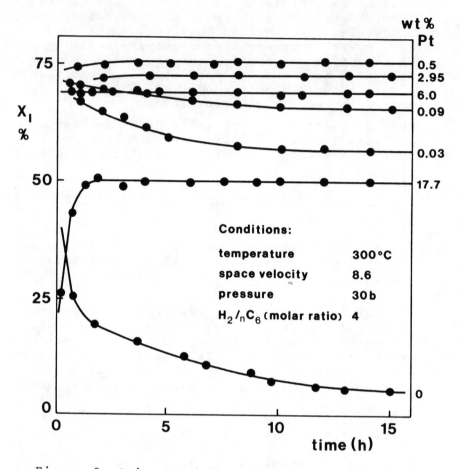

Figure 9. Aging period of Pt H Y catalysts for the hydroisomerization of n-hexane (T = 300°C,

$$P_t = 30 \text{ b}, \quad \frac{H_2}{n\,C_6} = 4 \text{)}$$

The behaviour of Pt H mordenite on the hydroisomerization of n-hexane is very similar to that of Pt H Y, concerning the aging period.

It would be interesting to compare the performances of Pt H Y and Pt H mordenite at 260°C, temperature, for which conversions are less than 10%, in order to select the best catalyst and also to determine the optimal platinum content.

As shown in Figure 10, at 260°C under a total pressure of 30 bars, we observe a sharp increase in the isomerization rate of n-hexane at low platinum contents. The maximum is reached for both catalysts somewhere in the range 2 to 3% Pt.

By comparing two curves we conclude that the Pt H mordenite presents a higher isomerizing activity, certainly due to its higher acidity (54).

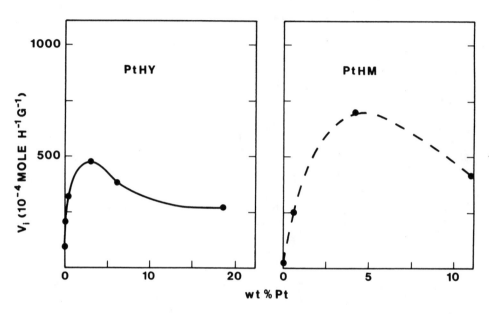

Figure 10. Comparison of the isomerizing activity as a function of the platinum content for Pt H Y and Pt H mordenite

$$(T = 260^{\circ}C, \ P_t = 30 \ b)$$

The decrease in the isomerizing activity at higher platinum contents can be explained (46) by the coke poisoning of the acid function. The higher formation of coke at high platinum contents is caused by the hydrogenolysis of n-hexane on the platinum, which becomes more important than hydrocracking (56, 57).

Other authors (55) have compared the performances of H mordenite and H Y loaded with a noble metal, and their results lead to the same conclusion, that the bifunctional catalysts based on H mordenite are more active for paraffin isomerization.

3.2 Reaction mechanisms of n-hexane hydroisomerization

Having discussed the influence of the platinum content on the selectivity and activity of Pt H zeolites we should next clarify the mechanism of hydroisomerization of n-hexane on metal acid bifunctional catalysts.

Weisz (58) proposed a mechanism as described below

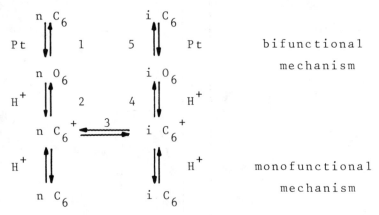

If the mechanism is purely acid the formation of the carbonium ion is the rate-determining step.

The C_6^+ is formed through the hydrogen transfer between the n-paraffin and a carbonium ion previously present or by a reaction on the Brönsted or Lewis sites of the zeolite (60).

For low metal contents, the hydrogenation and dehydrogenation reactions are the limiting step, the isomerization activity being proportional to the metal content. If the hydrogenating activity is high, which happens for higher metal content, the limiting step will be the skeletal isomerization of olefins on acid sites and the activity depends only on the acidity of the zeolitic catalysts (59). However, for the highest platinum contents we observed a decrease in the isomerizing activity, already considered above.

At low conversion, and neglecting the external and internal diffusional limitations, the isomerization rate of n-hexane can be expressed, according to the bifunctional mechanism, in which the limiting step is the acid isomerization of olefins, as follows:

$$r = k_3 \; C_m \; K_2 \; K_1 \quad \frac{P_{n\,C_6}}{pH_2 + K_1 \; K_2 \; P_{n\,C_6}} \tag{43}$$

where k_3 is the isomerization rate constant of carbonium ions, C_m the concentration of Brönsted sites of the zeolite and K_1 and K_2 the equilibrium constants of the dehydrogenation of n-hexane and of the carbonium ion formation.

The validity of this equation, and consequently of the bifunctional mechanism was tested (43) for the hydroisomerization of n-hexane on Pt H Y (6.0 wt %) under a total pressure of 40 bars, and a good agreement was found with the experimental values.

Kinetics studies of the n-hexane isomerization carried out by Braun and co-workers (61) have shown that no significant mass transfer limitations on the external and internal surface of pellets of Pt H mordenite (0.5% wt % Pt) occur. The experimental values obtained for effectiveness factors corresponding to pellet diameters in the range of 0.5 to 4 mm and at a temperature of 270ºC, lie between 0.9 and 1.0.

3.3 Activation energies of isomerization and selectivities as functions of the platinum content for Pt H Y and Pt H mordenite

As shown in Figure 11, the activation energies of isomerization for Pt H Y and Pt H M obtained in a range of 230 to 300ºC under a total pressure of 30 bars increase with the increase in the platinum content, reaching a plateau of 37 K cal/mole for low platinum contents (about 0.5 wt %). This behaviour can be explained through the bifunctional mechanism described above.

For the H Y and H M the limiting steps is the formation of the carbonium ion whose activation energy is low (60).

As the platinum content increases there is a change towards the bifunctional mechanism where the limiting step is the skeletal isomerization of intermediate olefins.

Braun et al. (61) obtained similar values as activation energies of isomerization for Pt H mordenite.

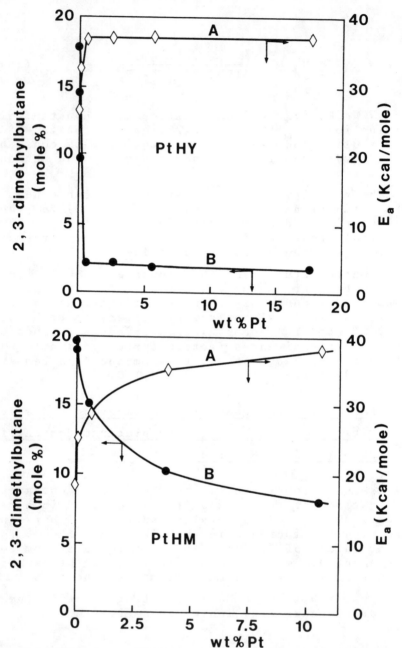

Figure 11. Activation energy of n-hexane hydroisomerization and selectivity of 2,3 - dimethylbutane on two series of PtHY and PtHM catalysts, as a function of the platinum content (T = 260°C, P_t = 30 bars, $\frac{H_2}{n\ C_6}$ = 4).

Plot A, activation energy; plot B, selectivity of 2,3--dimethylbutane.

The change of the selectivities of 2,3 - dimethyl butane as a function of platinum content follows an opposite trend to that obtained for the activation energies of isomerization.

In fact, for Pt H Y there is a strong decrease of the selectivity for 2,3 - dimethylbutane over the initial part of the curve, which then reaches a plateau for platinum contents > 0.5 wt %. As regards the other isomers, the 2- and 3- methylpentane are obtained as an equilibrium mixture and also very low percentages·of 2,2 - dimethylbutane which has a low rate of formation (62).

For Pt H M, the selectivity decrease for 2,3 - di methylbutane is much slower, and even for high platinum contents its selectivity is about 7%. We can assume (63) that the migration from the metal to the acid sites is slower than the isomerization and consequently the olefins have sufficient time to attain equilibrium on acid sites, thus obtaining the dimethylbutanes as primary products.

Conclusions

The analysis of the behaviour of the catalysts Pt H mordenite and Pt H Y in the hydroisomerization of n-hexane, led us to some important conclusions, namely, the fact that Pt H M has a higher isomerizing activity and selectivity for the dimethylbutanes of high octane number.

These properties may explain the preference that is given to the use of Pt H mordenite in the Hysomer Shell process.

The development of more active catalysts will allow the decrease of the hydroisomerization temperature and consequently to obtain a maximum yield of branched paraffins with high octane level.

Acknowledgments

Much of the work reported here, was carried out by F. Ramôa Ribeiro during his stay at the French Petroleum Institute. Special thanks are due to Dr. Ch. Marcilly (French Petroleum Institute) and Prof. Dr. M. Guisnet (University of Poitiers) for the helpful discussions on several aspects of this work.

566

References

1. J. McEvoy in "Catalyst for the Control of Automati ve Pollutants" Ed. Robert Gould, ACS Monograph 14$\overline{3}$, 1975, American Chemical Society, Washington

2. A.P.Bolton, in "Zeolite Chemistry and Catalysis" Ed. J.A.Rabo p.750, ACS Monograph 171, 1975, American Chemical Society, Washington

3. J.Oelderick, E.L.Mackor, J.C.Platteeuw, A.Van der Wiel, USP 3 201 494, Shell, International Research Maatschappij Fr. 2 005 043

4. W.N.Lyster, J.L.Hubbs, H.W.Prengle, AIChE J. 10, 1964, 907

5. J.H.Sinfelt, Advan. Chem. Eng. 5, 1964, 37

6. Oil Gas J. (Aug. 16, 1971), 67

7. A.H.Richardson, M.F.Olive, 68th National Meeting, AIChE, paper 53d, 1971

8. F.G.Ciapetta, J.B.Hunter, Ind. Eng. Chem. 45, 1953, 147

9. J.A.Rabo, P.E.Pickert, D.N.Stamires, J.Boyle, Int. Congr. Catal., 2nd, Paris, 1960, Nº104

10. J.A.Rabo, P.E.Pickert, R.L.Mays, Ind. Eng. Chem. 53, 1961, 147

11. V.Schomaker, Proc. Third Intern. Congr. Catalysis, 2, 1964, 1264

12. A.Voorhies, Ph.Bryant, AIChE J. 14, 1968, 852

13. Hydrocarbon Process, 53 (9), 1974, 212

14. H.W.Kouwenhoven. W.C.Van Zijll Langhout, Chem. Eng. Prog., 67, 1971, 65

15. F.Ribeiro, Ph. D. Thesis, Poitiers, France, 1980

16. D.W.Breck, "Zeolite Molecular Sieves", Wiley - Interscience, New York, 1974

17. C.V.McDaniel, P.K.Maher, Proc. Int. Congr. Mol. Sieves, 1st, London, 1967, 186-195

18. G.T.Kerr, J. Catal., 13, 1969, 114

19. U.S. 4.036. 739, 1971

20. H.W.Kouwenhoven, Adv. Chem. Series, 121, 1973, 529

21. U.S. 3.527.835, 1970

22. J.F. Le Page et al. in "Catalyse de Contact", Ed. Technip, Paris, 1978

23. F.Ribeiro, Ch.Marcilly, G.Thomas, C.R. Acad. Sc. Paris, 287 C, 1978, 431

24. F.Ribeiro, Ch.Marcilly, Rev. Inst. Fr. Pét., 34 (3), 1979, 405

25. F.Ribeiro, Ch.Marcilly, M.Guisnet, M.Portela, Proc. 3rd Int. Chem. Eng. Conference, Portugal, 1981, 2.106

26. F.Ribeiro, Ch.Marcilly, in "Recent Progress Reports, 5th Int. Conference on Zeolites (Napoli 1980)" (R.Sersale, C.Colella, R.Aiello, Eds.) p.135, Giannini, Napoli, 1981

27. F.Ribeiro, Ch.Marcilly, M.Guisnet, Rev.Port. Quím., in press

28. F.Ribeiro, Ch.Marcilly, M.Guisnet, Proc. 7th Iberoam. Symp. Catal.,Argentina, 1, 1980, 274

29. U.S. 3.354.077, 1967

30. U.S. 3.506.400, 1970

31. U.S. 3.591.488, 1971

32. C.V.McDaniel, P.K.Maher, in "Zeolite Chemistry and Catalysis", Ed. J.A.Rabo, p.285, ACS Monograph 171, 1975, American Chemical Society, Washington

33. J.Scherzer, J.L.Bass, J. Catal. 28, 1973, 101

34. P.Jacobs, J.B.Uytterhoeven, J. Catal. 22, 1971, 193

35. U.S. 3.897.327, 1975

36. R.A.Dalla Betta, M.Boudart, Proc. 5th Int. Congr., Miami Beach, 2, 1972, 1329

37. P.Gallezot, A.Alarcon Diaz, J.A.Delmon, A.J. Renouprez, B.Imelik, J. Catal., 39, 1975, 334

38. T.Kubo, H.Arai, H.Tominaga, T.Kunugi, Bull. Chem. Soc. Jap., 45, 1972, 607

39. F.Ribeiro, Ch.Marcilly, Proc. Vth Int. Symp. Heterog. Catal., Varna, Bulgaria, 1983 (submitted)

40. P.E.Eberly Jr., C.N.Kimberlin Jr., J. Catal., 22, 1971, 419

41. Shell International Research Mij., B.P. 1.189.850, (April 29, 1970)

42. F.Ribeiro, Ch.Marcilly, M.Guisnet, Proc. 8th Iberoam. Symp. Catal., Spain (Huelva), 1, 1982, 219

568

43. F.Ribeiro, Ch.Marcilly, M.Guisnet, J.Catal. $\underline{78}$, 1982, 267

44. P.G.Menon, G.F.Froment, J. Catal., $\underline{59}$, 1979, 1938

45. P.G.Menon, G.F.Froment, Applied Catal., $\underline{1}$, 1981,31

46. F.Ribeiro, Ch.Marcilly, M.Guisnet, E.Freund, H. Dexpert, "Catalysis by Zeolites", ed. B. Imelik, p.319, 1980, Elsevier Scientific Publishing Company, Amsterdam

47. R.Beecher, A.Voorhies, Ind. Eng. Chem., Prod. Res. Develop. $\underline{8}$, 1969, 366

48. N.L.Cull, H.H.Brenner, Ind. Eng. Chem., $\underline{53}$, 1961, 833

49. M.A.Lanewala, P.E.Pickert, A.P.Bolton, J.Catal., $\underline{9}$, 1967, 95

50. A.P.Bolton, M.A.Lanewala, J.Catal., $\underline{22}$, 1971, 419

51. Kh. M. Minachev, Neftekhimia, $\underline{4}$, 1964, 850

52. Kh. M. Minachev, Ya. I. Isakov, in "Zeolite Chemistry and Catalysis" Ed. J.A.Rabo, p.592, ACS Monograph 171, 1976, American Chemical Society, Washington

53. F.Chevalier, M.Guisnet, R.Maurel, 6th Proc. Intern. Congr. Catal., London, $\underline{2}$, 1976, 478

54. M.L.Poutsma, in "Zeolite Chemistry and Catalysis" Ed. J.A.Rabo, p.437, ACS Monograph 171, 1976, American Chemical Society, Washington

55. B.W.Burbridge, I.M.Keen, M.K.Eyles, Advan. Chem. Ser. $\underline{102}$, 1971, 400

56. H.Matsumoto, Y.Saito, Y.Yoneda, J.Catal., $\underline{22}$, 1971, 182

57. A.M.Gyul'maliev, I.I.Levitskii, E.A.Uddl'tzova, J.Catal., $\underline{58}$, 1979, 144

58. P.B.Weisz, in "Advances in Catalysis" (D.D.Eley, H.Pines and P.B.Weisz Eds.) $\underline{13}$, 1963, 137, Academic Press, New York

59. F.Chevalier, Ph. D. Thesis, Poitiers, France, 1979

60. F.E.Condon, P.H.Emett, $\underline{6}$, 1958, 83

61. G.Braun, F.Felting, H.Schoenberger, in "Molecular Sieves II", Ed. J.R.Katzer, p. 504, ACS Symposium Series 40, 1977, American Chemical Society, Washington

62. M.Guisnet, J.J.Garcia, F.Chevalier, R.Maurel, Bull. Soc. Chim., 1976, 1657

63. F.Ribeiro, Ch.Marcilly, M.Guisnet, J.Catal., 78, 1982, 275

ZEOLITES AS CATALYSTS IN XYLENE ISOMERIZATION PROCESSES

M. Guisnet and N.S. Gnep

Laboratoire Associé au CNRS - Catalyse Organique
Université de Poitiers, France

The o-xylene and above all p-xylene are basic substances essential in the organic chemical industry especially for the manufacture of plastifiants, resins, fibres and polyester films ; m-xylene, the isomer produced in the greatest quantity does not offer a very great interest and is most often converted into o- and p-xylenes(1,2). Xylenes are produced by catalytic reforming ; the C_8 aromatic cuts coming from the distillation of reformates contain xylenes in their equilibrium mixture (o-xylene $\simeq$ 25 %, m-xylene $\simeq$ 50 % and p-xylene $\simeq$ 25 %) and ethylbenzene in higher quantity than at equilibrium (10 to 40 % instead of 7 %). According to the economic context and the importance of the aromatic industrial plant, ethylbenzene can either be recuperated or transformed into xylenes. The aim of an isomerization unit will be therefore to obtain a maximum yield of o-xylene and above all of p-xylene from C_8 aromatic cuts containing or not ethylbenzene. Taking into account the very high m-xylene content in the equilibrium mixture of xylenes, the conversion, at each passage, of m-xylene is very reduced, thus the recycled quantity important (typical value recycled/fresh charge = 3).

1. MAIN INDUSTRIAL PROCESSES OF C_8 AROMATIC ISOMERIZATION

If xylene isomerization occurs on acid catalysts, ethylbenzene isomerization however is not observed on these catalysts. This reaction in fact demands the simultaneous presence of both an acid and a hydrogenating function (bifunctional catalysis). Therefore there are two types of isomerization processes, one allowing the isomerization of the three xylenes and the other the whole C_8 aromatic cut (2).

1.1 Processes of xylene isomerization

The reaction temperature depends directly on the acidity of the catalyst :
- ⩽ 150°C for very strong acids of Friedel Crafts type,
- from 250 to 450°C for zeolitic catalysts,
- from 380 to 500°C for medium strength acids such as halogenated alumina or silica-alumina.

According to whether the reaction temperature is or is not lower than the critical xylene temperature (about 345°C), the process can be operated
- either in liquid phase :
     e.g. Mitsubishi Gas Chemical Process using $HF-BF_3$; Low Temperature Isomerization from Mobil using ZSM-5 zeolite,
- or in gas phase :
     e.g. Isoforming from Esso Research and Engineering ; Isolene I from Toray.

The stronger the catalyst activity and above all the greater its life span (especially its resistance to coking) and its selectivity (limited secondary reaction of disproportionation, preferential orientation towards p-xylene) the better will be the process (2).

1.2 C_8 aromatic isomerization process (xylenes + ethylbenzene)

In order to allow ethylbenzene isomerization, all the processes use bifunctional noble metal acid catalysts on fixed beds, operating under hydrogen pressure (1 to 2 MPa) and at a temperature ranging from 380 to 460°C. In these conditions, C_8 naphtenes are formed ; they are recycled with unconverted C_8 aromatics (2).

Different acid supports can be used (1,2) :
- Silica-alumina (Octafining from Arco-Engelhard),
- Chlorinated alumina (Isomar from UOP),
- Fluorinated alumina (Isarom from IFP),
- Zeolitic catalysts in mordenite base (Isolene II from Toray, Octafining II from Arco-Engelhard, Aris from Veb Leuna Werke) or in ZSM-5 zeolite base (Mobil process).

Here again the aim is to obtain a very high degree of activity, a good stability and a high selectivity.

2. XYLENE ISOMERIZATION ON ACIDIC ZEOLITE CATALYSTS

Over acid catalysts, xylenes undergo several reactions : isomerization but also disproportionation into toluene and trimethylbenzenes and finally coke formation responsible for catalyst deactivation. The importance of these two latter reactions must evidently be limited ; moreover it would be interesting to orient xylene isomerization towards p-xylene formation. It will be shown here, on the basis of the mechanism of these various reactions,

what must be the physicochemical characteristics of zeolite catalysts in order to satisfy these requirements.

2.1 Isomerization mechanisms

Three mechanisms have been proposed to account for aromatic
hydrocarbon isomerization (3) : an intramolecular mechanism, an
intermolecular mechanism via transalkylation products and a dealkylation-alkylation mechanism. In the case of xylene isomerization,
the intramolecular mechanism is the most probable, although on
zeolite various authors have proposed mechanisms involving transalkylation products ; the dealkylation-alkylation mechanism is very
unlikely because it involves a very unstable methyl carbocation
(4,5).

The intramolecular mechanism is described in figure 1 ; the
selectivity of m-xylene isomerization (para/ortho ratio) should not
depend on the characteristics of acid centers which are present on
the catalyst since the formations of ortho and para isomers proceed
through the same steps. This effectively is the case (figure 2) :
the para/ortho ratio is the same (equal to 1.1 at 350°C) on
silica-alumina, fluorinated alumina and a series of stabilized Y
zeolites which differ by their acidity.

However, the formation of para isomer which has a smaller molecular size than the ortho isomer can be strongly favoured if the
zeolite has a pore size close to that of the xylenes. Figure 2
shows that ZSM-5 zeolite in particular is well adapted to obtain
selectively para isomer. Moreover it must be noted that notably
higher values of this ratio can be obtained on modified ZSM-5
zeolite (6).

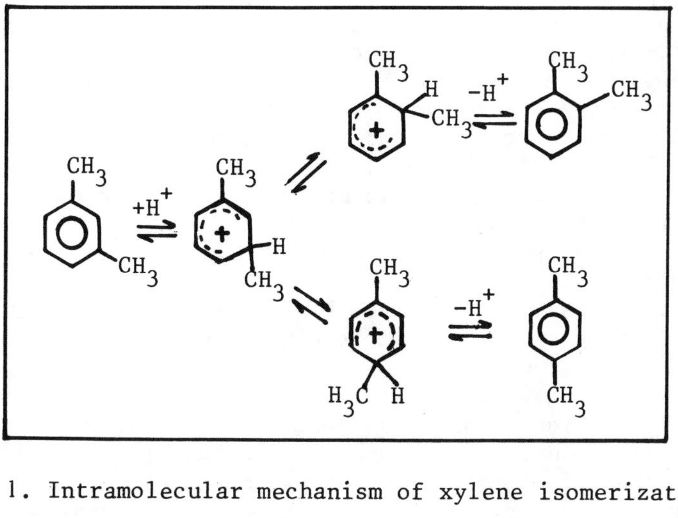

Fig. 1. Intramolecular mechanism of xylene isomerization

574

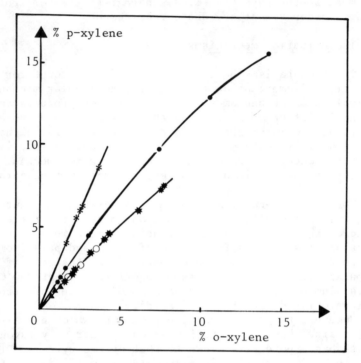

Fig. 2. m-Xylene isomerization on silica-alumina (O), fluorinated
alumina (▲), Y zeolite (✳), mordenite (●) and ZSM-5
zeolite (✱).

2.2 Disproportionation mechanisms

Besides the mechanism of dealkylation-alkylation which would
be very unlikely in the case of xylenes (cf. isomerization) two
other mechanisms have been proposed (3) :
  - the first is very similar to the dealkylation-alkylation
mechanism. The methyl group of a benzenium ion undergoes the nu-
cleophilic attack of an aromatic molecule and is transferred without
appearing in carbocation form ;
  - the second involves benzylic carbocations and diaryl-
methane intermediates (figure 3).
It may be noted that the first mechanism invokes the same benzenium
ion intermediates as the intramolecular isomerization mechanism,
while the second mechanism invokes different intermediates. Several
observations recently published favour the disproportionation
mechanism by the intermediate of benzylic carbocations :
  - the addition to o-xylene of alkanes capable of giving a
tertiary carbocation (methyl-2 pentane, methylcyclohexane...) causes
an important decrease in the disproportionation activity without

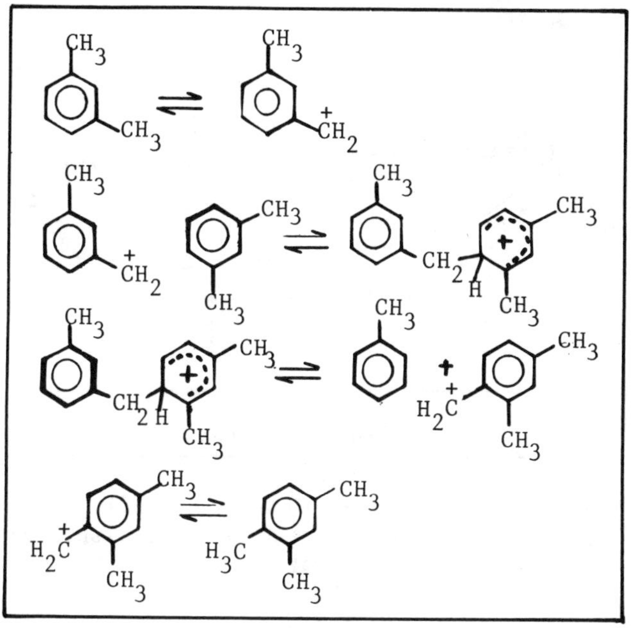

Fig. 3.   Disproportionation mechanism of xylenes via benzylic
carbocation intermediates.

affecting in a significant manner the isomerization activity
(figure 4). In order to explain this result, it is proposed that
the disproportionation intermediates but not those of isomerization
are consumed by reacting with the alkanes (7). This implies the
existence of different intermediates in the two reactions : the
disproportionation intermediate cannot therefore be the benzenium
ion intermediate of isomerization ; no reaction moreover is possible between this latter intermediate and alkanes. However, benzylic
carbocations can react with branched alkanes with hydride transfer
from the alkanes to the benzylic carbocations :

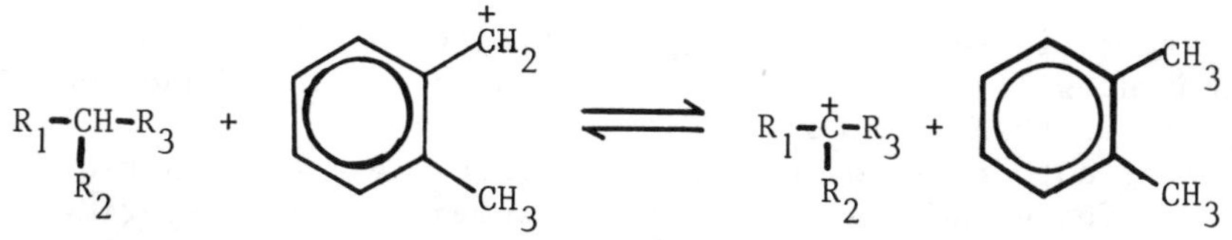

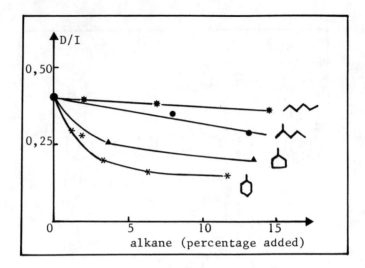

Fig. 4.   Transformation of o-xylene on Y zeolite at 350°C :
          Influence of alkane content on the disproportionation (D)
          to isomerization (I) rate ratio.

     - hydrogen under pressure inhibits toluene disproportionation
on mordenites (8). This inhibiting effect is explained by a decrease
of the concentrations of benzylic carbocation intermediates caused
by the following reaction (reverse reaction of benzylic carbocation
formation by an aromatic adsorption on a Brönsted acid site) :

$$H_2 \; + \; \underset{}{\overset{+}{C}H_2\text{-}C_6H_5} \; \rightleftharpoons \; CH_3\text{-}C_6H_5 \; + \; H^+$$

Here again no possible reaction exists between benzenium ions and
hydrogen.

2.3 Disproportionation/Isomerization rate ratio

     The ratio between the disproportionation (D) and the isomeri-
zation rates (I) depends a great deal on the acidity and on the
porosity of the zeolites .
     Most authors agree to recognize that these two reactions in-
volve different acid sites. However, some poisoning experiments by
pyridine carried out on Y zeolite show that active sites in disproportionation and isomerization of xylene (figure 5) differ very
slightly in acid strength (9). Yet these sites are certainly different in nature : the isomerization requires only one Brönsted acid
center (10,11) while bimolecular disproportionation is more

577

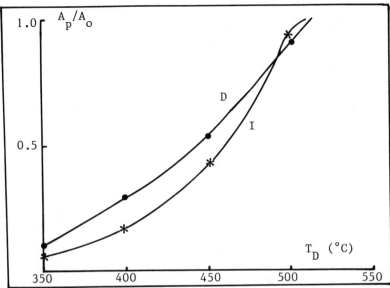

Fig. 5. Isomerization (I) and disproportionation (D) of o-xylene
on a pyridine poisoned Y zeolite : Ratio of the activities
of poisoned (A_p) and unpoisoned (A_o) catalysts for
different desorption temperatures (T_D).

demanding : the acid centers are probably constituted by a pair of
acid sites (12). In agreement with this hypothesis, it can be noted
that disproportionation is always more sensitive than isomerization
to zeolite treatments which modify the acid center concentrations
(exchange of protons by alkaline ions, chemical dealumination, calcination under wet air) (13). Thus the exchange of protonic mordenite by sodium ions which decreases the acid center concentrations
definitely increases the isomerization selectivity (14,15).

The D/I ratio depends on zeolite pore size and especially on
the space available in the vicinity of active centers. Actually,
the bimolecular intermediates and transition states of disproportionation, contrarily to those of intramolecular isomerization are
very bulky and their formation can be inhibited as a result of
steric constraints. This explains the D/I decrease with the intracrystalline cavity diameter (16) : it is notably smaller on mordenite than on Y zeolite and practically equal to zero on ZSM-5
zeolite. However, another explanation must be found for the low
value of the D/I ratio in the case of offretite. In this case,
diffusional limitations restricting the desorption of trimethyl-
benzene the size of which is greater than that of p-xylene are
responsible for the low disproportionation activity (15).

578

2.4 Activity and stability of catalysts

Figure 6 shows that zeolites are more active than amorphous
catalysts. Among the zeolites, mordenite presents a very high
initial activity but it is very rapidly deactivated at least under
atmospheric pressure. This deactivation is due to the formation of
very heavy polyaromatic compounds ("coke"). On the contrary, ZSM-5
zeolite is practically not deactivated. On this zeolite, steric
constraints inhibit the formation of highly bulky transition states
necessarily involved in the numerous bimolecular steps of coke formation. Coke however can only be formed on external surface of the
zeolite (17,18).

Mordenite stability can notably be improved (19,20). It is
particularly the case if they have been calcined under wet air since
this treatment eliminates preferentially the acid sites of great
coking activity (21). Mordenite stability moreover is actually improved by operating under high hydrogen pressure (20,22). Finally
the association of a hydrogenating function with an acid function
allow us to limit coke formation and mordenite deactivation ;
however the hydrogenating power of catalysts must not be very important so as to avoid secondary reactions of hydrogenolysis and
hydrocracking.

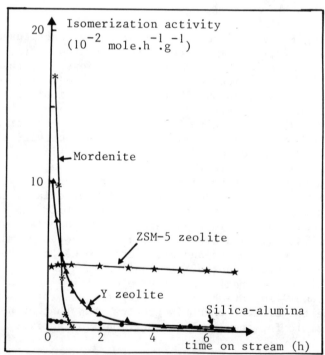

Fig. 6. o-Xylene isomerization on mordenite, Y zeolite, ZSM-5
zeolite and silica-alumina.

3. ISOMERIZATION OF C_8 AROMATIC HYDROCARBONS ON BIFUNCTIONAL
ZEOLITE CATALYSTS

Ethylbenzene requires for its isomerization bifunctional metal-
acid catalysts : thus P.B. Weisz (23) shows that platinum and
silica-alumina used separately have no activity while their mixture
is very active. On these catalysts, xylenes can be isomerized on
acid centers or again by the successive intervention of hydrogenating
and acid centers as is the case for ethylbenzene. Various secondary
reactions moreover can be observed :
   - xylenes and ethylbenzene disproportionation on acid centers,
   - dealkylation on metallic or acid centers,
   - hydrogenation into naphtenes on metallic centers,
   - hydrocracking through a bifunctional mechanism.

3.1 Ethylbenzene isomerization mechanism

A kinetic study of ethylbenzene isomerization (24) carried out
on platinum fluorinated alumina catalysts and on mixtures of pla-
tinum-inert alumina and fluorinated alumina has allowed us to
propose the reactional scheme shown in figure 7 :

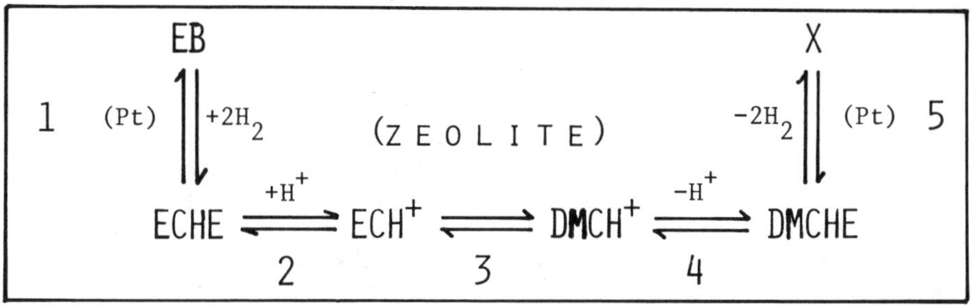

Fig. 7. Ethylbenzene isomerization reactionnal path.

EB : ethylbenzene, ECHE and ECH$^+$: olefins and carbocations with
an ethylcyclohexane skeleton, DMCHE and DMCH$^+$: olefins and carbocations with a dimethylcyclohexane skeleton, X : xylenes.

Ethylbenzene is hydrogenated on metallic sites into ethylcyclohexenes and ethylcyclohexane ; the highly active ethylcyclohexenes
are isomerized on acid centers into dimethylcyclohexenes which are
dehydrogenated into xylenes on metallic sites. Over industrial catalysts, the hydrogenation activity is sufficient for the hydrogenation and dehydrogenation reactions to be more rapid than the isomerization of the intermediate olefins ; the slow step of the
reactional process is then the rearrangement of ethylcyclohexylic

carbocations into dimethylcyclohexylic ones (step 3) e.g. in the ethylbenzene isomerization into o-xylene.

According to Nitta and Jacobs (25) the mechanism of figure 7 accounts equally for ethylbenzene isomerization on Pt-zeolite.

3.2 Selectivity of ethylbenzene transformation

This selectivity depends a great deal on the reaction temperature (24,25).

- At low temperature, aromatic hydrogenation into naphtenes is thermodynamically favoured. Taking into account the high hydrogenating activity of the catalysts the naphtene formation can become important ; therefore the operation temperature will be chosen so as to maintain the naphtene content below 10 % (2).

- At high temperature, disproportionation and dealkylation reactions become preponderant. Furthermore, deactivation becomes rapid.

The importance of the secondary reactions of disproportionation and dealkylation is all the greater than the acid centers of the catalyst are stronger ; therefore the most selective catalysts for ethylbenzene isomerization must be of medium acidity (24,25). This is not surprising since the acid strength required for olefin skeletal isomerization (limiting reaction in ethylbenzene isomerization) is not as great as that required for ethylbenzene disproportionation (26). Moreover, disproportionation being bimolecular probably requires a pair of acid centers while olefin isomerization being intramolecular requires only one Brönsted acid site. The catalysts should be all the more selective than the density of their acid centers is lower. Finally, like for xylene isomerization, zeolites which are not very active in disproportionation as a result of steric constraints (mordenite and above all ZSM-5) will present a great interest. This interest will be even greater if p-xylene isomer can be privileged for configurational reasons as is the case with ZSM-5 zeolites.

Conclusion

Numerous processes of C_8 aromatic isomerization using zeolites as catalysts have been recently developped. Zeolites such as mordenite and ZSM-5, while being very stable, present the advantage of being more active than amorphous catalysts. Moreover their porous structure give them a high selectivity : the secondary disproportionation reaction is very limited and the isomerization, at least in the case of ZSM-5 catalysts, is oriented towards the p-xylene formation.

References

1. Chauvel A., Lefebvre G. and Raimbault C., Production d'olé-fines et d'aromatiques (Publications de l'Institut Français du Pétrole, Technip 1980), p. 95.
2. Marcilly C., Techniques de l'Ingénieur (1983), in press.
3. Poutsma M.L., Zeolite Chemistry and Catalysis (J.A. Rabo Ed.) ACS Monograph 171 (American Chemical Society, Washington, 1971) pp. 431-528.
4. Lanewala M.A. and Bolton A.P., J. Org. Chem. 34 (1969) 3107.
5. Guisnet M., Gnep N.S., Bearez C. and Chevalier F., Catalysis by Zeolites (B. Imelik et al Eds.), Studies in Surface Science and Catalysis, Vol. 5 (Elsevier Scientific Publishing Company, Amsterdam, 1980) pp. 77-83.
6. Young L.B., Butter S.A. and Kaeding W.W., J. Catal. 76 (1982) 418.
7. Gnep N.S. and Guisnet M., React. Kinet. Catal. Lett. (1983) in press.
8. Gnep N.S. and Guisnet M., Applied Catalysis 1 (1981) 329.
9. Guisnet M. and Perot G., Symposium on Shape Selective Catalysis, 185th ACS National Meeting, Seattle, March 1983.
10. Ward J.W. and Hansford R.C., J. Catal. 13 (1969) 154.
    Hansford R.C. and Ward J.W., J. Catal. 13 (1969) 316.
11. Ward J.W., J. Catal. 13 (1969) 321 ; Ibid. 14 (1969) 365 ; Ibid. 17 (1970) 355 ; Ibid. 26 (1972) 451 ; Ibid. 26 (1972) 470.
12. Csicsery S.M. and Hickson D.A., J. Catal. 19 (1970) 386.
13. Martin de Armando M.L., Gnep N.S. and Guisnet M., J. Chem. Research (S) (1981) 8.
14. Ratnasamy P., Sivasankar S. and Vishnoi S., J. Catal. 69 (1981) 428.
15. Tejada G., Gnep N.S. and Guisnet M., to be published.
16. Weisz P.B., Proceedings of the 7th International Congress on Catalysis, Part A (Seiyama T. and Tanabe K. Eds., Kodansha LTD, Tokyo, 1980) pp. 3-20.
17. Butt J.B., Delgado-Diaz S. and Muno W.E., J. Catal. 37 (1975) 158.
18. Butt J.B., J. Catal. 41 (1976) 190.
19. Marcilly C., French Patents (1976).
20. Gnep N.S., Martin de Armando M.L., Marcilly C., Ha B.H. and Guisnet M., Catalyst Deactivation (Delmon B. and Froment G.F. Eds.), Studies in Surface Science and Catalysis 6 (Elsevier Scientific Publishing Company, Amsterdam, 1980) pp. 79-89.
21. Mirodatos C. and Barthomeuf D., J. Chem. Research (1981) 39.

22. Gnep N.S., Martin de Armando M.L. and Guisnet M., React. Kinet. Catal. Lett., 13 (1980) 183.
23. Weisz P.B., Advances in Catalysis 13 (Academic Press, London, 1962) p. 137.
24. Gnep N.S. and Guisnet M., Bull. Soc. Chim. de France, (1977) 429.
25. Nitta M. and Jacobs P.A., Catalysis by Zeolites (Imelik B. et al Eds.) Studies in Surface Science and Catalysis, Vol. 5 (Elsevier Scientific Publishing Company, Amsterdam, 1980) pp. 251-259.
26. Marsicobètre D., Gnep N.S., Guisnet M. and Maurel R., Rev. Port. Quim. 18 (1976) 316.

ION EXCHANGE SEPARATIONS WITH MOLECULAR SIEVE ZEOLITES

JOHN D. SHERMAN

MOLECULAR SIEVE DEPARTMENT, UNION CARBIDE CORPORATION,
TARRYTOWN TECHNICAL CENTER, TARRYTOWN, NEW YORK 10591, USA

ABSTRACT

Molecular sieve zeolite cation exchangers provide unique
combinations of selectivity, capacity and stability not available
in other ion exchangers. Commercial applications include
separations of radioisotopes, waste water ammonia removal and as
detergent builders. New zeolite products with still higher ammonium
ion selectivities and capacities and products with improved calcium
and magnesium exchange capabilities for detergent applications have
also been developed.

New applications in artificial kidney devices, radioactive
waste separations and disposal, heavy metals removal and animal
feeding are under development. Potential uses in metals recovery
and in separations and purifications of non-ferrous metals are
being explored. Applications have grown rapidly with increasing
awareness of zeolite properties and the imaginative consideration
of their potential uses in ion exchange separations.

1 BACKGROUND

Molecular sieve zeolites are crystalline, hydrated alumino-
silicates of (most commonly) Na^+, K^+, Mg^{++}, Sr^{++} and Ba^{++} cations.
The aluminosilicate portion of the structure is a 3-dimensional open
framework consisting of a network of AlO_4 and SiO_4 tetrahedra
linked to each other by sharing all the oxygens. Zeolites may be
represented by the empirical formula:

$$R_{2/n}^{n+} \cdot Al_2O_3 \cdot x\ SiO_2 \cdot y\ H_2O$$

In this oxide formula x is normally $>$ 2 since AlO4 tetrahedra are joined only to SiO4 tetrahedra;   n is the cation valance. The framework contains channels and interconnected voids occupied by the cations and water molecules. The cations are quite mobile and can usually be exchanged, to varying degrees, by other cations.

In 1858, Eichorn (70) showed zeolites are capable of reversible exchange of cations. This is a characteristic property of the zeolites. Extensive studies have also shown exchange selectivities in zeolites do not follow the typical rules and patterns exhibited by other organic and inorganic ion exchangers. Also, many zeolites provide combinations of selectivity, capacity and stability superior to the more common organic and inorganic cation exchangers.

The term "zeolite" properly refers to the <u>crystalline molecular sieve ion exchangers</u>. Unfortunately, some confusion has been created by improper use of the term. The accepted term for a synthetic <u>amorphous</u> aluminosilicate is "<u>permutite</u>". The zeolite cation exchangers discussed herein are of the <u>crystalline</u> framework alumino-silicate type with the scientifically correct designation "<u>zeolite</u>."

Union Carbide Corporation owns the rights to the trademark LINDE in the U.S.A., Canada and Mexico. Our references in this paper to LINDE are to the trademark owned by Union Carbide Corporation. We want to avoid any confusion with the LINDE trademark as owned by others elsewhere in the world.

Development of zeolites was preceded by development of the permutites which supplemented the use of natural microcrystalline aluminosilicates ("green sands" or glauconites), and led to significant expansion of the use of ion exchange, especially for water softening. A disadvantage of the permutites was their solubility, at extremes of pH, rendering it nearly impossible to regenerate them by H^+ exchange.

Although the process of ion exchange was discovered in 1850 (185, 192), it was not applied as an industrial separation process until 1905, when Gans (83) demonstrated it could be used as a unit process for both water softening and removal of certain metal ions, especially iron and manganese (56). From 1905 to 1935, the only ion exchangers available were the inorganics, usually aluminosilicates – either green sands or permutites, operated only in the neutral pH regions, using salt as the regenerant. In 1935, sulfonated coal ion exchangers which could be regenerated with acid were commercially developed. Condensation polymers with sulfonic and amino groups were developed shortly thereafter. In 1946, sulfonated and aminated copolymers of styrene and divinyl benzene became available. The development of these new organic ion exchangers with superior stability and ease of regeneration enabled a rapid expansion in the range of industrial applications of the ion exchange process (56).

Why, then, has there been a renewed interest in inorganic ion exchangers and particularly, zeolites, for ion exchange applications? The greater stability of some zeolites (vs. organic resin ion exchangers) under certain conditions and their high selectivity towards particular ions have combined to allow the development of new applications for the ion exchange process. This has not generally involved the displacement by zeolites of other exchangers in existing applications, but rather the development of entirely new applications for which the existing ion exchangers were not well suited. It is reasonable to expect that this trend will continue.

Zeolites did not find significant use commercially as ion exchangers until the early 1960's, due to lack of availability and lack of knowledge of their properties. Both barriers disappeared during the 1950's and 1960's. The discovery by R. M. Milton and co-workers at Union Carbide that zeolites could be synthesized at convenient conditions led to the discovery of dozens of new zeolite structures and assured their availability in commercial quantities. Extensive exploration by Union Carbide in the late 1950's also resulted in discovery in the Western U.S. of many deposits of natural zeolites of sufficient quantity and purity for commercial use.

Knowledge of the desirable ion exchange properties of zeolites also became available due to the reports of Ames (1-15), Amphlett (16), Barrer (26, 29, 30), Thomas (184) and others.

The first commercial uses were developed in the early 1960's by Ames et al. for processing radioactive wastes. Their stability in ionizing radiation and at elevated temperatures and pH levels, and excellent capacities and selectivities, caused certain zeolites to be uniquely suited for recovery and concentration of radioisotopes for long-term storage. Later discovery (also by Ames) that some zeolites had excellent NH_4^+ ion selectivity led to the development in the late 1960's / early 1970's of the second significant commercial application: the removal of NH_4^+ from municipal wastewater.

Other applications are being developed. Some are discussed in the present paper. Growing knowledge of zeolites and growing needs (in pollution abatement, energy production, agriculture, aquaculture, animal nutrition, metals processing and biomedical applications) promise many exciting new applications for these unique materials.

Background References: The structure, chemistry and use of zeolites was broadly reviewed by D. W. Breck (51), including a 64-page chapter on ion exchange suggested for background reading. Sherry (172, 174) reviewed ion-sieving and ion selectivity phenomena. Cremers (66) reviewed more recent studies. Others (16, 27-30, 38, 39, 96, 139) reviewed earlier work. Mercer and Ames reviewed radwaste and NH_4^+ removal applications (127). Zeolite ion exchange applications more broadly were reviewed by Sherman (166).

Barrer (40) reviewed equilibrium and kinetic aspects of zeolite ion exchange at the 5th International Conference on Zeolites. Rees will discuss binary and ternary cation exchange on zeolites at the coming 6th International Conference (141).

2 ION EXCHANGE PROPERTIES OF MOLECULAR SIEVE ZEOLITES

The above papers extensively review the basic ion exchange phenomena observed on zeolites. The present paper includes only a few examples. Typical properties are summarized in Table 1 (166).

TABLE 1. TYPICAL PROPERTIES OF THE MOST COMMON MOLECULAR SIEVE ZEOLITES

Zeolite type	Pore openings (hydrated form)	Typical SiO_2/Al_2O_3 mole ratio	Typical maximum theoretical cation exch. capacity (Na^+ form, anhydrous)
Analcime	2.6 Å	4	4.9 meq/g
Chabazite	3.7 × 4.2 Å, and 2.6 Å	4	4.9
Clinoptilolite	4.0 × 5.5 Å, and 4.4 × 7.2 Å, and 4.1 × 4.7 Å	10	2.6
Erionite	3.6 × 5.2 Å	6	3.8
Ferrierite	4.3 × 5.5 Å 3.4 × 4.8 Å	11	2.4
Mordenite*	6.7 × 7.0 Å 2.9 × 5.7 Å	10	2.6
Phillipsite	4.2 × 4.4 Å 2.8 × 4.8 Å 3.3 Å	4.4	4.7
LINDE A	4.2 Å into α cage 2.2 Å into β cage	2	7.0
LINDE F	~3.7 Å	2	7.0
Zeolite HS (Hydroxysodalite)	2.2 Å	2	7.0
LINDE L	7.1 Å	6	3.8
Large-port Mordenite	6.7 × 7.0 Å 2.9 × 5.7 Å	10	2.6
LINDE Omega	7.5 Å	7	3.4
Zeolite P (LINDE B)	3.1 × 4.4 Å 2.8 × 4.9 Å	3	5.8
LINDE T	3.6 × 5.2 Å	6.9	3.4
LINDE W	4.2 × 4.4 Å	3.6	5.3
LINDE X	7.4 Å into α cages 2.2 Å into β cages	2.5	6.4
LINDE Y	7.4 Å into α cages 2.2 Å into β cages	4.8	4.4

*The large pores are partially blocked in natural mordenite.

TABLE 2. ION SIEVING EFFECTS IN MOLECULAR SIEVE ZEOLITE ION EXCHANGE

Zeolite	Exchange proceeds		Exchange negligible		Size of largest pore openings**
	Cation	Diameter* (Å)	Cation	Diameter* (Å)	
Analcime	Rb^+	2.96 (X)	Cs^+	3.38 (X)	2.6 Å
		6.58 (H)		6.58 (H)	
Chabazite	Cs^+	3.38 (X)	$N(CH_3)_4^+$	6.94 (X)	3.7 X 4.2 Å
		6.58 (H)		7.34 (H)	
Clinoptilolite	Cs^+	3.38 (X)	$N(CH_3)_4^+$	6.94 (X)	4.4 X 7.2 Å
		6.58 (H)		7.34 (H)	
LINDE X	$N(CH_3)_4^+$	6.94 (X)	$N(C_2H_5)_4^+$	8.0 (X)	7.4 Å
		7.34 (H)		8.0 (H)	

*Data from Nightingale (130)
(X) = (Crystal radius) X (2)
(H) = (Hydrated radius) X (2)
**From crystallographic data.

Ion Sieving Effects: Ion sieving effects are observed with the
zeolites having the smallest pore openings and with the largest
cations, as illustrated in Table 2 (166). Some ions are exchanged even
though their hydrated diameters are much larger than the pore
openings; this requires an exchange of solvent molecules. Ions with
too large anhydrous diameters are totally excluded. Exchange rates
vary due to differences in energies required to exchange solvent
molecules and may be very slow at lower temperatures if the water of
hydration is strongly held: e.g., Ca^{++} enters the Type A zeolite
rapidly at 25 C, whereas Mg^{++} enters very slowly (see Figure 1)(167).

Figure 1.

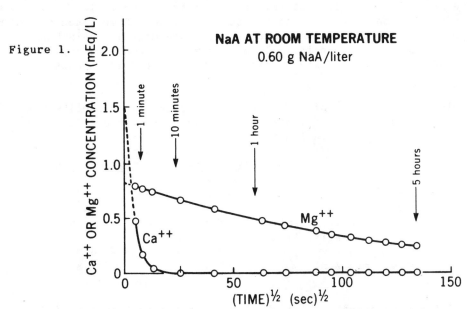

588

There are also ion sieving effects <u>inside</u> the zeolite cages in which larger ions are excluded from entering some of the inner cavities of the zeolite due to the small size of the entrances to these cavities. For example, at 25 C large ions do not replace all Na^+ ions in LINDE Type X or Y zeolites; 16 Na^+ ions per unit cell can not be replaced by La^{+++} or alkaline earth exchange of zeolite Y (170) or by exchange of Mn^{++}, Co^{++} and Ni^{+1} on zeolites X and Y (82). These 16 sodium ions are present in side cavities with small entrances which prevent the larger ions from entering at low temperature. The exchangeable ions are those present in the supercages.

<u>Effect of Ion Volume:</u> The size of the entering cation can also appreciably affect both the rate and extent of ion exchange. For example, in exchange of various alkyl ammonium ions in clinoptilolite at 60 C, complete replacement of all the Na^+ ions was accomplished by each of these cations, except for trimethyl-, i-propyl-, and n-propyl-ammonium ions, for which only partial replacement was accomplished. The water content of the zeolite decreased with increasing loading and with increased volume of the organic cation. The larger ions above could not enter channel systems controlled by diffusion through 8-rings, whereas they could enter those channels controlled by diffusion through 10-rings. Still larger ions such as tetramethyl- and t-butyl-ammonium ions were completely excluded from even the largest pores (32).

<u>Ion Exchange Capacity</u>: The ion exchange capacity of a zeolite ion exchanger is a function of its SiO_2/Al_2O_3 mole ratio, and also of its cation form. The maximum exchange capacities of common zeolites are given in Table 1. Not all the capacity is available to all ions due to the ion sieving effects discussed above. Comparison of LINDE Type A zeolite and a common strong acid type organic resin ion-exchanger is given in Table 3. As may be seen, the zeolite compares well in capacity on either a weight or volume basis.

Table 3. Comparison of LINDE Type A Molecular Sieve Zeolite with Common (Polystyrene Divinylbenzene Sulfonic Acid Type) Organic Resin Ion Exchangers (166).

	LINDE A zeolite	Organic Resin
Form	Mesh or Beads	Beads
Cation	Na^+	Na^+
Bulk Density (g/ml)	0.65	0.38
Total Capacity (max.)		
mEq / g	5.6 *	4.8
mEq / ml	3.6	1.8

* Based on anhydrous weight of 80% LINDE Type A zeolite / 20% bond.

Ion Exchange Selectivities: Zeolites commonly exhibit high selectivities for exchange among cations which will easily enter the zeolite pores. For example, LINDE Type A zeolite provides a striking selectivity for Ca^{++} over Na^+ compared with common organic resin strong acid type cation exchangers.

Each zeolite provides a different pattern of ion exchange selectivity. Selectivity series of increasing preference for exchange of different cations for the most common zeolites are summarized elsewhere (166). In general, common organic resin cation exchangers prefer ions of higher charge. This is also true of many zeolites (e.g., LINDE Type A and X zeolites). However, some zeolites show marked selectivity for some monovalent cations over common divalent cations. For example, LINDE W exchanges NH_4^+ in marked preference to Ca^{++} and prefers Na^+ over $Ca^{+\prime}$. Most zeolites exhibit selectivities for Ag, Tl and Pb exchange and many also are selective for exchange of other non-ferrous metal cations (e.g., Cd, Zn and Cu).

Theoretical analyses of ion exchange in glasses led Eisenman (71) to conclude that selectivity was controlled by the anionic field strength of the exchange site, and to predict ion exchange selectivity series. Sherry (172) showed ion exchange behavior of zeolites is in general agreement with these predictions.

The zeolite ion exchange affinity sequence is often found to be in accord with the hydrated ionic radius, so the affinity sequence is: $Li^+ < Na^+ < K^+ < Rb^+ < Cs^+$, and for divalent ions: $Mg^{++} < Ca^{++} < Sr^{++} < Ba^{++}$. Often a zeolite favors the least hydrated ion, while the solution phase favors the most highly hydrated ion. In other words, water molecules in solution compete with the zeolite for attraction of the cations.

In his theoretical analysis of ion exchange selectivity in glasses, Eisenman (71) separated the free energy of the ion exchange reaction into two parts. As applied to zeolites, this may be written as follows (173):

$$\Delta G^O = (G_Z^A - G_Z^B) - (G_S^A - G_S^B)$$

where Z = zeolite and s = solution phase. The first term on the right is the difference between the free energies of ions A and B in the zeolite and the second term is their difference in solution, which equals the differences in the free energies of hydration of the ions. If the electrostatic fields in the zeolite are strong, then the first term dominates and small ions are preferred by the zeolite even though such ions have large free energies of hydration.

On the other hand, large, wholly hydrated ions are preferred in "weak" field zeolites, i.e., those with lower aluminum content and, thus, lower framework charge density. Also, if the pore volume of the zeolite is smaller, then the degree of hydration of ions inside the zeolite will be less, also favoring exchange of ions with lower hydration energies.

For example, LINDE Type A zeolite with a high aluminum content and high void volume exhibits the selectivity series (51): $Cs^+ < Li^+ < Rb^+ < NH_4^+ < K^+ < Na^+$. On the other hand, chabazite (35,172), erionite (7, 13, 172), and phillipsite (6, 33, 161), which each have lower aluminum content and lower pore volume (vs Type A zeolite), exhibit the "weak field" selectivity series: $Li^+ < Na^+ < K^+ < Rb^+ < Cs^+$. Since most zeolites have lower aluminum contents and lower pore volumes, the weak field selectivity series is often observed. However, even among these zeolites the relative selectivities for the cations in the series differ greatly.

Decreasing the aluminum content of the zeolite also increases the average distance between the adjacent anionic (AlO_2^-) sites on the zeolite framework. This, in effect, increases the difficulty of a single divalent cation in "satisfying" the fields of two adjacent anionic sites. As a result, as expected, the preference of the zeolite for divalent cations decreases, and that for univalent cations increases, as the aluminum content decreases (173).

Ion Exchange Equilibrium Isotherms: It is convenient to express the results of selectivity measurements as ion exchange isotherms. For a reaction in which the exchanging ions A and B, have charges Z_A^+ and Z_B^+, respectively, the ion exchange isotherms are typically expressed as plots of the equivalent cation fraction, As, of the ion A in solution against the fraction Ac of the same ion in the ion exchanging zeolite. Typical isotherms for ion exchange on zeolite A are shown in Figure 2. An excellent discussion of

Figure 2. Isotherms for Exchange of Ca^{++} for Na^+ at Room Temperature on LINDE Type A Molecular Sieve Zeolite(167).

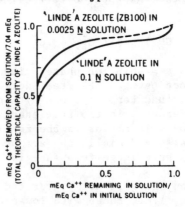

procedures for measurement and interpretation of ion exchange data on zeolites has been presented by Dyer, Enamy and Townsend (69).

Other variables Affecting Cation Exchange in Zeolites: The observed ion exchange selectivities and loadings on zeolites are dependent upon pH (H^+ is a competing cation), temperature and aqueous solution chemistry. The types and concentrations of anions, competing cations, solvent and complexing agents can each alter the quality of the ion exchange separation which can be achieved (via the effects of these variables upon the activities of the cations in solution). Fortunately, because of the rigidity of zeolite frameworks the effects of these variables upon the overall ion exchange performance of the zeolite are generally less complex and somewhat more predictable than with some organic resion ion exchangers in which complex sorption and swelling phenomena can occur, accompanied by changes in their ion exchange characteristics (96).

Complexing of the cation can markedly alter its exchange properties. For example, Ag^+ exchange is very much favored over Na^+, but Na^+ is preferred over the $Ag(NH_3)_2^+$ complexed ion on LINDE X zeolite. Thus, regeneration of the exchanged zeolites may be accomplished using complexing agents in the regenerant solution. Also, addition of complexing agents may allow zeolite ion exchange separations (e.g., for selective non-ferrous metal exchanges) not otherwise obtainable.

The effective ion exchange capacity of the zeolite may be greatest at elevated temperatures as in the admission of large ions at elevated temperatures (e.g., see 121). Also, in some cases, the ion exchange which has been carried out at higher temperatures may be partially irreversible at lower temperatures at which the large ions are not capable of diffusing through the smaller pores.

Ion Exchange Immiscibility Gaps: Unusual ion exchange behavior was observed by Olson and Sherry for Sr^{++}/Na^+ exchange of Type X zeolite (132) caused by a region of limited mutual solubility of the end members NaX and SrX. The ion exchange loading curves show a sudden jump when the Sr^{++} exchange levels on the zeolite reach 70%, and this is accompanied by an abrupt change in the unit cell dimensions of the zeolite crystals. Similar immiscibility gaps have been observed in other zeolites; e.g., see Barrer and Munday (34).

Exchange of Metal Ions: Exchange of multivalent metal ions is complicated by the need to maintain the pH levels in solution low enough to avoid the solubility limits of the metals and high enough to minimize proton exchange and hydrolysis of the zeolite. Since lower pH levels are required to provide adequate solubility of trivalent ions in aqueous solutions, there are few reports of such exchanges (principally of Y^{+++}, Ce^{+++}, La^{+++}), primarily on more acid-stable zeolites (LINDE X and Y, mordenite, clinoptilolite).

Exchange of LINDE Type A with multivalent ions presents the most difficult task experimentally, because of the limited stability of this zeolite at low pH levels. Breck et al. (48) reported successful exchange of NaA with Ba^{++}, Hg^{++}, Cd^{++}, Zn^{++}, Co^{++}, and Ni^{++}, but that Cu^{++} and Fe^{+++} exchange destroyed the structure and that no exchange could be observed with Ce^{+++}. Mercer and Ames (125) later reported the Ce^{+++} exchange of LINDE A zeolite, but without structural studies confirming retention of crystal structure.

Most recently, Wiers et al. (194) reported reversible Ca^{++}/Pb^{++} and Ca^{++}/Cd^{++} exchange, and successful but irreversible Cu^{++} exchange of $Ca^{++}A$, but observed complete crystal structure loss upon attempts at Hg^{++} exchange of $Ca^{++}A$. Even in methanol solution, Fe^{+++} nitrate and Al^{+++} nitrate exchange caused complete loss of crystallinity of $Na^{+}A$.

On the other hand, for LINDE Type X and Type Y zeolites, Na^{+}/Ce^{+++} (8, 10, 125), Na^{+}/Y^{+++} (146), and reversible Na^{+}/La^{+++} (170) exchanges have been reported.

The above has emphasized some pitfalls which may be encountered if zeolites are employed to separate metal ions. On the other hand, many zeolites are quite selective for exchange of particular metal ions and their use for pollution abatement, metals recovery and other unique separations offer promise for future applications.

Hydrogen Ion Exchange of Zeolites: When zeolites are contacted with solutions of lower pH values, proton exchange will occur. When ion exchange reactions are studied with solutions comprising salts which generate low pH solutions, the intended binary ion exchange may, in actuality, involve a ternary ion exchange with the hydronium ion H_3O^{+} as a third party to the exchange (e.g., 119–122, 131). The extent of proton exchange can be determined by computation from ion balances in solution or by chemical analyses of the zeolite.

Thermodynamic Treatment of Ion Exchange Equilibria and Selectivity

Numerous approaches to characterizing the equilibrium between ion exchange materials and aqueous solutions have been described in the literature. Gaines and Thomas (81) in 1953 proposed a rigorous thermodynamic treatment based upon the Law of Mass Action and treating the solvent as an independent variable. This approach appears to have the greatest value for analysis of the exchange behavior of zeolites.

Effect of Solution Normality on Ion Exchange Selectivity: Barrer and Klinowski (37) extended the results of Gaines and Thomas to evaluate quantitatively the changes in the shape and position of the ion exchange isotherms with variations in the total normality of the external electrolyte solution in contact with the zeolite. Their studies have shown that if Z_A is not equal to Z_B then, with increasing dilution, the isotherms become more and more rectangular and selective to the ion of higher valence (the "concentration / valency" effect), so that ion selectivity can arise universally from high dilution of the electrolyte solution, independently of the exchanger. They also demonstrated the use of this model to estimate ion exchange isotherms at different total solution normalities from a single isotherm at a single total solution normality (37).

Thermodynamic Treatment When Not All Ions are Exchangeable:
Sometimes not all the cations B^{Z_B+} in the starting zeolite can be replaced by entering cations A^{Z_A+}, due to steric limitations. This can occur if the entering cations are too large to penetrate into smaller cavities or too voluminous to allow entry of sufficient numbers for complete exchange –as in the case of $N(CH_3)_4^+$ exchange of NaX zeolite . In this case, the thermodynamic treatment must be modified to account for the incomplete exchange.

The proper calculational procedures for the case of incompete exchange, e.g., as employed by Barrer et al. (31) and Sherry (169, 171), involves the use of the equivalent cation fractions of the exchangeable cations only. An alternative approach proposed by Vansant and Uytterhoeven (189, 190) and discussed briefly by Breck (52) is incorrect, as has been discussed in great detail by Barrer et al. (36) and, more recently, by Dyer et al. (69).

Ternary Ion Exchange in Zeolites: Except for some early efforts by Mercer and Ames (15, 125), there has been very little published concerning the modelling of greater-than-binary ion exchange on zeolites. Recently, however, Rees et al. have described the ternary $Na^{++}/Ca^{++}/Mg^{++}$ exchange on LINDE Type A zeolite (41, 140, 141).

Fletcher, Townsend et al. have derived expressions for thermo-dynamic equilibrium constants for ternary ion exchange in zeolites (74-78, 187). Their most recent papers (78, 187) demonstrate their success with the $Na^+/ NH_4^+/ K^+$ -synthetic mordenite system.

3 ZEOLITE STABILITY

The molecular sieve zeolites have rigid, strong frameworks stable to high temperatures, oxidation/reduction, ionizing radiation, and not subject (as are many organic resin ion exchangers) to physical attrition due to osmotic shock. For the same reaons, the ion exchanger properties of the zeolites are relatively more constant and predictable over wide ranges of temperature, ionic strength, etc., than is often the case with other ion exchangers. Similarly, zeolite ion exchangers should not tend to adsorb organic molecules or ions and become "fouled" as readily as other ion exchangers.

594

Zeolites are also relatively stable at elevated pH levels (e.g., pH 7-11) at which other inorganic ion exchangers (e.g., zirconium phosphates, etc.) tend to lose functional groups due to slow hydrolysis (114). Zeolites are synthesized at elevated pH levels (e.g, pH 12-13+) and temperatures (e.g., 100-300°C) and are relatively stable at conditions only slightly less severe.

Practical Limits in the Use of Zeolites in Ion Exchange Separations: The chief limitation in the use of zeolites is due to their solubility in aqueous solutions. This limitation is not severe, except at extremes of pH. At the pH levels of natural surface waters (pH 6-10), most zeolites are relatively stable and will dissolve only very slowly.

Zeolite ion exchangers should not be employed below about pH 5-6 except for very brief exposures. Operation above pH ~6 is preferred. Proton exchange followed by slow hydrolysis of framework aluminum, leading to gradual loss of ion exchange capacity will occur at low pH levels. These reactions occur more readily on zeolites with low SiO_2/Al_2O_3 mole ratios and more rapidly at higher temperatures. If operation at low pH levels is required, actual experimental tests of zeolite stability should be made at conditions of intended use.

Zeolite Solubility: Zeolites are relatively stable over very broad ranges of conditions of interest for potential ion exchange applications, as evidenced by the presence of vast quantities of natural zeolites which were formed millions of years ago, as well as the current formation and persistence of vast quantities of certain zeolites (especially phillipsite and clinoptilolite) present in shallow sediments on the floors of the oceans. It is, therefore, worthwhile to consider the "equilibrium" solubility of zeolites, and to evaluate theoretically the effects of solution composition and pH on the solubility of the zeolite.

The basic approach involves the modelling of zeolite solubility, as indicated in the following equation for the case of a sodium-potassium-phillipsite. Based upon the free energy of formation of

$$(Na_{0.5} K_{0.5}) \cdot AlSi_3O_8 \cdot H_2O + 7 H_2O \rightleftharpoons$$
$$0.5 \ Na^+ + 0.5 \ K^+ + Al(OH)_4^- + 3 \ Si(OH)_4^0$$

this zeolite estimated (188) from literature data (160, 200), and assuming the modelling approach developed by natural water chemists, it is possible to calculate various combinations of pH, temperature, and concentrations (and activities) of silicon, aluminum and cation species in solution, which would exist in equilibrium with the zeolite. The results of such calculations [1] based upon the WATEQ

---

1) Sherman, J. D., Unpublished Studies, Union Carbide Corp. (1981)

Figure 3. WATEQ2 Calculation [1] of Si (mg / L) Concentrations in Eguilibrium with Phillipsite and Solutions Containing 0.066 N Na$^+$, 0.066 N K$^+$ and 100 μg Al / L at 25 C.

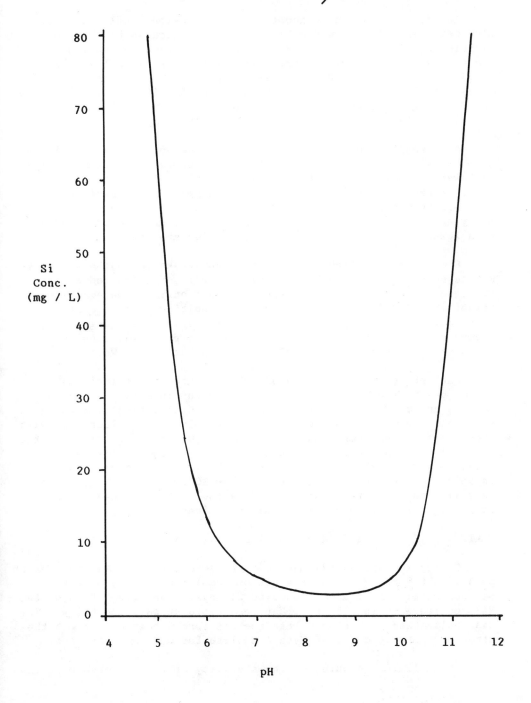

596

computer model (25, 188) are shown in Figure 3. As may be seen, such calculations predict that zeolites should exhibit low solubilities over a broad range of moderate pH levels ($\sim$pH 5.5–10.5).

In addition, higher concentrations of cations, silicates or aluminates in solution would be expected to decrease the rate of dissolution of the zeolite. Since the range of concentrations of these ions most commonly observed in natural water are 1 – 15 mg/l for Si, 0.03–0.05 mg/l for Al, with (typically) 0.5–2.0 mEq/L concentrations of cations, and pH levels of 5.0–10.5, it is expected that zeolites in contact with such waters will not be far from equilibrium, and will dissolve rather slowly.

Some applications can involve contacting very large quantities of water with rather small quantities of zeolite. For example, in wastewater treatment for NH_4^+ removal, simple calculations show that, if the effluent contained only $\sim$10 ppm of Si dissolved from the zeolite, then the entire zeolite bed would dissolve within about one year. In fact, actual experience has demonstrated loss rates much lower than this, even with regeneration at $\sim$ pH 12.

Similarly, in batch experiments, many weeks or months of contact may be required to reach "equilibrium" levels of dissolved Si in solution. Therefore, it is clear that, not only is the equilibrium solubility of zeolites very low but, in addition, the rate at which this equilibrium is approached is very low over broad ranges of pH. (However, at extremes of pH, e.g., pH 1-2 or pH 13-14, massive dissolution of most zeolites will occur in hours or days.)

Caullet, Guth and Way (59-62, 92) measured the solubility of 1-2 μm crystals of NaA and NaX for two months in $\underline{M}$ NaOH solutions at 25, 60, and 80 C. They observed congruent dissolution at these high pH levels (pH from 11.93 to 13.46). The solubilities increased with increasing pH and temperature, and the NaA was more soluble than the NaX zeolite. The concentration of Si in solution stabilized relatively rapidly (within $\sim$ 2 weeks), but the concentrations of Al in solution continued to rise even after 30-60 days in some cases. They also demonstrated that their data could be described by the following dissolution "equilibrium" equations:

$$NaAlSi_xO_{2(x+1)} + 2(x+1)\ H_2O \rightleftharpoons x\ Si(OH)_4 + Al(OH)_4 + Na^+$$

Cilley and Wiers (63) measured the acid uptake of NaA and CaA at pH 3-7 in the presence and absence of complexing agents over time periods ranging from a few hours to 21 days. The dissolution of the NaA zeolite was slow at pH 7, but very rapid at pH 5 and below. The CaA zeolite was more stable than the NaA form, in agreement wih the strong preference of LINDE Type A zeolite for Ca^{++} over Na^+ ions.

Some limited studies of the stability of 20 X 50 mesh particles of clinoptilolite were reported by Koon and Kaufman (112, 113). The weight loss of the zeolite was measured after cycling the zeolite between caustic and distilled water solutions at 2 hour intervals. If it is assumed the attrition occurred only during the high pH portion of the cycle, then the loss rates at pH levels of 11.5, 12.0 and 12.5 were 1.8, 3.0 and 4.8 %/day. Extrapolation suggests loss rates at pH 11, 10 and 9 of 1.1, 0.4 and 0.15 %/day are to be expected.

Taken together, these various published data suggest that zeolites differ substantially in their relative stability in, aqueous solutions but that, in general, dissolution should be slight at pH levels of approximately 6.0 to 10.5, and will increase rapidly as the pH is increased or decreased outside of this range. Higher temperature levels will, of course, also speed the dissolution.

Data from actual commercial applications of zeolites for ion exchange separations is very limited. In addition, the solutions being treated and regenerant solutions (which are often recycled) may contain silicate or aluminate ions, as well as exchangeable cations, and the presence of these will tend to slow down the rate of degradation of the zeolite or allow the degradation to proceed to "equilibrium" levels, and then proceed only slowly towards further degradation. Therefore, the lifetime of zeolite ion exchangers in commercial applications is often measured in years, rather than the days or months suggested by laboratory studies at more severe conditions, as outlined above.

On the other hand, the presence of complexing agents or other mechanisms which will remove aluminate and silicate ions from solution as they are formed will, of course, encourage the more rapid degradation of the zeolite. For example, Cilley and Wiers (63) reported that sodium tripolyphosphate, glycine, and citrate at pH 5 all accelerated the degradation of the NaA zeolite by a mechanism of aluminum extraction.

Other published literature on zeolite solubility is sparse. However, several conclusions may be reached from the published literature and our own studies [2], with regard to the patterns of degradation of zeolites at various pH levels. At very high pH levels, all zeolites tested dissolved congruently. At lower pH levels, e.g., 11-12, the rate of removal of silicon was much more rapid than that of aluminum, i.e., the zeolites dissolved incongruently, leaving behind a solid product containing a lower SiO_2/Al_2O_3 molar ratio than existed in the zeolite initially. At pH levels from 5-10, the rates of loss were very low. Loss rates gradually increased as the pH was increased or decreased beyond this range.

---

2) Unpublished studies of C. C. Chao, A. C. Frost, C. H. Nuermberger, R. J. Ross, and J.D. Sherman, assisted by J. Dubaniewicz and P. E. West, Union Carbide Corporation.

While the effect of pH on the stability of the zeolites is in good qualitative agreement with the theoretical solubility predictions discussed above, the zeolites appear to be more acid-stable than would be predicted by Figure 3. However, more detailed characterization of the surviving solid product is required to determine what fraction of the remaining solid retains its ion exchange capacity. Nevertheless, our studies have indicated that the zeolites do retain both their structure and their ion exchange capacity over broader ranges of pH than the equilibrium solubility model would suggest. The likely reason is that the kinetics of dissolution are extremely slow, except at extremes of pH.

Our own and other studies indicate that, generally speaking, the aqueous solution stability of a zeolite increases as its silica content increases. The higher silica content zeolites may generally be employed at process conditions involving more severe pH levels with lower rates of loss.

Unlike most organic resin cation exchangers, zeolites cannot be regenerated with strong mineral acids for use in demineralization applications. Instead, zeolites will be used in "ion interchange", as opposed to "demineralization" applications. So long as pH levels are not extreme, zeolites offer very attractive combinations of capacity, stability, and often extraordinary selectivities, as compared to the conventional organic resin or other inorganic ion exchangers.

4 PRESENT APPLICATIONS OF MOLECULAR SIEVE ZEOLITES IN ION EXCHANGE

The following sections discuss uses of zeolites in ion exchange by application type, with emphasis on the unique properties of the zeolites employed in each. Recent advances in the development of improved zeolites for use in ammonium ion exchange are described, as is also the newest application of zeolites in ion exchange: as builders in household detergents. Later sections briefly describe opportunities for other new ion exchange separation and purification applications based upon the unique properties of zeolites.

4.1 Ion Exchange of Cesium and Strontium Radioisotopes

Due to their stability in the presence of ionizing radiation, and in aqueous solutions at high temperatures, inorganic ion exchangers offer significant advantages in separation and purification of radioisotopes.

Exploratory studies of the ion exchange properties of these materials by Ames (1-13) revealed the high selectivities and capacities of several zeolites for cesium and strontium radioisotopes. Subsequently, several new processes were developed using these zeolites, as outlined in Table 4 (166).

Table 4. Cs and Sr Radioisotope Recovery and Purification

Service	Zeolite	Remarks and References
[137]Cs/High Level Wastes	LINDE AW-500 (IONSIV IE-95) (300 ft[3] bed)	First charge treated 3 million gal. then pressure drop increased (almost plugged) due to Al salt precipitation from the feed solution, 2nd charge treated several million gallons with no significant capacity loss (54, 127).
Cs/Rb,K,Na Purification of product from above	Large-pore Mordenite (1 l. bed)	Pilot plant; full-scale facility later operated at Hanford (47, 127).
[90]Sr,[137]Cs/ Low Level Waste Water in Fuel Storage Basin	Clinoptilolite (4 beds, 5.3 ft[3] each)	Nonregenerative use. Capacity: 12,000 gal wastewater treated / ft[3] of zeolite (90, 127, 195).
[137]Cs/ Evaporator Overheads & Misc. Wastes	LINDE AW-500 (IONSIV IE-95) (9.2 ft[3] bed)	Nonregenerative use. Capacity: 23-76,000 gal. wastewater treated / ft[3] of zeolite (90, 99, 127).
[137]Cs/ Process Condensate Wastewater	Large pore Mordenite	1.8 m deep layer of mordenite plus 0.6 m layer of organic resin ion exchanger. Regenerated with $NaNO_3$ solution (94, 127).

4.2 Zeolite Use at the Three Mile Island (TMI) Nuclear Power Plant

Zeolites have also been employed for the removal of radioactive cesium and strontium from contaminated waters from the accident at the TMI-2 nuclear power plant in Harrisburg, Pennsylvania. This led to increased interest in the radiation stability properties of zeolites. Brief early studies (79) had been noted by Mercer and Ames (127), but there had not been broad dissemination of such information.

Zeolite Stability in Ionizing Radiation: Roddy (145) reviewed the use of zeolites in radioactivity removal from liquid waste streams. He reported no deleterious effects on the structure or ion exchange properties of zeolites tested (Union Carbide's IONSIV IE-95 (AW-500), A-51 (4A), X-61 (13X) and Y-71 (Type Y) zeolites; Norton synthetic mordenite; and natural clinoptilolite] have been noted for cumulative exposures from 1×10^4 to ~1×10^8 Grays (1×10^6 to ~1×10^{10} rads). [Note: organic resin exchangers generally

show appreciable degradation when accumulated doses exceed
10^5-10^6 rad, and massive damage at doses of 10^9 rad (176).]

More recently, Bibler, Wallace, and Ebra (43) reported that
doses up to 3 X 10^{10} rads had no effect on the crystal structure of
Union Carbide's IONSIV IE-95 zeolite or on its ability to retain
Cs-137. They also studied the generation of gas due to radiolytic
decomposition of water contained in the zeolite. Pillay (136, 137),
reported studies of the effects 4.4 X 10^6 Gy doses on IONSIV IE-95
zeolite and observed that gas generation increased as the zeolite
water content increased. Bibler et al. (43) concluded that, if the
water is removed from the zeolite, Cs-137 can safely be stored on the
zeolite. Also zeolite incorporation into a glass (melting point
greater than 1000 C) should prevent hydrogen production because of
the low amount of water remaining in the zeolite in the glass.

Decisions were made, in fact, to employ Union Carbide's IONSIV
A-51 and IE-96 (a special Na^+ form of IONSIV IE-95) zeolite ion
exchangers for the clean-up of the contaminated water at Three Mile
Island and to ship the spent, radioisotope-loaded zeolite beds to
Hanford to be incorporated into glass waste forms for ultimate
disposal.

Zeolite Ion Exchange System at Three Mile Island: The TMI-2
accident began on March 28, 1979. The zeolite ion exchange system
for decontamination of the high activity level water in the Reactor
Containment Building (RCB) sump and reactor coolant system (RCS) at
TMI-2 was developed during early summer 1979. It is known as the SDS
(Submerger Demineralizer System)(142), since it is installed under
water in a spent fuel storage basin. Figure 4 (124) shows the SDS
flowsheet. Performance tests and design considerations for this
system are described (53, 57, 58, 65, 67, 98, 102, 107, 124, 175).
The compositions of the contaminated waters are shown in Table 5 (57).

Bench scale tests of the IONSIV IE-95 zeolite using RCB water
from TMI showed that cesium could be removed from the RCB water by a
factor of 10^4 for at least 1000 bed volumes (57) (42,000 Ci/liner)
without cesium breakthrough, but with significant breakthrough of
strontium, which could be removed by additional downstream liners.

Table 5. Composition of Contaminated Waters at Three Mile Island (57)

	Reactor Coolant System	Reactor Containment Building
Volume	~90,000 gallons	~700,000 gallons
Sodium	1350 ppm	1200 ppm
Boron	3870 ppm	2000 ppm
Cesium	1.5 ppm	0.8 ppm
Strontium	0.05 ppm	0.1 ppm

Figure 4.   Three Mile Island (TMI-2) SDS Radwaste System(124).

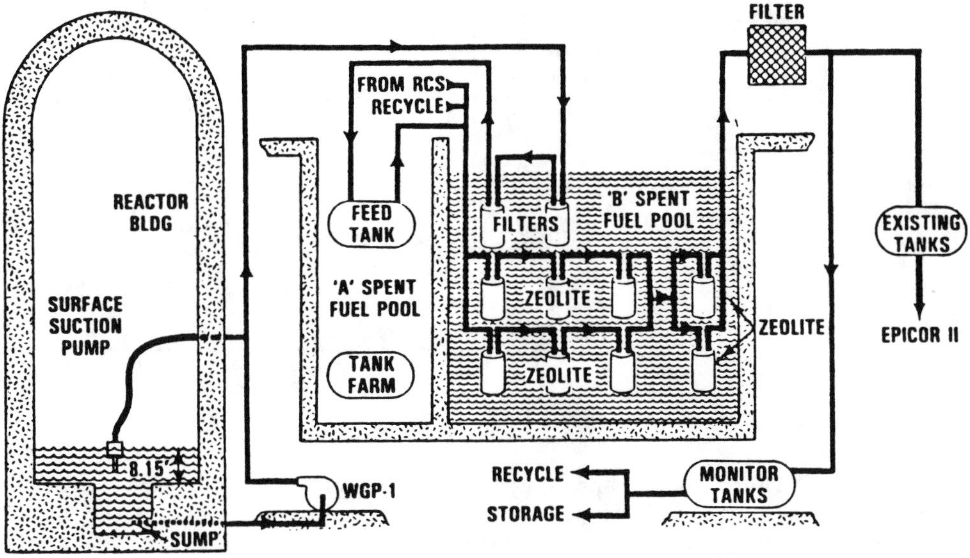

Each "liner" contains approximately 60 gal. of zeolite. Extrapolation indicated that increasing the loadings to 60,000 Ci, or more, of cesium per liner should be readily achievable and desirable to concentrate the wastes onto fewer liners.

The Cs^+/Na^+ and Sr^{++}/Na^+ ion exchange selectivities of the zeolites control their performance in this application.   The IONSIV IE-96 zeolite employed in this application has excellent Cs^+ selectivity, but lower Sr^{++} selectivity.   Therefore, although the Cs^+ radioisotopes would be held strongly by the IONSIV IE-96 zeolite, the Sr^{++} radioisotopes would be less strongly held and would break through the column first. Therefore, the procedure to load the zeolite columns to 60,000 Ci/liner would generate a number of additional liners loaded with strontium only.

During this same time period, we recommended (and later learned others also recommended) that Union Carbide's IONSIV A-51 zeolite ion exchanger be employed for removal of the strontium because of its excellent Sr^{++}/Na^+ ion exchange selectivity and capacity and its excellent stability in ionizing radiaton.   Tests using synthetic TMI water confirmed the excellent strontium selectivity of the IONSIV A-51, and laboratory tests at ORNL using actual TMI water confirmed that a mixture of IONSIV IE-96 to remove cesium and IONSIV A-51 to remove strontium would provide more efficient decontamination, and generate a much smaller volume of spent zeolite for subsequent

602

Figure 5. Comparison of Performance of Single Zeolite and Mixed Zeolite(65).

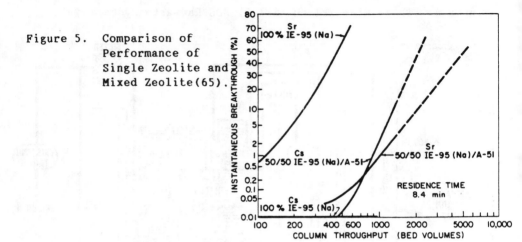

disposal. These tests and a computerized mathematical model used to evaluate the data and to predict the performance of the mixed bed SDS system will be described (102). It was concluded that a mixture of 5.5 ft^3 IONSIV IE-96 and 2.5 ft^3 IONSIV A-51 placed in each column "liner" should provide optimum performance.

A test with TMI-2 RCB water and a bed containing equal parts of IONSIV IE-96 and IONSIV A-51 zeolites provided the breakthrough curves in Figure 5 compared with those obtained when using only IONSIV IE-96 (65). The mixed zeolite bed capacity for strontium sorption is increased by a factor of about 10, even though strontium exchange is slower, as indicated by the lower slope of the break-though curve. The capacity for cesium is adequate for a throughput of up to about 2000 BV, representing a factor of 10 increase over the original design.

The TMI-2 SDS system employing the mixture of IONSIV IE-96 and IONSIV A-51 zeolites was operated from June 1981 to February 1982, to successfully process the ～650,000 gallons of RCB sump water with the results summarized in Table 6 (98). By use of the mixture of zeolites, the quantities of spent zeolite material loaded with

Table 6. SDS Effectiveness in Processing Reactor Building Sump Water

Radionuclide	Influent (μCi/ml)	Effluent (μCi/ml)	Decontamination Factor
Cs-134	13.1	0.0001	130,000
Cs-137	123	0.00086	140,000
Sr-90	5.14	0.0088	590

320,000 Curies of radioactivity removed from the sump was reduced to only 1550 kg of spent zeolites, corresponding to a volume reduction of 1460 (vs. the contaminated water volume). The spent zeolite beds will be stored for a period at TMI and then shipped to the Pacific Northwest Laboratory for tests to demonstrate the vitrification of the zeolites, on a production scale, using the 'in-can' melting process (175).

In light of the excellent performance of the mixed zeolite beds, the SDS system has subsequently been employed for decontamination of the Reactor Coolant System water. This will be continued to reduce radiation hazards during removal of the damaged fuel elements from the reactor to complete the TMI recovery project.

4.3 Ammonium Removal from Municipal Wastewater

In early studies of the use of ion exchange for wastewater ammonia removal, a permutite type exchanger and various organic resin ion exchangers were found to have poor selectivity for ammonium ions, resulting in unacceptably low ammonium loadings and low regeneration efficiency, corresponding high costs, and problems of brine disposal.

This situation was altered dramatically by a report by Ames (15) which presented data showing the superior ammonium ion selectivity of several zeolite ion exchangers tested. Clinoptilolite and Union Carbide's AW-400 were most promising. Subsequent pilot plant tests using clinoptilolite demonstrated ammonium removal greater than 95%. Regeneration was accomplished using a lime-salt solution which was effectively reused after ammonia was removed from the regenerant solution by air stripping and exhausted to the atmosphere. An improved process (the ARRP Process) for the rejuvenation of the regenerant solution at elevated pH levels, followed by the removal of the NH_3 from the air by acid scrubbing, and recycling the air to the stripper was developed later. All of these developments were reviewed in detail elsewhere (166).

Improved Ammonium Exchangers: Although clinoptilolite performs quite well in this service, exchangers with higher capacity should provide significantly improved overall process performance. Studies at Union Carbide Corporation led to the discovery that the synthetic LINDE IONSIV F zeolite is more effective than clinoptilolite in removing NH_4^+ from wastewater (50, 51). Laboratory tests confirmed the higher cycled NH_4^+ loadings of the LINDE F zeolite (166).

Further exploratory tests led to the discovery that zeolites of the Type W-merlinoite-phillipsite-gismondine-NaP group provide superior NH_4^+ exchange characteristics (162, 163, 168). Equilibrium isotherms, shown in Figures 6 and 7, confirmed the superior NH_4^+ exchange capacities of the LINDE F and W zeolites.

Figure 6.   Figure 7.

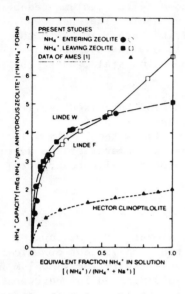
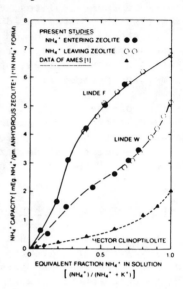

Figure 6. NH_4^+ Capacities in Competition with Na^+
            in 0.1 N Solutions at Room Temperature (168).

Figure 7. NH_4^+ Capacities in Competition with K^+
            in 0.1 N Solutions at Room Temperature (168).

Cyclic Column NH_4^+ Exchange Tests: Column tests compared
the performance of LINDE F, LINDE W and clinoptilolite at conditions
simulating removal of NH_4^+ from municipal wastewater (168).
Earlier studies (166) had also shown attractive NH_4^+ ion exchange
performance by other members of the Type W–phillipsite–gismondine
family of zeolites. Therefore, a sample of natural phillipsite was
also tested. The clinoptilolite and the LINDE F samples lost $\sim$ 50%
and 55%, respectively, of their initial dynamic NH_4^+ capacities in
just three cycles, believed due to the gradual buildup on these zeolite
of more strongly held K^+ and Ca^{++} ions which are not efficiently
removed in subsequent regenerations. The following steady-state, cycle
dynamic NH_4^+ (10% breakthrough) ion exchange capacities were
estimated. As may be seen, the LINDE W mesh provides $\sim$ 3 1/2 times
greater dynamic NH_4^+ ion exchange capacity as compared to
clinoptilolite (168):

IONSIV W Mesh Developmental Sample	$\sim$650 BV
Pine Valley Phillipsite	$\sim$550 BV
IONSIV F Mesh Developmental Sample	$\sim$350 BV
Hector Clinoptilolite	$\sim$190 BV

Other studies indicate that applications of IONSIV F and W zeolites for wastewater NH_4^+ removal should provide substantial reductions in overall process costs. Recent pilot scale tests for actual wastewater NH_4^+ removal using samples of IONSIV W mesh product prepared at semiworks scale confirmed the above results.

Commercial Wastewater NH_4^+ Removal Processes Employing Zeolites: The first large-scale installations began operation in California in 1978 and in Virginia in 1979. Both installations employ clinoptilolite in 20 X 50 mesh form and both employ the Ammonia Removal and Regeneration Process (ARRP Process) developed by the CH2M-Hill consulting firm, as reviewed elsewhere (143, 166, 181).

The Tahoe-Truckee plant encountered $CaCO_3$ "scale" formation problems in the ARRP system due to inadequate pH control. These problems were surmounted by modest alterations of equipment and operating procedures (138). Despite initial difficulties, the overall plant performance for the last 30 days of operation covered by the report (55) was quite satisfactory. The total nitrogen concentrations were reduced from 39.0 mg/l in the feed down to only 0.89 mg/l in the final effluent, for a 98% overall nitrogen removal.

Recent reports include mathematical modelling of column NH_4^+ removal (159, 160) and of biological regeneration (154-158) of clinoptilolite. Tests comparing clinoptilolite, erionite, mordenite and phillipsite (106) confirmed, in agreement with our studies, that phillipsite provided the best NH_4^+ capacity ($\sim$ 70% greater than the best clinoptilolite), but it was too friable for commercial use.

Liberti et al. have also disclosed an interesting combination of selective phosphate exchange on a weak anion resin exchanger and selective ammonium ion exchange on clinoptilolite in a process which recovers a slow release fertilizer, $MgNH_4PO_4 \cdot 6H_2O$, from the concentrated regenerant stream (44, 117, 118, 134). The value of the fertilizer should pay most of the cost to remove both ammonia and phospate, for an estimated overall cost only $\sim$ 30% of the cost of biological wastewater treatment for ammonia removal (44, 134).

Other recent studies include use of clinoptilolite for removal of NH_4^+ ions from drinking water (84). Tests will also be made of the treatment of municipal wastewater for potable water reuse in Denver, Colorado (20, 147).

4.4 Molecular Sieve Zeolite Builders in Detergents

The prime function of phosphates in detergents is to reduce the activity of the "hardness" ions, Ca^{++} and Mg^{++}, in the wash water by complexing. Zeolite ion exchangers in powder form can also provide this service by removing Ca^{++} and Mg^{++} from the solution and replacing them with "soft" ions such as Na^+ (42, 152, 182).

Heavy duty powder detergents employing the Na form of the LINDE A zeolite in low or zero-phosphate formulations are already being sold in several areas of the United States, Europe, and elsewhere. This use has grown rapidly and now consumes hundreds of millions of pounds of zeolites annually worldwide.

This application has been developed primarily by scientists at Henkel (42, 152, 182) in Germany and Procter and Gamble in the United States (151, 194). The literature on this application, especially the patent literature, is voluminous and specialized. It will not be reviewed in detail here. Instead the discussion will be limited to the related ion exchange behavior of zeolites.

Detailed studies of the ion exchange equilibria of Ca^{++} or Mg^{++} with Na^+ on the LINDE Type A zeolite have been reported by Barri and Rees (41, 140), as shown in Figure 8. Thermodynamic modelling generated the smoothed curves fitting this data, providing elegant proof of the "concentration/valency" effect discussed earlier. As may be seen, the LINDE Type A zeolite is very effective in removing calcium from solution. High calcium loadings on the zeolite may be achieved in equilibrium with solutions of very low residual calcium content, as desired for efficient soil removal with a laundry detergent.

Figure 8. Na-Ca and Na-Mg Binary Exchange Isotherms: O, 0.2 N; X, 0.1 N; $\bullet$, 0.05 N; $\square$, 0.01 N; $\blacksquare$, 0.005 N.

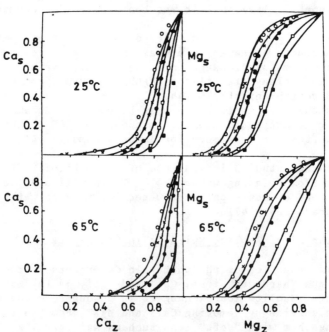

In order to provide maximum effectiveness, the zeolite should also perform its function very rapidly. The rate of Mg^{++} removal from a water of "average" hardness by LINDE A zeolite is much lower than that of Ca^{++} removal, and this difference becomes more pronounced at lower temperatures.

The results of ion exchange rate tests with NaA zeolite are shown in Figure 3 (167). As may be seen, the calcium is removed from solution extremely rapidly, but the magnesium is not; even after five hours magnesium exchange equilibrium has still not been attained.

Figure 9 (167) displays effects of temperature. The calcium removal is rapid at 20 C but it is even faster at the higher temperatures. At all temperatures the magnesium removal is relatively slow and equilibrium is not achieved at 10 minutes, even at 120 F. Therefore, the Type A zeolite is not well suited for magnesium removal for detergent applications, particularly at the cooler water washing conditions often encountered in the United States. Exploratory studies led to the development of a product, called LINDE ZB300 Zeolite Builder, which removes not only the calcium but also magnesium, rapidly and efficiently.

Figure 10 (167) shows results of room temperature tests for times of interest in actual laundry detergent uses, but with only magnesium present in the solution. As may be seen, the results for 30 seconds and 10 minutes on the X zeolite are very close together,

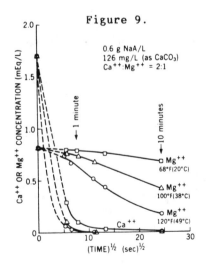 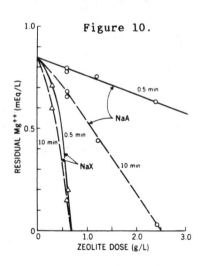

Figure 9. Simultaneous Ca^{++} and Mg^{++} Exchange by LINDE Type A Zeolite

Figure 10. Mg^{++} Exchange by LINDE Type A and Type X Zeolites at Room Temperature.

608

indicating that equilibrium is very rapidly achieved. Therefore, the X zeolite can be considered as an alternative to the A zeolite if magnesium removal as well as calcium removal is desired. With this in mind, the use of mixtures of the two zeolites was tested. One would expect such mixtures would provide performance intermediate between that of the two zeolites separately.

However, the performance of the mixtures was found to be substantially superior to that of the zeolites individually, revealing a synergistic effect, shown in Figure 11 (68, 167). The expected performance of the mixtures would lie on straight lines connecting the points on the right-hand and left-hand axes for the two pure zeolites. As may be seen, the mixtures performed far better than expected based upon a linear mixing of the two end-members. This improved performance is particularly noticeable at lower temperatures, but it is also observed at higher temperatures.

Such mixtures of detergent grade LINDE Type A and X zeolites are now sold commercially under the designation Union Carbide ZB300 Zeolite Builder. The performance of the ZB300 zeolite product compared with Type A zeolite is shown in Figure 12. As shown, substantially improved magnesium removal is obtained.

Figure 11.   Figure 12.

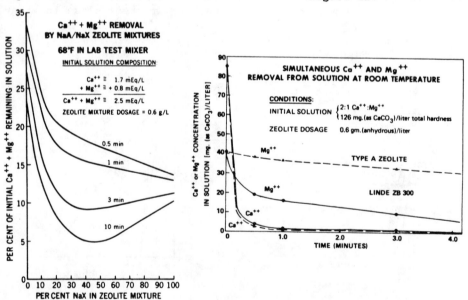

Figure 11. Ca^{++} and Mg^{++} Removal by NaX / NaA Zeolite Mixtures(167).

Figure 12. Simultaneous Ca^{++} and Mg^{++} Removal from Solution(167).

"Terg-O-Tometer" tests conducted using a variety of soil types, cloth types and skeletal detergent formulations showed that the ZB300 mixed zeolite product often provides detergency performance nearly identical to the ZB100 product (Union Carbide's detergent grade LINDE Type A zeolite). However, in a substantial number of other examples, the ZB300 product provides modest but significant improvements in soil removal performance, as reported earlier (167).

5 FUTURE APPLICATIONS

5.1 Radioactive Waste Storage

As discussed earlier, zeolites are employed in separations of long-lived Cs and Sr radioisotopes. These radioisotopes can also be retained on zeolites for long-term storage by ion exchange onto the zeolite, drying the zeolite to prevent excessive pressure after the container is sealed, and sealing the containers by welding (126).

Since zeolites contain alkali metal or alkaline earth oxides, alumina and silica (major constituents of many common glasses), heating to temperatures sufficient to cause destructon of the zeolite crystal structure can convert the zeolite to a glass. Addition of suitable flux calcining agents can allow this to be accomplished at lower temperatures. Leach rates for alkali and alkaline-earth elements from aluminosilicate glasses are extremely low (e.g., $10^{-7} gm/cm^2$-day). The chemical durability, low leach rates, and high thermal conductivity of glass combine to make this an ideal form for immobilizing radioactive wastes.

One process employs a hydrous metal oxide type cation exchanger (Na Ti_2O_5H) to trap 90_{Sr} and other radioisotopes from liquid wastes from fuel reprocessing, followed by a zeolite bed to trap the 134_{Cs} and 137_{Cs}. The ion exchangers are removed, blended, dried and hot-press- sintered to yield stable ceramic discs with low leach rates (18). The process has been tested at Hanford (19). Similarly, radioactive Cs, Sr and Pu were sorbed on LINDE AW-500 zeolite for final soldification in concrete or glass (101, 196).

5.2 Zeolite Enhanced Biological Nitrification

The removal of NH_4^+ from municipal waste water by use of zeolite ion exchangers in a physical-chemical treatment process involving alternating cycles of loading and regeneration was discussed earlier. In such cycles the zeolite is regenerated by mass-action or chemical driving forces. The NH_4^+ loaded on the zeolite may also be removed by biological nitrification:

$$NH_4^+ + 2O_2 \xrightarrow{\text{bacteria}} NO_3^- + 2H^+ + H_2O$$

In effect, the addition of a zeolite to the activated sludge will impart selective NH_4^+ exchange capabilities to the sludge, thus improving its ability to remove NH_4^+ from the waste water. The conversion of NH_4^+ by nitrifying bacteria will regenerate the zeolite. Such "zeolite enhanced biological nitrification" has been demonstrated by Sims (177) and Sims and Little (178). The ability of the zeolite to pick up NH_4^+ during peak load periods and subsequently gradually release it may provide higher overall nitrification rates and improved ability to handle shock loads.

5.3 NH_4^+ from Industrial and Agricultural Waste Water

In addition to treatment of municipal waste water, it is anticipated that zeolites will find use in removal of NH_4^+ from industrial and agricultural waste water streams. Here the availability of several different zeolites (each having NH_4^+ ion exchange selectivity) with different properties offers the possibility of selecting the optimum exchanger for a particular service on the basis of its ability to selectively remove NH_4^+ in the presence of different competing cations.

5.4 Regeneration of Artificial Kidney Dialysate Solutions

Hemodialysis treatment in artificial kidney systems involves the transfer of uremic wastes through suitable membranes by dialysis to a dialysate fluid while the small pores of the membranes prevent loss of desirable blood components. Rather large volumes (100-300 liters) of dialysate solution are required for a single treatment. Interests in reducing the size of the equipment required and in achieving portability have led to the development of a process to remove the waste products from the spent dialysate solution, so the dialysate solution can be continuously reused. Then, as little as 1-2 liters of dialysate can be adequate for one hemodialysis treatment.

Of all the uremic waste substances, urea is the most abundantly generated. Improved urea-binding sorbents have been sought for many years for dialysate regeneration in hemodialysis or peritoneal dialysis, to reduce the quantities of dialysate required and, thereby reduce the size and weight of such systems. Unfortunately, although activated carbon is an effective sorbent for many waste metabolites and drugs, it possesses insufficient capacity to remove urea effectively from dilute solutions. Other sorbents (resin ion exchangers, oxystarch, etc.) have also been studied, but a non-soluble, selective urea-binding sorbent with high urea sorption capacity at physiological pH levels has not been found.

In an alternative method urea is hydrolyzed to form ammonium ions which are then removed by ion exchange. Dialysis systems employing such a process were developed (88, 89, 123) and sold by CCI Life Systems, Inc., based upon use of an immobilized urease enzyme

catalyst to hydrolyze urea to NH_4^+, followed by zirconium
phosphate to remove the NH_4^+ cations, and hydrous zirconium
oxide to remove phosphate and fluoride anions. $Ca^{+'}$, $Mg^{''}$ and
K^+ removed by the zirconium phosphate must be added back in the
required, controlled concentrations to the regenerated dialysate.

The LINDE F and W zeolites discussed earlier provide unique
selectivities for NH_4^+ in the presence of other common alkali
and alkaline earth cations and should, therefore, provide improved
performance over zirconium phosphate in dialysate regeneration.
Andersson et al. (17) disclosed the use of a zeolite ion exchanger
for this purpose. Their data confirm our results (see above) showing
that several zeolite ion exchangers provide higher selectivities for
NH_4^+ over Na^+ and for Na^+ over Ca^{++} and $Mg^{+'}$, compared to
a zirconium phosphate exchangers. The highest NH_4^+ selectivities
were provided by phillipsite and LINDE F zeolite ion exchangers.
Both showed higher exchange capacities compared to clinoptilolite.

NH_4^+ Exchange for Urea Removal: Extension of the use of
these new IONSIV F and W ion exchangers to remove urea nitrogen is
quite straightforward. The NH_4^+ generated by urea hydrolysis is
exchanged onto the zeolite. The CO_3 simultaneously generated
must be neutralized by acid addition (directly or via use of a
suitable buffer) to control pH near physiological pH levels ($\sim$7.4
pH) and at suitable levels ($\sim$pH 7-8) to maintain the urease enzyme
activity. The system chemistry is shown below (164):

Urea Hydrolysis: urease enzyme
$$(NH_2)_2CO + 2 H_2O \xrightarrow{\hspace{3cm}} 2 NH_4^+ + CO_3^-$$

Neutralization of Ammonium Carbonate:
$$2 NH_4^+ + CO_3 \rightleftharpoons 2 NH_4^+ + HCO_3^-$$

NH_4^+ Ion Exchange:
$$(R_{2/n}^n) \text{ zeolite} + 2 NH_4^+ \rightleftharpoons (NH_4^+)_2 \text{ zeolite} + 2/n\ R^{n+}$$

The effective NH_4^+ binding capacities of IONSIV F-80 and
IONSIV W-85 zeolites in contact with blood or dialysate solutions
were estimated (164) to be quite high, even in the presence of
competing K^+, Ca^{++} and Mg^{++} cations. For example, the desired
removal of 12-30 gms urea/day for treatment of chronic renal disease
would require only 0.9-2.2 lbs/day of IONSIV NH_4^+ exchanger.

Later studies by Klein (103-105) and Gregonis and Walker (90)
confirmed the high NH_4^+ capacities of the IONSIV F and W zeolites at
conditions simulating their use in artificial kidney devices (104).

The electrolyte balances required for dialysate regeneration are
much more complex than can be provided by simple NH_4^+ exchange. As
noted above, it is necessary to neutralize the CO_3 generated by

urea hydrolysis and to maintain suitable pH levels. In addition, K^+
and phosphate removal in controlled amounts is desired and blood pH
and electrolyte (Na^+, K^+, Ca^{++}, Mg^{++}, phosphate, bicarbonate, etc.)
balances must be carefully maintained for patient wellbeing.

One approach to achieving these desired balances involves the
use of these NH_4^+ selective zeolites in the partially Ca^{++} exchanged
form (22, 23, 164, 165). The Ca^{++} from the zeolite can then serve
as a "buffer" to remove the excess CO_3^{--} and also, with proper
adjustment, provide a means of phosphate removal by formation of
calcium phosphate adsorption complexes and/or precipitates, as shown
schematically (and over-simplified) below:

$$CaNa \text{ zeolite} + (NH_4)_2CO_3 + X \ NaH_2PO_4 \longrightarrow$$
$$NH_4 \text{ zeolite} + CaCO_3 + Ca \text{ phosphate}$$

Animal tests using a novel sorbent--slurrry--reciprocating-
dialyzer (SSRD) developed by S. R. Ash (24) have demonstrated the
effectiveness of this approach, and have shown that the use of
zeolite ion exchangers can make truly portable artificial kidney
devices a reality (22). These studies demonstrated the feasibility
of a small dialysis device with urea conversion to $(NH_4)_2CO_3$
and the use of (Ca^{++}, Na^+ form) IONSIV F and W zeolite ion
exchangers for selective NH_4^+ and K^+ removal, and Ca^{++}
release, to provide, in actual dog tests, ion fluxes appropriate
for the uremic patient (22).

5.5 Aquaculture

Ammonia is extremely toxic to aquatic animals. In closed
systems (e.g., aquariums) and when extensive water reuse is
practiced in high density fish culture (hatcheries, fish farming),
the ammonia released directly by the fish, from their other
nitrogenous wastes such as urea, and from bacterial deamination of
protein in feed and wastes will quickly reach toxic concentrations
if not removed.

Microbiological filters may be used for this purpose.
However, nitrifying bacteria are easily inhibited or killed by
various stresses (low temperatures, sulfides, methanol, heavy
metals, antibiotics used to control disease outbreaks, shock loads,
etc.). Toxic levels of ammonia may be quickly reached before the
biological filter operation can be reestablished to required levels.

For these reasons, a number of investigators have studied NH_4^+
removal by ion exchange on zeolites as an independent standby backup
system (for emergency use when upsets occur, or during treatment of
diseased cultures), and/or for polishing treatment to remove ammonia
which escapes the bioligical filter, or for reliable ammonia removal
in lieu of biological filters.

Braico (46) compared ion exchange using a zeolite with other methods of ammonia removal for reuse of fish hatchery waters. He concluded that zeolite ion exchange offers advantages of lower cost, higher ammonia removals, a chemical process which is more controllable than existing biological processes, and lower land area requirements.

Konikoff (111) compared the performance of clinoptilolite ion exchange vs. biofilters for ammonia removal. Substantial biological nitrification occurred in the clinoptilolite system at the conditions studied and consequent nitrite toxicity problems were observed in both the biofilter and ion exchange systems. (This can likely be avoided.)

Johnson and Sieburth (100) tested LINDE AW-500 and clinoptilolite zeolites for ammonia removal from from an active, closed system for Chinook Salmon and from artificial sea water. They concluded that, in high salinity waters, a zeolite ion exchange column is desirable as a secondary or backup system to biological filters for use in low density closed aquaculture systems. However, in fresh water systems, it is feasible to use ion exchange alone (instead of biological nitrification) with the advantage of being ready at any time to remove considerable quantities of toxic ammonia. Total removal of the ammonia using zeolites also avoids the further complication of nitrate buildup resulting from the biological oxidation of ammonia.

Slone, Turner, and Jester (180) demonstrated the utility of zeolite removal of ammonia escaping from a biofilter in a closed "silo" type fish culture system.

Liao and Mayo (115, 116) have examined broadly the requirements and options in water reconditioning - reuse in intensified fish culture.

Peters and Bose (135) also reported studies of the use of zeolite ion exchange for ammonia removal in hatchery or aquaculture water reuse systems.

Numerous other studies have also been reported in the literature (129, 166, 179). These have included demonstrations of the utility of zeolites in reducing levels of toxic ammonia in the transport of ornamental freshwater fish (45).

Torii (186) reviewed uses of natural zeolites (clinoptilolite and mordenite) in Japan and estimated that 5-10 million pounds per year of clinoptilolite are employed in removal of ammonia from aquaculture (e.g., eels, carps or sweetfish) ponds or tanks.

Preliminary tests have shown that the LINDE NH_4^+ ion exchangers discussed above also provide significant improvements in performance in aquacultural and related applications.

5.6 Feeding of Ruminant Animals

The digestive systems of ruminant animals (cattle, sheep, goats, deer, buffalo, etc.) include a bacterial fermentation "vat" (the rumen) in which plant and other feed materials may be broken down into smaller molecules and in which amino acids and certain vitamins may be synthesized. These rumen bacteria can even be fed sources of inorganic nitrogen, such as ammonia or urea, and employ these non-protein nitrogen (NPN) feeds to produce amino acids, which are ultimately converted to animal protein.

Substitution of NPN compounds for a portion (or all) of the natural protein in the animals diet offers major economics in the cost of feed. However, the quantities of NPN which may be fed are limited by the need to keep the ammonia concentrations in the rumen below toxic levels. If large quantities of urea are fed, toxic levels of ammonia can be reached, since urea fed to the animal is quickly hydrolyzed (by urease enzyme), releasing ammonium ions:

$$(NH_2)_2CO + 2H_2O \longrightarrow 2 NH_4^+ + CO_3$$

White and Ohlrogge (193) have disclosed that ion exchangers may be introduced into the rumen prior to feeding of NPN compounds so that the NH_4^+ ions are partially exchanged onto the ion exchanger (to reduce the NH_4^+ concentration in the ruminal fluid) and thereafter slowly released by the regenerant action of saliva (including Na^+ and K^+ ions) entering the rumen. They note that zeolite ion exchangers, especially the LINDE F zeolite, provide outstanding NH_4^+ ion exchange performance for this application.

Watanabe (191) has reported studies in Japan of actual feeding of zeolites to cattle as a dietary supplement. Feeding of the zeolite to the animals provided improved feeding efficiency. Kondo et al. (110) have reported the benefits of zeolite feed supplements on calf growth.

5.7 Other Uses in Agriculture, Horticulture and Animal Feeding

Mumpton and Fishman (128) and Torii (186) reviewed agricultural uses of natural zeolites including extensive studies in Japan, and some also in the United States, of the feeding of zeolites to animals, use in odor control, use of zeolites as soil conditioners, in fertilizers, as carriers of fungicides and pesticides, and as NH_4^+ exchangers to prolong the life of cut flowers. There have been many studies (21, 72, 108, 109, 133) of feeding of zeolites to swine and poultry. Reported benefits have included: increased weight gain, increased feeding efficiency, reduced incidence of intestinal and other diseases and reduced death rates; and lower odor of animal excrement.

Although the exact mechanisms by which the benefits of the zeolites are achieved are not known in all of the agricultural applications, ion exchange properties are likely of great importance. The natural zeolites tested, principally mordenite and clinoptilolite (which are known to have selectivity for NH_4^+ exchange), provided reduced odor of animal droppings, improved retention of nitrogen fertilizer in the soil, and removal of NH_4^+ from water.

5.8 Metal Removal, Recovery and Separations

Many zeolites exhibit high selectivities for various heavy metals and are under consideration for use in recovery of precious and semi-precious metals and for removal of heavy metals from industrial and metals processing waste waters.

Because of their availability (especially in Japan), the zeolites clinoptilolite and mordenite have been studied for heavy metals (especially Cd, Cu, Pb, and Zn) removal from waste waters (64, 80, 91, 93, 148-150, 153, 183, 197-197). The high selectivities of several zeolite ion exchangers for Ag^+ also suggests their use for the recovery of silver from waste waters.

Separations and purifications of non-ferrous metals may also be accomplished by zeolite ion exchange. For example, Breck (49) reported the unique separation of Co^{++} and Ni^{++} on LINDE Type A zeolite. Many other separations of non-ferrous metals are also possible (264). Separations of both free and complexed ions may be accomplished (e.g., 49, 51), suggesting that zeolite ion exchangers may provide unique new separations and purifications in the processing of non-ferrous metals.

Acknowledgements

Mr. R. J. Ross and also Messrs. Y. J. Doerr, J. Dubaniewicz, C. H. Nuermberger, and R. E. West and Ms. C. S. Roth collaborated on the NH_4^+ exchange work and Dr. J. M. Bennett and Ms. J. P. Cohen performed related crystallographic studies. Dr. A. F. Denny and Messrs. A. J. Gioffre, Jr., and G. M. Straehle shared in various aspects of the detergent product development studies. Dr. A. C. Frost and Messrs. R. J. Ross, C. H. Barkhausen, P. E. West and Ms. J. E. Stern assisted in numerous other ways, all greatly appreciated. I also wish to acknowledge the many contributions to our ion exchange studies made by the late Dr. Donald W. Breck; we miss Don's warm friendship and stimulating discussions. Many thanks also to Ms. B. Bjorkman and Ms. G. Solano for their fine secretarial assistance.

Finally, and most of all, I wish to thank Carol, my wife, for her constant good cheer, patience and encouragement during this and many other such projects.

616

REFERENCES

(1) Ames, L.L., jr. Amer.Mineral. 45 (1960) 689.
(2) Ames, L.L., jr. and B.W. Mercer. Report No. HW-SA-2175, Hanford Lab. Oper. (1961).
(3) Ames, L.L., jr. Amer.Mineral. 46 (1961) 1120.
(4) Ames, L.L., jr. Amer.Mineral. 47 (1962) 1067.
(5) Ames, L.L., jr. Amer.Mineral. 47 (1962) 1317.
(6) Ames, L.L., jr. Amer.Mineral. 49 (1964) 127.
(7) Ames, L.L., jr. Amer.Mineral. 49 (1964) 1099.
(8) Ames, L.L., jr. J.Inorg.Nucl.Chem. 27 (1965) 885.
(9) Ames, L.L., jr. Amer.Mineral. 50 (1965) 465.
(10) Ames, L.L., jr. Can.Mineral. 8 (1965) 325.
(11) Ames, L.L., jr. Can.Mineral. 8 (1966) 572.
(12) Ames, L.L., jr. Can.Mineral. 8 (1966) 582.
(13) Ames, L.L., jr. Amer.Mineral. 51 (1966) 903.
(14) Ames. L.L., jr. J.Inorg.Nucl.Chem. 29 (1967) 262.
(15) Ames, L.L., jr. Proc. 13th Pacif. NW. Indus. Waste Conf., 6-7, Apr.,1967, Wash. State Univ. (Publ. by Tech. Ext. Serv., Wash. State Univ., Pullman, Wa., 1967), p. 135.
(16) Amphlett, C.B.,Inorganic Ion Exchangers (New York, Elsevier, 1964) Ch.3.
(17) Andersson, S., I. Grenthe, E. Jonsson and L. Nauclars. Ger.Offen. 2,515,212 (1975); Brit. Pat. 1,484,642 (1977).
(18) Anon. New Process Consolidates Radioactive Wastes, Chem.Eng. News (Jan. 12, 1976) 32-33.
(19) Anon. Consolidating Radioactive Wastes, Chem.Eng. (July 19, 1976) 158.
(20) Anon. Water Reuse Facility under Construction Near Denver, J. Am. Water Works Assoc. 73 (1981) 40.
(21) Araki, K. and S. Honda. Japanese Pat. 74,034,898 (1974).
(22) Ash, S.R., R.G. Barile, J.A. Thornhill, J.D. Sherman and N-H.L. Wang. Trans. Am. Soc. Artif. Intern. Organs 46 (1980) 111.
(23) Ash, S.R., Belg. Brevet d'Invention No. 874,429 (Aug. 8, 1979).
(24) Ash, S.R., P.G. Wilcox, D.L. Wright and D.P. Kessler. Mass Transfer in a Wearable Artificial Kidney. Presented at A.I.Ch.E. Mtg., New York City (Nov. 15, 1977).
(25) Ball, J.W., E.A. Jenne and D.K. Nordstrom. in Chemical Modelling in Aqueous Systems, E.A. Jenne, Ed., ACS Symp. Series 93 (Washington,D.C., Am. Chem. Soc.,1979), pp. 815-835.
(26) Barrer, R.M., J.Chem.Soc. (1950) 2342.
(27) Barrer, R.M. and J.D.Falconer. Proc.Roy.Soc. A236 (1957) 227.
(28) Barrer, R.M., Proc.Chem.Soc.(London) (1958) 99.
(29) Barrer, R.M., Brit.Chem.Eng. 4 (1959) 267.
(30) Barrer, R.M., Chem.Ind. (1962) 1258.
(31) Barrer, R.M., L.V.C.Rees and M.Shamsuzzoha. J.Inorg.Nucl.Chem. 28 (1968) 629.
(32) Barrer, R.M., R. Papadopoulos and L.V.C.Rees.Ibid.,29(1967)2047
(33) Barrer, R.M. and B.M. Munday, J.Chem.Soc. A1971 (1971) 2904.
(34) Barrer, R.M. and B.M. Munday, Ibid., (1971) 2914.

(35) Barrer, R.M. and J. Klinowski, Ibid., (1972) 1956.
(36) Barrer, R.M., J. Klinowski and H.S. Sherry. J.Chem.Soc.,
     Faraday II, 69 (1973) 1669.
(37) Barrer, R.M. and J. Klinowski. J.Chem.Soc.,Far.Trans.I 70
     (1974) 2080.
(38) Barrer, R.M. Bull.Soc.Fr.Mineral.Cristall., 97 (1974) 89.
(39 Barrer, R.M., in L.B. Sand and F.A. Mumpton, eds., Natural
     Zeolites, Occurrence, Properties, Use (New York, Pergamon
     Press, 1978) pp. 385-395.
(40) Barrer, R.M. Proc.5th Int.Conf.On Zeolites, Naples, Italy,
     (1980), L.V.C.Rees, Ed., p.273.
(41) Barri, S.A.I. and L.V.C.Rees. J.Chrom. 201 (1980) 21.
(42) Berth, P., G. Jakobi,E. Schmadel, M.J. Schwuger and C.H.
     Krauch. Angew. Chem. Intern. Edit., 14(2) (1975) 94.
(43) Bibler, N.E., R.M. Wallace and M.A. Ebra. Effects of High
     Radiation Doses on LINDE IONSIV IE-95. Presented at
     A.I.Ch.E.Mtg., Aug. 16-19, 1981, Detroit, Mich.
(44) Boari, G., L. Liberti, R. Passino and D. Petruzelli. Water
     Res.15 (1981) 337.
(45) Bower, C.E. and D.T. Turner, Progr. Fish Cult. 44 (1982) 19.
(46) Braico, R.D., Ammonia Removal from Recycled Fish Hatchery
     Water, M.S.Thesis, Montana State Univ. (1972).
(47) Bray, L.A. and Fullam, H.T., Molecular Sieve Zeolites I (2nd
     Int. Conf.) (1970), p.450.
(48) Breck, D.W., W.G. Eversole, R.M. Milton, T.B. Reed and T.L.
     Thomas. J.Am.Chem.Soc. 78 (1956) 5963.
(49) Breck, D.W., J.Chem.Ed. 41 (1964) 678.
(50) Breck, D.W., U.S.Patent 3,723,308 (Mar.27 1973).
(51) Breck, D.W. Zeolite Molecular Sieves.Structure, Chemistry
     and Use (New York, Wiley-Interscience, 1974).
(52) Breck, D.W., Ibid., p.535.
(53) Brooksbank, R.E., et al., DOE-SDS-Task Force, U.S. Dept. of
     Energy Report DOE/NE-0012 (1981).
(54) Buckingham, J.S., U.S. Atomic Energy Corp. Doc. No.
     ARH-SA-49 (1970)
(55) Butterfield, O.R., T.J. Kennedy and C.F. Woods. Start-up and
     Operation of the Tahoe-Truckee Sanitation Agency, Advanced
     Wastewater Treatment Plant, 51st Ann.Conf., Water Pollu.
     Control Fed., Anaheim, Ca., (Oct.1-6,1978).
(56) Calmon, C., Ion Exchange for Pollution Control (Boca Raton,
     Florida, CRC Press, 1979) Vol. I, pp. 3-21.
(57) Campbell, D.O., E.D. Collins, L.J. King and J.B. Knauer. Oak
     Ridge Natl. Lab. Rept. ORNL/TM-7448 (1980).
(58) Ibid., Oak Ridge National Lab. Rept. ORNL/TM-7756 (1982).
(59) Caullet, P., J.A. Guth, A. Kalt, E. Nanse and R. Wey,
     C.R.Acad.Sci., Paris, 287 (1978) 763.
(60) Caullet, P., J.A. Guth and R. Wey. Ibid., 288 (1978) 1.
(61) Caullet, P., J.A. Guth and R. Wey. Ibid., 288 (1978) 1059.
(62) Caullet, P., J.A. Guth and R. Wey. Bull.Mineral., 103 (1980)
     330.

(63)   Cilley, W.A. and B.H. Wiers. Zeolite A Proton Exchange and
       Dissolution, Paper R6, Discussions, 4th Internatl.Conf.on
       Zeolites, Chicago, (1977).
(64)   Chelischev, N.F., N.S. Martynova, L.K. Fakina and B.G.
       Berenshtein, Doklady Akad. Nauk. SSSR 217 (1974) 1140.
(65)   Collins, E.D., D.O. Campbell, L.J. King, and J.B. Knauer,
       A.I.Ch.E. Symp. Series No.213, vol 78 (1982) pp. 9-15.
(66)   Cremers, A., Ion Exchange In Zeolites, in Molecular Sieves
       II, ACS Symposium Series 40 (Washington, D.C., Am. Chem.
       Soc., 1977) 179.
(67)   D'Ambrosia, J.T., A.I.Ch.E.Symp.Series No.213, 78 (1982)
       27-32.
(68)   Denny, A.F., A.J. Gioffre and J.D. Sherman. U.S. Patent
       4,094,778 (1978).
(69)   Dyer, A., H. Enamy and R.P. Townsend, Sep. Sci. Tech.
       16(2) (1981) 173.
(70)   Eichorn, H. Pogendorf Ann. Phys. Chem. 105 (1858) 126.
(71)   Eisenman, G. Biophys. J. 2 (1962) 259.
(72)   England, D.C. and P.B. George, Rept. 17th Swine Day,
       Spec.Rept. 447 (Agric. Expmt. Sta., Oregon State Univ.,
       1975).
(73)   Fletcher, P. and R.P. Townsend. Proc. 5th Int. Conf. on
       Zeolites, Naples, Italy (ed.L.V.C.Rees) (1980), p.311.
(74)   Fletcher, P. and R.P. Townsend. Ibid., 77 (1981) 955.
(75)   Fletcher, P. and R.P. Townsend. Ibid., 77 (1981) 965.
(76)   Fletcher, P. and R.P. Townsend. Ibid., 77 (1981) 2077.
(77)   Fletcher, P. and R.P. Townsend. Ibid., 79 (1983) 419.
(78)   Fletcher, P. and R.P. Townsend. Paper Submitted To Phil.
       Trans. Roy.Soc.,(London).
(79)   Fullerton, R., U.S. Atomic Energy Comm. Doc. No. HW-69256
       (1961).
(80)   Fujimori, K. and Moriya, Y., Asahi Garash Kogyo Giutsu
       Shoreikai Kenkyu Hokoku 23 (1973) 243.
(81)   Gaines, G.L. and H.C. Thomas, J. Chem. Phys. 21 (1953) 714.
(82)   Gallei, E., D. Eisenbach and A. Ahmed, J. Catalysis 33
       (1974) 62
(83)   Gans, R., Jahrb. Konig. Preuss. Geol. Landesanstalt (Berlin)
       25 (1905) 2; (1906) 2763; Chem. Ztg. (1907) 28.
(84)   Gaspard, M., A. Neveu and G. Martin, Water Res. 17 (1983)
       279.
(85)   Gleason, G.H. and A.C. Loonam. Sewage Works J. 5(1) (1933)
       61.
(86)   Gleason, G.H. and A.C. Loonam. Sewage Works J. 6(3) (1934)
       450.
(87)   Golden, T.C. and R.G. Jenkins. J.Chem.Eng.Data 26 (1981) 366.
(88)   Gordon, A., M. Popovtzer, M. Greenbaum, L. Morantz, M.
       McArthur, J.R. De Palma and M.H. Maxwell. Proc.Conf., 5th,
       1968, Excerpta Med. Found., (1969) pp. 86-96.
(89)   Greenbaum, M., L. Morantz, A. Gordon, M.H. Maxwell and M.
       McArthur, Abstracts, N.I.H. Chronic Uremia Program,

P.B. 179667 (1969).

(90) Gregonis, D. and J. Walker, A Membrane System to Remove Urea from the Dialyzing Fluid of the Artificial Kidney, Proc.11th Annual Contractors Conf., AK-CUP, N1AMMD, Natl. Inst. Health, U.S. Dept. H. E. W., Washington, D.C. (1979) p. 162.

(91) Gushima, Y., Japan. Patent 74 05,889 (Jan.19, 1974).

(92) Guth, J.A., P. Caullet and R. Wey, Proc.5th Int. Conf. Zeolites, L.V.C. Rees, ed., (Philadelphia, Heyden, 1980) 30-39.

(93) Hagiwara, Z. and M. Uchida, in L.B. Sand and F.A. Mumpton, eds., Natural Zeolites, Occurrence, Properties, Use (New York, Pergamon Press, 1978) 463-470.

(94) Hanson, G.L., Proc. Natl. Conf. Complete Water Reuse, Washington, D.C., A.I.Ch.E.-E.P.A., 360-366(1973).

(95) Hawkins, D.B. and J.H. Horton, U.S.Atomic Energy Comm. Doc. No. PP-1245 (1971).

(96) Helfferich, F., Ion Exchange (New York, McGraw-Hill, 1962), 10-13, 122-123, 133, 145-146, 160-161, 185-193.

(97) Hertzenberg, E.P. and H.S. Sherry, in Adsorption and Ion Exchange with Synth. Zeolites, W.H.Flank, ed. (Washington, D.C., Am. Chem. Soc., 1980) 187.

(98) Hofstetter, K.J., C.G. Hitz, T.D. Lookabill and S.J. Eichfeld, Submerged Demineralizer System Design, Operation and Results, Presented at Am. Nucl. Soc. Decontamination Conf., Niagara Falls (Sept., 1982).

(99) International Atomic Energy Agency, Vienna, Technical Report Series No.136 (1972) 61 and 64; and No.78 (1967) 73.

(100) Johnson, P.W. and J.McN. Sieburth, Aquaculture 4 (1974) 61.

(101) Kelley, J.A., W.H. Hale, J.A. Stone and J.R. Wiley, A.I.Ch.E. Symp. Series No.154, 72 (1976) 128.

(102) King, L.J., D.O. Campbell, E.D. Collins, J.B. Knauer and R.M. Wallace, Evaluation of Zeolite Mixtures for Decontamination of High-Activity-Level Water in the Submerged Demineralizer System (SDS) Flowsheet at the Three Mile Island Nuclear Power Station, Unit 2, to be presented at 6th Int. Zeolite Conf., Reno, July 10-15, 1983.

(103) Klein, E., F.F. Holland and K. Eberle, Trans. Am. Soc. Artif. Organs 24, (1978) 127.

(104) Klein, E. and F.F. Holland. Hemoperfusion Kidney Liver Support Detoxif. Proc. Int. Symp. (1980) 63-79.

(105) Klein, E., F.F. Holland, R.P. Wendt, H. Gidden and K. Eberle, Report PB 81-122681, Publ.by NTIS, U.S. Dept. Commerce (Nov.7, 1980).

(106) Klieve, J.R. and M.J. Semmens. Water Res. 14 (1980) 161.

(107) Knauer, J.B., D.O. Campbell, E.D. Collins and L.J. King, Oak Ridge Natl. Lab. Rept. ORNL/TM-8333 (1982).

(108) Komakine, U.S.Patent 3,836,676 (1974).

(109) Kondo, T. and B. Wagai. Experimental Use of Clinoptilolite-tuff as Dietary Supplement for Pigs, Yotonkai (May, 1968) 1-4.

(110) Kondo, K., et al. Effect of Zeolite on Calf Growth, Chikusan

No Kenikyu 23 (1969) 987.

(111) Konikoff, M., Comparison of Clinoptilolite and Biofilters for Nitrogen Removal in Recirculating Fish Culture Systems, Ph.D. Thesis, Southern Illinois Univ. (1973).

(112) Koon, J.H. and W.J. Kaufman, U.S. Environ. Protection Agency Report No.17080 DAR (Sept.,1971).

(113) Koon, J.H. and W.J. Kaufman, J. Water Pollu. Cont. Fed. 47 (1975) 448.

(114) Larsen, E.M. and D.R. Vissers, J. Phys. Chem. 64 (1960) 1732.

(115) Liao, P.B. and R.D. Mayo, Aquaculture 1 (1972) 317.

(116) Liao, P.B. and R.D. Mayo, Aquaculture 3 (1974) 61.

(117) Liberti, L., Water Res. 13 (1979) 65.

(118) Liberti, L., N. Limoni, R. Passino and D. Petruzelli, in Pawlowski, Ed., Physicochemical Methods for Water and Wastewater Treatment (Elmsford, New York, Pergamon Press, (1980) pp.73-85.

(119) Maes, A. and A. Cremers, in W.M.Meier, J.B. Uytterhoeven, eds., Molecular Sieves, 3rd Int. Conf., Adv.Chem.Series 121 (Washington, D.C., Am.Chem.Soc., 1973) p. 230.

(120) Maes, A. and A. Cremers, in J.B.Uytterhoeven, ed., Proc.3d Int. Conf. on Molecular Sieves (Leuven Univ. Press, 1973) 192.

(121) Maes, A. and A. Cremers, J. Chem. Soc., Far. Trans. I 71 (1975) 265.

(122) Maes, A., J. Verlinden and A. Cremers, in R.P. Townsend, ed., The Properties and Applications of Zeolites, Special Publ. No.33 (London, The Chemical Society, 1979) p. 269.

(123) Marantz, L.B., M. Greenbaum and M.J. McArthur, Ger.Offen. 2,100,961 (July 15, 1971).

(124) Mc Goey, R.J. and C. Hitz, Processing Three Mile Island Unit 2 Accident Radioactive Waste Waters, Presented at Am. Inst. Chem. Engrs. Meeting, Detroit, Mich. (Aug.19, 1981).

(125) Mercer, B.W. and L.L. Ames, Unclassified Hanford Laboratories Report HW-78461,(1963).

(126) Mercer, B.W. and W.C. Schmidt, U.S. Atomic Energy Comm. Accession No. 14466, Report No. RL-SA-58 (1965).

(127) Mercer, B.W. and L.L. Ames, in L.B. Sand and F.A. Mumpton, eds., Natural Zeolites, Occurrence, Properties, Use (New York, Pergamon Press, 1978) pp. 451-462.

(128) Mumpton, F.A. and P. Fishman, J. Animal Sci. 45(5) (1977) 1188.

(129) Mumpton, F.A., in L.B. Sand and F.A. Mumpton, eds., Natural Zeolites, Occurrence, Properties, Use (New York, Pergamon Press, 1978) pp. 3-27.

(130) Nightingale, E.R., J. Phys. Chem. 63 (1959) 1381.

(131) Nikashina, V.A., J. Chrom. 120 (1976) 155.

(132) Olson, D.H. and H.S. Sherry, J. Phys. Chem. 72 (1968) 4095.

(133) Onogi, T., Experimental Use of Zeolite Tuffs as Dietary Supplements for Chickens, Report Kamagata Stock Raising Institute (1966) 7-18.

(134) Passino, R. and L. Liberti. Resource Recov. Conserv. 6 (1981)

263.

(135) Peters, M.D. and R.J. Bose, Tech. Rep. Fish Mari. Serv.
(Canada) (1975) 535-546.

(136) Pillay, K.K.K., A.I.Ch.E. Symp. Series No. 213, 78 (1982) 33

(137) Pillay, K.K.K., Report NE/RWM-80-3, Penn.State Univ.,
University Park, Pa. (Oct., 1980).

(138) Prettyman, R.F. and R.F. Woods, Ammonia Removal at Tahoe-
Truckee - First Year in Review, 21st Northern Regional Conf.,
Calif. Water Pollu. Control Assoc., San Francisco, Oct.18,
1979.

(139) Rees, L.V.C., Ann. Rept. Progr. Chem. Sect. A, 67 (1970) 191.

(140) Rees, L.V.C., in R.P. Townsend, ed., The Properties and
Applications of Zeolites, Special Publ. No.33 (London, The
Chemical Society, 1980) 218.

(141) Rees, L.V.C., Binary and Ternary Cation Exchange in Zeolite A,
Paper to be presented at the 6th Internatl.Conf.on Zeolites,
Reno, Nevada, July 10-15, 1983.

(142) Reust, R.R., CNSI/TMI Water Cleanup System Process
Description, AGNS Draft Report, Allied General Nuclear
Services(Aug., 1979).

(143) Robbins, M.H., Jr. and G.A. Gunn, Proc.Water Reuse Symp.;
Water Reuse - from Research to Applicn., Vol.2 (A.W.W.A.
Res.Foundn. (1979) 1311.

(144) Robie, R.A. and D.R. Waldbaum, Thermodynamic Properties of
Minerals and Related Substances at 298.15 K(25.0 C) and One
Atmosphere (1.013 Bars) Pressure and at Higher Temperatures,
U.S. Geol.Survey Bull. (1968) 1259.

(145) Roddy, J.W., Oak Ridge Natl. Lab. Rept. ORNL/TM-7782 (Aug.,
1981).

(146) Rosolovskaya,E.N., K.V. Topchievka and S.P. Dorozhko, Russ.
J.Phys.Chem. 51 (1977) 861.

(147) Rothberg, M.R., S.W. Work, K. Linstedt and E.R. Bennett,
Proc. Water Reuse Symp.; Water Reuse - from Research to To
Applicn., Vol.1 (1979) A.W.W.A.Res.Foundn., pp.105-138.

(148) Sanga, S., A. Kurata and Takahushi, Mizu Shori Gijutsu
14(11)(1973)1151.

(149) Sato, I., Chicka Shigen Chosajo Hokoku (Hokkaido) 47 (1975)
63.

(150) Sato, E., Sen'i Kako 27(12) (1975) 714.

(151) Savitsky, A.C., Soap Cosmet.Chem.Spec. 53(3) (March, 1977) 29.

(152) Schwuger, M.J. and H.G. Smolka. Physicochemical Aspects of
Phosphate Substitution by Heterogeneous Ion-exchangers in
Detergency, Presented at 49th Natl.Colloid Symp., Clarkson
College, Potsdam, N.Y. (June, 1975).

(153) Semmens, M.J. and M. Seyfarth, in L.B. Sand and F.A. Mumpton,
eds., Natural Zeolites, Occurrence, Properties, Use (New
York, Pergamon Press, 1978) pp. 517-526.

(154) Semmens, M.J., J.T. Wang and A.C. Booth, J.Water Pollu.
Control Fed. 49 (1977) 2431.

(155) Semmens, M.J., The Feasibility of Using Nitrifying Bacteria to

Assist the Regeneration of Clinoptilolite, 32nd, Indus.Waste Conf., Purdue Univ., 1977 (publ.1978).

(156) Semmens, M.J. and P.S. Porter, J. Water Pollu. Control Fed. 51(1979) 2928.

(157) Semmens, M.J. and R.R. Goodrich, Environ. Sci. Technol. 11 (1977) 255.

(158) Semmens, M.J. and R.R. Goodrich, Ibid., 11 (1977) 260.

(159) Semmens, M.J., A.C. Booth and G.W. Tauxe, J. Environ. Eng. Div., Proc. Am. Soc. Civil Engrs. 104, No.EE2, 231 (1978).

(160) Semmens, M.J., J.R. Klieve, D. Sonabrich and G.M. Tauxe, Water Res. 15 (1981) 655.

(161) Sherman, J.D. and R.J. Ross, Union Carbide, Unpublished Results (1971, 1973)

(162) Sherman, J.D. and R.J. Ross, Separation of Ammonium Ions from Aqueous Solutions, Ger.Offen. 2,531,338 (Feb.12,1976); Brit.Pat. 1,510,018 (May 10, 1978); U.S.Patent 4,344,851 (Aug. 10, 1982).

(163) Sherman, J.D. and Ross, R.J., LINDE IONSIV W-85 - Phillipsite - Gismondine Molecular Sieve Zeolites for Ammonium Ion Exchange, Bulletin F-4094a, Union Carbide Corp., New York (1977).

(164) Sherman, J.D., LINDE IONSIV Zeolite NH_4^+ Exchangers for Artificial Kidney Applications, in Artificial Kidneys, Artificial Liver, Artificial Cells, T.M.S.Chang, Ed., Plenum Press, New York (Feb., 1978).pp.267-274.

(165) Sherman, J.D., Removal of Uremic Substances with Zeolite Ion Exchangers, European Patent Appl.81106583.8, (Aug.25, 1981).

(166) Sherman, J.D., Ion Exchange Separations with Molecular Sieve Zeolites, in Adsorption and Ion Exchange Separations, J.D.Sherman, ed., A.I.Ch.E.Symp.Series , 74, No.179 (1978).pp.98-116.

(167) Sherman, J.D., Denny, A.F. and Gioffre, A.J., Soap Cosmet. Chem. Spec., 33-40 (Dec., 1978) 64-68 .

(168) Sherman, J.D. and Ross, R.J., in L.V.C.Rees, ed., Proc. 5th Intl. Conf. Zeolites, (Philadelphia, Heyden, 1980) 823.

(169) Sherry, H.S., J. Phys. Chem. 70 (1966) 1158.

(170) Sherry, H.S., J. Coll. Interface Sci 28 (1968) 288.

(171) Sherry, H.S., J. Phys. Chem. 72 (1968) 4086.

(172) Sherry, H.S., Ion Exchange (ed.Marinsky) Vol 2, Chapt.3 (New York, Marcel Dekker, 1969) 89.

(173) Sherry, H.S., Ion Exch. Process Ind., Intl. Conf., July, 1969, Soc. Chem Ind. (1969), p.329.

(174) Sherry, H.S., in E.M. Flanigen, L.B. Sand, eds., Molecular Sieve Zeolites 1, (Washington, D.C., Am. Chem. Soc., 1971) 89.

(175) Siemens, D.H., D.E. Knowlton and M.W. Shupe, A.I.Ch.E. Symp. Series No.213, Vol 78 (1982), pp. 41-44.

(176) Simon, G.P., Stability of Ion Exchangers in Ionizing Radiation, in Ion Exchange for Pollution Control, C. Calmon and H.Gold, eds., Vol. 1, (Boca Raton, Florida, CRC Press, 1979) 55-70.

(177) Sims, R.C., Environ. Sci. Notes, 9, 2 (1972).
(178) Sims, R.C. and L.W. Little, Environ. Letters 4(1) (1973) 27.
(179) Sims, R.C. and E. Hindin, in Chemistry of Wastewater
       Technology, A.J.Rubin, ed. (Ann Arbor, Mich., Ann Arbor
       Science, 1978) 305.
(180) Slone, W.J., in Proc. Bio-Engrg. Symp. for Fish Culture, L.J.
       Allen and E.C.Kinney, eds. (Bethesda, Md., Fish Culture Sect.,
       Am. Fisheries Soc., 1981) 104-115.
(181) Smith, S.A., R.L. Chapman, O.R. Butterfield, Proc. Water Reuse
       Symp.; Water Reuse - from Research to Application, Vol.2
       (1979). A.W.W.A. Res. Foundn., pp. 1435-1445.
 (182) Smolka, H.G. and M.J. Schwuger, in L.B. Sand and F.A. Mumpton,
       eds., Natural Zeolites, Occurrence, Properties, Use (New York,
       Pergamon Press, 1978) 487-493.
(183) Tanaka, Y. and S. Koide, Bosei Kanri 18(10) (1974) 11-17 .
(184) Thomas, T.L., Process for Cation Separation Using Zeolite
       Materials, U.S.Patent 3,033,641 (May 8, 1962).
(185) Thompson, H.S., J. R. Agric. Soc. Engl., 11, 68 (1850).
(186) Torii, K., in Natural Zeolites, Occurrence, Properties, Use,
       L.B.Sand, F.A.Mumpton, eds., (Pergamon Press, New York, 1978),
       p.441.
(187) Townsend, R.P., P. Fletcher and M. Loisidou, Prediction of
       Multicomponent Ion Exchange Equilibria in Zeolites, to be
       presented at 6th Internatl. Conf. on Zeolites, Reno, Nevada,
       July 10-15, 1983.
(188) Truesdell, A.H. and B.F. Jones, U.S. Geol. Survey J. Res., 2
       (2) (1974) 233.
(189) Vansant, E.F. and J.B. Uytterhoeven, Trans. Far. Soc. 67
       (1971) 586.
(190) Vansant, E.F. and J.B. Uytterhoeven, Ibid., 67 (1971) 2961.
(191) Watanabe, Report on Use of Zeolite Tuff as Dietary Supplement
       for Cattle, Report of Okayama Pref. Fed. Agric. Coop. Assoc.
       (Apr., 1971).
(192) Way, J.T., J. R. Agric. Soc. Engl., 11, 313 (1850); 13, 123
       (1852).
(193) White, J.L. and A.J. Ohlrogge, Canadian Patent 939,186
       (Jan.1, 1974)
(194) Wiers, B.H., R.J. Grosse and W.A. Cilley, Environ. Sci.
       Technol. 16 (1982) 617.
(195) Wilding, M.W. and D.W. Rhodes, U.S. Atomic Energy Comm. Doc.
       No. IDO-14624 (1963).
(196) Wiley, J.R. and R.M. Wallace, Savannah River Lab. Rept.
       DP-1388.
(197) Yokota, F., Y. Tanaka and H. Fukaya, Aichi-ken Kogyo Shidesho
       Hokoku 10 (1974) 66-70.
(198) Yokota, F. and T. Kato, Kagaku Kojo 18(10) (1974) 75-79.
(199) Yoshida, H., A. Kurata and S. Sanga, Mizu Shori Gijutsu
       17(3) (1976) 219-226.
(200) Zen, E-an., Am. Mineral. 57 (1972) 524.

SELECTIVE ADSORPTION PROCESSES : N-ISELF

Philippe JACOB

Process Engineer - FEYZIN Refinery. ELF-AQUITAINE

Part I : A SURVEY OF THE N-ISELF PROCESS

I - GENESIS OF THE N-ISELF PROCESS : the ELF-SOLAIZE Research Center and the preparatory chromatography.

I.1. Presentation of the Elf-Solaize Research Center.
I.2. The stress put by ELF on preparatory chromatography.

II - WHAT IS INTERESTING ABOUT N-ISELF PROCESS :

II.1. Gasoline manufacturing in a refinery.
II.2. How to fractionate light gasoline .

III - PROCESS OPERATION :

III.1. Principle of operation.
III.2. Characterization of the separation.
III.3. Yielding of the process.
III.4. Adsorbant regeneration.
III.5. Energy consumption.

I - GENESIS OF THE N-ISELF PROCESS

I.1. Presentation of the Elf-Solaize Research Center (CRES) :

In the Elf-Aquitaine organization, the ELF-SOLAIZE Research Center is linked to the Refining-Marketing Direction of operations, which finances most of its researches, the other ones being paid out directly by SNEA, according to the own scopes of the CRES. It regroups more than 350 people distributed among six sections, whose main activity areas are :

. processes
. lubes and additives
. environment
. analysis methods
. chemistry and catalysis
. energy.

Born 13 years ago, an age that for a research center as for a man is the age of discretion, the CRES may already be credited with a lot of realizations :

. all ELF and ANTAR trade-marks lubes
. the polymer-asphalt STYRELF
. the oleophile drums for treating oily water
. the cloud point additives, the combustion additives
. the anti-fouling SPIRELF spring for protection of exchanger pipes.
. the various processes of preparatory chromatography
. and of course the N-ISELF process.

I.2. The stress put by ELF on preparatory chromatography :

Soon after its founding, the Elf Solaize research center developped a lot of preparatory chromatography processes. The principle of these processes is very plain : it's that of chromatography used as an analysis method : the mixture that one wants to separate is periodically introduced into an adsorbant packed column, which selectively retains the components to separate.

In this area, the stress was put on improvement of those key-points that simultaneously allow a good separation, a high percentage of recovery, and a high efficiency, i.e. :

. the periodic injection system of the feed to be separated, when the feed must be vaporized before entering the separation column, in order to prevent preferential departing.

. the product trapping and recovery system that prevents the pollution of a product by another one and allows to
recycle the intermediate fractions.

All these abilities that met in CRES for the stress on preparatory chromatography partly explain the genesis of the N-ISELF
process.

II - WHAT IS INTERESTING ABOUT THE N-ISELF PROCESS

II.1.Gasoline manufacturing in a refinery

The manufacturing of motor-spirits uses some rather different
basis :

. the reformat, proceeded from hydrotreated heavy
naphta, is the main basis, its octane number being very high.

. the alkylat resulting of the combination of isobutane and isobutene.

. the fluid catalytic cracker gasoline.

. the steam-cracker pyrolisis gasoline, the light
aromatic compounds of which having been previously extracted.

In addition to these natural basis come some fatal basis, the
octane number of which is much worst : these are light gasolines coming either from topping or from catalytic cracking,
the octane number of which is much lower (particularly the
motor octane number).

Let us notice that this light gasoline is the natural feedstock of the steam cracker, too.

We know why light gasoline is neither quite satisfactory as a
gasoline basis nor quite proper to be transformated into ethylene. That is due to the nature of the mixed hydrocarbons,
those being either N-paraffins with a linear structure and
which are a good feedstock for a steam cracker or I-paraffins
or naphtenes having a good octane number but leading in a
steam cracker to a poor ethylene yielding.

Here appears the interest of light gasoline fractionating :
one would obtain a N-paraffin rich cut that could be used
according to its degree of purity:

- as a steam-cracker feedstock (cf. fig. 1)

628

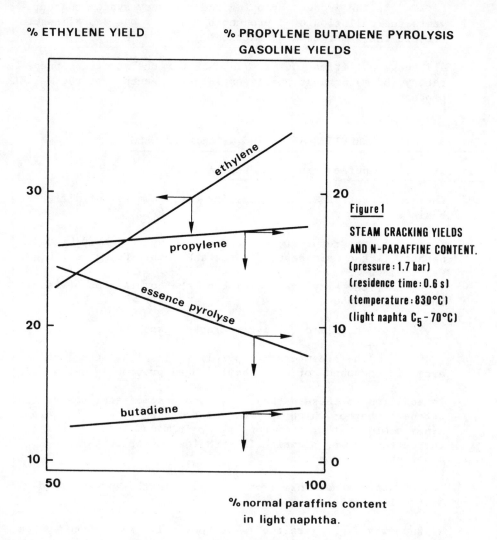

% ETHYLENE YIELD

% PROPYLENE BUTADIENE PYROLYSIS
GASOLINE YIELDS

ethylene

propylene

essence pyrolyse

butadiene

% normal paraffins content
in light naphtha.

Figure 1

STEAM CRACKING YIELDS
AND N-PARAFFINE CONTENT.
(pressure : 1.7 bar)
(residence time : 0.6 s)
(temperature : 830°C)
(light naphta C_5 - 70°C)

     - as an isomerisation feedstock in order to
give back a gasoline basis,

     - as it appears or re-fractionated into nC_5,
nC_6, nC_7 as a solvent.

the resulting cut, being iso-paraffin rich, may very
easily be integrated to gasolines (cf. fig. 2).

Figure 2

VARIATION OF CLEAR RESEARCH OCTANE NUMBER
WITH CONTENT OF NORMAL PARAFFINS.

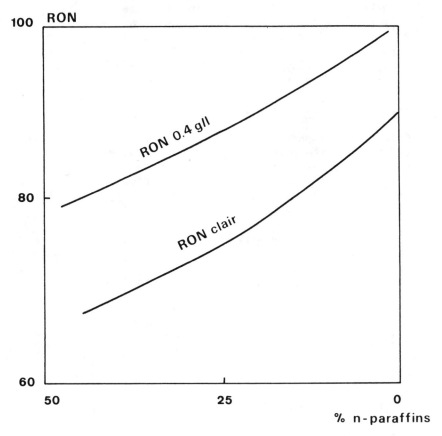

% n-paraffins

II.2.How to fractionate light gasoline ?

Isomers have very closed boiling points (for example 36°C for
N-pentane and 27.9°C for iso-pentane) what leads for an atmospheric distillation to a considerable number of trays. Moreover
it would be necessary to design an unit for each wanted separation deisopentaniser, deisohexaniser, etc...), all this
bringing about a high investment and a very important energy
consumption.

In order to reduce all these drawbacks in the light gasoline
distillation, the ELF-AQUITAINE research teams met a quite
different approach to the problem.

So was the N-ISELF process born.

III - PROCESS OPERATION

The N-ISELF process makes use of the principle of gas-solid chromatography. The solid is a molecular sieve used as an adsorbent. The solid, because of its crystaline structure, only absorbs molecules of equivalent diameter less than 5 Å. Normal paraffins are thus adsorbed, while iso-paraffins, naphtenes and aromatics are not. These molecular sieves were designed specially for the N-ISELF process.

III.1. Principle of separation (cf. fig. 3)

Most of the separation processes by adsorption makes use of adsorption-desorption cycles which are determined by varia- tion of a physical parameter such as total pressure or tempe- rature. On the contrary the N-ISELF process is isobar and rather isotherm. The active parameter is the partial pressure of the n-paraffins.

Periodically some vaporized hydrocarbons, carried by the hydrogen stream, comes into contact with the sieve-bed. Parts of adsordables molecules (n-paraffins) are retained. The other ones then flow out of the bed, followed by those adsorbed molecules which are desorbed by hydrogen.

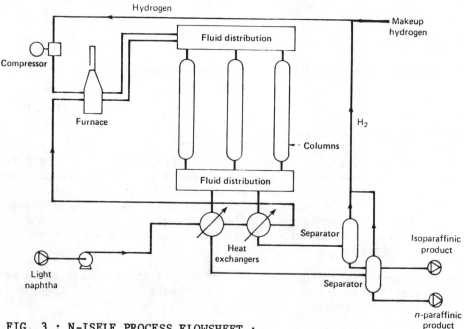

FIG. 3 : N-ISELF PROCESS FLOWSHEET :

At each injection process, one iso-paraffin rich mixture and one n-paraffin-rich mixture then leave the bed. The products are then cooled and separated from hydrogen gas and then recycled by a compressor.

The industrial process has several beds in parallel. The number of beds is determined by the ratio : cycle time/injection time of the load on a given bed. Switching between columns is done by a valve system managed in real time by a programm clock so as the process becomes pseudo-continuous.

The process includes many operating variables such as : total pressure, temperature, cycle time, timing structure of the different sievings. This confers on the N-ISELF unit a large operating flexibility allowing it to produce two cuts more or less enriched with varying yields.

III.2.Characterization of the separation :

Separation may be characterized by a factor varying from 0 to ∞, which is the product of two ratios :

- the ratio : iso-content in the iso-cut/iso-content in the N-cut,

- the ratio : N-content in the N-cut/N-content in the iso-cut.

For a given cut, we may represent on the same figure the percentage of iso-paraffin in the iso-cut and its RON in terms of the yield of the iso-cut.

For instance a feedstock containing 50 % of iso-paraffins has a RON of about 69. A separation factor of about 30 allows us to reach 50 % of iso-cut with a clear octane number of 78, and containing about 82 % of iso-paraffins (cf. fig. 4).

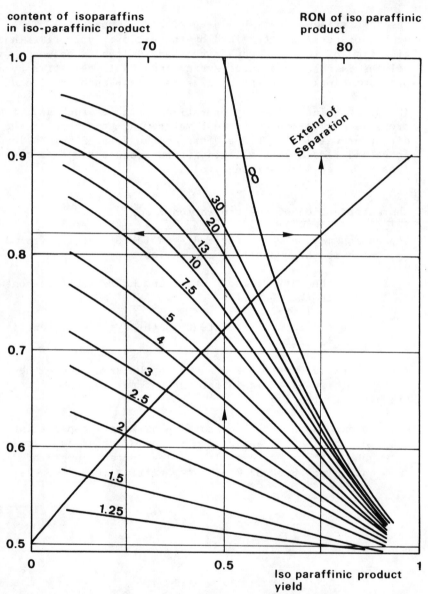

content of isoparaffins
in iso-paraffinic product

RON of iso paraffinic
product

FIG. 4 : CHARACTERIZATION OF THE SEPARATION

III.3. Yielding of the process :

Operation is characterized by the type of load, the properties of both the N- and iso-cuts, operating variables, and treatment costs.

For any given feedstock, the yields in iso-cut (and consequently in N-cut) may be selected between 25 % and 75 %, according to the designed purity. The N-ISELF process may for instance yield high purity n-paraffins. These are indeed always present in effluents, because they desorb slowly, whereas the iso-paraffin peak may cross the bed practically without being distorded. It is then possible, by varying the cut points, to easily modify the product yieldings and purities.

This flexibility is all the more operational that, as soon as the process was designed, a high level automatization was achevied to pilot feedstock injection and products recovery.

III.4. Adsorbent regeneration :

The sieve efficiency decreases over time. This deactivation may be taken up by modifying the cycle time and cut points, but if one wants to maintain the whole separation range, an "in situ" burn off regeneration must be achieved every 1600 hours for each adsorbent bed. The useful life of adsorbers regenerated in this way is about three years.

III.5. Energy consumption :

As the feedstock, recycle gas and product flows are almost stationary, the thermal integration in this process is analogous to that of feed/effluent processes like hydrotreating for example : the feedstock is preheated by the hot streams coming from the separation columns. The fuel consumption is strongly reduced : about 1 TEP for 100 Tons of feed.

The N-ISELF process being isobar, the recycle compressor has only to compensate for pressure drop in the streams. Its consumption is about 20 KWH/Ton.

At last, one must stabilize the products flowing from the separators, what means a 3 tons consumption of steam for 100 tons of feed.

Part 2 : MODELIZATION AND OPTIMIZATION

I -   MATHEMATICAL MODEL OF AN ADSORPTION COLUMN

II - PROCESS SIMULATION

III - COMPUTER SIMULATION PROGRAMM CAPABILITIES

IV - USE FOR DESIGNING A NEW UNIT OR OPTIMIZING AN
      EXISTING UNIT ON A NEW FEEDSTOCK.

I - MATHEMATICAL MODEL OF AN ADSORPTION COLUMN

The model used here belongs to the MCE type (Mixing cells in Cascade with mass exchange).

Adsorption of each n-paraffin is represented by the empirical isotherm of PETERSON & REDLICH taking into account the component coupling :

$$q_n = \frac{A_n P_n}{1 + \Sigma_j B_j P_j^{g_j}} \qquad\qquad n = 1, N$$

The hypothesis for this model are the following :

- negligible pressure drop,
- constant temperature,
- internal adsorption kinetics with a linear potential.

When the material balance equations for each component are written, they are linearized around an operational point selected to represent the average conditions of the adsorbent saturation during one cycle. This linearization is justified by the facts that isotherm curvature is slight at high temperature and that recycled gas contains n-paraffins too, so that there is still partial pressure of n-paraffins beyond which isotherm is practically linear.

The linearized equations are solved in the Laplace space in order to obtain a transfer matrix :

$$G(\gamma) = \left[g_{k1} (\gamma) \cdot e^{i\alpha_{k1}} \right] \qquad\qquad \begin{matrix} k = 1, N+I \\ 1 = 1, N+I \end{matrix}$$

One then computes the amount of each component present in a given cut leaving the unit between the $\emptyset_1$ and $\emptyset_2$ instants. On that purpose a trapping function is defined, such as :

$$p(t) = 1 \qquad\qquad \emptyset_1 < t < \emptyset_2$$
$$p(t) = 0 \qquad\qquad \text{if not}$$

The k component flow in the considered cut is given by :

$$S_k(t) = p(t) . \sum_{m=1}^{N+I} h_{km}(t) \otimes e_m(t)$$

In that equation, h_{km} is an impulsional response, $e_m(t)$ being the input signal of the m component and $\otimes$ the convolution product.

These functions being periodic, with a T period (the cycle time), they can be decomposed into Fourrier series and the wanted quantity :

$$m_k = G . \int_0^T s_k(t) \, dt \qquad (G = \text{recycle gas flow})$$

is nothing but the constant term $S_k(o)$ of the Fourrier transformation equation, multiplied by G :

$$m_k = G \sum_{n=-\infty}^{n=+\infty} P(-n) . \sum_{m=1}^{N+I} G_{km}(n) . E_m(n)$$

II - PROCESS SIMULATION

The N-ISELF process is simulated from the above described model by equalizing the ratio : cycle time/injection time with the number of parallel columns. Each component amount in the N-cut and the Iso-cut is computed, as well as the composition and flow of the hydrogen gas coming from the separators. The recycle flow is adjusted by an hydrogen make-up. Computing is iterative, transfer functions are modified at each step until convergence is achieved (cf. fig. 1).

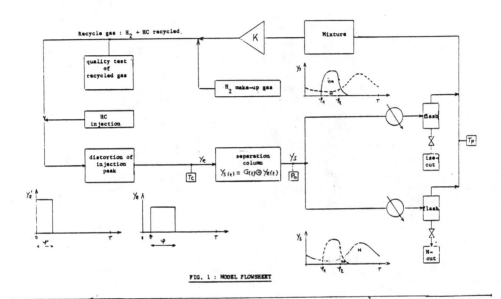

FIG. 1 : MODEL FLOWSHEET

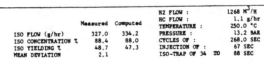

	Measured	Computed	H2 FLOW :	1268 M³/H
			HC FLOW :	1.1 g/hr
ISO FLOW (g/hr)	327.0	334.2	TEMPERATURE :	250.0 °C
ISO CONCENTRATION %	88.4	88.0	PRESSURE :	13.2 BAR
ISO YIELDING %	48.7	47.3	CYCLES OF :	268.0 SEC
MEAN DEVIATION	2.1		INJECTION OF :	67 SEC
			ISO-TRAP OF 34 TO	88 SEC

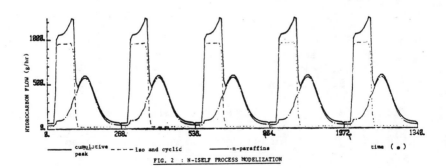

cumulative peak ——— --- iso and cyclic ——— n-paraffins   time (•)

FIG. 2 : N-ISELF PROCESS MODELIZATION

III - COMPUTER SIMULATION PROGRAMM CAPABILITIES

The programm is designed for simulating adsorption of n-paraffins from C3 to C10. It can handle feeds including as many as 50 components. Each adsorbent is characterized by 3 parameters taking into account its particularities and correcting the empirical isotherm and the kinetics laws :

- an adsorption factor,
- a kinetics factor,
- an aging factor.

These adjusting factors are determined for each adsorbent from the results obtained on a pilot in order to adjust computing to achieved measures.

Besides it, the N-ISELF simulation modulus can be integrated to a general programm of chemical engineering simulation, so allowing its upstream or downstream integration into a refining schema, specially when the N-cut that it produces is refractionated into nC5, nC6, nC7, or when its feed needs to be pretreated.

IV - USE FOR DESIGNING A NEW UNIT OR OPTIMIZING AN EXISTING UNIT ON A NEW FEEDSTOCK :

This option allows us to maximize a criterion (a technical or economical one) with respect to a constraint on product specification (n-paraffin content in the N-cut for instance) :

The variable may be :

1. cycle time,
2. pressure,
3. recycle flow,
4. column volume (adsorbent mass),
5. column number.

The data are : feedstock composition, temperature and specifications of a key-product (N-cut or iso-cut). Respect of specifications is obtained by computing optimal cut-points. Selection criteria may be :

1. Operating cost by ton of feedstock,
2. Operating cost by ton of key-product,
3. Actualized profit,
4. Key-product yield,
5. Profitability rate,

6. Pay-out time,
7. Capital investment per ton of key-product.

These various options allow us to quickly determine the capital investment and operating costs of a new unit, as well as its main characteristics (column number and sizing) and operating variables.

The programm gives the intermediate stream characteristics (colums effluents, recycle gas) and so allows us an exact sizing of the equipments (heat-exchangers, furnaces, compressor).

On an existing unit it allows for example to quickly determine the best cycle time, recycle flow and cut points for a given feed, in order to maximize the key-product yield.

Part III : INTEGRATION OF THE N-ISELF PROCESS IN THE REFINING AND PETROCHEMICAL INDUSTRY : CASE STUDIES.

INTRODUCTION

I -   INTEGRATION OF THE N-ISELF PROCESS IN A REFINERY.

   I.1.   Material Balance.
   I.2.   Incidence on the production of premium grade gasoline.
   I.3.   Characteristics of the feedstock.
   I.4.   Profitability.
   I.5.   Case in which the N-cut is sold as a solvent base.

II - A 100 000 TONS/YEAR N-ISELF UNIT FOR N-HEXANE PRODUCTION.

   II.1.   Feed characteristics.
   II.2.   Product characteristics.
   II.3.   Description of the unit.
   II.4.   Material balance (kg/hr).
   II.5.   Profitability.

III - STEAM-CRACKER FEED PROCESSING

   III.1. Naphta processing.
   III.2. Material Balance.
   III.3. Profitability.

INTEGRATION OF THE N-ISELF PROCESS IN THE REFINING AND
PETROCHEMICAL INDUSTRY : CASE STUDIES :

INTRODUCTION

The cost of processing is low for a refining process, above
all when consideration is given to the economies with respect
to refining and petrochemicals.

This type of unit may be located in a refinery or in a
chemical plant.

The main points which are sources of profit in refining are
as follows :

. with an n-iself and a fixed production rate, the pool NOR
  improves and the reformer, a unit which converts 5 to 10 %
  of its feed into gas, may be taken partially out of action.

. the iso-paraffin produced by the n-iself process improve
  the susceptability of gasoline to lead.

. in some refineries, the n-iself process increases the motor
  octane number which is lowered by the F.C.C. gasolines.

. n-iself provides "light" NOR and increases the volumetric
  fuel yield of the crude oil processed.

In many cases which have been studied, utilisation is justified by these advantages themselves from the point of view
of gasoline pool and is amplified if the advantages from the
point of view of petrochemicals are taken into account.

. improvement of ethylene yield,
. improvement of propylene yield,
. improvement of butadiene yield,
. reduction of coking,
. reduction of fuel consumption.

I - INTEGRATION OF THE N-ISELF PROCESS IN A REFINERY

I.1.Material balance :

Using two feedstocks of light and heavy Arab origin, the
material balance of the N-ISELF unit is the following :

from light Arab	Feedstock (C5 - 70°)	Iso-cut	N-cut
density	0.663	0.663	0.663
RON (0.4)	77	.90	-
flow (t/h)	13	6.527	6.442
yield	-	50.2 %	49.6 %

from heavy Arab			
density	0.665	0.665	0.665
RON (0.4)	84.2	93.8	-
flow (t/h)	13	6.527	6.442
yield	-	50.2 %	49.6 %

Gain in octanes : iso-cut/feedstock :

> from light Arab + 13 points
> from heavy Arab + 9.6 points.

Taking into account the potential of the C5-70° cut contained in each of the two crude oils (3.4 % on light Arab, 3.8 % on heavy Arab) and a treatment of 60 % light Arab and 40 % heavy Arab, the following is retained :

iso-cut (density = 0.663   feedstock (density = 0.663
        (RON (0.4) = 91.6   (RON (0.4) = 80

I.2. Incidence on the production of premium grade gasoline

Without N-ISELF,   N and B quantities of naphta and butane enter into the composition of the premium grade gasoline S.

With N-ISELF,   the supplementary iso-cut which is going to enter into the composition of the premium grade gasoline will give rise to a supplementary quantity of Δ S premium grade to the detriment of the ΔN and ΔB quantities of naphta and butane.

We will therefore have the folowing equivalence :

$$S = iso - \Delta N - \Delta B$$

The characteristics of the previous components are :

	density	TV (bar)	RON (0.4)
iso-cut	0.663	0.72	91.6
naphta	0.663	0.68	80
butane	0.580	4.4	99
premium grade gasoline	(summer	0.65	98
	(winter	0.8	98

50 kt/year of iso-cut are available (75.4 k m^3).

Let us suppose that half of this cut will enter into premium grade gasoline of the winter and summer quality. For a year, the following is found :

$$(\Delta N = - 26.8 \text{ k } m^3 \text{ of light naphta in the pool}$$
$$(\Delta S = + 75.4 \text{ k } m^3 \text{ of iso-cut in the pool}$$
$$(\Delta B = - 0.2 \text{ k } m^3 \text{ of butane in the pool.}$$

This leads to :

. the usage of 48,380 m^3 of naphta to make 48,380 m^3 of premium grade gasoline, releasing 200 m^3 of butane.

Taking into account the paritary densities, the usage of 32,100 tons of light naphta leads to producing 36,200 tons of premium grade gasoline and releasing 100 tons of butane.

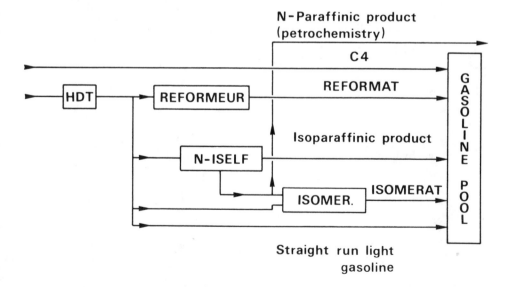

Figure 1.

I.3. Characteristics of the feedstock :

The study on the unit has been carried out with two reference
feedstock and a capacity of 100,000 tons/year for an operation
of 7700 hours without regeneration.

Feedstock characteristics (hydrotreated light gasolines direc
straight run from atmospheric distillation) :

	Feedstock I C5 - 70°C From heavy Arab	Feedstock II C5 - 70°C From light Arab
Density at 15°C	0.645	0.645
TVR	0.720	0.720
% distilled at 70°C	100	100
RON clear	69	63.1
RON + 0.4 g Pb/1	84.2	77
% n-paraffins	51	56.36
% Naphta + aromatics	4	5.57
% Iso-paraffins	45	38.07
S contents	< 1 ppm	< 1 ppm
Water contents	20 to 30 ppm	
MON clear	68	62
MON + 0.4 g Pb/1	83.3	76

Composition (mole %) :

N-butane	0.44	0.13
Isopentane	14.73	12.53
N-pentane	33.16	34.74
Cyclo C5	1.71	1.54
3 methyl C5	29.89	23.07
N-hexane	17.38	20.42
Cyclohexane	2.09	2.83
Benzene	0.29	0.77
Isoheptane	0.29	2.48
N-heptane	0.02	1.07

I.4. Profitability

Total investments : the project considered requires :

43 MF   (May 1982)
investment

Variable costs (for 13 t/h) :

. electricity : 435 KW
. fuel gas :   1400 thermies
. steam :   1 t/hr
. water :   300 m^3/hr

426.16 F/hr

Regeneration (4 per year) :

. electricity : 59 600 KW
. fuel gas :   36 000 thermies
. water :   57 600 m^3
. air :   8 000 m^3
. nitrogen :   13 200 m^3
. sieve :   101 kf/year
                     3 kf/year

193.432 F/year

Variable annual costs for 7,692 hr/year

Variable costs :   3.27 MF
Regeneration costs :   0.19 MF

3.46 MF

Annual standing charges

1 man on duty :   1.00 MF
Staff insurance   1.00 MF

2.00 MF

Costs per year :   5.46 MF

Annually the usage of 32,100 t/year of naphta leads to
producing 30,200 t/year of paritary density premium grade
gasoline and to releasing 100 t/year of butane.

Due to unwanted coastal traffic, differentials must be introduced with respects to the long term prices :

- 9.7 F/t for premium grade gasoline
+ 230 F/t for butane
- 38.5 F/t for naphta.

This means :

supplementary premium grade gasoline : + 77.02
naphta consumed :   - 61.06
butane released :   + 10.18

SUPPLEMENTARY VALORIZATION : 20 MF

Annual costs :   5.46 MF

Pay out = $\dfrac{43}{14.54}$ = 3.0 years

sensitivity :

. investments : + 20 %   pay out : 3.3 yrs
. octane gain : 10 % lower hoped for :   pay out : 3.5 yrs
. naphta price (gap between naphta and premium
  grade gasoline reduced by 10 %)   pay out : 3.3 yrs

I.5. Case in which the n-cut is sold as a solvent base

40,000 T/yr of base sold at the price of premium grade gasoline (cf. West Germany)

Δ between naphta and premium grade : + 225.7 T/t
- 40 F/t for transport

Supplementary annual valorization :

40,000 x 185 = 7.4 MF

MBA : 20 + 7.4 - 5.6 = 21

Pay out = $\dfrac{43}{21}$ = 2 years.

II - A 100 000 TONS/YEAR N-ISELF UNIT FOR N-HEXANE PRODUCTION

II - 1. Feed characteristics :

Raffinate :

- composition :

	design feed % weight :	fluctuation % weight :
. Butane	1	0.15 - 2.5
. Isopentane	4	2.2 - 4.7
. N-pentane	5	2.8 - 6.8
. Isohexane	40	35 - 45
. N-Hexane	24	20.5 - 24.5
. Iso-heptane	13	9.3 - 18.6
. N-heptane	8	5.7 - 11.4
. Octanes	2.5	1.5 - 3.5
. Aromatics	2.5	1.5 - 3.5
. Sulphur ppm	1 - 2	

- Bromine number   0.05 - 0.10 g Br/100 g
- Glycol (ethylen glycol)   maximum : 50 ppm
- Pressure   minimum : 2 eff. bar
- Temperature   25°C minimum
- Water satured (no free water)

Hydrogen gas :

- composition (% volume)
  - H_2 96 % minimum
  - Methane   4 % maximum
  - CO   5 ppm maximum
  - CO_2 30 ppm maximum

- pressure   19 eff. bar minimum
          21 eff. bar maximum

- temperature   ambient

UNIT CAPACITY

Rated capacity of the unit is 100 000 T/yr feed, with an operation factor of 7800 hr/yr.

II.2. Product characteristics

N-Hexane cut

- composition :

. N-hexane content	85 % weight
. Aromatics content	200 ppm
. Sulphur content	2 ppm
. Density at 20°C	0.660
. Color (H_2SO_4)	1
. Alcohol (OH)	25 ppm
. non-volatiles	5 mg/100 ml
. Bromine number	0.015 Br_2/100 g
. Ether	100 ppm
. Colors (hazene)	20
. aspect	limpid, transparent

II.3. Description of the unit

The N-hexane cut production unit includes the following sections :

- a section for separating the iso-cut and the N-cut on a molecular sieve (N-ISELF process).
- a hydrogenation section of the N-cut.
- a fractionating section.

Separation of the iso-cut and the n-cut :

The raffinate is fed to the N-ISELF unit by the feed pump, and then mixed to a little part of the recycle hydrogen rich recycle gas.

This mixture is preheated in the feed/effluent heat exchanger, then it is heated up to the separation temperature in the furnace.

The remaining recycle gas is preheated in the recycle/iso

effluent heat exchanger, and then heated up to a temperature higher than the separation column temperature in the furnace. Indeed, the furnace insuring both the duties : feed and recycle preheating, is controlled by the feed outlet temperature, whereas the recycle temperature at the separation column inlet is controlled by mixing some too hot gas coming out of the furnace and some cold gas taken from the compressor outlet.

The recycle gas is used as a carrier gas and is permanently injected, with a flow control, into the five separation columns.

The feed is intermittently, cyclically, injected into each of the five separation columns (during 1/5th of cycle-time into each).

The hydrocarbons contained in the feed are adsorbed on the molecular sieve and selectively eluated by hydrogen.

First, the iso-paraffins are desorbed and directed to the iso stream (during 1/3rd of cycle-time).

During the end of the cycle, N-paraffins come out of the column and are directed to the N-stream, thanks to a timing system on the output valve.

The use of five columns allows the continuous injection of the feed, one column after the other.

The iso-paraffins are directed (with 1/3rd of the recycle gas) to the separation drum, after cooling down with the recycle gas, and then in a water cooling system.

The N-cut, mixed with 2/3rd of the recycle gas, is directed to the hydrogenation reactor.

N-cut hydrogenation

In order to meet the aromatic content in the N-hexane cut specifications, the N-cut is hydrogenated.

The N-cut is cooled down by heat exchange with the feed, then in a cooling system with water ; the gas and liquid are then separated in a separation-drum.

The vapors coming from both separators and the make-up gas are recovered by a compressor and form the recycle gas.

Fractionating

The N-cut feeds a first fractionating column, called the depentanizer, where three products are withdrawn : on top a vapor distillate which is sent to the fuel-gas network, in the side-stream the n-C5 cut, and at the bottom the n-C6 cut. This column is reboiled with eff. 7 bar steam.

The n-C5, n-C6, n-C7 cuts are cooled down in water heat-exchangers and sent to the storage capacities.

The Iso-cut, after preheating with the stabilized iso-cut, feeds a stabilization column where a vapor distillates is drawn on top, and then sent to the fuel-gas gathering system, and where the stabilized iso-cut is drawn at the bottom. This column is reboiled with eff. 7 bar vapor. A condenser, inte-grated in the column top, insures the reflux.

After heat-exchange with the feed of the stabilization column, the iso-cut is recovered at the bottom by a pump and then cooled down in a water heat-exchanger and sent to the storage facilities.

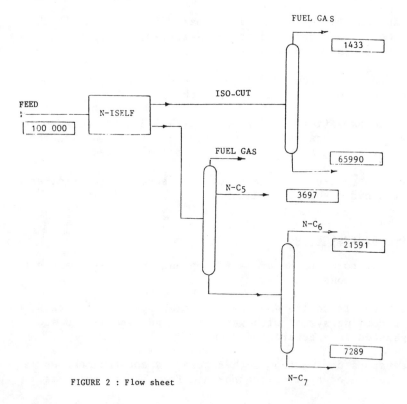

FIGURE 2 : Flow sheet

II.4. Material balance in kg/hr

	FEED	H$_2$ make-up gas	nC$_5$ cut	nC$_6$ cut	nC$_7$ cut	Iso-cut	Fuel gas
Hydrogen		3,71					3,4
Methane		1,25					1,2
Iso butane	128,2		4,1			2,0	121,9
Iso-pentane	512,8		37,5	1,1		436,4	37,1
n-Pentane	641,0		372,9	41,7		211,2	15,3
Iso-hexane	5 128,1		10,1	215,2	0,7	4 909,0	1,1
n-Hexane	3 076,8		49,4	2 495,8	34,4	497,0	
Iso-heptane	1 666,6			4,7	45,9	1 615,2	
n-Heptane	1 025,5			9,4	845,3	170,8	
Octane plus	320,5			0,1	8,2	311,8	
Aromatics	320,5			0,5		306,6	
TOTAL	12 820,0	4,96	474,0	2 768,5	934,5	8 460,0	180,0

II.5. Profitability

Balance	Kg/h	% N
- Feed	12,820	37
	(100,000 t/yr)	
- Iso-cut	8,460	89,7
- NC$_5$-cut	474	
- NC$_6$-cut	2,768	
- NC$_7$-cut	934,5	
$\triangle$ fuel-gas	183,0	

Hypothesis

. Iso-cut : Introduction of the iso-cut into the motor-fuel stock brings about the production of 47,512 t/yr of paritary density premium grade gasoline and 130 t/yr of C4 by use of 42,497 t/year of iso-cut.

. N-cut : the N-cuts, i.e. 32,578 t/yr, are sold on the solvent market at premium grade gasoline price.

CAPITAL INVESTMENT

(including hydrotreatment and additional offsites facilities)

. Variable charges

	Consumption	Price	Time/yr	Total
Electricity	339 KW	0.24 F/KWh	7800 hrs	0.635 MF
Steam	2.9 t/hr	97 F/t	7800 hrs	2.194 MF
Fuel-gas	0.17t/hr	1498 F/t	7800 hrs	1.986 MF
Water	$285m^3$/hr	0.1 F/m^3	7800 hrs	0.222 MF
TOTAL			7800 hrs	5.037 MF/yr

. 4 regenerations each year   0.2   MF/yr

. Standing charges   2   MF/yr

. Iso-cut valorization   20.5   MF/yr

. N-cut valorization   7.4   MF/yr

. Losses   2.6   MF/yr

. Gross Profit : 20.5 + 7.4 - 2. - 0.2 - 5. - 2.6 = 18.1

. Pay-out : 40/18.1 $\sim$ 2.2 years

III. STEAM-CRACKER FEED PROCESSING

III.1. Naphta processing (cf. fig. 3)

In a first stage, long cut naphta is splitted into heavy naphta and light naphta which is then desulfurized by hydro-treating.

This light hydrotreated naphta is then fed to the N-ISELF process which produces two cuts :

- a N-cut used as a steam-cracking feed.
- an Iso-cut used in the motor-fuel stock.

The naphta preparation is needful because :

. N-ISELF is specially adapted to light naphta.
. Light naphta must be desulfurize in order to remove sulphur containing components and olefins traces, which are impurities harmful to N-ISELF.

Of any way, the iso-cut must be desulfurized.

. enrichment with heavier n-paraffins and with C7, C8, doesn't bring about any noticeable increase in the olefin yield during steam-cracking of the so processed naphta. This conclusion arises from extended studies carried in our Company for the operation of our steam-crackers.

So we set up a process in which only light hydrotreated naphta is processed by N-ISELF.

III.2. Material balance

The long cut naphta feed is 1.36 mt/yr. Preparation of this feed leads to the following quantities :

. long cut naphta feed :	1.36 mt/yr
. light desulfurized naphta :	0.57 mt/yr
. heavy naphta :	0.79 mt/yr
. light iso-naphta :	0.31 mt/yr
. light N-naphta :	0.26 mt/yr

The N-ISELF intended for this project is then 0.57 mt/yr.

654

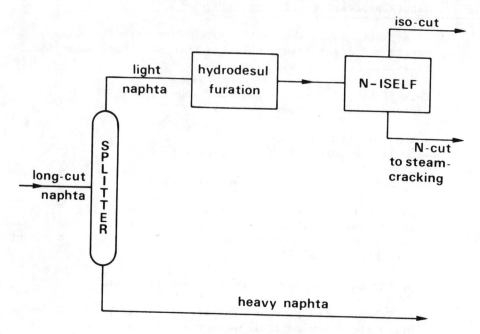

FIG. 3 : REFINING FLOWSHEET

III.3. Profitability

Capital investment :

. N-ISELF	5 70 kt/yr	150.8 MF
. Hydrotreating	570 kt/yr	45 MF
. Splitter	1,360 kt/yr	29.5 MF
		225.3 MF

Costs :

Standing yearly costs :

N-ISELF : 3 men/yr	7.5 MF
Hydrotreating : 1 man/year	2.8 MF
Splitter : 1/4th man/year	2.25 MF
	12.55 MF

Variable yearly costs :

N-ISELF :	16.47 MF
Hydrotreating :	9.7 MF
Splitter :	10.55 MF
	26.72 MF

Profitability :

. Iso-cut : 310 kt of iso-cut will give an increasing and a lightening in the motor-fuel stock, this being done by the process described in the case 1.

Valorization will then be of 99.2 MF.

. N-cut : enrichment with N-paraffins brings about better olefin-yields, hence a better naphta valorization. This leads to a 44.5 MF additional valorization of products coming from the 260 kt naphta steam-cracking (cf. detailed economical study in the appendix).

. Summing-up : the additional valorization is then :

$$99 + 39 = 138 \text{ MF}$$

for a capital investment of 225.3 MF and variable and standing costs of 39.7 MF, i. e. a pay-out of 2.28 years.

A P P E N D I X

PROCESSING COSTS

Hydrotreating :

. Capacity : 0.57 MT/yr

        investment   45   MF

. Standing costs :

1 man/year	1.0 MF/yr	
Maintenance (3 % of capital investment)	1.2	
Overhead expenses	0.6	
	2.8	2.8 MF

. Variable costs :

Fuel :	0.0105 at 1.100 F/t	11.55
Electricity :	8 KWH/t at 0.26 F/KW	2.08
Steam t/t :	0.03 to 97 F/t	2.91
Water m^3/hr :	0.5 to 0.01	
		17.00F/t

i. e. yearly :   9.7 MF

Splitter

. Capacity : 1.36 mt/yr

        investment   29.6 MF

. Standing costs :

0.25 man/yr	0.25 MF/yr	
Maintenance (3 % I)	1	
Overhead expenses	1	
	2.25 MF/yr	2.25MF

. Variable costs :

Steam 0.1 t/t at 92 F/t, i. e. 9.76 F/t

i. e. yearly   10.5 MF/year

PRESSURE SWING ADSORPTION

C.N. Kenney and N.F. Kirkby

Department of Chemical Engineering, Cambridge, U.K.

1. : INTRODUCTION

1.1

Pressure swing adsorption (PSA) is a process for separating gas mixtures based upon recently developed adsorbent solids capable of selectively retaining specific gases. The major operating parameter in a PSA system is pressure; most commercial units operate at, or near, ambient temperature. A typical PSA cycle is shown in Fig. 1, the basic steps of which may be summarised as follows:

(1) Feed gas is compressed into an adsorbent fixed bed, the adsorbable components are adsorbed and adsorbate-free gas accumulates at the closed end of the bed.

(2) At the upper operating pressure, adsorbent-free gas (usually the "product") is withdrawn from the far end of the bed whilst feed flow is maintained. Three distinct regions form in the bed:

a) Near the feed entry the bed is saturated with adsorbate and the gas phase is of feed composition.

b) Near the exit the bed is still adsorbate-free, and the gas phase has product composition.

c) The region between the above two is called the mass transfer zone or adsorption front. This is where adsorption is occurring and gas phase composition changes rapidly with axial position. The adsorption front moves alsong the bed as more feed is introduced. Eventually the bed would be completely saturated, the mass transfer zone having reached the end of the bed, at which point feed is said to "breakthrough" into the product stream.

(3) To regenerate the saturated bed the next step involves reducing the operating pressure (Blowdown). The adsorbate is largely

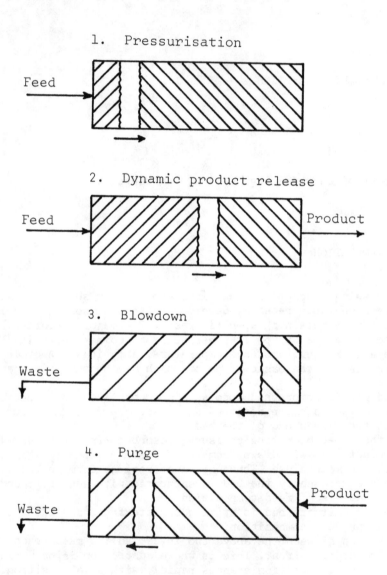

Fig. 1 : Basic steps of a PSA cycle

desorbed into the gas phase and released as a waste stream.
(4) To complete the regeneration of the bed the adsorbent is
purged with product quality gas, usually counter-currently from
the product end. Again distinct regions form and eventually purge
gas would breakthrough into the waste and further purging would
needlessly discharge potential product.

Thus PSA has similarities to other fixed bed , and chromatographic separation processes. Calculating the balance between
feed flows and part-cycle times is the crucial problem and no
comprehensive solution exists. Most existing plants have been
designed from empirical data, or grossly simplified mathematical
models.

1.2 Uses of PSA/Competing Processes

There are four major areas in which PSA is applied:
i)   Air drying
ii)   Hydrogen recovery
iii) Air separation
iv)   Exotic separations

1.2.1 Air drying. This is the oldest use of PSA and probably the
most widely studied. Competing methods for removing water from
ambient air are compressor/chiller systems, which are less
energy efficient than PSA, or thermal swing adsorbers (where
regeneration is achieved by heating and purging at constant
pressure) which have the disadvantage that their thermal inertia
necessitates long time cycles and therefore low throughput per
unit bed volume.

1.2.2 Hydrogen recovery applications. In commercial terms this
is the largest group of applications; since 1966 Union Carbide
alone have built over 40 major units. Hydrogen is an expensive
feedstock and recovery systems have attracted much attention.
The processes competing with PSA are based on palladium diffusion
or are cryogenic processes.

In contrast to these , PSA has proved itself to have higher
reliability and more flexibility. It generally requires less
maintenance and can produce hydrogen at very high purity. PSA
technology has recently been applied most successfully to H_2
recovery from ammonia synthesis loop purge gas streams producing
up to 10% reductions in feedstock/fuel costs.

1.2.3. Air separation. PSA can be used to produce either oxygen
or nitrogen from air. Low volume requirements (< 200 Nm^3/hr) or
oxygen include:
i)   Biological treatment of municipal and industrial wastewater.
ii)   Feed to ozone generators for wastewater treatment.

iii) Oxygen for rivers and reservoirs, especially for fish farming.
iv) Bleaching of chemical pulp and treatment of black liquor in
     the paper industry.
v)   Nonferrous metal smelting.
vi) Medical applications, domiciliary O_2 generators etc.
vii) Chemical oxidation processes.
viii)Enriched O_2 combustion atmosperes to enhance fuel economy.

Nitrogen applications centre around inert gas requirements;
inert flushing gas for petroleum tanks, plant flushing prior to
start-up etc. For low volume applications, PSA has specific
advantages over cryogenic methods and/or road tanker delivery
systems:
i)   fast and simple start-up and shut-down;
ii)   simple, reliable and maintenance-free operation;
iii) low operating costs;
iv)   low capital costs, especially as special materials of con-
      struction are not required;
v)   highly flexible production capability minimises storage
      capacity and/or allows efficient operation even with large
      demand fluctuations;
vi)   being frequently skid mounted, construction and installation
      costs are minimised.

1.2.4 Other separations.
(1) Gaseous isotope separations in the nuclear reprocessing
    industry.
(2) Helium recovery from blast furnace gases.
(3) Hydrocarbon separations, eg. paraffins from aromatic hydro-
    carbons.

1.2.5. PSA disadvantages. PSA technology is not usually competitive
at high throughputs. Hydrogen process adsorbents have been most
successfully developed, having extremely low affinity for hydrogen
in comparison with the common impurities encountered. However, in
air separation particularly, product to waste ratios are of the
order of 1:10 or worse; at high product flows very large
quantities of waste gas have to be pumped, increasing costs
dramatically.
Optimal cycle studies should offset these adsorbent
limitations but such studies are complicated both experimentally
and theoretically especially with the models available.

2.: PSA PROCESSES

2.1 Historical Background

The mention of the adsorption of gases and vapours on solids derives
from dates back to Biblical times. Langmuir (1) derived a theory
of adsorption phenomena based on assuming that only a monolayer

forms, that adsorption is localised and the heat evolved is
independent of surface coverage. This led to the Langmuir isotherm,
still widely used to quantify the amount adsorbed as a function
of pressure at constant temperature. McBain (2) coined the term
"molecular sieve" to define porous solids which exhibit the
property of acting as sieves on a molecular scale. Barrer (3)
published data showing the selective adsorptive properties of
chabazite for air and correctly attributed the higher quadrupole
moment of nitrogen as being the origin of its high affinity.

The adsorptive properties of various forms of carbon and
coke became apparent, particularly as the result of research at
the German Mining Research Institute. As a result of collaboration
Dr. Ing H. Kahle at Linde obtained a German Patent in 1942
entitled "Adsorption Process" which led to the SORBOGEN1 process
for H_2O and CO_2 removal from air, by 1954, using a pressure swing
cycle, Kahle (4)). However, most authors attribute the invention
of PSA to Dr. C.W. Skarstrom, working at Esso Research and
Engineering, the first major patent being given in 1960 (see
Skarstrom (5)). In these two decades there were several important
developments; synthetic zeolite could be manufactured, and natural
zeolite doctored by ion exchange, much improving selectivity and
total capacity. In Germany especially, control of pore shapes,
and size distributions led to some highly efficient carbon
molecular sieves. The Skarstrom process was probably the first
.to exploit these developments. These ideas spread rapidly and
there are now over 400 patents held world-wide by some seven or
eight major companies.

2.2 Adsorbents

The two major types of adsorbents used, are both broadly
referred to as molecular sieves: molecular sieving carbon (MSC)
and zeolites. MSC manufacturers have declined requests to
provide adsorbent for this research, so only a brief review of
MSC properties is given.

2.2.1 Zeolite structure and properties. The natural and synthetic
zeolites are characterised by being alumino-silicate crystallites
of complex molecular structures. Breck (6) gives an account of
all major varieties, structures, properties, etc. Types A and X
are predominantly used for PSA.

The cubic crystallites have roughly spherical cavities
(α-cages) interconnected two windows per cavity, typical dimensions
being 11,4 Å internal diameter for α-cages and 4.5 Å diameter
windows (5A Zeolite). In the preparation, metal ions can be
exchanged with the walls of the cages or windows (4A Zeolite),
thereby regulating the dimensions and creating electrostatic fields
within the cavities. Selective adsorptive properties result from

two factors: large molecules are not capable of passing into the α-cages, the cut-off being determined by the window dimensions; secondly, molecules capable of entering the α-cages will be selectively retained, preference being given to molecules with the highest quadrupole moment.

The normal method of expressing the adsorptive capacity is by relating amount adsorbed per unit mass of adsorbent to pressure, for pure gases at constant temperature. Typical adsorption iso-therms for nitrogen and oxygen on a 5A zeolite are shown in Fig. 2. Pure nitrogen is more strongly adsorbed than oxygen at both temperatures shown, but this information does not show how mixtures of these gases will be adsorbed. Ruthven (7) published an isotherm equation, derived from statistical thermodynamics specifically for 5A zeolites and generalised this into a multi-component theory (Ruthven, Loughlin and Holborow (8)) which has been shown to work well for binary mixtures of gases (Ruthven (9).

For molecules that cannot enter the α-cages, modelling the diffusion mechanisms has been studied and much useful data, correlations and models published (Ruthven, Derrah and Loughlin (10)) for hydrocarbon diffusion and Ruthven and Derrah (11) for mono- and diatomic gases onto 4A and 5A zeolite.

2.2.2 Molecular sieving carbons. Commercially available MSC are manufactured from coke by processes which narrow the pore size distribution. Juntgen et al (12) and (13) of Bergbau-Forschung GmbH have developed activated carbons that are remarkably oxygen specific, and are the basis of a nitrogen process (Knoblauch (14)). The activated carbons used contain bottle and slit shaped pores. Separation of mixtures occurs not because of equilibrium capacity differences but because of wide differences in the rate of up-take. (See Fig. 3). Reyhing (15) and Leitgeb (16) discuss this difference in some detail, and the implications for future research on molecular sieves and process design.

Walker Lamond and Metcalfe (17) have made MSC's by carbonising and pore-blocking thermosetting organic polymers. Subsequently Nandi and Walker (18) compared these adsorbents with coke based MSC and carbons obtained from coconut shells. Nandi and Walker (19) describe an oxygen separation process but unlike the Bergbau-Forschung process these seem to be equilibrium isotherm dominated. They do point out, however, that carbon has a major advantage over zeolite because zeolites are highly hydrophilic and feed air must often be dried before being fed to air separation beds; this is not necessary with MSC.

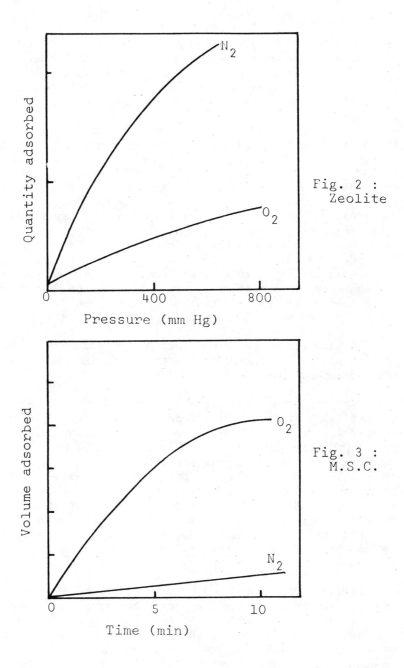

Fig. 2 :
Zeolite

Fig. 3 :
M.S.C.

Zeolite and molecular sieve carbon selectivity for oxygen - nitrogen.

2.3 Industrial Processes

In view of the very large number of processes patented (over 400) it would be a major undertaking to review a fully representative cross-section. Wagner and Stewart (18) and Anon (19) publish a broad survey of PSA systems applied to hydrogen separations, the latter showing some rare economic data.

For oxygen production an introduction to the patents is given by Lee and Stahl (20) in which processes were reviewed in six groups: Esso Research and Engineering, Union Carbide, L'Air Liquide, Bayer/Mahler, Nippon Steel and W.R. Grace processes. To up-date this list the BOC and Air Products processes should be added (see Smith and Armond (21) and Sircar and Zondlo (22)). All these processes use zeolite to preferentially adsorb nitrogen from air to produce up to 95% oxygen. A major distinction that should be drawn between these cycles is that several (Batta Union Carbide, and L'Air Liquide) use a product release that is not accompanied by simultaneous feed. This type of process follows the Kahle system and is called "non-dynamic" product release. Where feed is introduced and product withdrawn simultaneously the cycles are said to use dynamic product release.

To illustrate the common part cycles and multibed arrangements an Esso patent is discussed whilst Table 1 is used to summarise the remainder.

2.3.1Air production. Skarstrom (5) patented a process shown schematically in Fig. 4 . The process is superatmospheric in that it operates above atmospheric pressure, blowdown being to atmospheric pressure. Pressurisation is followed by product release, and then depressurisation to atmosphere. Bed B is operated a half cycle out of phase, releasing product while Bed A is depressurised, and repressurising. This is the simplest scheme that will give a virtually constant flow of product. A purge stream is added at the end of depressurisation which reduces the product flow but does not take it all. At atmosperic pressure after purging, each bed is opened direct to the high pressure feed to repressurise the bed.

Berlin (23) introduced "bed pressure equalisation" (BPE) to repressurise partially the purged bed from the exhausted bed. This stage has the advantage of reducing the rather violent repressurisation of the bed coming on stream, and consequently reduces problems of adsorbent attrition, and fatigue of the mechanical structure. However, this step further disrupts the product flow rate and to compensate Marsh (24) introduced a storage vessel to conserved compressed gas for the purge step.

PRIMARY PRODUCT

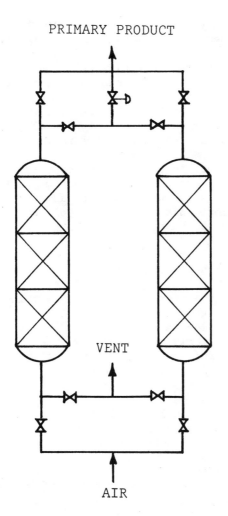

VENT

AIR

Fig. 4 : Process schematic for E.R.&E. U.S. Patent
2,944,627 (C.W. Skarstrom, July 12, 1960)

Table 1 is a highly compressed summary of the other seven major company processes. The following points should also be made:

(1) The mid-bed line on the L'Air Liquide processes makes them remarkably versatile so that cycle rearrangements can be employed to give large turndown ratios.

(2) The Bayer sieves, at least until recently, were the most selective for air separation, and do not need clay binders to hold the zeolite crystallites in pellet form (see Heinze et al (25)).

(3) The BOC cycles do not involve a feed compressor, the feed is sucked into the beds by the product compressor. Waste is withdrawn typically with a water-ring pump and as a result energy consumption is kept low. This subambient cycle works on the steepest parts of the isotherms and this enhances the efficiency further.

(4) The Air Products system is the only one which produces both oxygen and nitrogen. A hybrid 5A zeolite is used called Norton Zeolon 900-Na, in very small pellets (1/16 inch) which ensures that very narrow mass transfer zones develop within the beds. 13X zeolite is used in the pre-treatment columns.

(5) The Bendix Corporation (see Bendix (26)) have adapted these two bed cycles, particularly the Batta (27) version, for respiratory support systems for medical use. Other companies have also shown interest in this potentially large and lucrative market (Armond (28)).

In a paper published in 1977 (29) the same authors made plant design considerations on an oxygen from air process using a 5A molecular sieve in two adsorbent beds as an illustration. They considered the removal of moisture from the feed, cycle time, bed fluidisation, choice of adsorbent, power consumption, effect of ambient temperature and process equipment. The influence of each of these factors on the cost of the plant and its operational costs and effectiveness are discussed.

2.4 Nitrogen Production

Following on from the work done by Jüntgen (13) using a coke adsorben K. Knoblauch (14) recently published an article advocating the use of such an adsorbent coke for the production of N_2 with a purity of between 97 and 99.9% by volume. The plant is the classical pressure swing process for continuous operation with one reactor being loaded while the other is being regenerated. The optimum period for loading and regeneration is 60 seconds with desorption at a pressure of 70 Torr. Higher N_2 purity is achieved at the expense of gas product quantity though this may be compensated for by increasing the operating pressure. The unit is versatile enough to produce O_2 as a main product by dividing the desorption step into 3 stages. The system is aimed at small volume users and is available in standard sizes from 10 to 1000 m^3/h.

The Linde BF process is described by Dr. Jörg Reyhing (30) and gives a brief history of PSA investigations at the Linde Division of Union Carbide. The paper is mainly concerned with a process for the production of about 99% N_2 from air at a pressure of between 2 and 8 bar using two 0.5 m^3 reactor vessels. Operation is in the conventional manner and various relationships between N_2 product quantity, air product quantity and specific energy requirement with feed O_2 content and process pressure as parameters are given. In a short article by J.C. Davis (31) a summary is given of the so-called Lindox route for waste-water treatment and a discussion of 3-bed and 4-bed PSA systems.

3.: HYDROGEN PRODUCTION AND OTHER GAS SEPARATIONS USING PSA

3.1 Hydrogen Production

The purification of industrial hydrogen containing gases was historically the next commercial application of PSA to follow the purification of air feed to cryogenic plants, and "heatless drying". The first successful PSA hydrogen unit was started up in 1966; since that time over 40 have been designed by Union Carbide alone. Pure hydrogen is in growing demand on the industrial market in the food, metallurgical, glass, electronics and chemical industries. Since the traditional production methods by electrolysis or cracking involve some disadvantages in terms of cost price and output, the purification of hydrogenated mixtures available at a relatively low price is becoming more important, especially since it is nearly always cheaper to recover available H_2 than to manufacture it. Alexis (32) pioneered the first work on purifying low grade hydrogen streams which indicated that the method could competitively produce 95% to 98% H_2 when compared to the palladium diffusion or cryogenic processes. A typical H_2 plant with a PSA purification system is discribed by Raghuraman and Johansen (33) and features the following process steps: feed desulphurization, steam reforming, high temperature shift conversion, purification in a PSA system. The first 3 steps are the same as for a conventional plant, whereas low temperature shift, CO_2 removal, and methanation are replaced by one process unit - the PSA system. In such a system, hydrogen purities of up to 99.999 mole % are reported by Heck and Johansen (34) using a naphtha feed. Activated carbon and zeolite molecular sieves are commonly employed as adsorbents and hydrogen is very weakly adsorbed at ambient temperatures, thus lending itself particularly well to production at high purity by the PSA route.

Katira et al. (35) patented a system containing four beds in parallel which has two internal recovery steps. The inlet stream containing 98 mole % H_2 and 2 mole % methane was fed to the system at pressures ranging from 150 to 400 p.s.i.g. and produced 99.999

mole % hydrogen. Activated carbon was found to be the most
economic adsorbent for this separation and economic optimization
performed showed that the ratio of the amount of purge gas to
amount of feed gas should be kept as low as possible without
decreasing the product purity. W. Wolf, working in the Linde
Division, published a review of Union Carbide's PSA hydrogen (36)
purification process covering work done by Stewart and Heck(37)This
is basically the same system patented by Katira et al(38)Wolf point
out that PSA compared with the conventional scheme provides:
(1) 10% reduction in feedstock plus fuel costs
(2) 5% to 7% lower total production costs
(3) higher reliability
(4) less maintenance
(5) more flexibility
(6) higher product purity
(7) lower capital costs for small H_2 producing units but slightly
    higher costs for larger plants.
Katira and Stewart have reviewed the performance of the Union
Carbide.'s units both from a mechanical and process standpoint and
this shows PSA hydrogen systems to be generally proven in commercial practice with the expectation of further and wider use in
larger applications e.g. pipeline H_2. Details have been given by
K. Knoblauch of Bergbau-Forschung and S. Dunlop of Petrocarbon of
a PSA-based process designed to recover H_2 from ammonia purge
streams(39)This PSA unit was installed after the ammonia absorber
which washes NH_3-containing purge gas from the synthesis loop.
Feed to the PSA unit therefore consists of ammonia, water, methane,
argon and nitrogen, which is produced as a pure product. In the
zeolite bed reactors the optimum adsorption pressure was found to
be between 20 and 25 bar with a regeneration pressure of 3 bar.
The system has the ability to change the operating conditions, by
removing more product, to produce lower purity hydrogen for recycle rather than 99.9% H_2 for sale. Hence the plant can adapt to
a fluctuating market.

This method of hydrogen recovery from an ammonia plant purge
stream with subsequent sale or recycle back to the synthesis loop
for conversion to more ammonia has been shown by Petrocarbon
Developments (40) and I.C.I. to be highly profitable using a
cryogenic recovery process. The application of PSA to this
operation or similarly to methanol synthesis purging gas is of
great interest at the moment. The latter has been studied by
Eluard and Simonet (41) using an "Air Liquide" molecular sieve -
"Alite". They showed the efficiency of H_2 extraction to range
between 60% and 93% depending upon the type of cycle and the gas
stream treated. An average efficiency of about 80% was achieved
with corresponding high degrees of purity (above 99.5% in every
case). In the systems using vacuum desorption the purities
obtained were the highest, at 99.99%+. The authors commented upon
the high degree of success that L'Air Liquide has attained using

PSA in their four units which separate:
(1) 99.95% pure H_2 from cracked ammonia
(2) 99.95% pure H_2 from a mixture of 85% H_2, 15% N_2
(3) 99.95% pure helium from a mixture of He, 10% air
(4) extra pure H_2 from reforming gas.

3.2 Other Gaseous Mixture Separations

Pressure swing adsorption as a method of gas separation has not been confined to the gas mixtures already mentioned. In 1974 Bird and Granville (42) published a paper on the separation of nitrogen from helium using PSA which produced 98.0 to 99.99 vol. % helium using activated carbon beds and a short cycle regime at ambient temperature. This operation was effective in its separation, but no economic considerations have been.

Other PSA separations which have been less widely investigated include paraffins from cyclic hydrocarbons, nitrogen and methane, n-tetradecane from iso-octane/n-tetradecane mixtures, and hydrocarbon mixtures.

4.: THEORY

The central difficulty of a quantative description of PSA is its discontinuous nature, and whilst in principle this should not alter the fundamental equations, it does mean that separation solutions must be sought for each new set of initial and boundary conditions that arise from one process step to another.

We shall consider only two theoretical approaches to modelling PSA. In the first a relatively simplified model is adopted for the adsorption of a single gas in an inert carrier which emphasise the movement of zones in pressurisation and purging. Since one column is in practice purged with enriched gas from another column there is an optimum purge: too much purge wastes enriched gas and too little fails to regenerate the bed. In the second analysis detailed (equilibrium) modelling of a single bed is given. This shows how the shape of the mass transfer zone varies with operating parameters and in particular gives a quantitative description of the separation which occurs in the pressurisation step

4.1 Single Adsorbed Gas

Shendalman and Mitchell (43) worked on the idealised system of 1.09% CO_2 in He on silica gel where only the CO_2 adsorption need be considered. The two bed process involved the original Skarstrom cycle of pressurisation to 59 psia, dynamic product release, depressurisation to 21 psia and purging from product counter currently. This is shown in Fig. (5).

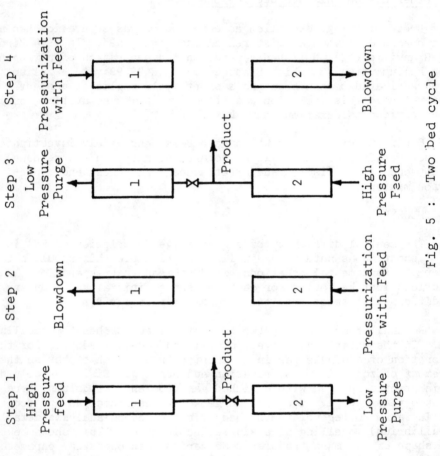

Fig. 5 : Two bed cycle

They made the following assumptions:
1) Instantaneous equilibrium with constant partition coefficient
   k. (Linear CO_2 isotherm with no co-adsorption or hysterisis.)
2) Non-adsorbing carrier gas present in large excess.
3) Isothermal operation.
4) One-dimensional system in plug flow with no axial diffusion.
5) Negligible spatial pressure gradients.
6) Ideal compressible gas behaviour.

These assumptions result in a component mass balance of the form:

$$\varepsilon \left\{ \frac{dc}{dt} + \frac{d(vc)}{dz} \right\} + (1 - \varepsilon) \frac{dn}{dt} = 0 \qquad (1)$$

and an overall continuity equation:

$$\frac{dP}{dt} + \frac{d(vP)}{dz} = 0 \qquad (2)$$

Employing the equilibrium relationship, N = KC, and assuming
ideal gas behaviour result in the single equation

$$P(\varepsilon + (1 - \varepsilon)k) \frac{dy}{dt} + Pv\frac{dy}{dz} + (1 - \varepsilon)ky \frac{dP}{dt} = 0 \qquad (3)$$

which is quasi-linear and by the method of characteristics yields
the following pair of equations.

$$\frac{dt}{P(\varepsilon + k (1 - \varepsilon))} = \frac{dz}{Pv} = \frac{-dy}{(1 - \varepsilon)ky(dP/dt)} \qquad (4)$$

From these equations Shendalman and Mitchell trace the trajectories
of constant composition in the distance-time plane. During the
constant pressure steps these trajectories have constant velocity,
whereas during repressurisation and depressurisation, since the
pressure drop across the bed is neglected, the equation (2)
implies linear velocity gradients, equation (5)

$$\frac{dv}{dt} = \frac{-1}{P} \frac{dP}{dt} \qquad (5)$$

It was then observed that because blowdown and repressurisation
occupy small fractions of the total cycle time, and depend only
on initial and final states these steps can be considered

instantaneous. By then considering the relative penetration depths (the distance a concentration front travels during a process step) of the product and purge steps, the critical purge rate, necessary to prevent feed breakthrough was calculated. Additionally, performance predictions for the steady-state (performance after a large number of cycles) could be made at any purge to feed ratio.

Their results for the comparison of theory and experiment were disappointing; the theory consistently predicted 30% lower concentrations in the product than shown by experiments. Further, the predicted purge to feed ratios were not confirmed. Errors were primarily attributed to the linear isotherm assumption, although rate and dispersion processes were also acknowledged as being important.

Chan et al (45) extended this analysis by considering the blowdown and repressurisation steps; following on from Shendalman and Mitchell's analysis they define a parameter, β, such that,

$$\beta = \frac{\varepsilon}{\varepsilon + k(1 - \varepsilon)} \tag{6}$$

where β represents the ratio of the superficial to concentration front velocity. As a result the characteristic equations become:

$$\frac{dz}{dt} = \beta v \tag{7}$$

$$\frac{dy}{dt} = (\beta - 1) \frac{1}{P} \frac{dP}{dt} y \tag{8}$$

The penetration distances of the fronts during the constant pressure came directly from equation (7)

$$L_h = \beta v_h \Delta t_h \tag{9}$$

$$L_1 = \beta v_1 \Delta t_1 \tag{10}$$

where v_h and v_1 are constant linear velocities and the Δt, half-cycle times. For the steps where pressure is changing, the change in composition and position of the characteristics are given by

$$z_h = z_1 \, \{\tfrac{P_1}{P_h}\}^\beta \tag{11}$$

$$y_h = y_1 \, \frac{(P_1)^\beta}{(P_h)} \tag{12}$$

Shendalman's analysis can now be revised, the net displacements of fronts in the high and low pressure beds being given by

$$\Delta L_1 = L_1 - L_h \, \{\tfrac{P_1}{P_h}\}^\beta \tag{13}$$

$$\Delta L_h = L_1 \, \{\tfrac{P_1}{P_h}\}^\beta - L_h \tag{14}$$

The critical purge to feed ratio, γ, being when $\Delta L_1 = \Delta L_h = 0$, whence,

$$\gamma = \frac{V_1}{V_h} = \frac{L_1}{L_h} = \{\tfrac{P_h}{P_1}\}^\beta \tag{15}$$

and there is no net movement of the concentration front from one cycle to the next. When $L_1/L_h > \gamma$ and $L_1 < L$ they state that:

$$\lim_{N \to \infty} y_N^{Pr} = 0 \tag{16}$$

Chan (45) analysed these expressions in some detail and compared them with the Shendalman and Mitchell's experimental results. The conclusions are that by treating pressurisation and blowdown in the above manner an exact mass balance is obtained, and that gas composition for these two steps, both in the columns and

in that released, are functions of the position of the characteristics within the column.

4.2. Binary Mixture

Flores-Fernancdez (1978) applied the general method outlined
above to model an air separation process. A single bed was
considered and as a result the theory was not extended to
consider penetration depths. Equations were derived in which the
oxygen isotherm was considered linear and the nitrogen was given
by the Langmuir equation. Subscript A refers to oxygen and B to
nitrogen.

The analysis recognises that substantial quantities of
both gases are absorbed, unlike simplified treatments of
chromatography where constant mass flows are assumed.

$$N_B = \frac{K_B P_B}{1 + K_B P_B} \qquad\qquad (17)$$

Equation (17) is identical with the linear isotherm when $K_B = 0$.
Like the previous analysis two mass balance equations arise
for the fluid and solid phases.

Consider experimental results for pressurisation, shown in
Fig. (9) in which Y_A, the oxygen gas concentration is given
for a bed closed at the upper (exit) end plotted against position
in the bed, $\bar{Z} = Z/L$; the final pressure inside the column (P_2)
as the parameter. The solid lines represent the best curves
drawn through the experimental data; the broken lines give the
calculated behaviour. It can be seen that as the pressure inside
the column is increased the gas phase is enriched in oxygen;
the oxygen enrichment in the upper part of the bed, where a
plateau is formed, is greater than in the lower part of the bed.
The plateau progressively moves upwards and is further enriched
in oxygen by increasing the pressure in the system. The region
before the plateau indicates the penetrations of the feed gas;
the plateau itself represents the initial gas in the bed which
has been forced upwards by the feed gas and has been enriched
in oxygen. The concentration in the lower part of the bed is
slightly dependent on the feed penetration depth (i.e., final
pressure in the bed). Also the concentration profiles are
basically independent of the pressurisation time.

This behaviour can be understood using a linear equilibrium
model for which the adsorption isotherm assumes the equilibrium

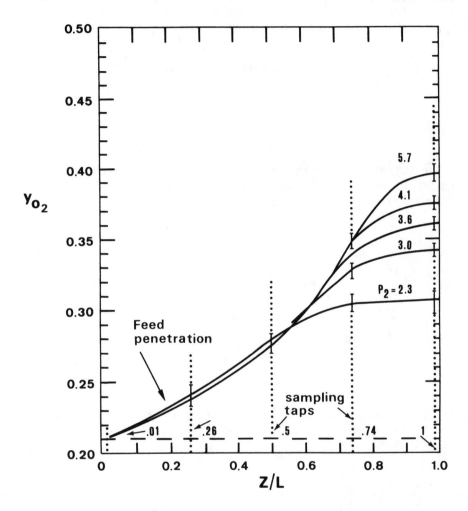

Fig. 9 : Experimental concentration profiles for the air pressurisation runs.

$N^x = f(c) = K_i(p_i)$ with N^x_i the equilibrium number of moles of adsorbate (O_2 or N_2) per unit mass of solid; P_i is the partial pressure of the species i in the gas phase and K_i is a Henry type adsorption constant. In this work, at $295°C$;

$$K_A(O_2) = 0.129 \times 10^{-8} \text{ kmol/(kg adsorbent) } N/m^2$$

and

$$K_B(N_2) = 0.301 \times 10^{-8} \text{ kmol/(kg adsorbent) } (N/m^2).$$

The rates of adsorption of oxygen (r_A) and nitrogen (r_B) are given by the equilibrium relations expressed by Equations (18) and (19):

$$r_A = \frac{\partial}{\partial t} (N^*_B) = K_A \frac{\partial P_A}{\partial t} \tag{18}$$

$$r_B = -\frac{\partial}{\partial t} (N^*_B) = K_B \frac{\partial (P-P_A)}{\partial t} \tag{19}$$

$$-\frac{\partial(q)}{\partial Z} = \frac{\varepsilon}{RT} \frac{\partial P_A}{\partial t} + w(r_A + r_B) \tag{20}$$

(20) is a total material balance on a section of bed of thickness ΔZ, when the total pressure in the bed varies as a function of time. For component (A)

$$-\frac{\partial q_A}{\partial Z} = \frac{\varepsilon}{RT} \frac{\partial P_A}{\partial t} + w(r_A + r_B) \tag{21}$$

with ε the fractional void space in the packed bed

By substituting Equation (18) into Equations (20) and (21) the following expressions result

$$-\frac{\partial q}{\partial Z} = a_2 \frac{\partial P}{\partial t} + a_3 \frac{\partial P_A}{\partial t} \tag{22}$$

and

$$- \frac{\partial(y_A q)}{\partial Z} = a_1 \frac{\partial P_A}{\partial t} \qquad (23)$$

where $a_1 = \epsilon A/RT + w\, K_A$; $a_2 = \epsilon A/RT + w\, K_B$; $a_3 = w(K_A - K_B)$.

The oxygen bed capacity coefficients, a_1 and a_2, represent the maximum oxygen capacities respectively of the bed in the gas phase and on the solid phase per unit bed length per unit pressure. The boundary conditions are :-

$q(t, Z = L) = 0.0$ (no flow at the closed end)

$y_A(t = 0, Z) = y_A^o$ (uniform initial concentration)

$y_A(t, Z = 0) = y_{Af}$ (constant feed gas composition)

$P = p(t)$ (total pressure as a function of time)

(24)

Equations (22) and (23) may be combined to eliminate the partial derivative $(\partial P_A/\partial t)$; thus

$$- \frac{\partial q}{\partial Z} + \frac{a_3}{a_1} \frac{\partial(y_A q)}{\partial Z} = a_2 \frac{\partial P}{\partial t} \qquad (25)$$

Since the coefficients (a_1, a_2 and a_3) are constant, and the total pressure derivative $(\partial P/\partial t)$ is only a function of time (i.e., no pressure drop along the bed); Equation (25) can be integrated in the Z direction, from any position inside the bed, $Z = Z'$, to the closed end, $Z = L$, to give

$$q = \frac{a_1 a_2 (L - Z)}{a_1 - a_3 y_A} \frac{\partial P}{\partial t} = \frac{a_2 (L - Z)}{1 - (1-\alpha)y_A} \frac{\partial P}{\partial t} \qquad (26)$$

where α is conveniently defined as $\alpha = a_2/a_1$ and $(1-\alpha) = a_3/a_1$.

The above equation satisfies the flow boundary condition for a closed end as expressed by Equation (24). Therefore the

independent time function, f(t), which is obtained from the
integration of the partial differential equation, Equation (25),
is not included because it is equal to zero. Maximum flow is
attained at the bed entrance; where $Z = 0$ and $y_A = y_{Af}$.

The material balance on A, Equation (23), can be rewritten
as

$$\frac{\partial y_A}{\partial t} + \frac{q}{(a_1 - a_3 y_A)P} \frac{\partial y_A}{\partial Z} + \frac{a_3(1 - y_A)y_A}{(a_1 - a_3 y_A)P} \frac{\partial P}{\partial t} = 0 \quad (26\,a)$$

where the coefficients are related to the parameter 'α' by

$$\alpha = a_2/a_1 \text{ and } (1 - \alpha) = a_3/a_1.$$

Equation (26) is a quasilinear partial differential equation
of the first order and can be solved either by the method of
characteristics or by an explicit method of finite difference
approximations. The characteristic equations are: -

$$\frac{\partial Z}{\partial t} = \frac{q}{(a_1 - a_3\ y_A)P} = \frac{\alpha(L-Z)}{(1-(1-\alpha)y_A)2} \frac{1}{P} \frac{\partial P}{\partial t} \quad (27)$$

$$(28)$$

$$\frac{dy_A}{dt} = \frac{-a_3(1 - y_A)y_A}{(a_1 - a_3 y_A)P} \frac{\partial P}{\partial t} = \frac{-(1-\alpha)(1-y_A)y_A}{(1 - (1-\alpha)y_A)} \frac{1}{P} \frac{\partial P}{\partial t}$$

where q is given by Equation (25) and α is the bed fractionation
factor ($\alpha = a_2/a_1$). Note that the characteristics are
independent of the pattern selected to pressurise the column.
Equation (27) and (28) for the linear equilibrium adsorption
of a binary gas mixture are similar to the equations presented
by Shendalman and Mitchell (1972) for the linear equilibrium
adsorption of a single component from a dilute solution, but
here the superficial velocity gradient in the bed is non-linear
and is a function of the concentration in the gas phase.
Equation (28) can be integrated from y_{A1} (the concentration
in the bed at $\bar{Z} = \bar{Z}_1$ and $P = P_1$) to Y_{A2} (the concentration
in the position $\bar{Z}=\bar{Z}^1$ at $P = P_2$) when the pressure changes
from P_1 to P_2 (i.e., the time varies from t_1 to t_2); where
$\bar{Z} = Z/L$.

The result is:

$$y_{A2}/(1-Y_{A2})^{\alpha} = (y_{A1}/(1-y_{A1})^{\alpha})\ (P_2/P_1)^{(\alpha-1)} \tag{29}$$

Equation (27) can be combined with Equation (28) to obtain a relation between y_A and $\bar{Z}$. The resulting ordinary differential equation can be integrated from $\bar{Z}_1$ to $\bar{Z}_2$, and from y_{A1} to y_{A2}. The result is given by

$$\frac{1 - \bar{Z}_2}{1 - \bar{Z}_1} = \frac{1 - (1-\alpha)y_{A2}}{1 - (1-\alpha)y_{A1}}\ \frac{1 - y_{A2}}{1 - y_{A1}}^{1/(\alpha-1)} \frac{y_{A1}}{y_{A2}}^{\alpha/(\alpha-1)} \tag{30}$$

where the distance Z is measured from the bed inlet. For the special cases '$y_{A1} = 0$' and '$y_{A1} = 1$', the concentration remains constant and the characteristics are displaced according to the relations

$$(1 - \bar{Z}_2)\ /\ (1 - \bar{Z}_1) = (P_1/P_2)^{\alpha} \qquad , \quad \text{for } y_{A1} = 0 \tag{31}$$

and

$$(1 - \bar{Z}_2)\ /\ (1 - \bar{Z}_1) = (P_1/P_2)^{1/\alpha} \qquad , \quad \text{for } y_{A1} = 1 \tag{32}$$

which are obtained by direct integration of Equation (27).

If the characteristics cross, then there is the formation of a shock wave. It can be shown from the material balances across the shock wave that the velocity of transition of the shock is given by

$$\frac{dZ}{dt} = \frac{\alpha(L - Z)}{(1 - (1-\alpha)y_{Aa})\ (1 - (1-\alpha)y_{Ab})}\ \frac{1}{P}\ \frac{\partial P}{\partial t} \tag{33}$$

where y_{Aa} and y_{Ab} are the concentrations ahead and before the shock wave. The flow rates ahead and before the shock wave are related by:

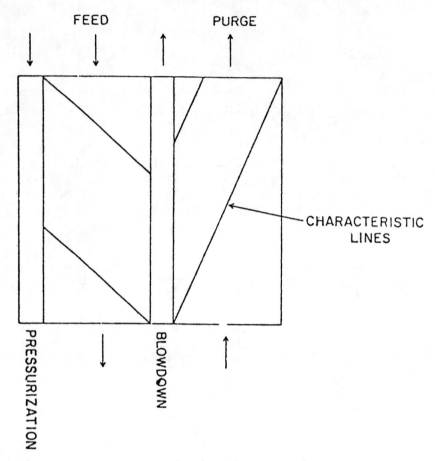

Fig. 6 : Characteristic lines - single gas

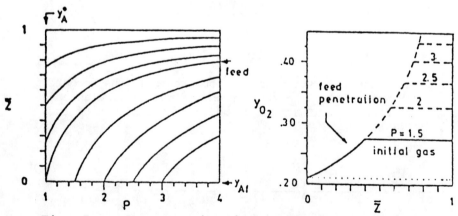

Fig. 7 : Characteric lines and concentration
profiles during pressurisation.

Uniform displacement of initial gas : $y_A^O = y_{Af}^O = 0.21$

$$q_a/q_b = (1 - (1-\alpha)y_{Ab}) / (1 - (1-\alpha)y_{Aa}) \qquad (34)$$

The integration of Equation (33) can be performed numerically with the aid of Equations (29) to (32).

Note that three cases may arise in the solution of the above equations; for the pressurisation step these depend on the relative value of the initial concentration in the bed (y_A^o) relative to the feed concentration (y_{Af}^o), as depicted in figure 7. The characteristics are uniformly distorted by pressure increase when the initial and feed concentrations are equal; figure where $y_A^o = y_{Af} = 0.21$ and $\alpha = 2.15$. The characteristics cross and the discontinuity originated at $\bar{Z} = 0$ and $t = 0$ (or $P-P^o$) is propagated as a shock wave when $y_A^o > y_{Af}$; figure 8 where $y_A^o = 0.5$, $y_{Af} = 0.21$ and $\alpha = 2.15$.
The characteristics emerging from the origin diverge and the discontinuity is propagated as a simple wave when $y_A^o > y_{Af}$; figure 8 where $y_A^o = 0.1$, $y_{Af} = 0.21$ and $\alpha = 2.15$. The concentration profiles are also shown in figure 8.

An 'approximate' solution to the equations (25) and (26) can be more simply obtained numerically by considering the bed as made up of a series (n) of well mixed cells. The partial differential equation (26) leads to a set of ordinary differential equations soluble by Runge-Kutta methods. A refinement is to allow for the non-linearity of the isotherm for nitrogen writing the isotherm

$$N^x = K_i (P_i) \text{ as}$$

$$N_B^x = \frac{K_B P_B}{1 + K_B' P_B}$$

with $K_B = 0.301 \times 10^{-8}$ kgmole/(kg ad.)(N/m^2) and

$$K_B' = 7.5 \times 10^{-7} \ (N/m^2)^{-1}.$$

The problem now becomes more complicated since the coefficients a_2 and a_3 of (22) now become $a_{2L} = a_1 - a_{3L}$ with a_{3L} given by

682

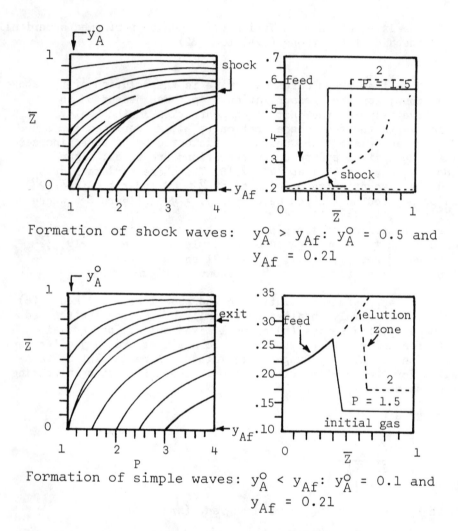

Formation of shock waves: $y_A^o > y_{Af}$: $y_A^o = 0.5$ and
$y_{Af} = 0.21$

Formation of simple waves: $y_A^o < y_{Af}$: $y_A^o = 0.1$ and
$y_{Af} = 0.21$

Fig. 8 : Characteristic lines and concentration
          Profiles during pressurisation.

$$a_{3L} = w \{K_A - K_B/(1 + K'_B \, p(1-y_A))^2\} \tag{35}$$

and the modified equation (24)

$$- \frac{\partial q}{\partial Z} + \frac{a_{3L}}{a_1} \frac{\partial(y_A \, q)}{\partial Z} = a_{2L} \frac{\partial P}{\partial t} \tag{36}$$

is coupled to (23) with non-linear coefficients a_{2L} and a_{3L}.

Corresponding characteristic equations can be derived, although the details are not given here.

4.2.2 Pressure cycling. Consider now a single bed in which the three steps of pressurisation , product release and depressurisation occur, each for a time λ.
The dimensionless cycling time, θ_c, is defined by

$$\theta_c = t/\lambda - 3(n_c - 1) \tag{37}$$

where λ is the step time and n_c is the number of cycles; θ_c is therefore the range 0 to 3.
In the three step process examined the equations and boundary conditions are

a) For Pressurisation ; $0 < \theta_c < 1$ and equation (24) holds

$$y_A(\theta_c = 0, Z) = Y_{A1} (Z)$$

but for the first cycle y_{A1} (Z) = 0.21

$$Y_A(\theta_c, Z = 0) = y_{Af} = 0.21$$

$$q(\theta_c, Z = L) = 0 \tag{38}$$

$$\frac{\partial P}{\partial t} = +P_1 (t)$$

where the positive sign (+) in the last equation indicates that the pressure in the system increases with time.

$$1 < \theta_c < 2$$

b) Product Release (step no. 2), the pressure in the system is considered constant, $P = P_h$, and the above equations reduce to

$$- \frac{\partial q}{\partial Z} = a_{3L} \ P \ \frac{\partial y_A}{\partial t} \qquad (39)$$

and

$$- \frac{\partial (y_A \ q)}{\partial Z} = a_1 \ P \ \frac{\partial y_A}{\partial t} \qquad (40)$$

$$y_A(\theta_c = 1, Z) = y_{A2} \ (Z)$$

$$y_A(\theta_c, Z = 0) = Y_{Af} = 0.21 \qquad (41)$$

$$q(\theta_c, Z = L) = q_p = \text{constant}$$

$$P = P_h = \text{constant}$$

c) Depressurisation

$$2 < \theta_c < 3$$

$$y_A(\theta_c = 2, Z) = y_{A3} \ (Z)$$

$$\frac{\partial y_A}{\partial t} (\theta_c, Z = L) = - \frac{a_{3L} \ (1-y_A)y_A}{(A_1 - a_{3L} \ y_A)} \ \frac{1}{P} \ \frac{\partial P}{\partial t} \qquad (42)$$

$$q(\theta_c, Z = L) = 0$$

$$\frac{\partial P}{\partial t} = - P_3 \ (t)$$

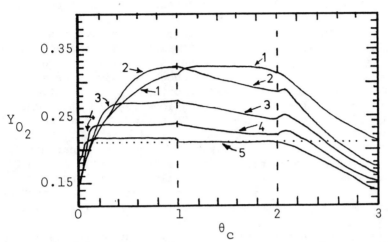

Fig.10 : Experimental concentration profiles

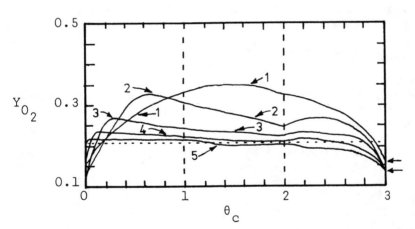

Fig.11: Predicted concentration profiles for pressurisation, product release and depressurisation.

Numbers 1-5 indicate sample points

where the negative sign (-) in Equation (42) indicates that the pressure in the system decreases with time.

The solution of the above equations has again been obtained numerically by using a first order difference formulae to represent the partial derivatives. In contrast to the pressurisation study the present system is not really amenable to solution by the method of characteristics because of the concentration changes from cycle to cycle before the 'steady-state' is attained. The characteristic lines during the pressure changes are not straight lines and therefore the task involved in tracking the shock and simple waves becomes cumbersome or impractical after obtaining the solution for the first step of the first cycle. Equations (24) to (30) were programmed on CSMP3 (Continuous System Modelling Program) following the guidelines described above and were solved on a 370 IBM computer. From figs.10,11 it is evident that the non-linear model provides an accurate representation of the experimental oxygen content of the product in the case of the longest cycling time (λ = 42.25) and the figure shows that the detailed concentration in the column can also be modelled effectively.

In summary, it appears that the nitrogen-oxygen separation process can probably be adequately represented by an 'approximate' solution of an equilibrium model taking into account the non-linearity of the nitrogen adsorption isotherm. This analysis points the way to allowing for non-equilibrium effects and modelling a multibed process.

4.3 Other Modelling Methods

Limitations of space make it impossible to give an account of other methods of solution. They have been summarised in Table 2.

5.: THE ECONOMICS OF PSA

There are many different types of cryogenic process and, also several different types of molecular sieve PSA schemes. The most economical process to obtain a given product must be selected considering factors such as:
    Quantity of products
    Consumption of energy
    Investment costs
    Purity of products
    State of products
    Number of products
    Flexibility of the plant
    Cost of maintenance
    Admissable dimensions of the plant.
The consumption of energy for the production of oxygen, nitrogen and argon is of great industrial interest. Compared with cryogenic

methods the increase of energy consumption at decreasing plant size (< 3,000 nm^3/hour) is mainly due to the increasing cold losses. The decrease in energy at increasing plant size (> 3,000 nm^3/hour) is mainly due to the increasing isothermal efficiency of the air compressors. In general the energy consumption for producing oxygen by the adsorption process either with zeolite molecular sieves or coal molecular sieves is relatively high and it seems that it cannot be decreased by increasing the size of the adsorption plants and it is apparent that it is better suited to small-volume users. This works to the advantage of PSA since cryogenic plants do not scale down well below 30 tons/day.

Usually though plants with a lower specific power consumption (cryogenic and carbon sieve) will be more expensive to build. The economics of purifying hydrogen by PSA are determined by the cost of the equipment and the cost of the feed, since no virtually utilities are required. The equipment cost is a function of the throughput, type and quality of the impurities to be removed and adsorbent properties. Feed costs may be from zero for a gas being flared but appreciable if the feed has fuel value or an alternative use. Nontheless, the PSA H_2 process is a simple plant with no rotary equipment, leading to low capital expenditure and more reliable plant operation. PSA has advantages in that it can handle broad ranges of impurities, low H_2 content in the feed gas and is capable of producing a higher % purity H_2 than cryogenics. Bergbau-Forschung and Petrocarbon see a major advantage of the PSA units over cryogenic systems in the fact that capital costs fall steadily as they are scaled down, because of this economic advantage of scale PSA purge stream purification units are of particular interest to operators of plants producing a small scale purge stream.

A summary of the existing applications of PSA and cryogenic systems was included in Wolf's paper in 1976. This indicates main applications of PSA still lie with air separation and the purification of hydrogen containing streams. Further improvements and applications are foreseeable in the future of PSA and will be due to the use of better adsorbents and newer cycles. Better control components could lead to further improvements. The result will be that the economics of PSA will become more attractive and more industrial gases will be separable by PSA due to capital cost reduction and higher product yield.

REFERENCES

1. Langmuir, I., J.Am.Chem.Soc. 38, 221 (1916)

2. McBain, J.W., The Sorption of Gases and Vapours by Solids, Chapter 5. Routledge & Sons, London (1932).

3. Barrer, R.M., Proc.Roy.Soc.(London) A167, 392 (1938).

4. Kahle, H., Chem.Eng.Techn, 75 (1954).

5. Skarstrom, C.W., US Patent 2,944,627 (1960).

6. Breck, D.W., Zeolite Molecular Sieves, J.Wiley & Sons, NY (1973).

7. Ruthven, D.M., Nature, Phys.Sci. 232 (29): 70 (1971).

8. Ruthven, D.M., Loughlin, K.F., Holborow, K.A., Chem.Eng.Sci. 28, 701 (1973).

9. Ruthven, D.M., AIChEJ 22 (4): 753 (1976).

10. Ruthven, D.M., Derrah, R.I., Loughlin, K.F., Canad.J.Chem. 51 (21): 3514 (1973).

11. Ruthven, D.M., Derrah, R.I., J.Chem.Soc.Faraday Trans. I, 71 (10): 2031 (1975).

12. Juntgen, H., Knoblauch, K., Schroter, H.J., Berichte der Bunsen-Gesellschaft fur physikalische Chemie, 79 (1975) 9, 8240826.

13. Juntgen, H., Knoblauch, K., Munzner, H., Schilling, H.D., 5th Int.Conf. on Magnetohydrodynamic Electrical Power Generation, Munich, 19-23 April 1971.

14. Knoblauch, K., Chem.Eng. (NY) 85 (25): 87 (1978).

15. Reyhing, J., Linde AG Werksgruppe TVT Munchen, Oct. (1975).

16. Leitgeb, P., Linde AG Werksgruppe TVT Munchen, Oct. (1975).

17. Walker, P.L., Lamond, T.G., Metcalf, J.E., Proc.2nd Int. Carbon and Graphite Conference (1966).

18. Nandi, S.P., Walker, P.L., Fuel, 54 (1975).

19. Nandi, S.P., Walker, P.L., Separ.Sci., 11: 441 (1976).

20. Lee, H., Stahl, D.E., AIChE Symp.Ser. 69(134): 1 (1973).

21. Smith, K.C., Armond, J.W., Cryotech '73 Proc. (pub. 1974): 101 (1974).

22. Sircar, S., Zondlo, J.W., US Patent 4,013,429 (1979).

23. Berlin, N.H., US Patent 3,280,536 (1969).

24. Marsh, W.D., US Patent 3,142,547 (1964).

25. Heinze, G., Mengel, M., Reiss, G., German Patent 2,329,210 (1976).

26. Bendix Corporation, UK Patent 1,467,288 (1977).

27. Batta, L.B., US Patent 3,717,974 (1973).

28. Armond, J.W., US Patent 4,065,272 (1976).

29. Smith, K.C. and Armond, J.W., UK Chem.Soc. Autumn Meeting (Aberdeen), (1977).

30. Reyhing, J., Linde Aktiengesellschaft, October (1975) Munchen.

31. David, J.C., Chem.Engng, $\underline{79}$, 88-89, October, (1972).

32. Alexis, R.W., Chemical Engineering Progress, $\underline{63}$, (5), 69-71, (1967).

33. Raghuraman, K.S. and Johansen, T., Processing, 10-11, October, (1978).

34. Heck, J.L. and Johansen, T., Hydrocarbon Processing, $\underline{57}$, (1) 175-7, January, (1978).

35. Katira, C.H., Doshi, K.J. and Stewart, H.A., Paper 38a presented at 68th National Meeting of the AIChE, Houston (1971).

36. Wolf, W., The Oil and Gas Journal, $\underline{74}$, (8), 88 (1976).

37. Stewart, H.A. and Heck, J.L., Chemical Engineering Progress 78, September, (1969).

38. Katira, C.H. and Stewart, H.A., Cryotech 73 Proceedings, 78-84, (1974).

39. Petrocarbon Developments seminar, London, June 1979, in The Chemical Engineer, 395 (1979).

40. Banks, R., Chemical Engineering, October 10, 1977.

41. Eluard, R. and Simonet, G., Chimie et Industrie - Genie Chimique, $\underline{103}$, 15, (1970).

42. Bird, G. and Grancille, W.H., Advances in Cryogenic Engineering, $\underline{19}$, 463-73, (1974).

43. Shendalman, L.H. and Mitchell, J.E., Chem.Eng.Sci. $\underline{27}$, 1449, (1972).

44. Flores-Fernandez, G., Ph.D. dissertation, Cambridge University, (1978).

45. Chan, Y.N.I., Hill, F.B. and Wong, Y.W., Report BNL-25398, Brookhaven National Laboratory, Upton, NY, USA (1978).

46. Szolcsanyi P., Horvath, G., Kotsis, L., Szanya, T., Proc. Chem.Equip.Des.Automn, S-J, 5th Congr.CHISA, Prague, Czechoslovakia (1975).

47. Sebastian, D.J.G., Private Communication (1978).

48. Garg, D.R. and Ruthven, D.M., Chem.Eng.Sci. $\underline{29}$, 571 (1974).

49. Turnock, P.H. and Kadlec, R.H., AIChE Journal, $\underline{17}$, 335 (1971).

50. Kowler, D.E. and Kadlec, R.H., AIChE Journal, $\underline{18}$, 1208 (1972).

51. Wagner, J.L. and Stewart, H.A., AIChE Symp.Series, $\underline{69}$ (134): 1 (1973).

52. Anon, Nitrogen, $\underline{121}$, 37 (1979).

NOMENCLATURE

a_1 oxygen bed capacity coefficient, eqn.(23): $kmol/(m)(N/m^2)$.

a_2 nitrogen bed capacity coefficient, eqn.(23): $kmol/(m)(N/m^2)$.

a_3 coefficient defined as $a_3 = a_1 - a_2 = w(K_A - K_B)$.

c   concentration in the gas phase: $kmol/m^3$.

K_A, K_B adsorption equilibrium constants: $kmol/(kg\ adsorbent)(N/m^2)$.

L   bed length: m.

ΔL net front displacement: m.

$\bar{n}$ number of cells.

n_c number of cycles in pressure swing process.

n   molar density of adsorbed phase per unit volume of bed: $kmol\ m^{-3}$

N   moles of gas adsorbed per unit mass solid: $kmol/kg$ adsorbent

P   total pressure: N/m^2 or bar.

P_L partial pressure of component 'L': N/m^2.

q   molar flow rates: kmol/s.

r   rate of adsorption: $kmol/(kg\ adsorbent)(s)$.

R   gas law constant.

t   time: s.

T   temperature: $^\circ K$.

v   interstitial or front velocity: ms^{-1}.

w   weight of adsorbent per unit length of bed: kg/m.

y   mole fraction in gas phase.

Z   axial position from inlet: m.

$\bar{Z}$ dimensionless position in bed (Z/L).

α bed fractionation factor (a_2/a_1).

β ratio of interstitial velocity to front velocity.

ϵ external porosity of bed.

θ_c dimensionless cycling time.

λ duration of step in cycle: s.

a,b   conditions ahead and before shock wave.

A   oxygen : B nitrogen.

f   pertains to feed.

h   high pressure in cycle.

o   initial condition.

TABLE 1

COMPANY	BEDS	OPERATING PRESSURE RANGE	CYCLE FEATURES
Union Carbide	4	Super-ambient	Pressurise - Dynamic product release - BPE - Provide purge - complete blow-down - receive purge - BPE - etc.
	>2	Super-ambient	Simultaneous pressurisation from feed and product ends, non-dynamic product release.
L'air Liquide	1 to 10	From Super- to Sub-ambient	1) Each bed has a third pipe connected halfway up bed 2) Non-dynamic product release through a second bed at lower pressure 3) Vacuum applied at mid-point of bed 4) Air feed drying section of silica gel, to product zeolite.
Bayer-Mahler	3	From Super- to Sub-ambient	1) Used special 4A sieves (Ca^{2+} and Sr^{2+} exchanged) 2) Each bed has drying section 3) Vacuum regeneration including drying.
Nippon Steel	3	From Super- to Sub-ambient	1) Cycles similar to Bayer-Mahler above 2) Natural zeolite used: Calcareous Tufa.
W.R.Grace & Co.	2 + 1 tank	Super-ambient and Super-sub-ambient	Tuned up Esso Process 1) Dynamic product release 2) Bed pressure equalisation 3) Vent to tank 4) Blowdown to atmosphere or vacuum pumped.
BOC Techsep	2	Sub-ambient	1) Composite beds: silica gel drying, 5A zeolite 2) Dynamic product release - feed sucked through by product compressor 3) Vacuum regeneration, repressuri-sation from product.

TABLE 1   continued

COMPANY	BEDS	OPERATING PRESSURE RANGE	CYCLE FEATURES
BOC Techsep	3	Sub-ambient	As above, with 4) Breakthrough from product release fed to regenerated bed and released as second-out product.
Air Products and Chemicals	2 + 2 tanks	Super-sub-ambient	Produces both oxygen and nitrogen Offspec. O_2 product and N_2 product collected separately, and used to cross purge beds in a 4-part cycle. Separate dryers required using 13x zeolite. 4 pumps used, for recompression of purge gases.

694

TABLE 2	Summary of Numerical Solutions
AUTHOR	TYPE OF SOLUTION AND ASSUMPTIONS
Szolcsanyi *et al* (1975) (46)	Explicit finite difference scheme. 1) One dimensional flow - plug flow no radial dispersion 2) No axial diffusion 3) Hydrodynamics dominated by valve formulae at inlet and outlet, and Darcy's law within the bed 4) Equilibrium adsorption is maintained 5) Linear isotherms 6) Ideal gas behaviour 7) Isothermal operation 8) Simple initial and boundary conditions.
Sebastian (1978) (47)	Explicit finite difference scheme, convergence contained by monotonicity conditions. Vacuum desorption step only. A mathematical paper. 1) Plug flow with axial diffusion 2) Ergun equation for pressure drop across bed 3) Isothermal operation/ideal gas behaviour 4) Unspecified rate of adsorption functions 5) Complex boundary conditions.
Garg & Ruthven (1973) & (1974) (48)	Implicit finite difference scheme of the Crank-Nicholson type. Constant pressure adsorption process considered only. 1) Plug flow with axial diffusion 2) No spatial pressure gradients 3) Isothermal operation/ideal gas behaviour 4) Non-linear mass transfer model separately considering micro and macro pore diffusion.
Kowler & Kadlec (1972) and Turnock & Kadlec (1971) (49), (50).	Process optimisation on a cell model. 1) Isothermal operation/ideal gas behaviour 2) Darcy's Law pressure drop across bed 3) Viscosity of the gas is composition independent 4) Plug flow 5) Instantaneous equilibrium 6) Freundlich isotherm applies 7) Relative volatility relates adsorptive capacity and is composition invariant. 8) The equilibrium amount adsorbed is independent of composition 9) Each cell is ideally mixed and at constant pressure

LIST OF PARTICIPANTS

BELGIUM

Ph. Bodart - Facultés Univ. de Namur - B 5000 Namur

J. Costa - Univ. Liège - B 4000 Liège

J. Martens - Katholieke Univ. Leuven - B 3030 Leuven

CANADA

D. Ruthven - Univ. New Brunswick - Fredericton

F. Smith - Univ. Prince Edward Island - Charlotte

DENMARK

N. Blom - Haldor Topse A/S - DK 2800 Lyngby

J. Larson - Technological Institut - Denmark

FRANCE

F. Fajula - Univ. Montpellier - 34075 Montpellier

N. Gnep - Univ. Poitiers - 86022 Poitiers

J. Lucien - Shell Recherche - 76530 Grand Couronne

T. Labourel - Elf Recherche - 69360 St. Symphorien
            d'Ozon

GERMANY

D. Arntz - Degussa - D 6450 Hanau 1

M. Baacke - Degussa - D 6450 Hanau 1

GREECE

D. Zamboulis - Univ. Thessaloniki - Thessaloniki

ICELAND

G. Einarson - Technological Institute - Reykjavik

ITALY

P. Ciambelli - Univ. Napoli - 80134 Napoli

P. Corbo - Univ. Napoli - 80134 Napoli

I. Ferino - Univ. Cagliari - 09100 Cagliari

A. La Ginestra - Univ. Roma - 00185 Roma

G. Gubitosa - Donegani Research Institute - 28100
            Novara

R. Maggiore - Univ. Catania - 95125 Catania

R. Monacci - Univ. Cagliari - 09100 Cagliari

P. Porta - Univ. Roma - 00185 Roma

A. Villanti - Anic/Chisec - 20097 S. Donato Milanese

ISRAEL

M. Steinberg - Univ. Hebrew - Jerusalem

NETHERLANDS

C. W. Engelen - Univ. Eindhoven - 5600 MB Eindhoven

E. Groenen - Koninklijke-Shell Laboratorium - 1031 CM
            Amsterdam

H. Okkersen - Dow Chemical - Terneuzen

F. Roozeboom - Esso Chemie - 3000 HE Rotterdam

W. Van Erp - Koninklijke-Shell Laboratorium - 1003
            Amsterdam

NORWAY

G. Boe - ELKEM - N 4620 Vagsbygd

G. Haegh - Central Institute Industrial Research -
            - N Oslo 3

K. Kinnari - Statoil - N 4001 Stavanger

S. Kolboe - Univ. Oslo - N Oslo 3

O. Onsager - Univ. Trondheim - N 7034 Trondheim

J.Raeder - Univ. Oslo - N Oslo 3

SPAIN

A. Corma - Instituto Catalisis Petroleoquimica -
      - Madrid 6
J. Juan Aguera - ENPETROL - Cartagena
A. Lopez Agudo - Instituto Catalisis Petroleoquimica -
      - Madrid 6
C. Ballesteros Martin - Univ. Madrid - Madrid 3
A. Lucas Martinez - Univ. Madrid - Madrid 3
J. Pajares - Instituto Catalisis Petroleoquimica -
      - Madrid 6
A. Villarroya Palomar - Industrias Quimicas del Ebro -
      - Zaragoza

TURKEY

E. Alper - A. U. Fen Fakultesi - Ankara

SWITZERLAND

G. Gut - ETH-Zentrum - CH 8092 Zurich

UNITED KINGDOM

S. Fegan - Univ. Edinburgh - Edinburgh EH93JJ
A. Hope - Univ. College London - London WCIH OAJ
D. Rawlence - Joseph Crosfield Sons - Warrington
            WA5 1AB
M. Sanders - Univ. College London - London WCIH OAJ
D. Swindells - Univ. Aberdeen - Old Aberdeen AB92UE
D. Whan - Univ. Edinburgh - Edinburgh EH93JJ

U.S.A.

L. Sand - Worcester Polytechnic Institute - Worcester

698

PORTUGAL

M. J. Afonso - Inst. Superior Técnico - 1000 Lisboa

J. Bordado - Quimigal - Barreiro

M. Brotas - Faculdade de Ciências - Lisboa

J. Caeiro - Petrogal - Lisboa

C. Costa - Faculdade de Engenharia - 4099 Porto Codex

M. C. Dias - Inst. Superior Técnico - 1000 Lisboa

J. Figueiredo - Faculdade de Engenharia - 4099 Porto
      Codex

F. Freire - Inst. Superior Técnico - 1000 Lisboa

C. Henriques - Inst. Superior Técnico - 1000 Lisboa

J. Justino - Inst. Superior Técnico - 1000 Lisboa

F. Lemos - Inst. Superior Técnico - 1000 Lisboa

M. A. Lemos - Inst. Superior Técnico - 1000 Lisboa

L. Sousa Lobo - Univ. Nova de Lisboa - 2825 Monte da
      Caparica

J. M. Loureiro - Faculdade de Engenharia - 4099 Porto
      Codex

M. A. Mendes - Petrogal - Lisboa

M. F. Martins - Inst. Superior Técnico - 1000 Lisboa

L. Palha - Inst. Superior Técnico - 1000 Lisboa

M. J. Pires - Inst. Superior Técnico - 1000 Lisboa

M. F. Portela - Inst. Superior Técnico - 1000 Lisboa

M. F. Ribeiro - Inst. Superior Técnico - 1000 Lisboa

M. C. Rodrigues - Inst. Superior Técnico - 1000 Lisboa

A. N. Santos - Univ. Nova de Lisboa - 2825 - Monte da
      Caparica